VESTIGIA MATHEMATICA

VESTIGIA MATHEMATICA

Studies in medieval and early modern mathematics in honour of H.L.L. Busard

edited by

M. Folkerts and J.P. Hogendijk

Rodopi

Amsterdam - Atlanta, GA 1993

ISBN: 90-5183-536-1 (CIP)

Printed in The Netherlands

Contents

Introduction 7

H. L. L. Busard: Publications 11

P. Bockstaele	A challenge to the mathematicians of the University of Leuven as a New Year's gift for 1639	15
H. J. M. Bos	Johann Molther's 'Problema Deliacum', 1619	29
S. Brentjes	Varianten einer Ḥaǧǧāǧ-Version von Buch II der *Elemente*	47
C. Burnett	Ocreatus	69
A. Djebbar	Deux mathématiciens peu connus de l'Espagne du XIe siècle: al-Mu'taman et Ibn Sayyid	79
Y. Dold-Samplonius	The volume of domes in Arabic mathematics	93
M. Folkerts	Die *Rithmachia* des Werinher von Tegernsee	107
J. P. Hogendijk	The Arabic version of Euclid's *On Division*	143
B. Hughes	Robert Recorde and the first published equation	163
W. Kaunzner	Über die beiden nachgelassenen mathematischen Handschriften von Adam Ries	173
P. Kunitzsch	"The Peacock's Tail": On the names of some theorems of Euclid's *Elements*	205
R. Lorch	Abū Kāmil on the pentagon and decagon	215
J. van Maanen	The 'double-meaning' method for dating mathematical texts	253
G. Molland	Roger Bacon's *Geometria Speculativa*	265
B. A. Rosenfeld	"Geometric trigonometry" in treatises of al-Khwārizmī, al-Māhānī and Ibn al-Haytham	305
C. J. Scriba	Zur Aufgabe 86 des Byzantischen Rechenbuchs Cod. Vindob. Phil. Gr. 65	309
J. Sesiano	La version latine médiévale de l'Algèbre d'Abū Kāmil	315
A. Simi, L. Toti Rigatelli	Some 14th and 15th century texts on practical geometry	453

Addresses of the authors 471

Introduction

Hubertus Lambertus Ludovicus Busard was born on 21 August 1923 in Sittard (The Netherlands, Province of Limburg). He had three sisters and no brothers. After finishing grammar school in Sittard with certificates α (1942) and β (1943), he studied mathematics, physics and astronomy at the University of Utrecht. In 1949 he graduated in mathematics. In 1951 he was appointed as a lecturer in mathematics and physics at the Technical College in Venlo, a city in the South of the Netherlands, very close to the German border. He taught there until his retirement in 1984. Since 1951 he has been happily married to Yvonne (née Bollen). They have two children and five grandchildren.

It was E. J. Dijksterhuis (1892-1965), professor of history of science in Utrecht (since 1953) and Leiden (since 1955), who first interested Busard in the history of science, and he became Dijksterhuis' last student. Busard was awarded his doctorate in science (Dr.rer.nat.) in Utrecht in 1961 for a thesis on Nicole Oresme's *Quaestiones super geometriam Euclidis*, which he edited for the first time (**1**). This dissertation was the beginning of Busard's involvement in Western medieval mathematics. The only articles he has published outside the field of medieval mathematics are on Viète (**3**, 1964) and Mydorge (**6**, 1965); they were related to studies by J. E. Hofmann, who invited him to his meetings on history of science at the Mathematical Research Institute Oberwolfach (Black Forest). Connected with his thesis are two articles on infinite series in the Middle Ages (**2**, **4**) and another (**18**) on the sources of Oresme.

Almost all Busard's publications are editions of Latin mathematical texts. They cover all branches of medieval western mathematics. In geometry he edited the influential treatise of Dominicus de Clavasio (**5**, 1965) and three texts on surveying which were translated by Gerard of Cremona in the 12th century (Abū Bekr, *Liber Saydi Abuothmi*, *Liber Aderameti*: **8**, **11**, 1968-69), Ahmed ibn Jusuf's book on similar arcs, also translated from the Arabic by Gerard (**19**, 1973), the treatise on isoperimetry translated directly from the Greek (**27**, 1980), and John of Gmunden's work on trigonometry (**23**, 1971). Arithmetical texts edited by Busard are the well-known treatise on the computation with fractions by Johannes de Lineriis (**9**, 1968), the *Arithmetica speculativa* by Johannes de Muris, which is closely connected with Boethius' work on arithmetic (**14**, 1971), and the highly interesting and influential treatise on theoretical arithmetic by Jordanus de Nemore, who was one of the most important mathematicians in the Latin West during the Middle Ages (**32**, 1991; the main results in **34**, 1992). Yet another work by Jordanus was edited by Busard, namely the treatise on proportions, which appeared together with a similar work attributed to Campanus (**17**, 1971). In the same year he published an edition of Albert of Saxony's treatise on proportions (**12**). Some results of Busard's early research became part of a small booklet on mathematics in the Middle Ages which is based on a paper given in Paris (**10**, 1969). In 1974 Busard published the text of an algebraic chapter in Johannes de Muris' *De arte mensurandi* (**23**). Later he decided to edit the whole text of this compilation. He has now finished this work, and it will be published soon (**37**).

Although the editions mentioned above are very substantial contributions to the history of medieval Western mathematics, Busard has done even more: perhaps the most important part of his *œuvre* consists of his numerous articles and books on the Latin Euclid. Before Busard's time only the reworking by Campanus (ca. 1255) had been printed – in editions of the 15th and 16th centuries –, and almost all information on the Latin Euclid in the Middle Ages depended on an article by Marshall Clagett in 1953 (The Medieval Latin Translations from the Arabic of the *Elements* of Euclid, with Special Emphasis on the Versions of Adelard of Bath, in: *Isis*, 44 (1953), 16-42). It was Busard who put our knowledge of the history of this most important Greek mathematical text in the Latin West on a solid base: in the last 25 years he has produced critical editions of six complete versions of the *Elements* which originated in the 12th and 13th centuries, and in addition he has edited some scholia and commentaries connected with Euclid's main work (**20**, **21**, **22**) – to say nothing of his articles on the Euclid tradition in the West (**15**, **25**, **26**, **30**, **35**). From Busard's research we know that Euclid's *Elements* were translated in the 12th century from the Arabic three times: by Adelard of Bath (the so-called "Adelard I" version), Hermann of Carinthia and Gerard of Cremona. Based upon Adelard's and Hermann's translations is the version which Clagett called "Adelard II", but which, as Busard showed, was probably written by Robert of Chester. This reworking was the most influential of these texts: it became the basis of the Campanus edition ("the medieval Euclid") and of another version, unknown to Clagett, upon which the version sometimes attributed to Albertus Magnus depends. Except for the Campanus text, Busard has edited all of them using all known manuscripts ("Adelard I": **29**, Hermann: **7**, Gerard: **28**, Version II: **33**, the anonymous redaction based upon it: **36**). Besides the translations from the Arabic and the later reworkings, there is also a translation of the *Elements* made directly from the Greek in the middle of the 12th century, and this translation, too, has been edited by Busard (**31**). Therefore, we now have editions of almost all Latin texts, apart from the so-called "Adelard III" version and some special redactions of which only one manuscript is known. It is almost incredible that this huge task has been performed by only one man: Dr. Busard. He has done for the Latin Euclid what Marshall Clagett has done for Archimedes in the Middle Ages, and no other historians of mathematics in this century have edited such a large amount of important material from the Middle Ages as Clagett and Busard have done. It is to their credit that our knowledge of the history of mathematics in the Middle Ages has advanced so considerably and that almost all important Latin texts are now available in modern editions.

Busard's work on the history of mathematics was honoured when he was elected as corresponding member of the Académie Internationale d'Histoire des Sciences in 1971 and as effective member in 1978.

Until his retirement in 1984, Busard did all his scientific work in his spare time. The example of Busard shows that a high scientific output need not conflict with family commitments. In fact, his harmonious family life has been the inexhaustible source of energy for his work. Busard has always been a very cooperative and helpful person, as most of the contributors to this volume know from personal experience.

The plan to honour Busard on his 70th birthday with a volume of studies dates from 1990, and it was secretly discussed at a conference on medieval mathematics in Wolfenbüttel (this was somewhat difficult because Busard was of course present). It was very easy to find enthusiastic authors. Many of the articles in this volume of studies are in the style of Busard's own work, and some of these articles are the fruit of hundreds or even thousands of hours of labour. Together they reflect the high esteem of their authors for Dr. Busard.

We thank the Department of Mathematics of the University of Utrecht for generously providing access to its computer facilities, and we are grateful to Gerhard Brey (München), Benno van Dalen, Wilberd van der Kallen and André de Meijer (all of Utrecht) for their help in solving computer problems, to Prof. Klaus Lagally (Stuttgart) for providing his program ArabTEX for typesetting Arabic, and to the Managing Director of Rodopi Publishers (Amsterdam-Atlanta), F. van der Zee, for his willingness to publish this volume of studies.

We would like to conclude this introduction by wishing Dr. Busard and his family many more happy and fruitful years.

The editors:

M. Folkerts J. P. Hogendijk

Dr. H. L. L. Busard at work on an edition of Euclid's *Elements*

H. L. L. Busard: Publications

(**1**) Nicole Oresme, *Quaestiones super geometriam Euclidis*, Leiden (E.J. Brill), 1961 (XIV + 179 pp.).

(**2**) Über unendliche Reihen im Mittelalter, in: *L'Enseignement Mathématique*, tome 8, fasc. 3-4 (1962), 281-290.

(**3**) Über einige Papiere aus Viètes Nachlaß in der Pariser Bibliothèque Nationale (mit Wiedergabe des bisher ungedruckten Textes aus nouv. acqu. lat. 1643), in: *Centaurus*, **10** (1964), 65-126.

(**4**) Unendliche Reihen in *A est unum calidum*, in: *Archive for History of Exact Sciences*, **2** (1965), 387-397.

(**5**) The Practica geometriae of Dominicus de Clavasio, in: *Archive for History of Exact Sciences*, **2** (1965), 520-575.

(**6**) Über die Verwandlung eines Quadrats in ein regelmäßiges Vieleck und die Konstruktion dieser Vielecke über einer gegebenen Linie bei Claude Mydorge, in: *Janus*, **52** (1965), 1-39.

(**7**) *The Translation of the* Elements *of Euclid from the Arabic into Latin by Hermann of Carinthia (?)*: books I-VI, *Janus*, **54** (1967), 1-140, and published separately, Leiden (E. J. Brill), 1968 (142 pp.); books VII-IX, *Janus*, **59** (1972), 125-187; books VII-XII, Amsterdam (Mathematisch Centrum), 1977 (198 pp.).

(**8**) L'algèbre au moyen âge: Le "Liber mensurationum" d'Abû Bekr, in: *Journal des Savants*, 1968, 65-124.

(**9**) Het rekenen met breuken in de middeleeuwen, in het bijzonder bij Johannes de Lineriis, in: *Mededelingen van de Koninklijke Vlaamse Academie voor Wetenschappen, Letteren en Schone Kunsten van België*. Klasse der Wetenschappen, Jaargang 30, nr. 7, Brussels 1968 (36 pp.).

(**10**) *Quelques sujets de l'histoire des mathématiques au moyen-âge*, Paris 1969 (Université de Paris, Palais de la Découverte, D 125) (32 pp.).

(**11**) Die Vermessungstraktate *Liber Saydi Abuothmi* und *Liber Aderameti*, in: *Janus*, 56 (1969), 161-174.

(**12**) Der Tractatus proportionum von Albert von Sachsen, in: *Österreichische Akademie der Wissenschaften, math.-nat. Klasse, Denkschriften*, 116. Band, 2. Abhandlung, Wien 1971 (pp. 43-72).

(**13**) Der Traktat De sinibus, chordis et arcubus von Johannes von Gmunden, in: *Österreichische Akademie der Wissenschaften, math.-nat. Klasse, Denkschriften*, 116. Band, 3. Abhandlung, Wien 1971 (pp. 73-113).

(**14**) Die "Arithmetica speculativa" des Johannes de Muris, in: *Scientiarum historia*, **13** (1971), 103-132.

(**15**) [Commentary to:] J. E. Murdoch, The Medieval Euclid: Salient aspects of the translations of the *Elements* by Adelard of Bath and Campanus of Novara, in: *XIIe Congrès International d'Histoire des Sciences, Paris 1968, Actes*, tome I B, Paris (Albert Blanchard) 1971, 88-90.

(**16**) Der *Codex orientalis 162* der Leidener Universitätsbibliothek, in: *XIIe Congrès International d'Histoire des Sciences, Paris 1968, Actes*, tome III A, Paris (Albert Blanchard) 1971, 25-31.

(**17**) Die Traktate *De Proportionibus* von Jordanus Nemorarius und Campanus, in: *Centaurus*, **15** (1971), 193-227.

(**18**) Die Quellen von Nicole Oresme, in: *XIII. Internationaler Kongreß für Geschichte der Wissenschaft, UdSSR, Moskau, 18-24 August 1971, Colloquium: Wissenschaft im Mittelalter; Wechselbeziehungen zwischen dem Orient und Okzident*, Moskau (Nauka) 1971 (55 pp.). Separately published in: Janus **58** (1971), 161-193.

(**19**) Der *Liber de arcubus similibus* des Ahmed ibn Jusuf, in: *Annals of Science*, **30** (1973), 381-406 (with P. S. van Koningsveld).

(**20**) Über einige Euklid-Kommentare und Scholien, die im Mittelalter bekannt waren, in: *Janus*, **60** (1973), 53-58.

(**21**) Über einige Euklid-Scholien, die den Elementen von Euklid, übersetzt von Gerard von Cremona, angehängt worden sind, in: *Centaurus*, **18** (1974), 97-128.

(**22**) Ein mittelalterlicher Euklid-Kommentar, der Roger Bacon zugeschrieben werden kann, in: *Archives Internationales d'Histoire des Sciences*, **24** (1974), 199-218.

(**23**) The second part of chapter 5 of the *De arte mensurandi* by Johannes de Muris, in: *For Dirk Struik*, ed. R. S. Cohen et al., Dordrecht (D. Reidel) 1974, 147-167.

(**24**) Zum Gedenken an Prof. Dr. J. E. Hofmann, in: *RETE*, **2** (1974), 298-302.

(**25**) Über die Übermittlung der *Elemente* Euklids über die Länder des Nahen Ostens nach West-Europa, in: *XIVth International Congress of the History of Science, Tokyo & Kyoto, Japan, 19-27 August, 1974. Proceedings No. 2*, Tokyo 1975, 31-34.

(**26**) Über die Überlieferung der *Elemente* Euklids über die Länder des Nahen Ostens nach West-Europa, in: *Historia Mathematica*, **3** (1976), 279-290.

(**27**) Der Traktat *De isoperimetris*, der unmittelbar aus dem Griechischen ins Lateinische übersetzt worden ist, in: *Mediaeval Studies*, **42** (1980), 61-88.

(**28**) *The Latin translation of the Arabic version of Euclid's* Elements *commonly ascribed to Gerard of Cremona*, Leiden (New Rhine Publishers), 1983 (XXVIII pp. + 503 coll.).

(29) *The First Latin Translation of Euclid's* Elements *Commonly Ascribed to Adelard of Bath*, Toronto (Pontifical Institute of Mediaeval Studies), 1983 (VI + 425 pp.).

(30) Some Early Adaptations of Euclid's *Elements* and the Use of its Latin Translations, in: *Mathemata. Festschrift für Helmuth Gericke*, ed. M. Folkerts and U. Lindgren, Stuttgart (Franz Steiner), 1985, 129-164.

(31) *The Mediaeval Latin Translation of Euclid's Elements Made Directly from the Greek*, Stuttgart (Franz Steiner), 1987 (411 pp.).

(32) *Jordanus de Nemore, De elementis arithmetice artis. A Medieval Treatise on Number Theory. Part I: Text and Paraphrase, Part II: Conspectus Siglorum and Critical Apparatus*, Stuttgart (Franz Steiner), 1991 (372 pp., 188 pp.).

(33) *Robert of Chester's (?) Redaction of Euclid's* Elements: *the so-called Adelard II Version*, 2 Vols., Basel, Boston, Berlin (Birkhäuser), 1992 (950 pp.) (with M. Folkerts).

(34) The *Arithmetica* of Jordanus Nemorarius, in: *Amphora. Festschrift für Hans Wussing*, ed. S. Demidov, M. Folkerts, D. Rowe, C.J. Scriba, Basel, Boston, Berlin (Birkhäuser), 1992, 121-132.

In press:

(35) *Lateinische Euklidübersetzungen und -bearbeitungen aus dem 12. und 13. Jahrhundert*. To be published in the Wolfenbüttel colloquium on history of medieval mathematics.

(36) *A Thirteenth-Century Adaptation of Robert of Chester's Version of Euclid's* Elements. To be published in: Algorismus, München (Institut für Geschichte der Naturwissenschaften).

(37) *Johannes de Muris,* De arte mensurandi. *A Geometrical Handbook of the Fourteenth Century.* To be published with Franz Steiner Verlag (Stuttgart).

Dr. Busard contributed the following articles to C.G. Gillispie (ed.), *Dictionary of Scientific Biography*, New York, Charles Scribner's sons, 1970-1976: Bouvelles, Charles (vol. II, 360-361); Buot, Jacques (vol. II, 592-593); Carcavi, Pierre de (vol. III, 63-64); Clavius, Cristoph (vol. III, 311-312); Deparcieux, Antoine (vol. IV, 38-39); Despagnet, Jean (vol. IV, 74-75); Frenicle de Bessy, Bernard (vol. V, 158-160); Guldin, Paul (vol. V, 588-589); Hardy, Claude (vol. VI, 112-113); Henry of Hesse (vol. VI, 275-276); La Faille, Charles de (vol. VII, 557-558); Lansberge, Philip van (vol. VIII, 27-28); Le Paige, Constantin (vol. VIII, 250); Pitiscus, Bartholomeo (vol. XI, 3-4); Roomen, Adriaan van (vol. XI, 532-534); Ver Eecke, Paul (vol. XIII, 615-616); Viète, François (vol. XIV, 18-25).

Dr. Busard contributed the following articles to the *Lexicon des Mittelalters*, München/Zürich (Artemis), 1980 - ... :
Bartholomaeus von Parma: Bd. 1, Sp. 1496; Bonatti, Guido v. Forli: Bd. 2, Sp. 402; Bonfils (Immanuel ben Jakob): Bd. 2, Sp. 411; Bradwardine, Thomas: Bd. 2, Sp. 538-539; Domninos von Larissa: Bd. 3, Sp. 1226; Dorotheus von Sidon: Bd. 3, Sp. 1321-1322; Geometrie/Erdmessung: Bd. 4, Sp. 1271-1273; Gerhard von Brüssel: Bd. 4, Sp. 1317; Hermann von Carinthia: Bd. 4, Sp. 2166; Johannes (Danck) de Saxonia (J. Danekow de Magdeborth): Bd. 5, Sp. 568; Johannes de Muris (Jehan de Murs): Bd. 5, Sp. 591; Jordanus Nemorarius (J. de Nemore): Bd. 5, Sp. 628.

A challenge to the mathematicians of the University of Leuven as a New Year's gift for 1639

Paul Bockstaele

In the Albert I Royal Library in Brussels, there is, bound together with mathematical and astronomical works,[1] a rare document that throws some light on the practice of mathematics at the University of Leuven in the first half of the seventeenth century. It is a broadside in which an anonymous author offers three mathematical problems to the mathematicians of the Leuven University as a New Year's Gift for 1639 (Figure 1). The title is: *Strenae Mathematicae, Ex Scientia occultorum Numerorum desumptae, omnibusque Mathematicis huius almae Universitatis Lovaniensis anno 1639, Kalendis Ianuarij propositae.* It was printed in Leuven by Cornelius Coenesteyn, probably in 1638. What follows is an analysis of this document and an investigation of the identity of the author. First we give a translation of the paper. The mathematical symbols used in it have been left unchanged.

Mathematical new year's gifts

Selected from the Science of the Hidden Numbers[2] and offered to all the Mathematicians of the salutary University of Leuven on the first day of the year 1639.

First

In the introduction to his *Methodus Polygonorum,* the most noble and illustrious Adrianus Romanus presents to all the mathematicians of the entire world a problem derived from this science. For three examples he gives the solution. It is asked that these solutions be proven, and if errors occur in the formulation or in the solution, to indicate and correct them.

Second

Given four straight lines A, B, C, D whereof, if A is 1(1), B is equal to $5(1) - 5(3) + 1(5)$. If B is 1(1), then C is $3(1) - 1(3)$ and D is $9(1) - 30(3) + 27(5) - 9(7) + 1(9)$. C and D form together with the hypotenuse 2 a right triangle. It is asked that the aforesaid straight lines be expressed in radical signs. However, if one takes a fifth straight line E to be 1(1), then A is $5(1) - 5(3) + 1(5)$. It is asked how great is E, so close that it neither falls short of, nor exceeds [the value] by 1/10000000000000000 in absolute numbers.

[1] Inc. B 625.

[2] By the "scientia occultorum numerorum" is meant algebra.

STRENÆ MATHEMATICÆ,

Ex Scientia occultorum Numerorum desumptæ, omnibusq; Mathematicis huius almæ Vniuersitatis LOVANIENSIS anno CIↃ. IↃC. XXXIX. Kalendis Ianuarij propositæ.

PRIMA.

SVmmus Vir ac Clariss. D. D. ADRIANVS ROMANVS in Præfatione Methodi Polygonorum, omnibus totius orbis Mathematicis Problema, ex hac scientia deductum, proposuit, cuius tria exempla ipse soluit. Hæ igitur Solutiones demonstrandæ proponuntur; &, si alibi in proponendo aut soluendo error subsit, hic detegendus atque emendandus petitur.

SECVNDA.

SVnt quatuor Rectæ, A. B. C. D. quarum prima A si fuerit I (1), B erit 5 (1) — 5 ③ + I (5). Quod si B fuerit I (1). erit C 3 (1) — I (3) D verò 9 (1) — 30 (3) + 27 (5) — 9 (7) + 1 (9). C verò & D cum hypothenusa 2 triangulum constituunt rectangulum. Quæruntur Rectæ prædictæ A. B. C. D. in numeris Radicalibus adæquatis. Si verò assumatur quinta E fueritq; I (1) tum A sit 5 (1) — 5 (3) + 1 (5). Quæritur, quanta sit E? Tam propè ne $\frac{1}{10000000000000000}$ aut deficiat aut excedat in numeris absolutis.

TERTIA.

OMnium Æquationum Algebraicarum, à plurimis profundioris doctrinæ Viris hactenus tantopere desiderata, Vniuersalis Methodus Analytica petitur.

SED

Quis leget hæc? vel Duo vel Nemo,
Turpe & miserabile.

Typis CORNESTENII.

Figure 1

THIRD

A general analytic method for the solution of algebraic equations, which is so much desired by many learned men, is asked for.

BUT

Who will read this? Maybe two, maybe nobody.
A shame and a pity.

Printed by COENESTENIUS

With three algebraic problems, the unknown author challenges the Leuven mathematicians to demonstrate their learning and knowledge. However, he did not expect much of a result. The first problem was taken from a work of Adriaan van Roomen,[3] who taught medicine and mathematics in Leuven from around 1586 to 1592. At the beginning of 1593, he left for Würzburg to teach medicine at the just established University. As a mathematician, he was interested in the compilation of goniometric tables and everything related to it: the calculation of the sides of regular polygons, the n-section problem of an angle, squaring the circle, and so on. And these are precisely the problems raised in the work cited in the *Strenae*. It appeared in 1593 under the title of *Ideae Mathematicae Pars Prima, sive Methodus Polygonorum qua laterum, perimetrorum et arearum cujuscunque polygoni investigandorum ratio exactissima et certissima; una cum circuli quadratura continentur.*[4] In the introduction, van Roomen told how he had devoted himself for years to the finding of a general rule for the calculation of the sides of regular polygons. He had discovered three methods, one of which made use of algebraic equations. For all regular polygons from the triangle to the 80-sided polygon, he had already formulated the matching equation. He hoped to publish the results of his research in some ten books. The first four constitute his *Ideae Mathematicae Pars Prima.* The rest of this planned work never appeared, apart from a few fragments published under other titles. In these first four books, van Roomen gives the sides of the regular convex 3-, 4-, 5-, and 15-sided polygon inscribed in a circle with radius 1, and of polygons that result by repeated doubling of the number of sides, accurate to 32 decimal places. He continued the calculations to the polygon with 15×2^{60} sides and derived the number π to 16 decimal places.

At the end of the *Praefatio* appears the problem to which the pamphlet refers (Figure 2). Under the title *"Mathematical problem presented for solution to all the mathematicians of the world"*, van Roomen asks that the following equation of the 45th degree be solved:

[3]On Adriaan van Roomen or Adrianus Romanus (1561-1615) see P. Bockstaele in *Nationaal Biographisch Woordenboek*, vol. 2 (Brussels 1966), s. v. Roomen, col. 751-765.

[4]Printed in Leuven by Joannes Masius. A part of the edition, however, was taken over by the Antwerp printer and publisher Joannes van Keerbergen. In some copies, therefore, Masius's name and address are replaced by *Antwerpiae, Apud Ioannem Keerbergium.*

PROBLEMA MATHEMATICVM

omnibus totius orbis Mathematicis ad conſtruendū propoſitum.

SI duorum terminorum prioris ad poſteriorem proportio ſit, ut 1 (1) ad 45 (1) -- 3795 (3) + 9,5634 (5) — 113,8500 (7) + 781,1375 (9) — 3451,2075 (11) + 1,0530,6075 (13) -- 2,3267,6280 (15) + 3,8494,2375 (17) -- 4,8849,4125 (19) + 4,8384,1800 (21) -- 3,7865,8800 (23) + 2,3603,0652 (25) — 1,1767,9100 (27) + 4695,5700 (29) -- 1494,5040 (31) + 376,4565 (33) -- 74,0259 (35) + 11,1150 (37) — 1,2300 (39) + 945 (41) -- 45 (43) + 1 (45), deturque terminus poſterior, invenire priorem.

Exemplum primum datum.

SIt terminus poſterior *r bin.* 2 + *r bin.* 2 + *r bin* 2 + *r* 2. quęritur terminus prior. SOLVTIO. Dico terminū priorem eſſe *r bin.* 2 - *r bin.* 2 + *r bin.* 2 + *r bin.* 2 + *r* 3.

Exemplum ſecundum datum.

Sit terminus poſterior *r bin.* 2 + *r bin.* 2 - *r bin.* 2 - *r bin.* 2 - *r bin.* 2 - *r* 2. quæritur terminus prior. SOLVTIO. Terminus prior eſt *r bin.* 2 - *r bin.* 2 + *r bin.* 2 + *r bin.* 2 + *r bin.* 2 + *r* 3.

Exemplum tertium datum.

Sit terminus poſterior *r bin.* 2 + *r* 2, quæritur terminus prior.
SOL. Terminus prior eſt *r bin.* 2 - *r quadrin.* 2 + *r* $\frac{3}{16}$ + *r* $\frac{15}{16}$ + *r bin.* $\frac{5}{8}$ - *r* $\frac{5}{64}$

Si in numeris abſolutis ſolinomijs id proponere libuerit: Sit poſterior terminus $r\,3\,\frac{4142,1356,2373,0950,4880,1688,7242,0969,8078,5696,7187,5375}{10000,0000,0000,0000,0000,0000,0000,0000,0000,0000,0000,0000}$.

Quæritur terminus prior. SOLVTIO. Terminus prior erit

$$r\,\frac{27,4093,0490,3522,5243,1015,8831,2112,6838,8180}{10000,0000,0000,0000,0000,0000,0000,0800,0000,0000}$$

EXEMPLVM QVÆSITVM.

SIt poſterior terminus *r trinomia* 1 $\frac{3}{4}$ -- *r* $\frac{5}{16}$ -- *r bin.* 1 $\frac{7}{8}$ -- *r* $\frac{45}{64}$ quæritur terminus prior. Hoc exemplum omnibus Mathematicis ad conſtruendum ſit propoſitum. Non dubito quin *Ludolf van Cellen* ejus ſolutionem, ſaltem in numeris ſolinomijs ſit inventurus.

M E-

Figure 2

$$45x - 3795x^3 + 95634x^5 - 1138500x^7 + 7811375x^9 - 34512075x^{11} + 105306075x^{13} - 232676280x^{15} + 384942375x^{17} - 488494125x^{19} + 483841800x^{21} - 378658800x^{23} + 236030652x^{25} - 117679100x^{27} + 46955700x^{29} - 14945040x^{31} + 3764565x^{33} - 740259x^{35} + 111150x^{37} - 12300x^{39} + 945x^{41} - 45x^{43} + x^{45} = b$$

for

$$b = \sqrt{1\frac{3}{4} - \sqrt{\frac{5}{16}} - \sqrt{1\frac{7}{8} - \sqrt{\frac{45}{64}}}}.$$

This is, at first sight – even for an exceptionally gifted calculator like van Roomen – an impossible task. It is obvious that the meaning of the question cannot be the solution of an arbitrarily chosen nth-degree equation. The way to the solution must lie somewhere else, namely, in the recognition of the geometric problem that is offered under the guise of this algebraic equation. Perhaps to give his readers a hint, van Roomen gives three examples:

Example 1: If, in the given equation,

$$b = \sqrt{2 + \sqrt{2 + \sqrt{2 + \sqrt{2}}}},$$

then the solution is

$$x = \sqrt{2 - \sqrt{2 + \sqrt{2 + \sqrt{2 + \sqrt{3}}}}}.$$

Example 2: If

$$b = \sqrt{2 + \sqrt{2 - \sqrt{2 - \sqrt{2 - \sqrt{2 - \sqrt{2}}}}}},$$

then the solution is

$$x = \sqrt{2 - \sqrt{2 + \sqrt{2 + \sqrt{2 + \sqrt{2 + \sqrt{3}}}}}},$$

Example 3: If one takes $b = \sqrt{2 + \sqrt{2}}$, then the solution is

$$x = \sqrt{2 - \sqrt{2 + \sqrt{\frac{3}{16}} + \sqrt{\frac{15}{16}} + \sqrt{\frac{5}{8} - \sqrt{\frac{5}{64}}}}}.$$

In the last example, van Roomen gives the value of b and of the solution also in decimals, b accurate to $1/10^{48}$, the solution to $1/10^{40}$.

Among the problems that captured the attention of mathematicians in van Roomen's time were the squaring of the circle and the compilation of goniometric tables. In this, the calculation of the sides of regular polygons as a function of the radius of the circumscribed circle occupied an important place. For those working actively on it, the complicated radical expressions in van Roomen's problem were easily recognizable as sides or diagonals of regular polygons inscribed in a circle with radius 1. Once one has grasped this, one has the first indication about the direction in which the solution has to be sought.

Let us analyse the three examples van Roomen gives. As abbreviations, we will use the following symbols:
s_n for the side of the regular n-gon inscribed in a circle with radius 1;
$s_{n,p}$ for the diagonal of the regular n-gon that subtends an arc of $p\times 360°/n$;
chd α for the chord of arc α;
arc a for the arc subtended by chord a.

In Example 1, the second member b of the equation appears to be the chord of the supplement of arc s_{32} which is the diagonal $s_{32,15}$:

$$b = \text{chd }(180°\text{- arc } s_{32}) = s_{32,15} = \text{chd } 168°45'.$$

As a solution, van Roomen gives s_{96} = chd 3°45′. And it turns out that $45 \times 3°45' = 168°45'$. A solution of the equation is thus the chord of the forty-fifth of arc b.

In the second example, van Roomen gives as the solution the side s_{192} of the 192-sided polygon, which is the chord of 1°52′30″. As the value for b, one now expects the chord of $45 \times 1°52'30'' = 84°22'30''$, which is

$$\text{chd }(90°\text{- arc } s_{64}) = s_{64,15} = \sqrt{2-\sqrt{2-\sqrt{2+\sqrt{2+\sqrt{2}}}}}.$$

Van Roomen apparently made a mistake here, for the value given for b is $s_{128,43}$ = chd 120°56′15″.

In the third example, b = chd (180°- arc s_8) = $s_{8,3}$ = chd 135°. As a solution, van Roomen gives s_{120} = chd 3°. Because $45 \times 3° = 135°$, one finds here, too, that the given solution is the chord of the 45th part of arc b.

The problem hidden behind van Roomen's equation thus comes down to this: Given the chord of an arc, find the chord of the forty-fifth of that arc. Once one grasps this, it is no longer so difficult to solve the problem. What he gives as the value for b is s_{15} = chd 24°. The solution will thus be

$$x = s_{675} = \text{chd }(\tfrac{1}{45}\cdot 24°) = \text{chd } 32' = 2 \sin 16'.$$

In contrast with the chords from the examples, this can no longer be expressed by radicals. An approximate value is given by the sine table.

Van Roomen's challenge to all the mathematicians of his time did not remain without response. François Viète, *maître des requêtes* and counselor to Henry IV of France, published in 1595 in Paris his ***Ad problema quod omnibus Mathematicis totius orbis construendum proposuit Adrianus Romanus Responsum.*** Already in the 1570s, Viète had worked on the problem of the angle division and had obtained some remarkable results. We are particularly interested here in the formulas in which he expressed the chord of an uneven multiple of an arc α in terms of the chord of α. If the radius of the circle is 1, and if chd $\alpha = x$, then one has:

$$\begin{aligned}
&\text{chd } 3\alpha = 3x - x^3\\
&\text{chd } 5\alpha = 5x - 5x^3 + x^5\\
&\text{chd } 7\alpha = 7x - 14x^3 + 7x^5 - x^7\\
&\text{chd } 9\alpha = 9x - 30x^3 + 27x^5 - 9x^7 + x^9\\
&\text{etc.}
\end{aligned}$$

Viète also tells how the coefficients can be calculated for an arbitrary n. For n=45 the formula becomes:

$$\text{chd } 45\alpha = 45x - 3795x^3 + 95634x^5 - \ldots - 45x^{43} + x^{45},$$

which is nothing other than van Roomen's famous equation. Viète immediately recognized it. In his answer, he gives not only the solution chd 32′ = 2 sin 16′, but also the 22 other positive roots:

$$x = 2 \sin\,(16' + k \cdot \tfrac{360°}{45}),\ k = 1, 2, 3, \ldots, 22.$$

After this long excursus let us read again the first problem of the Leuven pamphlet. It is asked to prove the three examples van Roomen gives, and to indicate and correct any oversights or errors in the formulation or in the solution. From the way the question is formulated, it is clear that the author of the paper knew Viète's *Responsum.* Indeed, in Viète we find criticisms of the way in which van Roomen formulated the problem as well as corrections of the examples and supplements to the solutions. Viète begins by noting "that van Roomen's problem is ridiculous, if it is not corrected."[5] For van Roomen formulated it as follows: when $a : b = x : (45x - 3795x^3 + \ldots + x^{45})$, then calculate a when b is given. Correctly, Viète pointed out that this problem is totally undetermined, because, with b given, one can choose x arbitrarily and from it then determine a.[6]

We have already pointed out that the value given for b is wrong in van Roomen's second problem. Viète also observed this. As a correction he proposed

$$b = \sqrt{2 - \sqrt{2 - \sqrt{2 + \sqrt{2 + \sqrt{2 + \sqrt{2}}}}}} = s_{128,31} = \text{chd } 87°11'15'',$$

[5]"Γελοῖον autem est Adriani Problema, nisi emendetur." ***Responsum,*** f. 2r.

[6]This curious interpretation of equations as proportions van Roomen took from Stevin's ***Arithmetique,*** p. 264-266.

which, however, is also wrong.

Moreover, he had still other objections to the examples. According to him, they contributed nothing to the finding of the solutions because, by the purely analytical presentation of the problem, its geometrical origin remained entirely hidden. The last criticism concerns the incompleteness of van Roomen's solution: he only gives one root each time.

The second problem from the *Strenae* is a system of five equations with five unknowns in which equations appear of the second, third, fifth and ninth degree. For the indication of the powers of the unknown, the author uses the same sybols as Simon Stevin did in his *Arithmetique* (Leiden, 1585) and van Roomen in his *Ideae mathematicae*.[7] (1) or 1(1) represents the first power of the unknown x, (2) the second power, and so on; 1(3) means x^3, 3(5) is $3x^5$. etc. Transcribed in our modern symbols, the problem looks like this: calculate A, B, C, D and E if they satisfy:

(1) $B = 5A - 5A^3 + A^5$
(2) $C = 3B - B^3$
(3) $D = 9B - 30B^3 + 27B^5 - 9B^7 + B^9$
(4) $C^2 + D^2 = 4$ (C and D are sides of a right triangle with hypotenuse 2)
(5) $A = 5E - 5E^3 + E^5$.

From the formulation, it is clear that we are dealing here with a geometric problem and not a purely algebraic problem. One recognizes immediately three of the multiple-angle formulas of Viète given above. This leads us to interpret A, B, C, D and E as chords in a circle with radius 1. Equations (2) and (3) indicate that C and D are the chords of, respectively, the threefold and the ninefold of the arc subtended by B:

$$\text{arc } C = 3 \text{ arc } B \text{ and arc } D = 9 \text{ arc } B.$$

From (4), it follows that arc C + arc $D = 180°$. Combined with the preceding, this gives

$$12 \text{ arc } B = 180°, \text{ arc } B = 15°, B = s_{24}.$$

Starting from $s_6 = 1$, one can find by repeatedly doubling the number of sides:

$$B = s_{24} = \sqrt{2 - \sqrt{2 + \sqrt{3}}}.$$

C is the chord of $3 \times 15° = 45°$, which is $s_8 = \sqrt{2 - \sqrt{2}}$. From (4), D is now immediately found: $D = \sqrt{2 + \sqrt{2}}$. From (1), arc $B = 5$ arc A. Because

[7] Stevin used a circle within which the exponent was written. Probably for typographic reasons, the circle was replaced by parentheses in the pamphlet.

$B = s_{24}$, $A = s_{120}$. Proceeding from s_{15}, by doubling the number of sides three times, one finds;

$$A = s_{120} = \sqrt{2 - \sqrt{2 + \sqrt{\frac{3}{16}} + \sqrt{\frac{15}{16}} + \sqrt{\frac{5}{8} - \sqrt{\frac{5}{64}}}}}.$$

Equation (5), finally, indicates that arc $A = 5$ arc E. Because A is the side of a regular 120-gon, E is the side of the 600-gon:

$$E = s_{600} = \text{chd } 36' = 2 \sin 18'.$$

The third New Year's gift is undoubtedly the most remarkable one. After 16-th century mathematicians like dal Ferro, Tartaglia, Cardan and Ferrari had succeeded in working out general methods for the solution of third- and fourth-degree equations, the demand became ever more urgent for a generally applicable way to solve algebraic equations, in connection with the problem of the angle n-section, among other reasons. It is this demand that our anonymous author articulated in his third New Year's gift. An adequate answer he certainly did not receive. The last words of the pamphlet, indeed, indicate that he did not have any great expectations and that he did not have a very high opinion of the knowledge and learning of the Leuven mathematicians of that time. Moreover, one may ask whether he himself grasped the fundamental difference between the first two, originally geometric problems, and the last purely analytical or algebraic problem.

This analysis of the *Strenae* says nothing yet about its author. For the time being, a lack of information makes it impossible to postulate a name with any certainty. What follows are only possibilities and assumptions based on what we know about the teaching of mathematics in Leuven in the first half of the 17th century.

Already from the 15th century on, the principles of arithmetic and geometry were taught on a modest scale in each of the four pedagogies or colleges that together formed the Leuven faculty of arts. Since 1563 moreover, there had been a public chair for mathematics. Around 1590, a group of students had formed around Adriaan van Roomen. These students who helped him with his calculations, were a kind of working community that undoubtedly stimulated the interest in mathematics in Leuven. However, the departure of van Roomen from Leuven in 1593, together with the perilous situation of the city and the university due to war and occupation had seriously retarded this development. The public chair for mathesis, which remained unoccupied for years, received a new holder only in 1610: Joannes Storms or Sturmius (1559-1650) from Mechelen. Presumably in 1579, he was admitted to the degree of magister in arts in the pedagogy *De Lelie*, where he was assigned the teaching of philosophy six years later. At the same time he began to study medicine. On 9 November 1593 he was admitted to the degree of doctor in medicine. Thus he must certainly have had Adriaan van Roomen as professor and, perhaps, also attended his mathematics lessons. When he was appointed to the chair of mathematics, Sturmius was already fifty years old. He kept this subject in hand for thirty-six more years. In the meantime, in 1634, when he was already seventy-five, he had himself appointed professor in medicine. Sturmius's long professorate probably hindered the study of mathematics in Leuven for years. Around 1638, the year that is of interest here, hardly two or three students attended the public lectures on mathematics. Sturmius became too old to do his job properly. It is not impossible that the scornful comments in the pamphlet and the displeasure that echoes through them were prompted by Sturmius's teaching. A candidate for the chair of mathematics, gradually having become tired of waiting, sent his challenge to the Leuven mathematicians to remind them in that way of the miserable condition of mathematical education.

Among the Leuven professors who manifested some interest in mathematics and astronomy were Ericius Puteanus (1574-1646), successor of Justus Lipsius, and Libert Froidmont or Fromondus (1587-1653). Neither of them, however, comes into consideration as the author of the mathematical pamphlet. Rather, both its style and its contents indicate that the author would have to be sought among the students of Adriaan van Roomen or among those interested in his work. Of those who studied mathematics with him, he mentioned four in the introduction of his *Ideae Mathematicae*.[8] First, there were two medical students, Bernandus Lordel from Mouzon and Joannes vanden Weege from Boxtel, who often helped him with his calculations. He was convinced that they, when they completed their studies, would again devote themselves fully to mathematics. Then he named Thomas Fienus, who was already established in Antwerp as a doctor of medicine and was engaged primarily with astronomy and the construction of sundials. The last student van Roomen mentioned was Cornelis Opmeer from Delft, who, although he studied law, was a faithful attendant in his classes on astronomy. Of these four, Fienus later worked in Leuven, and also vanden Weege, at least if we may identify him with a van der Wegen who stayed in Leuven after 1630. In 1593, Fienus was appointed to the medical faculty, but he died

[8] "...aliquos è meis discipulis praeterire non possum, imprimis duos artium Magistros & Medicinae studiosos, qui in laboribus hisce calculi ferendis, non raro socios sese adjunxerunt M. BERNARDUM LORDEL Mosomensem, & M. IOANNEM vanden WEEGE Boxtalensem, qui hoc tempore non exiguos in Mathesi fecere progressus; atque non dubium est, quin post Medicinae (cui uti par est peculiariter incumbunt) studium, totis viribus contendent, ut ad divinae hujus scientiae culmen perveniant. His quoque adjungere libet D. THOMAM FIENUM Antwerpiensem, qui jam Lovanio relicto, Antwerpiae Medicum agit, primoque mobili, sed peculiariter horologiographiae incumbit. Nec quoque praeterire possum D. CORNELIUM OPMEER Delphium, qui licet juri operam daret, me tamen Mathesin profitentem audivit libentissimè; ita ut in primi mobilis doctrina progressus fecerit non vulgares. Hi ergo & ipsi Mathesin aliquando propriis ornabunt promovebuntque laboribus." *Ideae Mathematicae*, f. **ij verso.

already in 1631, which rules him out as author of the *Strenae*. Van Roomen's Joannes vanden Weege was admited on 2 May 1591 to the degree of *artium magister* and then commenced his medical studies. In August 1593 he signed, together with five other medical students including Lordel, a petition to the Burgomaster and the Council of the City of Leuven to support the candidacy of Fienus for an appointment in the medical faculty.[9] He signed the petition as "Joannes vander Wegen, artium doctor". After the completion of his studies, he returned to the region of his birth. He settled in 's-Hertogenbosch, where he worked primarily as "engineer and mathematician" of the Archduke Albert and Archduchess Isabella. As an architect, he built a powderhouse for the city in the years 1618-1619. He was involved in the defense of 's-Hertogenbosch against the troops of Frederick Henry in 1629.[10] After the capitulation of the city on 14 September 1629 we lose track of him. Did he, together with the bishop, the governor and the garrison, leave the city? Did he afterwards settle in Leuven? On 27 November 1631, a Joannes van der Weghe from 's-Hertogenbosch re-registered at the University of Leuven as a *medicinae studiosus*.[11] After an absence from Leuven, one could, by means of a *reintitulatio*, be taken up again in the university community and enjoy its privileges. It is possible that he was the former student of Adriaan van Roomen who, while waiting for the chair of mathematics occupied by Sturmius to become vacant, hoped to obtain permission to give private lessons. We cannot be certain. The re-registration could also have been that of Joannes van der Weghen from Boxtel, who, ten years previously, on 16 December 1621, was matriculated as a student in the pedagogy *De Valk*.[12] What is certain is that after 1631, a van der Wegen, who had the reputation of a good mathematician, was active in Leuven and even was paid for giving lessons in mathematics. This appears from a few remarks in passing in the correspondence of Descartes and Puteanus, who mention a certain van der Wegen.

Immediately after the appearance of his *Discours de la Méthode*, Descartes sent copies to several scholars, including the Leuven professor Plempius.[13] Plempius mentioned on 15 September 1637 the receipt of three copies, of which he had already given one to Fromondus and another to father Fournet.[14]

For Descartes' *Dioptrique* he had only praise, and he admired *Les Météores*, although he did not share all of Descartes' opinions. The *Géométrie*,

[9] Leuven, Stadsarchief, no. 3990.

[10] H. Bots, I. Matthey, M. Meyer: *Noordbrabantse studenten 1550-1750* (Bijdragen tot de Geschiedenis van het zuiden van Nederland, XLIV), Tilburg, 1979, p. 758.

[11] "Joannes van der Weghe, Sylveducensis, maiorennis, medicinae studiosus, reintitulatus." A. Schillings: *Matricule de l'Université de Louvain*, vol. V, p. 244, no. 115.

[12] "Joannes van der Weghen, Boxtellanus." *Ibid.* p. 95, no. 228.

[13] Vopiscus Fortunatus Plempius was born in Amsterdam and studied in Leuven and in Italy. He became professor in medicine at Leuven in 1633.

[14] Probably François Fournet S. J., born in 1581. He taught theology in Douai, where he died in 1638. For Plempius' letter, see *Oeuvres de Descartes*. Publiées par Ch. Adam et P. Tannery. Vol. 1, Paris 1897, pp. 399-400.

however, he left untouched, for he was a stranger to this field. He promised to pass the work on to others who were interested in such matters. In his answer on 3 October, Descartes wrote the following:[15]

"I know that my *Géométrie* will have only a few readers, for it requires readers not only who are versed in everything that is known up to now in geometry and algebra, but also who are very diligent, very clever, and very observant, because I have omitted everything I judged to be not unknown to others and tried to say or at least touch on much in few words. As I have learned, there are two such people with you: Wendelinus[16] and Vander Wegen." Descartes asked Plempius to provide him with their opinion or that of others of his *Géométrie*. No response ever came.

Six years later we encounter Van der Wegen once again in letters from Puteanus to the astronomer Van Langren.[17] On 25 April 1644 Puteanus wrote the following from Leuven: "Our good lord Wendelinus keeps him in his Herk and his presbytery: what he does, I do not know. It saddens me when I think that such a man is not heard in our academy: he could teach the course in mathesis, which is here now mute. Van der Wegen is here as though he were not here. He draws pay so that he would also teach mathematics, but he thinks that it is sufficient to live here."[18]

Sturmius still occupied the chair for mathematics, but did poorly. Was van der Wegen perhaps paid as his *suppleant* but did he, too, neglect this task? He must have been around seventy at the time if he was the Joannes van der Wegen who studied mathematics under van Roomen.

Concerned over the teaching of mathematics at the university, Puteanus tried about a year later, on 4 February 1645, to persuade Van Langren to establish himself in Leuven to teach mathematics. He wrote: "It would be

[15] "Non ignoro Geometriam meam paucissimos lectores habituram; nam cùm ea scribere neglexerim quae ab alijs sciri suspicabar, & paucissimis verbis multa (imo omnia quae unquam in illâ scientiâ poterunt inveniri) vel complecti vel saltem attingere sim conatus, lectores non modo peritos eorum omnium quae hactenus in Geometriâ & Algebrâ cognita fuere, sed etiam valdè laboriosos, ingeniosos & attentos desiderat. Duos apud vos esse Wendelinum & Van der Waegen. Gratissimum mihi erit, si per te quidquid vel illi vel alij quilibet de ea iudicabunt, accepero." *Oeuvres de Descartes, op. cit.* Vol. 1, Paris, 1897. Letter LXXXVII, p. 411.

[16] Godfried Wendelen, born in Herk-de-Stad in 1580, probably studied at the Leuven faculty of arts. From 1598 to 1612, he lived in the Provence. After his ordination to the priesthood in 1620, he was parish-priest in turn in Geet-Bets and in Herk-de-Stad. He died in Ghent in 1667.

[17] Michel Florent van Langren, mathematician, astronomer and geographer, was born around 1600 and died in Brussels in 1675. In 1629, he was appointed "cosmographer and mathematician of His Majesty in Flanders".

[18] "Onsen goeden heer Wendelinus houdt hem in syn Herck, en syne Pastorie: wat hy doet, en weet ick niet. Bedroeve my, als ick dencke, dat alsulcken man in onse Academie niet gehoort en wordt: wie soude bequaem syn tot de lesse van Mathematica die nu alhier stom is? Vander Wegen is hier, al of hy hier niet en waere. Hy treckt gagie, dat hy oock soude de leeren de Mathematica: dan hem dunckt, dattet genoech is, alhier te woonen." J.J. Moreau: *Honderd veertien Nederlandse brieven van Erycius Puteanus aan de Astronoom Michael Florent van Langren.* Antwerpen 1957, Letter LXXXIX, p. 140.

good if you came to live here in Leuven, both for your name and for the increase of your profit. You would here always have several young men who wish to learn mathematics and would pay well. It is an honor to live in an Academy in which excellent men used to be, but where now there is nobody else but Van der Wegen, who is whole by himself."[19]

From Descartes' letter to Plempius one can deduce that van der Wegen had a reputation as a mathematician and resided in Leuven in 1637. Mathematical education was then already a dismal affair. The almost eighty-year-old Sturmius still occupied the public chair for mathematics. One can imagine that, in such circumstances, a gradually aging Van der Wegen, who had perhaps been hoping for years to succeed Sturmius, vented his exasperation and disappointment in the form of a challenge to the Leuven mathematicians. Direct evidence that this was the case, however, is lacking. After the reading of Puteanus' comments, moreover, it is just as plausible that Van der Wegen was one of the mathematicians aimed at in the pamphlet.

Whatever the case may be, it is certain that in the course of 1639 an attempt was made to reorganize the situation in Leuven somewhat. At the end of 1639, Gerard van Gutschoven[20] was appointed *suppleant* to Sturmius, with the result that the number of students in the mathematics course again rose. Van Gutschoven, who was born in Leuven in 1615, studied there and was admitted to the degree of licentiate in medicine in September 1635. He did not last long as a *suppleant* for Sturmius. Already in 1640, he left for Antwerp, leaving the mathematics chair as good as vacant. Presumably, between 1640 and 1643, he stayed for a time in the Northern Low Countries, where he met Descartes and became one of his most fervent followers. In 1643 or 1644 he returned to Leuven, where he established himself as a physician. Finally, in 1646, Sturmius resigned as professor of mathematics. Gerard van Gutschoven was appointed his successor. From 1659 to his death in 1668 he also taught anatomy, surgery and botany in the faculty of medicine.

With van Gutschoven we have another possible candidate for the authorship of the anonymous pamphlet. As a young scholar, well prepared in mathematics, he undoubtedly hoped to receive an appointment in Leuven before too long. One can well imagine that he, in his impatience, wanted to stress how urgent reform was by means of a mathematical New Year's gift.

And so we conclude our suppositions. Van der Wegen, van Gutschoven, or someone else? Perhaps a chance will provide the solution to this bibliographical riddle.

[19] "Dattet goet waer dat V. L. hier tot Loven quam woonen. Soo om naems, als profyte vermeerderinge. Hier soude V. L. altyt te doen hebben met verscheyde jonge heeren die Mathematica souden willen leeren, ende wel betaelen ... Het streckt oock tot eer, in een Academie te woonen; inde welcke eertyts fraeye mannen geweest syn, ende nu niemant en is dan Vander Wegen, dewelcke gans op hem selven is." J. J. Moreau, *op. cit.,* Letter XCI, p. 143.

[20] On Van Gutschoven see G. Vanpaemel in *Nationaal Biografisch Woordenboek*, vol. 13, Brussels, 1990, s. v. Gutschoven, col. 347-354.

Johann Molther's 'Problema Deliacum', 1619

Henk J. M. Bos

Introduction

In 1619 Johann Molther published at Frankfurt a book about the duplication of the cube, entitled *Problema Deliacum.* It attracted little interest at the time and references to the book in historical works are very scarce.[1] Yet for my studies on the concept of construction I have found the book revealing, and the occasion of the Festschrift for Dr Busard provides a welcome occasion to present a short note on its content, purpose and interest.

About Johann Molther himself little seems to be known.[2] He was born on March 28, 1591 in Grünberg in Hessen (Germany) as son of Johann Molther (1561-1618), pastor and professor of theology and Hebrew at Marburg university. About Molther senior's life we know more than about his son's because the latter included in his *Problema Deliacum* a long funereal poem for his father, detailing the qualities and the career of the deceased. Johann junior studied medicine and became professor of medicine at Marburg University in 1621. His date of death is unknown. Besides the *Problema Deliacum* he published an astronomical treatise and three medical disputations.[3]

In the style of his time, Molther did not spare words in phrasing the title of his study:

The Delian problem of doubling the cube, that is, given any solid, to

[1]Marin Mersenne expounded Molther's construction of two mean proportionals (see below, p. 43) in his *Harmonie universelle contenant la théorie et la pratique de la musique* Paris, 1636 (also fasimile ed. Paris 1965), p. 68; N.Th. Reimer mentioned Molther in his *Historia problematis de cubi duplicatione sive de inveniendis duabus mediis proportionalibus inter duas datas*, Göttingen, 1798; G.J. Toomer mentioned Molther's book in his article on Nicomedes in the *Dictionary of Scientific Biography* (ed. C.C. Gillispie, New York 1970-1980) vol. 10, pp. 114-116. I have found no other references.

[2]My sources for the biographical information are:
Zedler, J.H. (ed.), *Grosses vollständiges Universallexikon aller Wissenschaften und Künste* (64 vols, 4 suppl. vols), Leipzig 1732-1754 (reprint Graz, 1961-1964), vol. 21 (1739) col. 955;
Jöcher, C.G., *Allgemeines Gelehrtenlexicon*, 4 vols, Leipzig 1750-1751;
Adelung, J.C., Rotermund, H.W., Günther, O. *Fortsetzungen und Ergänzungen zu Jöcher's allgemeinem Gelehrtenlexicon*, 7 supplement vols, Leipzig 1784, 1787, Delmenhorst 1810, Bremen 1813, 1816, 1819, Leipzig 1897 (facsimile reprints of these volumes: Hildesheim (Olms), 1960-1961), vol. 3, col. 604 and suppl. vol. 4 col. 1960.

[3]Jöcher (note 2), suppl. vol. 4, col. 1960. The title of the astronomical treatise is given there as "Methodus erigendorum thematum astronom" (sic), Frankfurt 1618. I have not tried to trace these publications.

> make a similar solid in a given ratio, by means of the second Mesolabe,[4] by which two continuously proportional means can be taken. Now at last easily and geometrically solved after innumerable attempts of the most eminent mathematicians. The history of the problem is given first and some results are added about the trisection of an angle, the construction of a heptagon, the quadrature of the circle and two very convenient designs of proportional instruments.[5]

Such wordiness, and especially the claim to have "at last" solved the problem of duplicating the cube, are likely to draw a smile from the modern reader; the tradition of mistaken solutions of the problem, announced with similar pomp, is an old one. Molther's solution, however, was not mistaken, in particular he did not claim, as several others had done before him, to have solved the problem by straight lines and circles (ruler and compass[6]), which cannot be done. And although the mathematical content of his book is definitely unoriginal, there are aspects of his arguments which I find to be of some interest, especially because they relate to questions debated by some contemporaries of higher mathematical standing, such as Viète, Kepler and Descartes.

Duplication of the cube, mean proportionals and neusis

A mean proportional between two magnitudes a and b is a term of a geometrical sequence with a and b as first and last terms respectively. The simplest case occurs when the sequence has three terms: a, x, b, with

$$a : x = x : b \,;$$

x is then called the mean proportional or the geometric mean of a and b. The construction, by straight lines and circles, of the geometric mean between two

[4]The term Mesolabum was used in classical Greek geometry to denote an instrument for constructing mean porportionals. In the sixteenth and seventeenth centuries the term, also spelled *mesolabium*, no longer had a strictly instrumental connotation; it meant in general the art of constructing mean proportionals. Molther, however, used it in the classical sense. He called 'second' mesolabe the instrument for constructing *two* mean proportionals.

[5]*Problema Deliacum de cubi duplicatione, hoc est de quorumlibet solidorum, interventu Mesolabii secundi, quo duae capiantur mediae continue proportionales sub data ratione similium fabrica. Nunc tandem post infinitos praestantissimorum mathematicorum conatus expedite et geometrice solutum. Ubi historia problematis praemittitur, et simul nonnulla de anguli trisectione, heptagoni fabrica, circulique quadratura et duabus commodissimis instrumentorum proportionum formis inserentur.* The book was published at Frankfurt, "Typis ac sumptibus Antonii Hummii", in 1619. I have consulted the copy in the Bayerische Statsbibliothek at Munich (4° Math P 237) and a xerox of the copy in the library of Brown University, Providence, R.I., kindly provided by Prof. G.J. Toomer.

[6]In the following I use the terminology 'by straight lines and circles' rather than 'by ruler and compass' for constructions according to the Euclidean postulates because Euclid did not mention instruments to perform these constructions.

line segments is given in Euclid's *Elements* (II-13 and VI-14). The case of two mean proportionals arises when the sequence has four terms: a, x, y, b. Thus the problem of constructing two mean proportionals is: Given two line segments a and b, it is required to construct two line segments x and y such that a, x, y and b form a geometric sequence, i.e.:

$$a : x = x : y = y : b \,.$$

This problem was formulated and solved in classical Greek geometry. Tradition has it that the problem arose in connection with one of the three 'classical problems', namely the duplication of the cube: To construct a cube twice as large (in content) as a given cube. In algebraic terms, if the edge of the given cube is called a, it is required to find the edge x of a cube determined by $x^3 = 2a^3$. If we write $y = x^2/a$, the equation implies $a : x = x : y = y : 2a$, so x is the first of two mean proportionals between a and $2a$. Greek geometers also realized this fact (be it not exactly along the algebraic line above) and knew that hence a method for constructing two mean proportionals would imply a method to duplicate the cube.

Molther also reduced the duplication problem to a construction of two mean proportionals, for which he used an expedient called 'neusis'.[7] Greek geometers from the classical period had found that several problems which could not be constructed by straight lines and circles, could be reduced to the so-called 'neusis' problem. This problem was:

> **Problem** (neusis)
> Given: two straight lines l and m (see Figure 1), a point O and a line segment a.
> Required: to find a line through O, intersecting l and m in points A and B respectively such that $AB = a$.

The problem requires the segment a to be placed between the lines l and m in such a way that it points in the direction of O.[8] There were variants of the problem in which one or both of the straight lines were replaced by circles. In particular cases (for instance when the distances of O to l and m are equal) the neusis construction can be achieved by straight lines and circles only, but not in general. I shall use the term 'construction by neusis' for a construction which reduces a problem to a neusis problem, and I shall call a 'neusis procedure' any procedure to actually construct the solution of a neusis problem. In his book Molther solved the problem of two mean proportionals by a neusis construction and he explained a neusis procedure.

[7]On neusis in classical geometry see for instance *The works of Archimedes* (ed. T.L. Heath), New York (Dover), 1953 (reprint of ed. Cambridge 1897-1912), pp. c-cxxii, and W.R. Knorr, *The ancient tradition of geometric problems*, Boston etc. (Birkhäuser), 1986, pp. 178-187 and index *s.v.* 'neusis'.

[8]This explains the name of the problem: the segment has to be placed such that it 'verges' towards P; neusis derives from the Greek verb 'neuein' which means 'to verge'.

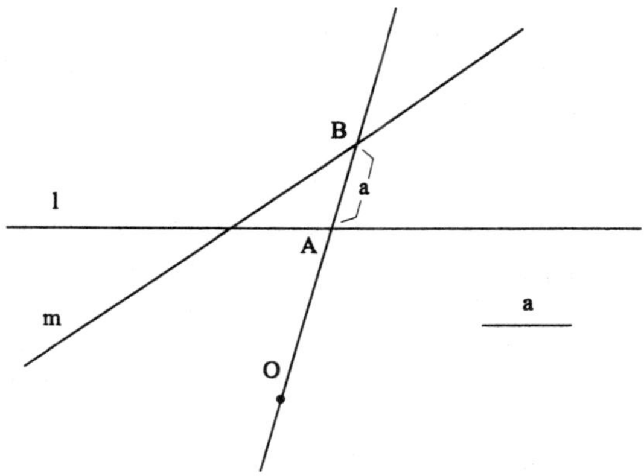

Figure 1

Neusis: the classical heritage

The mathematical techniques which Molther applied in his neusis construction were of classical origin and well known at his time. The two main sources from which early modern mathematicians learned about neusis and similar constructions and procedures were Eutocius' commentary to Proposition II-2 of Archimedes' *Sphere and Cylinder*, available since the beginning of the sixteenth century,[9] and Pappus' *Collectio*, first edited in print in 1588.[10]

In Proposition II-2 of the *Sphere and Cylinder* Archimedes assumed without further explanation the possibility of constructing two mean proportionals. In his commentary Eutocius gave twelve different constructions of this problem ascribing them to Plato, Heron, Philo of Byzantium, Apollonius, Diocles, Pappus, Sporus, Menaechmus, Archytas, Eratosthenes and Nicomedes.[11] The one by Nicomedes was a neusis construction. Several other solutions from Eutocius' list employed constructions like the neusis;

[9]The constructions which Eutocius mentioned were published in Georgius Valla, *De expetiendis et fugiendis rebus opus*, Venice 1501, book XIII, Ch. II, fols Uv^r-Xiv^r. This was a rather unsatisfactory version of the text. Johannes Werner gave a better edition in his *Libellus super viginti duobus elementis conicis (-) commentarius (-) cubi duplicatio (-)* Nürnberg, 1522, fols Civ^r - Hiv^r. The commentary of Eutocius became available in printed form in the Basel 1544 edition of Archimedes' *Opera* by Th. Geschauff. At the moment the text is easiest accessible through Vereecke's French translation: *Les oeuvres complètes d'Archimède suivies des commentaires d'Eutocius d'Ascalon* (tr. P. Vereecke), Paris, 1921.

[10]*Pappi Alexandrini mathematicae collectiones* (tr. ed. F. Commandino), Pesaro 1588. French translation: Pappus, *La collection mathématique* (tr. Paul Vereecke), Paris, 1933.

[11]There is one unattributed solution, mentioned after the one by Menaechmus.

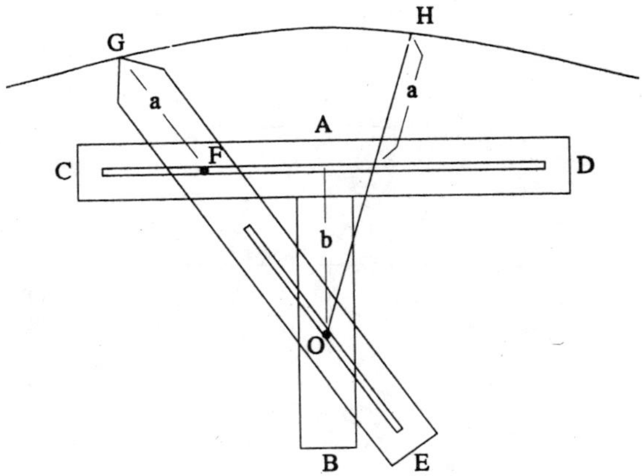

Figure 2

these were to be performed by shifting a marked ruler over the figure until, by trial and error, the correct position was found. Eutocius also mentioned a curve invented by Nicomedes by which a neusis problem could actually be solved. The curve has become known as the 'Conchoid of Nicomedes'. It is the curve traced by an instrument as in Figure 2, consisting of a system of perpendicular rulers AB, CD and a movable ruler EG. CD and EG have slots along their central lines; at O on AB and at F on EG pins are fixed which fall in the slots as shown. The distances $FG = a$ and $AO = b$ are constant (Nicomedes probably considered an instrument in which these distances were adjustable). If the ruler EG is moved the point G describes the conchoid. The pins and the slots ensure that in all its positions EG passes through the center O while F remains on the line CD. Thus any point H on the conchoid has the property that its distance to the base line CD measured along the line HO is equal to a. Eutocius showed that if for any given center, base line and segment the conchoid of Nicomedes can be drawn, the neusis problem can be solved (see Figure 3): draw the conchoid with respect to center O, segment a and base line l, let its intersection with m be B, then OB is the required line, intersecting l in A, with $AB = a$.

In his *Collectio* Pappus gave a classification of geometrical problems based on the curves used in their construction. He distinguished 'plane', 'solid', and 'line-like' problems. Plane problems were those constructible by the Euclidean means of straight lines and circles; 'solid' ones were those which, although not plane, could be constructed by the intersection of straight lines, circles and one or two conic sections. If the construction of a problem required more complicated curves than conic sections, they were 'line-like'.[12]

[12] *Collectio* (cf. note 10) book III, introduction to Proposition 5 and book IV, introduction to Proposition 34.

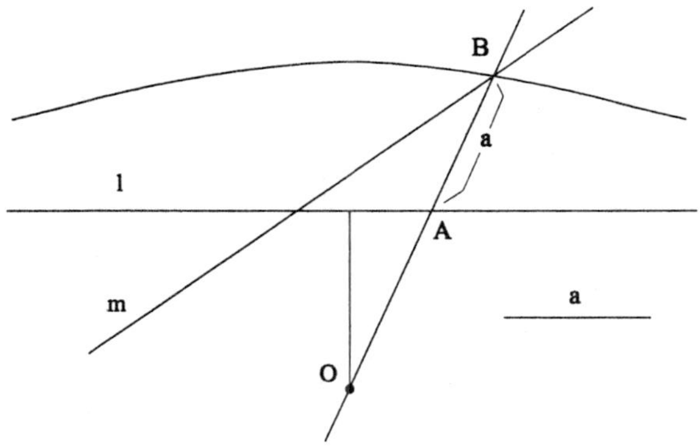

Figure 3

Pappus also discussed the neusis and its use in solving problems such as trisecting an angle or finding two mean proportionals. He was aware that the neusis problem was not plane; he proved that it was 'solid' by providing a construction by means of the intersection of a hyperbola and a circle.[13] The *Collectio* also contained a trisection by neusis.[14]

Thus the classical sources were quite informative about propositions and techniques relevant to the neusis procedure. They were, however, inconclusive with respect to the question whether procedures similar to the neusis, employing other means of construction than straight lines and circles and thereby being outside the Euclidean bounds of geometrical procedure, were legitimate. Eutocius' list of constructions suggested that the ancient mathematicians had not reached a consensus about which method was the most proper for solving the problem of mean proportionals. Eutocius himself hardly discussed the question of the geometrical legitimacy of the constructions; as far as he expressed preferences, they concerned the practical ease of the constructions, not their theoretical exactness. Pappus' classification of problems and his treatment of the neusis suggested that construction by intersection of conic sections should be considered as more geometrical than by using a conchoid or another neusis procedure.

[13] *Collectio* book IV Proposition 31.
[14] *Collectio* book IV Proposition 32.

Neusis: early modern interest

In the 1590's Viète gave the neusis construction a prominent place in his "Restored mathematical analysis or new algebra".[15] In his *Supplementum Geometriae* ("The Supplement of Geometry") of 1593[16] he gave neusis constructions for both the two mean proportionals problem and the trisection problem. He also showed that any geometrical construction problem which, translated into algebraic terms, led to an equation of the third or fourth degree, could be reduced to either a trisection or a determination of two mean proportionals. Thus a very large class of geometrical problems, including all the solid problems known at that time, could be reduced to a neusis.[17] Viète highlighted this result by stating that the Euclidean postulates, on which constructions by circles and straight lines were based, should be supplemented by a new one, postulating that to every center, base line and segment a neusis could be performed. This new postulate was the 'supplement of geometry' in the title of his book of 1593; Viète claimed that, with geometry thus supplemented, "no problem would be left unsolved".[18]

Legitimation

Viète formulated the neusis as a postulate; he did not propose any particular procedure for performing the construction. Nor did he argue explicitly why it should be allowed to accept neusis as a postulate; apparently he considered the postulate sufficiently legitimated by the fact that it made 'all' problems solvable. Molther was primarily interested in this legitimation question and he did not accept Viète's answer. He was not alone in this interest; the status and acceptability of constructions beyond those by straight lines and circles was widely discussed in early modern geometrical studies. I mention two examples to give some background to Molther's arguments.

Some years before Viète, Clavius had proposed another means to extend the boundaries of geometry so as to include such hitherto unsolvable

[15]In 1591 Viète published in Tours his *Isagoge in artem analyticen*; it was the first of a series of treatises which he then planned to publish and which were to form what he called the *Opus restitutae mathematicae analyseos seu algebra nova.* He did not complete this project. The *Isagoge* was republished in Viète, François, *Opera mathematica* (ed. F. van Schooten), Leiden, 1646 (facsimile reprint Hildesheim 1970), pp. 1-12. An English translation can be found in F. Viète, *The analytic art, nine studies in algebra, geometry and trigonometry* (tr. T.R. Witmer), Kent (Ohio), 1983, pp. 11-32.

[16]Tours 1593; it was the second of the series of treatises mentioned in the previous note. It is on pp. 240-257 of van Schooten's edition and on pp. 388-417 of Witmer's translation.

[17]In fact, Viète's result implies that *all* solid problems in Pappus' sense, that is, all those that can be solved by the intersection of conics, can be constructed by neusis, because such problems can be reduced to equations of degrees not higher than four. However, Viète did not explicitly draw that conclusion.

[18]"Nullum non problema solvere" (Van Schooten edition p. 12, see note 15). Remarkably, Viète did not discuss the problems leading to equations of higher degree than four, although he knew that the higher-order angular sections were of that class.

problems. He did so in a treatise on a special curve, namely the 'Quadratrix', inserted in the second edition of his Euclid.[19] From Pappus' *Collectio* Clavius had learned about that curve and its properties; in particular the fact that if the quadratrix were given, trisection, other angular sections and even the quadrature of the circle could be constructed. Ancient geometers, however, had expressed doubts about the legitimacy of this use of the curve because its definition, by a combination of motions, seemed to presuppose the quadrature of the circle, so that the curve's use in solving this problem would involve a *petitio principii*. Clavius provided a pointwise construction of the quadratrix which he considered to be fully geometrical. He then claimed with quite some emphasis (which he mitigated somewhat in later editions) that through his construction the use of the quadratrix in geometry was legitimized and the problems of the quadrature of the circle and the angular sections were truly geometrically solved.[20]

Viète and Clavius argued for an extension of the domain of geometry. Their view was opposed by others, notably by Kepler. In the same year as *Problema Deliacum* appeared, Kepler published his *Harmonices Mundi*.[21] The work contained a spirited defence[22] of Euclid's geometry against all classical and modern geometers who tried to extend the means of geometrical construction beyond straight lines and circles. According to Kepler line segments constructed by other means than straight lines and circles were not knowable, they fell outside the sphere of exact, genuinely geometrical knowledge. Also ratios involving such line segments were beyond the bounds of geometry. Kepler's reasons for this orthodox and purist conception of geometry were philosophical; in his view the Creator had shaped the world according to harmonious, knowable ratios and these were precisely the ones constructable by strictly Euclidean means; assuming that there were more such ratios would destroy the structure and the divine rationality of the creation.

Kepler was not the last mathematician to address the issue of acceptability of constructions beyond straight lines and circles. Indeed much of what Descartes' presented in his *Géométrie* of 1637 was directly or indirectly inspired by that question.[23]

---

[19] *Euclidis elementorum libri XV accessit XVI de solidorum regularium cuiuslibet comparatione* (ed. C. Clavius), 2 vols, Rome, 1589, vol. 1, pp. 894-918: "De mirabilia natura lineae cuiusdam inflexae, per quam et in circulo figura quotlibet laterum aequalium inscribitur, et circulus quadratur, et plura alia scitu iucundissima perficiuntur"; Clavius inserted the treatise also in his *Geometria Practica*, Rome, 1604 (pp. 359-370).

[20] Cf. "(-) admittere (-) descriptionem hanc nostram quadratricis lineae ut geometricam" (p. 898 in Euclid edition 1589, cf note 19); in the version of the *Geometria Practica* of 1604 Clavius added 'quodammodo' in front of geometricam (p. 362, cf note 19).

[21] Johann Kepler, *Harmonices mundi libri V*, Linz, 1619, in *Gesammelte Werke* (ed W. von Dyck e.a.), München 1937–, vol. 6 (1940). There is a German translation: J. Kepler, *Weltharmonik* (tr. M. Caspar), München, 1973 (repr. of edition 1939).

[22] Book I, *Werke* vol. 6, pp. 13-64, see especially the Prooemium, pp. 15-20.

[23] Cf. my articles: "On the representation of curves in Descartes' Géométrie", *Archive*

The Problema Deliacum

In his *Problema Deliacum*, then, Molther took a position in an existing mathematical debate. He proposed the use of neusis constructions to solve problems that could not be solved by straight lines and circles, and he presented in particular one such construction for two mean proportionals. The proposal was not new, as we have seen, and his construction was a modification of existing ones. Nor did Molther explicitly present them as original. What he did claim, however, as his own new and original contribution was his argument that the neusis construction was geometrical. In his opinion a convincing argument for the geometrical legitimacy of neusis constructions had not yet been given, and so, by presenting such an argument, he could claim to be the first to have really solved the problems of doubling the cube and constructing two mean proportionals. Before analysing Molther's arguments I give a brief survey of the content of the work.

The book opens with a dedicatory poem addressed to Maurice, landgrave of Hessen,[24] and a preface.[25] The main text starts with a history of the problems of cube duplication and constructing two mean proportionals.[26] Then follow four chapters, the first on the neusis postulate itself,[27] the second on the construction of two mean proportionals by means of the neusis,[28] the third on a particular problem in solid geometry which depends on the construction of two mean proportionals,[29] and the fourth on some related subjects.[30] The above mentioned funereal poem for Molther's father ends

for history of exact sciences, **24**, 1981, pp. 295-338; and "The structure of Descartes' Géométrie", in *Descartes: il metodo e i saggi; Atti del convegno per il 350o anniversario della publicazione del Discours de la Méthode e degli Essais* (ed. G. Belgioioso e.a., 2 vols, Florence, 1990) pp. 349-369.

[24]Pp. 3-5.

[25]Pp. 6-9.

[26]Pp. 10-28.

[27]"Postulatum genuinum. Lineae rectae è puncto ad duas lineas requisita applicatio." Pp. 29-50.

[28]"Mesolabii secundi expositio" pp. 51-69. The chapter ends (pp. 63-69) with the remark that the first of two mean proportionals between 1 and 31, that is $\sqrt[3]{31}$, provides a good approximation for π; this is the result on the quadrature of the circle announced in the title of the book; Molther gives the values as 314138+ and 314159+ respectively (for radius 100000).

[29]"Fabrica solidorum sub data ratione similium" pp. 69-72. This was a standard problem in the early modern geometrical literature, namely: given a solid A and a ratio $a : b$, to construct a solid B similar to A and such that the volumes of A and B are in the ratio $a : b$.

[30]Pp. 73-88. Molther discussed some aspects of the determination of mean proportionals in practice, both by numbers and by instruments. He also explained the two proportional instruments announced in the title of his book. They were variants of the proportional compass. In one of them (pp. 83-86) the two sets of scales normally drawn on the two legs of the compass were drawn parallel but at some distance on a single flat surface; Molther claimed that this device, because of the absence of a hinge, was easier to make and as accurate in use as the usual proportional compass. The second (pp. 86-88) was practically

the book.[31]

In the historical section Molther critically reviewed earlier constructions of mean proportionals. He dealt with the constructions from Eutocius' list and with a number of more recent ones (Cusanus, Ramus, Oronce Fine, Peletier, Viète, Clavius, Villalpandus, Salignac, and Metius). In Molther's opinion, all these constructions failed to be truly geometrical. Some of them used curves which were traced by machines or constructed pointwise; neither method could be accepted as fully geometrical. Other constructions were unpractical, merely approximative or just false. Among all the constructions Molther preferred the one by Nicomedes which used the conchoid, because that curve could be traced more easily than the conics or the cissoid (another curve used in a construction of two mean proportionals), and also because its pointwise construction (here he referred to Clavius, who gave such a construction of the conchoid in his *Geometria Practica*[32]) was easy and practical. Yet, curves as the conchoid were not traced in a truly geometrical way and it was still an open problem

> how one should geometrically achieve the placing of the lines in these required positions, in one immediate action, with no other instrument than those the geometer is absolutely allowed to use, and with such truth and precision that the procedures indeed can bear the test of reason's criticism.[33]

The postulate

At the end of the historical introduction Molther announced his own method for finding two mean proportionals in the following terms:

> We, however, have finally realized that the matter of this difficulty, investigated through many centuries, a stumbling block for the most ingenious of mortals, is really so smooth, easy, obvious, ready and evident, that, because it meets the very terms for a legitimate postulate, it has by right to be counted as the next postulate, so that it does not at all require a belabored construction and proof, as difficult problems do, but that, as a principle clear in itself, it needs only a simple explanation, after which anyone can understand it and give it its due assent.[34]

---

the same as the usual instrument but Molther proposed a different use of it by which it was not necessary to have the scales on both legs of the compass.

[31] Pp. 89-102.

[32] *Geometria Practica* (see note 19), pp. 301-304.

[33] "(-) quomodo eiusmodi linearum requisitae applicationes Geometrice sine alio quam Geometrae absolute concesso instrumento, mox prima actione praestarentur tanta veritate ac praecisione ut etiam rationis censuram sustinerent." P. 25.

[34] "At vero nos rem istam exploratae per plurima secula difficultatis, in qua mortalium ingeniosissimi haesitarunt, ita expeditam, facilem, obviam, parabilem promtamque dudum

So Molther claimed that the neusis construction could be accepted as a postulate in geometry on the same level as the traditional postulates that canonized the usual Euclidean constructions, and that therefore the duplication problem could indeed be solved geometrically by means of neusis. This, as we have seen, was precisely what Viète had done in his *Supplementum Geometriae*, be it that he had not explicitly justified the postulate status for the neusis. For Molther that was the heart of the matter, which may explain how easily he dismissed Viète's work in his introduction:

> With all his subtlety Viète gathered nothing that can stand the test of criticism.[35]

The crucial part of Molther's reasoning, then, was his argument why the neusis construction was as obviously possible and acceptable in geometry as the construction of straight lines and circles by ruler and compass. The first chapter of his book was devoted to this argument. It opened with the formulation of the neusis postulate:

> Let it be postulated that, given two lines and a point in position in the same plane, a line can be drawn from that point such that the segment intercepted on that line by the two given lines is equal to another straight line given in length.[36]

The justification of the postulate

To justify the status of postulate Molther explained a procedure for the neusis which, he claimed, was legitimately geometrical. The procedure may be summarized as follows:

Procedure[37] (Neusis, Molther)

Given: two straight lines l and m (see Figure 4), a point O and a segment a.

Required: to find a line through O, intersecting l and m in A and B respectively and such that $AB = a$.

Procedure:

animadvertimus, ut quia hasce Postulati legitimi conditiones obtinet, Postulatus sit proxima meritoque annumeranda, adeo ut nequaquam ceu Problema contentiosum anxiam constructionem et demonstrationem requirat, sed tanquam Principium per se manifestum, levi contenta sit explicatione, qua adhibita à quolibet capi et assensum meriri possit." P. 27.

[35] "Subtillissimus Vieta nihil quod censuram sustineret venatus est" p. 26.

[36] "Postuletur, duabus lineis punctoque in eodem plano situ datis, ut è puncto isto linea recta applicetur, cuius portio a lineis illis intersecta alteri rectae longitudine datae sit aequalis," p. 29.

[37] Pp. 31-36.

1. Take a ruler and mark points C and D on it with distance a.
2. Move the ruler over the plane in such a way that it slides along the point O and that the point C on the ruler is always on the given line l.
3. Stop that motion at the moment that the point D is on the line m; then draw a straight line along the ruler; call A and B its intersections with l and m respectively.
4. OAB is the required line.

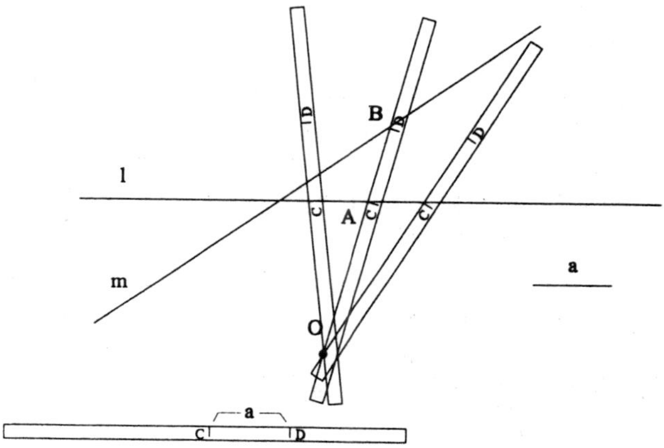

Figure 4

Molther analyzed this procedure at great length, arguing that each of its steps was legitimately geometrical. Since the first Euclidean postulate ascertained the tracing of straight lines, the ruler, by which this was done, was a geometrical instrument. Moreover such rulers could be made very precise, for instance by making them of metal, or by folding a piece of paper. The segment $CD = a$ could be marked on the ruler by a compass, which was also a geometrical instrument. The required movements of the ruler could be performed with great precision. This precision was guaranteed by our senses, which could judge whether the ruler remained along O during the motion and whether the point C moved along the given line l. It was also by the senses that the geometer decided to stop the motion at the moment that the point D lay on the second given line m. Molther argued that both motion and the testimony of the senses were implicitly assumed in the usual Euclidean postulates:

> For we have to use our sense to observe and acknowledge whether or not the ruler is placed in the way as postulated: because what can be clear by itself need not be made known through any demonstration. For at other times we acknowledge in no other way than by sense whether or not a ruler is duly placed along two points, from

> one of which to the other a straight line can be drawn according to the postulate; whether the given interval with which a circle is to be described is justly contained in a compass; and whether the one leg of the compass is rightly placed in the given center around which the circle is to be drawn. Indeed one sees immediately and with the same ease whether the ruler is along the point [O] and at the same time whether the point [C] on the ruler is at the line [l] and the point [D] on it at the line [m].[38]

Thus if one accepted the common Euclidean postulates one implicitly granted a legitimate place in geometry to motion and the testimony of the senses, and thereby to the neusis procedure.

Pure geometry

Although the starting point of Molther's legitimation argument was the precision of the neusis procedure as effectuated in practice, he did acknowledge geometry as a pure science, remote from the practice of drawing figures on paper by a step of abstraction or idealization. However, he argued that also if one considered pure geometry as an action of the mind alone, based on postulates, constructions still had to be performed in the mind by an inner sense, and this was done by procedures idealized from the actual physical construction procedures. Indeed the analysis of the role of motion and the senses in the actual physical procedure served to help the inner sense to perform the neusis abstractly in the mind as easily as it performed the mental operations corresponding to the use of ruler and compass. Therefore neusis should be accepted as a postulate in pure abstract geometry. Molther formulated this argument as follows:

> But also if one would judge that geometry it its most pure form should be practiced by action of the mind alone and based on its postulates, one detaches by mathematical abstraction the ideas of a material ruler and of a compass, and grasps them in one's mind, so that in the fantasy they serve as ruler and compass, guided by an interior sense. And thus it will be easy to imagine in thought the process of which we have shown how it is performed in reality.[39]

---

[38] "Sensu enim advertendum et agnoscendum sitne Regula ita ut postulatur applicata: quia nulla demonstratione id innotescat, quod per se liquidum esse potest. Quemadmodum alias haud aliter quam sensu cognoscimus sitne Regula ad duo puncta, è quorum uno ad alterum trahi postulatur Recta, debite applicata: sitne intervallum pro Circulo describendo datum Circini apertura iuste comprehensum: sitne pes Circini alter in centro dato circa quod gyranda est Peripheria, recte positus. Nempe eadem facilitate protinus cernitur, sitne Regula iuxta punctum A simulque Regulae punctum C in linea l, et Regulae punctum B in linea m." Pp. 33-34, in the translation I changed the letters to fit my figure.

[39] "Si quis autem existimet oportere puram puram [sic] Geometriam sola mentis actione, etiam secundum Postulata sua, exerceri; is quoque Regulae Circini materialis ideas

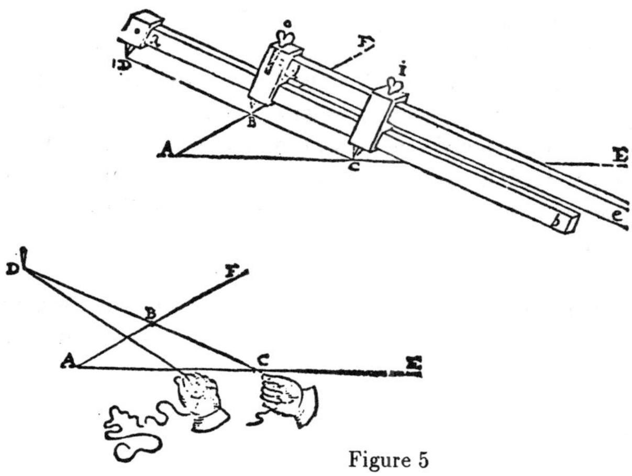

Figure 5

Molther's neusis instrument

To show the practical feasibility of neusis construction, Molther finally described two ways to actually perform the neusis, one by an instrument and one by a procedure involving a string. Figure 5 shows the drawings in his book.[40] The neusis was performed between the lines AE and AF from the point D; the given line segment was BC. The instrument was a combination of two rulers ab and de sliding along each other with adjustable points and pins. The distance oi along the ruler de could be made equal to the given segment. Ruler ab was placed with the pin at a in D and then turned around D while ruler de was made to slide along ab in such a way that the point corresponding to i moved along AE. The movement was stopped when the point corresponding to o was on the other line AF; at that position the ruler ab gave the required line.

The string procedure used a cord on which two points B and C were marked, with distance equal to the given segment. The cord, while being kept straight, was moved along the plane such that it wrapped around a pin in the given pole D and such that C followed the given line AE; the required position of the line was reached at the moment that B on the cord coincided with AF.

The construction of two mean proportionals

After these theoretical and practical arguments about the status and the execution of the neusis, Molther turned to the actual construction of solid

⟨afairesei⟩ Mathematica abstrahat et mente complectatur, ut in Phantasia per sensum interiorem Regulae ac Circini opera faciant. Sic enim proclive fuerit cogitando illud fingere, quod quomodo reipsa praestetur monstravimus." P. 36, text between ⟨ ⟩ in Greek.

[40]Pp. 38 and 39.

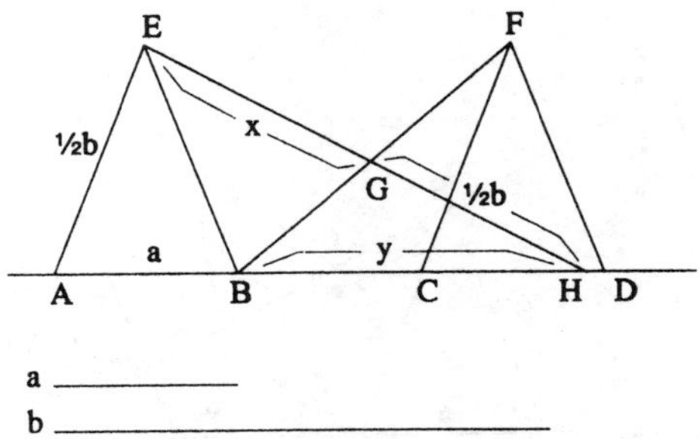

Figure 6

problems, by neusis but also otherwise. As most of these constructions are not particularly original or informative,[41] I shall only discuss the one for two mean proportionals.

The second chapter of *Problema Deliacum* was devoted to the *Mesolabium* proper, that is, to the construction of two mean proportionals. The construction was as follows:

> **Construction**[42] (two mean proportionals by neusis, Molther)
>
> Given: segments a and b ($a < b$) (see Figure 6). It is required to find their two mean proportionals x and y.
>
> Construction:
>
> **1.** Mark points A, B, C, D along a straight line such that $AB = BC = CD = a$.
>
> **2.** Construct two congruent isosceles triangles AEB and CFD, with $AE = EB = CF = FD = \frac{1}{2}b$; draw BF.
>
> **3.** By neusis (Molther here refers to his postulate), construct a line through E, intersecting the lines BF and AD in G and H respectively such that $GH = \frac{1}{2}b$.
>
> **4.** The required mean proportionals are $x = EG$ and $y = BH$.

The proof Molther gave for the correctness of this construction may be summarized as follows:

---

[41]Molther concluded his first chapter with a trisection by neusis (pp. 40-45) and a construction of the regular heptagon by means of a curve (introduced for this special purpose and constructed pointwise) which he called the "Helix (heptagoonografousa)" (pp. 46-48). The trisection is a variant of (and obviously inspired by) the neusis construction that occurs in Pappus' *Collectio* (cf. note 10); Molther referred to Clavius' *Geometria Practica* of 1604, where the same construction was given (on pp. 399-400, cf. note 19). Molther's change of the original construction was a slight practical improvement.

[42]Pp. 51-55.

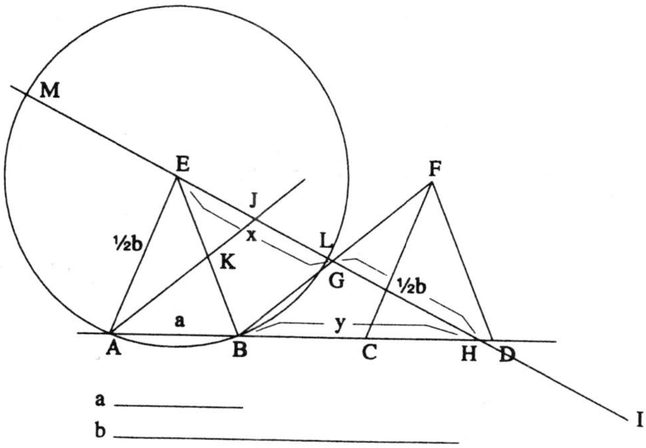

Figure 7

Proof[43]

1. Draw (see Figure 7) a circle with center E and radius $EA = \frac{1}{2}b$; prolong EH to both sides, take I on it with $GH = HI$ (and therefore $GI = b$); mark its intersections L and M with the circle, $LM = b$; draw a line through A parallel to BF, call its intersections with EG and EB J and K respectively.
2. Prove by similar triangles that $EK = KB$ and $EJ = JG$.
3. Note that, because $GH = \frac{1}{2}b = EL$, we have $EG = LH$, $MG = IL$ and $IE = HM$.
4. Prove by means of Euclid III-36 that $HL : HB = HA : HM$.
5. Because AJ is parallel to BF it follows that $AB : JG = BH : GH$; hence $AB : 2JG = BH : 2GH$; now $AB = a$, $2JG = EG = x$, $BH = y$ and $2GH = GI = b$, therefore $a : x = y : b$.
6. By similar triangles and the equalities of **2.** prove that $HA : EI = AB : EG$ and (using **4**) that $HA : EI = EG : HB$; as $AB = a$, $EG = x$ and $HB = y$ it follows that $a : x = x : y$.
7. From **5** and **6** it follows that $a : x = x : y = y : b$ as required.

If we compare Molther's construction and proof with those of Nicomedes (as given by Eutocius[44]) and of Viète (in his *Supplementum Geometriae*[45]) it appears that there is little originality in Molther's version. Without going into details here I may summarize the dependencies between the three solutions thus: Essentially they are the same, in particular with respect to the center,

[43]Pp. 55-58.
[44]Vereecke's translation (cf. note 9) pp. 618-620.
[45]Proposition 5, in Witmer's translation (cf. note 15) pp. 392-394.

lines and segment of the neusis. Viète simplified Nicomedes' proof by removing a number of inessential auxiliary elements in his figure; he introduced the isosceles triangle with sides a and $\frac{1}{2}b$ and the circle with radius $\frac{1}{2}b$ and chord a. Molther introduced more auxiliary lines, in particular the second isosceles triangle, which do not serve much function.

Conclusion – the interpretation of exactness

The justification of geometrical construction procedures beyond the use of straight lines and circles was a common concern in the early modern period. Mathematicians were unwilling to accept the restrictions of the Euclidean postulates, but the introduction of other means of construction than straight lines and circles evoked the question what exactness meant and where the borderline lay between legitimate, exact, or at any rate acceptable means of construction and inexact, unacceptable procedures. I term this issue the 'Interpretation of Exactness'.[46]

Mathematicians may adopt various strategies when confronted with the necessity to find new interpretations of exactness. They may, like Kepler, refer to authority and tradition and refuse to accept new interpretations. They may, as Descartes did in his *Géométrie*,[47] base a new interpretation of exactness on a philosophical analysis of pure understanding. They may, like Viète, motivate their choice implicitly by the quality of the resulting theory. They may also, like Clavius and like Molther, take practice as the starting point of their interpretation of exactness and argue for the legitimation of geometrical procedures by idealizing the criteria of precision that apply in practice.

All these strategies were tried out in the early modern period. They form, I find, a fascinating field of historical enquiry, not least because the various ideas developed in response to the questions concerning exact, acceptable constructions, although themselves of meta-mathematical nature, decisively influenced the development of mathematics in the early modern period. Although the idealization of practice, such as adopted by Clavius and Molther, appealed to a strong common experience of the practitioners of geometry, it was the least successful among the strategies mentioned above. Its arguments failed to convince and, contrary to the approaches exemplified by Viète and Descartes respectively, it led to no new mathematical results or techniques. Not surprisingly, then, Molther's *Problema Deliacum* met with little response and exerted no noticeable influence. Still, I think the book is of interest, because Molther worked out his legitimation in greater detail than we usually find. Thereby the *Problema Deliacum* offers us an instructive archetype of the approach to the interpretation of exactness which favours

[46] Cf. my 'On the interpretation of exactness', to be published in the Proceedings of the fifteenth international Wittgenstein symposium, held at Kirchberg am Wechsel (Austria) August 1992.

[47] Cf. the articles cited in note 23.

the idealization of practice.

Acknowledgements. I am grateful to G.J. Toomer for providing a Xerox copy of the *Problema Deliacum* and to Kirsti Andersen for helpful comments on an earlier version of this text.

Varianten einer Ḥaǧǧāǧ-Version von Buch II der *Elemente*

S. Brentjes

In MS Paris, BN, Persan 169H (im weiteren mit **P** abgekürzt) ist ein Fragment von Buch II der *Elemente* in arabischer Übersetzung enthalten, das sich auf Grund eines Zeugnisses von Ibn aṣ-Ṣalāḥ († 1153) in MS Oxford Marsh 720 als ein Abkömmling einer Version von al-Ḥaǧǧāǧ b. Yūsuf b. Maṭar (fl. zwischen 786 und 833) darstellt. Seine Analyse zeigt, daß es vermutlich auf der Version II dieses Übersetzers basiert. Allerdings ist es offenkundig keine reine Kopie dieser Version, sondern stellt anscheinend eine Bearbeitung derselben dar.[1]

In den Handschriften Escorial ar. 907 (im weiteren mit **E** abgekürzt), Rabāṭ Ḥasanīya 53 (im weiteren mit **R** abgekürzt) und Rabāṭ Ḥasanīya 1101, die arabische Fassungen der *Elemente* überliefern, sind u. a. in Buch II zu den Sätzen 1 bis 9 demselben Übersetzer zugeschriebene Zitate enthalten, die die Theoreme in einer vom jeweiligen Haupttext der Manuskripte deutlich abweichenden Form wiedergeben.

Nach der mittelalterlichen arabischen Überlieferung hat al-Ḥaǧǧāǧ b. Yūsuf zwei arabische Versionen der *Elemente* angefertigt. Nach Ibn an-Nadīm († 990) soll es sich dabei um zwei Übersetzungen gehandelt haben, während andere Quellen wie das Vorwort von MS Leiden 399.1 (im folgenden mit **L** abgekürzt), dem Kommentar Abū l-Faḍl an-Nairīzīs († 905 oder vor 922), oder eine Glosse in einem Exemplar der Rezension der *Elemente* von Naṣīr ad-Dīn aṭ-Ṭūsī (1201-1273) von einer Übersetzung und einer Bearbeitung sprechen.[2] Zwei explizit al-Ḥaǧǧāǧ b. Yūsuf zugeschriebene Fragmente müssen deshalb die Frage aufwerfen, wie sie sich zu diesen beiden Ḥaǧǧāǧ-Versionen verhalten und wie ihr Verhältnis untereinander beschaffen ist. Das Ziel des vorliegenden Artikels besteht deshalb in der Analyse der Beziehungen zwischen diesen beiden fragmentarischen Ḥaǧǧāǧ-Versionen von Buch II.

Da **P** bereits als vermutlicher Abkömmling der Ḥaǧǧāǧ-Version II identifiziert werden konnte, zerfällt die Aufgabe dieses Artikels in zwei Fragen: Könnten die Zitate in **E/R** Reste der Ḥaǧǧāǧ-Version I darstellen oder sind sie ebenfalls Abkömmlinge der Ḥaǧǧāǧ-Version II? Analog zu der bei der Identifikation von **P** aufgetretenen Grundproblematik steht darüber hinaus als dritte Frage zur Debatte, ob sich für die in **E/R** enthaltenen Zitate über deren Zuschreibung an al-Ḥaǧǧāǧ b. Yūsuf hinaus Argumente für eben diese

[1]Siehe dazu Brentjes 1991.

[2]Ibn an-Nadīm 1871/72, S. 265; *Codex Leidensis 399.1*, 1897, S. 4; MS Princeton, Yahuda Collection, 3770, f. 42[b].

Zuschreibung finden lassen. In einem ersten Schritt sollen deshalb die Zitate von **E** unter Berücksichtigung eventueller Abweichungen in **R** ediert und übersetzt werden. Eine Untersuchung der Beziehungen zwischen den drei Handschriften **E**, **R** und MS Rabāṭ Ḥasanīya 1101 kann dagegen hier nicht erfolgen, da letztere zum Zeitpunkt der Untersuchung nicht zur Verfügung stand.

1. Edition und Übersetzung der Ḥağğāğ-Zitate in E und R

II,1 **E**:[3]

وفى نسخة الحجاج يقول كل خطين يقسم احدهما
باقسام كم كانت فان الذى يكون من ضرب احد
الخطين فى الآخر مثل الذى يكون من ضرب الخط
الذى لم يقسم فى جميع اقسام الخط المقسوم .

Abweichungen in **R**:

–

Und in der Version von al-Ḥağğāğ wird gesagt: Teilt man von zwei (beliebigen) Strecken eine der Strecken in Teile, wie es (kommt), dann ist dasjenige, das aus der Multiplikation einer der beiden Strecken mit der anderen (entsteht), gleich demjenigen, das aus der Multiplikation der Strecke, die nicht geteilt worden ist, mit der Gesamtheit der Teile der geteilten Strecke (entsteht).

II,2 **E**:[4]

وفى نسخة الحجاج يقول كل خط يقسم اقساما فان الذى
يكون من ضرب الخط فى جميع اقسامه مثل الذى
يكون من ضرب الخط كله فى نفسه . وذلك ما اردنا ان
نبين .

Abweichungen in **R**:

statt نفسه: مثله;
ohne die Schlußformel.

Und in der Version von al-Ḥağğāğ wird gesagt: Eine (beliebige) Strecke möge in Teile geteilt werden. Dann ist das, was aus der Multiplikation der Strecke mit der Gesamtheit ihrer Teile (entsteht), gleich dem, was aus der Multiplikation der Strecke als Ganzes mit sich selbst (entsteht). Das ist es, was wir erklären wollten.

---

[3]f. 16[b], 23-26.

[4]f. 17[a], 3-6.

II,3 **E**:[5]

<table>
<tr><td dir="rtl">وفى نسخة الحجاج يقول كل خطين يقسم بقسمين فان
الذى يكون من ضرب جميع الخط فى احد القسمين
مثل الذى يكون من ضرب احد القسمين فى الآخر
ومن ضرب القسم الذى ضرب فيه الخط فى نفسه .</td><td>Abweichungen in R:
–</td></tr>
</table>

Und in der Version von al-Ḥağğāğ wird gesagt: Eine (beliebige) Strecke möge in zwei Teile geteilt werden. Dann ist das, was aus der Multiplikation der Gesamtheit der Strecke mit einem der beiden Teile (entsteht), gleich dem, was aus der Multiplikation eines der beiden Teile mit dem anderen und der Multiplikation des Teiles, mit dem die Strecke multipliziert wurde, mit sich selbst (entsteht).

II,4 **E**:[6]

<table>
<tr><td dir="rtl">وفى نسخة الحجاج يقول كل خط يقسم بقسمين فان
الذى يكون من ضرب جميع الخط فى نفسه مثل الذى
يكون من ضرب كل قسم فى نفسه ومن ضرب احد
القسمين فى الآخر مرتين . وقال ايضا وهذا لك استبان
السطحين الذين يكونان فى سطح مربع ويقطعهما
القطر مربعان .</td><td>Abweichungen in R:
fügt كله nach جميع الخط ein, hat statt هذا: هنا und fügt المتوازيين الاضلاع nach السطحين ein.</td></tr>
</table>

Und in der Version von al-Ḥağğāğ wird gesagt: Eine (beliebige) Strecke möge in zwei Teile geteilt werden. Dann ist das, was aus der Multiplikation der Gesamtheit der Strecke mit sich selbst (entsteht), gleich dem, was aus der Multiplikation eines jeden Teils mit sich selbst und der Multiplikation eines der beiden Teile mit dem anderen zweimal (genommen entsteht). Und es wird auch gesagt: Das ist für dich eine Erklärung, daß die beiden Flächen, die in der quadratischen Fläche sind und die beide die Diagonale schneidet, zwei Quadrate sind.

[5] f. 17^b, 3-6.
[6] f. 18^b, 3-5.

II,5 **E**:[7]

وفى نسخة الحجاج يقول كل خط يقسم بنصفين ، ثم
ايضا يقسم بقسمين (مختلفين) [8] فان الذى يكون من
ضرب القسم الاطول فى القسم الاصغر ومن فضل
نصف الخط على القسم الاصغر فى نفسه هو مثل الذى
يكون من ضرب نصف الخط فى نفسه .

Abweichungen in **R**:

addiert nach بقسمين: مختلفين, hat statt اصغر: اقصر.

Und in der Version von al-Ḥaǧǧāǧ wird gesagt: Eine (beliebige) Strecke möge in zwei Hälften und dann auch in zwei (verschiedene)[8] Teile geteilt werden. Dann ist das, was aus der Multiplikation des längeren Teils mit dem kleineren Teil und aus (der) des Überschusses der Hälfte der Strecke über den kleineren Teil mit sich selbst (entsteht), gleich dem, was aus der Multiplikation der Hälfte der Strecke mit sich selbst (entsteht).

II,6 **E**:[9]

وفى نسخة الحجاج يقول كل خط يقسم بنصفين ، ثم
يزاد فى طوله خط آخر فان الذى يكون من ضرب
جميع ذلك كله فى الخط المزيد ومن ضرب الخط الاول
فى نفسه جميعا هو مثل الذى يكون من ضرب نصف
الخط الاول اذا اضيف اليه الخط المزيد ، ثم يضرب
كله فى نفسه .

Abweichungen in **R**:

–

Und in der Version von al-Ḥaǧǧāǧ wird gesagt: Eine (beliebige) Strecke möge in zwei Hälften geteilt werden. Dann werde zu ihrer Länge eine andere Strecke hinzugefügt. Dann ist das, was aus der Multiplikation der Gesamtheit dessen als Ganzes mit der hinzugefügten Strecke und aus der Multiplikation der Hälfte der ersten Strecke mit sich selbst zusammen(genommen entsteht), gleich dem, was aus der Multiplikation der Hälfte der ersten Strecke (entsteht), wenn (mit) ihr die hinzugefügte Strecke verbunden wird (und) dann das Ganze mit sich selbst multipliziert wird.

[7] f. 19ª, 1-3.
[8] korrigiert nach **R**
[9] f. 19ª, 22-26.

II,7 **E**:[10]

وفى نسخة الحجاج يقول كل خط يقسم بقسمين فالذى
يكون من ضرب الخط فى نفسه وضرب احد القسمين
فى نفسه مجموعين مثل الذى يكون من ضرب الخط
فى القسم الذى ضرب فى نفسه مرتين مع ضرب
القسم الباقى فى نفسه .

Abweichungen in **R**:

fügt nach الخط im ersten und zweiten Teil der Behauptung كله und vor dem zweiten ضرب: من ein.

Und in der Version von al-Ḥağğāğ wird gesagt: Eine (beliebige) Strecke möge in zwei Teile geteilt werden. Dann ist das, was aus der Multiplikation der Strecke mit sich selbst und der Multiplikation eines der beiden Teile mit sich selbst, beides summiert, (entsteht), gleich dem, was aus der Multiplikation der Strecke mit dem Teil, der mit sich selbst multipliziert wurde, zweimal (plus) der Multiplikation des übrigen Teils mit sich selbst (entsteht).

II,8 **E**:[11]

وفى نسخة الحجاج يقول كل خط يقسم بقسمين ، ثم
يزاد فى طوله الخط مثل احد القسمين فان الذى يكون
من ضرب جميع ذلك فى نفسه مثل الذى يكون من
ضرب الخط الاول فى القسم المزيد اربع مرات وضرب
القسم الآخر في نفسه جميعا . وذلك ما اردنا ان نبين .

Abweichungen in **R**:

es fehlt am Ende in der Schlußformel: ان نبين.

Und in der Version von al-Ḥağğāğ wird gesagt: Eine (beliebige) Strecke möge in zwei Teile geteilt werden. Dann möge zur Länge der Strecke (etwas, das) einem der beiden Teile gleich ist, hinzugefügt werden. Dann ist das, was aus der Multiplikation der Gesamtheit dessen mit sich selbst (entsteht), gleich dem, was aus der Multiplikation der ersten Strecke mit dem hinzugefügten Teil viermal (genommen) und der Multiplikation des anderen Teils mit sich selbst (entsteht), zusammen(genommen). Das ist es, was wir erklären wollten.

[10] f. 19^b, 19-22.
[11] f. 20^b, 6-10.

II,9 **E**:[12]

وفى نسخة الحجاج قال كل خط يقسم بقسمين
متساويين ، ثم ايضا بقسمين مختلفين فان الذى يكون
من ضرب القسمين المختلفين كل واحد منهما فى نفسه
مجموعين هما مثلا ما يكون من ضرب الخط فى نفسه
ومثلا ما يكون من ضرب فضل نصف الخط على
القسم الاقصر فى نفسه جميعا . وذلك ما اردنا ان نبين .

Abweichungen in **R**:

Ausfall von بقسمين vor متساويين; fügt يقسم vor ايضا ein; ohne Schlußformel.

Und in der Version von al-Ḥaǧǧāǧ sagt man: Eine (beliebige) Strecke möge in zwei gleiche Teile, dann auch in zwei verschiedene Teile geteilt werden. Dann ist das, was aus der Multiplikation der beiden verschiedenen Teile, eines jeden der beiden mit sich selbst, beides summiert, (entsteht), gleich zweimal dem, was aus der Multiplikation der Strecke mit sich selbst (entsteht), und zweimal dem, was aus der Multiplikation des Überschusses der Hälfte der Strecke über den kürzeren Teil mit sich selbst (entsteht), zusammen(genommen). Und das ist es, was wir erklären wollten.

2. Vergleich zwischen P und E/R

Vergleicht man die Formulierung der Zitate in **E/R** mit **P**, so ist die Übereinstimmung in der syntaktischen Rahmenstruktur beider Versionen, d. h. in der Konstruktion *kull ... fa-inna ...*, sowie in ihrer prinzipiellen Ausfüllung unübersehbar. Die weitgehend geringfügigen Abweichungen zwischen **E**, **R** und **P** zeigen an, daß zum einen **E** und **R** in Bezug auf die Zitate wohl auf denselben Archetypus zurückgehen und daß zum anderen **P** und die Zitate von **E** aus der gleichen Grundquelle stammen. Das bedeutet, daß beide Versionen in der Tat auf ein und denselben Erstautor zurückgehen. Wenn also bei einer der beiden Versionen akzeptiert wird, daß sie auf der Arbeit von al-Ḥaǧǧāǧ b. Yūsuf beruht, dann muß das auch bei der anderen Version angenommen werden. Die Abweichungen zwischen den beiden Versionen lassen sich in mehrere Klassen unterteilen, die im folgenden diskutiert werden sollen.

[12]f. 21[a], 18-22.

2.1 Wesentliche Abweichungen zwischen E/R und P

Der erste Typ von Abweichungen wird durch jene Differenzen zwischen **P** und **E/R** konstituiert, die sich zur Unterscheidung der beiden Grundversionen und damit zu ihrer Charakterisierung benutzen lassen. Dazu gehören die Beschreibung von Quadraten und Rechtecken sowie die Verwendung von *miṯlihī* versus *nafsihī*. **P** beschreibt Quadrate stets durch *talbīn a fī miṯlihī/ nafsihī* [ziegelförmige (Fläche) *a* mit dem ihm Gleichen/mit sich selbst] und Rechtecke durch *talbīn a fī b* [ziegelförmige (Fläche) *a* mit *b*]. **E/R** verwendet für Quadrate immer *allaḏī yakūnu min ḍarb a fī nafsihī/ miṯlihī* [dasjenige, das aus der Multiplikation von *a* mit sich selbst/mit dem ihm Gleichen (entsteht)] und für Rechtecke immer *allaḏī yakūnu min ḍarb a fī b* [dasjenige, das aus der Multiplikation von *a* mit *b* (entsteht)]. **E** hat stets *nafsihī* und **R** hat nur in Satz II,2 *miṯlihī*; sonst aber stimmt es in dieser Hinsicht mit **E** überein. **P** dagegen hat in den Sätzen II,2 und II,3 *nafsihī*; in II,4 kommt *nafsihī* im Theorem vor, *miṯlihī* dagegen in der Behauptung. In den Sätzen II,5 bis II,10 und II,12 bis II,14 erscheint ausschließlich *miṯlihī*. In II,11 liegt eine zu II,4 analoge Situation vor: *nafsihī* tritt in der Aufgabenstellung auf und *miṯlihī* in der Konstruktion.

All das vermittelt den Eindruck, daß die ursprüngliche *talbīn*-Version mit *miṯlihī* gearbeitet hat, während die *ḍarb*-Version konsequent *nafsihī* verwendet. Letzteres wird auch durch die Sekundärüberlieferung der *ḍarb*-Version z. B. in der ersten *Risāla* der *Rasāʾil iḫwān aṣ-ṣafāʾ* und der Epitome der *Elemente* von Ibn Sīnā († 1036) bekräftigt.[13] Diese wesentlichen Differenzen zeigen, daß in **P** und **E/R** zwei (sprachlich) verschiedene Versionen der *Elemente* vorliegen.

Allerdings erhebt sich mit diesem Ergebnis die folgende Frage: ist überhaupt eine der beiden Fassungen **P** bzw. **E/R** direkt auf al-Ḥaǧǧāǧ b. Yūsuf zurückführbar oder stellen beide Versionen von anderen Autoren ausgeführte Bearbeitungen eines von der Ḥaǧǧāǧ-Version II abstammenden Archetypus dar? Die dritte Möglichkeit, beide Versionen als echte Ḥaǧǧāǧ-Texte zu akzeptieren, kollidiert mit der mittelalterlichen arabischen Überlieferung, wonach es, wie bereits erwähnt, nur zwei (echte) Ḥaǧǧāǧ-Versionen gegeben haben soll.

Direkte Anhaltspunkte für eine Antwort auf diese Frage stehen mir zur Zeit nicht zur Verfügung. Zwei indirekte Anhaltspunkte liefern jedoch zum einen die Sprache der beiden Versionen und zum anderen die arabische Sekundärüberlieferung.

Hinsichtlich der sprachlichen Form von **P** und **E/R** ist aufgrund anderer mathematischer sowie nichtmathematischer Quellen zu konstatieren, daß **P** ganz offensichtlich eine Sprachform verkörpert, die sich nicht in die mathematische Terminologie einzubürgern vermocht hat. Sie enthält mit *talbīn* einen

[13] Siehe *Rasāʾil iḫwān aṣ-ṣafāʾ*, 1306 H., S. 29-31 und Ibn Sīnā 1976, S. 69-86.

Ausdruck, der einen Syriazismus darstellen könnte,[14] was sein Verschwinden aus der mathematischen Terminologie erklären könnte, wenn man das Schicksal anderer Syriazismen z. B. in der philosophischen Literatur bedenkt. Wenn eine der beiden Versionen tatsächlich von al-Ḥaǧǧāǧ stammen sollte, dürfte es deshalb **P** sein, da im ersten Drittel des 9. Jh., als al-Ḥaǧǧāǧ seine zweite Version angefertigt hat, Syriazismen in der arabischen wissenschaftlichen Literatur noch verbreitet gewesen sind.

Außerdem zeigen Anmerkungen von Ibn al-Haiṯam in seinem *Šarḥ al-muṣādarāt kitāb Uqlīdis*, daß die von ihm benutzte Version der *Elemente* in den Büchern V und VII weitere Syriazismen enthalten hat. Andere von ihm erwähnte sprachliche Fassungen der fraglichen Termini lassen sich entweder aufgrund der uns vorliegenden Handschriften der *Elemente* als zur Isḥāq/Ṯābit-Tradition gehörig identifizieren oder anderen mathematischen Texten des 9. Jh. zuweisen.[15] Syriazismen sind dadurch als ein Element einer der Ḥaǧǧāǧ-Versionen indirekt bestätigt. Allerdings enthält Ibn al-Haiṯams Kommentar in Buch II keine der beiden Beschreibungen für Quadrate und Rechtecke von **P** oder **E/R**, sondern vielmehr eine der griechischen Edition (im weiteren mit **G** abgekürzt) angemessene geometrische Ausdrucksweise. Da Ibn al-Haiṯam nur die Definitionen (und in Buch I die Axiome und Postulate) bespricht, **P** und **E/R** aber die beiden Definitionen von Buch II ausgelassen haben, kann dieser Befund nicht für eine Klärung des Verhältnisses von **P** und **E/R** zu al-Ḥaǧǧāǧ herangezogen werden.

Bereits erwähnte Texte der arabischen Sekundärüberlieferung der *Elemente* wie die erste *Risāla* der *Rasā'il iḫwān aṣ-ṣafā'* oder die Epitome von Ibn Sīnā basieren auf einer *ḍarb*-Version. Der Erfolg der *ḍarb*-Version wird verständlich, wenn man bedenkt, daß die von ihr benutzte Sprache die der Arithmetiker und Algebraiker ist. Es ist aus meiner Sicht deshalb wenig plausibel anzunehmen, daß ein Bearbeiter diese weitverbreitete Terminologie durch die Ausdrucksweise von **P** ersetzt haben sollte, falls al-Ḥaǧǧāǧ seine Version II bereits in der Sprache von **E/R** verfaßt hätte. Die Ablösung der archaischen Sprache von **P** durch die Sprache von **E/R** dagegen paßt erheblich besser in das Bild von der historischen Entwicklung der Mathematik im islamischen Mittelalter, über das wir zur Zeit verfügen.

Andererseits liefern die algebraischen Arbeiten von M. b. Mūsā al-Ḫwārizmī (ca. 780-ca. 850) und Ibn Turk (ca. um 850)[16] Beispiele dafür, daß die Sprache von **E/R** bereits in der ersten Hälfte des 9. Jh. verwendet worden ist. Das wesentliche Element dieser Sprache besteht aus der Sicht der Überlieferung der *Elemente* in der Tatsache, daß eine arithmetische Terminologie zur Beschreibung geometrischer Objekte und ihrer Beziehungen bzw. Konstruktionen benutzt wird. Al-Ḫwārizmīs *Algebra* enthält in ihrem

[14] Vergleiche dazu Payne-Smith 1883, Bd. 2, fasc. 6, Sp. 1886f.; Levy 1879, Bd. 2, S. 468, Bd. 3, S. 122; Freytag 1837, S. 83; Ullmann 1983, Bd. II,1, S. 155, 178-180; *Lisān al-ʿarab* 1357 H./1956, Bd. 13, S. 375.

[15] Djebbar 1990; Brentjes 1992.

[16] Sayılı 1962, S. 144.

geometrischen Kapitel außerdem verschiedene Anleihen aus den *Elementen*.[17]

U. a. erscheinen in ihm die Sätze I,47 (zweimal, davon einmal paraphrasiert) sowie II,12 und II,13 (paraphrasiert). Die sprachliche Form, in der al-Ḫwārizmī I,47 angibt, unterscheidet sich in den beiden Fällen, in denen diese Aussage in seiner Arbeit auftritt, voneinander, hängt aber über die Verwendung von *ḍarb* und *nafsihī* sowohl miteinander als auch mit **E/R** zusammen. Da al-Ḫwārizmī aus zeitlichen Gründen nur eine der Ḥaǧǧāǧ-Versionen (oder eventuell eine ihrer Bearbeitungen), nicht aber die Übersetzung von Isḥāq b. Ḥunain oder deren Korrektur von Ṯābit b. Qurra benutzt haben kann, enthält die Formulierung von I,47 in beiden Fällen erwartungsgemäß weitere sprachliche Elemente wie *aṭwal/aqṣar* [länger/kürzer] oder *miṯl* [gleich] und die explizite Auffordung zur Summation der beiden Kathetenquadrate, die für **P** und **E/R** charakteristisch sind.[18]

Diese Übereinstimmung stellt ein zusätzliches Kriterium für die Ableitung von **P** und **E/R** von al-Ḥaǧǧāǧ dar. Die sprachliche Verwandtschaft al-Ḫwārizmīs mit **E/R** beschränkt sich im ersten Fall des Auftretens von I,47 jedoch nicht auf *ḍarb*, *nafsihī*, *miṯl*, *aṭwal/aqṣar* und die explizite Summationsaufforderung wie im zweiten Fall und bei den paraphrasierten Sätzen II,12 und I,13. Vielmehr erscheint dort auch die Einbettung von *ḍarb* durch *allaḏī yakūnu min* wie in **E/R** sowie die für **P** und **E/R** typische syntaktische Rahmenstruktur:[19]

واعلم انّ كلّ مثلّث قائم الزاوية فانّ الذى يكون من ضرب الضلعين الاقصرين كلّ واحد منهما في نفسه مجموعين مثل الذى يكون من ضرب الضلع الاطول فى نفسه .

> Und wisse, daß für jedes rechtwinklige Dreieck die (Summe) dessen, was aus der Multiplikation der beiden kürzeren Seiten, eine jede von den beiden mit sich selbst, (entsteht), gleich dem ist, was aus der Multiplikation der längsten Seite mit sich selbst (entsteht).

Dennoch scheint es sich offensichtlich nicht um ein vollständiges, direktes Zitat aus den *Elementen* zu handeln, da zum einen die Hypotenuse abkürzend als *aḍ-ḍilʿ al-aṭwal* [längste Seite] bezeichnet wird, während sie in der Glosse zu I,46[20] in **L**, die in der Sprache von **P** formuliert ist, *watar az-zāwiya al-*

[17]Bislang ist die Benutzung der *Elemente* durch al-Ḫwārizmī stets bestritten worden. Dabei hat man sich aber ausschließlich auf die Betrachtung des algebraischen Kapitels seiner Arbeit beschränkt. Die Untersuchung des Umfangs der Benutzung der *Elemente* durch al-Ḫwārizmī und die Frage, welche Version dieses Werkes er verwendet haben dürfte, wird eine gesonderte Studie zum Gegenstand haben.

[18]Siehe dazu Brentjes 1991.

[19]al-Ḫwārizmī 1968, S. 57.

[20]Bekanntermaßen erscheint dieses Theorem in den Ḥaǧǧāǧ-Varianten nicht als I,47, sondern als I,46.

qā'ima [Sehne (über) dem rechten Winkel] und in **G** ἡ τὴν ὀρθὴν γωνίαν ὑποτείνουσα πλευρά [die dem rechten Winkel gegenüberliegende Seite] genannt wird[21] und zum anderen die Reihenfolge der Satzteile gegenüber dieser Glosse und **G** vertauscht ist.

Die Interpretation dieses Fragmentes, seiner Beziehungen zur Arbeit von al-Ḥaǧǧāǧ sowie zu der Version von **E/R** ist überaus schwierig. Ein entscheidendes Moment ist die Frage, wie man den Charakter der Ḥaǧǧāǧ -Version II bestimmt – wie Ibn an-Nadīm als Übersetzung oder wie das Vorwort von **L** als Bearbeitung. Unabhängig von der Bewältigung dieser Probleme, die in einer anderen Studie versucht werden soll, zeigt das Fragment aber klar an, daß zumindest Bruchstücke einer *ḍarb*-Fassung der *Elemente* bereits in den 20er Jahren des 9. Jh. im Umlauf waren.

Ein zweiter wichtiger Hinweis der Sekundärüberlieferung findet sich in der Formulierung von II,5 in einer *Risāla fī masā'il al-ǧabr wa'l-muqābala* in MS Paris, BN, ar. 4946, d.h. erneut in einem algebraischen Text. Dieser Text gibt II,5 wie folgt an:[22]

(نريد ان نبيّن البرهان على مسائل الجبر المقترنة بمقدّمة واحدة وهي انّ) كلّ
عدد قسم بقسمين وبنصفين فانّ ضرب احد القسمين في الآخر ونصف الفضل
ما بين القسمين في مثله مجموعين مثل نصف العدد في مثله .

(Wir wollen den Beweis der zusammengesetzten algebraischen Probleme durch eine einzige Proposition erklären, (und zwar):) Eine (beliebige) Zahl möge in zwei Teile und in zwei Hälften geteilt werden. Dann ist das Produkt des einen der beiden Teile mit dem anderen und der Hälfte des Überschusses dessen, was zwischen den beiden Teilen ist, mit dem ihr Gleichen zusammengenommen gleich der Hälfte der Zahl mit der ihr Gleichen.

Seine sprachliche Verwandtschaft mit der Ḥaǧǧāǧ-Tradition ist unverkennbar. Vergleicht man dieses Fragment mit **E/R** und **P**, so fällt die enge Beziehung zu **P** ins Auge. Obwohl das Fragment arithmetisch formuliert ist, benutzt es in fast wörtlicher Übereinstimmung mit **P** im ersten Teil der Behauptung *aḥad al-qismaini fī l-āḫar* [einen der beiden Teile mit dem anderen] (**P**: *aḥad al-muḫtalifaini fī l-āḫar* [einen der beiden verschiedenen <Teile> mit dem anderen]). Aus dieser Beziehung läßt sich ableiten, daß in der Grundversion von **P** *aḥad al-qismaini al-muḫtalifaini fī l-āḫar* [einen der beiden verschiedenen Teile mit dem anderen] gestanden hat.

Eine zweite mit **P** in dieser Hinsicht übereinstimmende Fassung von II,5

[21] *Codex Leidensis 399.1*, 1897, S. 172; *Euclidis Elementa* 1969, S. 63.
[22] MS Paris, BN, ar. 4946, f 1[b], 1-4.

findet sich in der ersten *Risāla* der *Rasāʾil iḫwān aṣ-ṣafāʾ*:[23]

(مسألة) كلّ عدد قسم بنصفين ثمّ بقسمين مختلفين فانّ ضرب احد المختلفين في الآخر وضرب التفاوت في نفسه مساو لضرب نصف ذلك العدد في نفسه .

(Problem) Eine (beliebige) Zahl möge in zwei Hälften, dann in zwei verschiedene Teile geteilt werden. Dann ist das Produkt des einen der beiden verschiedenen <Teile> mit dem anderen und das Produkt der Differenz mit sich selbst gleich dem Produkt der Hälfte jener Zahl mit sich selbst.

E/R weicht hiervon mit *al-qism al-aṭwal fī l-qism al-aṣġar/aqṣar* [dem längeren Teil mit dem kleineren/kürzeren Teil] signifikant ab. Für diesen Umstand gibt es zwei Erklärungsmöglichkeiten: entweder hat es zwei verschiedene *ḍarb*-Versionen gegeben, die sich zwar beide letztendlich von al-Ḥaǧǧāǧ herleiten, aber von verschiedenen Archetypen angefertigt worden sind, oder **E/R** verkörpert in Bezug auf die kritische Passage eine spätere Bearbeitungsstufe als die beiden obigen Fragmente.

Ein weiteres wesentliches Moment ist das attestierte Überwiegen der *ḍarb*-Version, da Ibn an-Nadīm bekanntlicherweise berichtet, daß man sich (zu seiner Zeit, d. h. im 10. Jh.) auf die zweite Ḥaǧǧāǧ-Version gestützt habe.[24] Daraus ergibt sich die Frage, ob er vielleicht eine *ḍarb*-Version als zweite Ḥaǧǧāǧ-Übersetzung aufgefaßt hat.

2.2 Wesentliche Abweichungen zweiter Art zwischen E/R und P

Die folgenden substantiellen Unterschiede konstituieren eine zweite Klasse, da mit ihrer Hilfe der in **E/R** und **P** verkörperte momentane Stand der Überlieferung der beiden Grundversionen, auf denen sie basieren, charakterisiert werden kann. Es zeigt sich, daß dabei beide Texte wechselseitig Aufschlüsse über die Entfernung der jeweiligen anderen Handschrift von ihrer Grundversion erbringen. Die Unterschiede dieser Klasse betreffen die Sätze II,3, 5, 6 und II,7.

In II,3 stimmt das Zitat von **E/R** mit **G** in der Reihenfolge der beiden Flächen im zweiten Teil der Behauptung überein, d. h. hat *Rechteck plus Quadrat*, während **P** diese Reihenfolge umkehrt. In **P** folgt diese Umkehrung aus der gewählten Ausdrucksweise für das Quadrat, das als *talbīn ḏālika l-qism fī nafsihī* [ziegelförmige (Fläche) jenes Teils mit sich selbst] beschrieben wird. Diese Beschreibung ist aber nur dann eindeutig, wenn sie auf *talbīn*

---

[23] *Rasāʾil iḫwān aṣ-ṣafāʾ*, 1306 H., S. 30, 5f.

[24] Ibn an-Nadīm 1871/72, S. 265.

al-ḫaṭṭ fī aḥad al-qismaini [ziegelförmige (Fläche) der Strecke mit einem der beiden Teile], nicht aber, wenn sie auf das zu bildende *Rechteck* folgt. Das Zitat in **E/R** kann dagegen das *Quadrat* an der ursprünglichen Stelle plazieren, da es die Quadratseite als *ḏālika l-qism allaḏī ḍuriba fīhi l-ḫaṭṭ* [jener Teil, mit dem die Strecke multipliziert wurde] beschreibt. Diese Formulierung stellt aber offensichtlich keine Übersetzung der griechischen Ausdrucksweise τὸν ἀπὸ τοῦ προειρημένου τμήματος τετράγωνον [das Quadrat über dem genannten Teil] dar, während die Ausdrucksweise von **P** sehr wohl als Übersetzung des griechischen Ausdrucks akzeptiert werden kann. Damit stellt sich die Frage, ob II,3 von **E/R** zur originalen Fassung der *ḍarb*-Version gehört hat.

Ein Anhaltspunkt für eine Antwort darauf läßt sich aus der Sekundärüberlieferung gewinnen. Vergleicht man nämlich II,3 mit den beiden bereits oben benutzten Texten der Sekundärüberlieferung, so erhält man folgendes Bild:

– Satz II,3 in der Edition des *Kitāb aš-Šifā* von Ibn Sīnā ist offenkundig korrupt, aber es ist dennoch klar, daß wie in **E/R** zuerst das *Rechteck* und dann das *Quadrat* gebildet wird:[25]

اب قسم بقسمين على ه فضرب اب في احدهما وليكن جب الّذى هو اب في
به المساوي ا جب مساو لضرب اج في جب الّذى هو به في نفسه .

ab werde in *h* in zwei Teile geteilt. Dann werde *ab* mit einem von ihnen multipliziert. Sei das *ǧb*. Dasjenige, was *ab* mal *bh* gleich *ǧb* ist, ist gleich dem Produkt *aǧ* mal *ǧb*, welches *bh* ist, mit sich selbst.

Die Lücke besteht vor *fī nafsihī*, da nicht mitgeteilt wird, was mit sich selbst multipliziert werden soll. Fußnote 4 zu diesem Satz in der Edition vermittelt aber den Eindruck, daß der Ausfall nicht auf das zugrundegelegte Manuskript zurückzuführen ist, sondern auf einen Druckfehler, denn inbezug auf eine der Handschriften wird angegeben: *wa-ǧb ... nafsihī: wa-ǧb allaḏī fīhi ab fī nafsihī.*[26] Von dem zitierten *wa-ǧb ... nafsihī* gibt es im oben wiedergegebenen Satz aber nur das *fī nafsihī*. Es scheint also ein *wa-ǧb allaḏī ḍuriba fīhi ab* im Druck ausgefallen zu sein. Damit ist die Existenz der Vorgehensweise von **E/R** zumindest für das 10. Jh. attestiert. Die Fußnoten in der Edition des *Kitāb aš-šifā'* deuten außerdem darauf hin, daß Ibn Sīnā mit einer Version der *Elemente* gearbeitet hat, die die Quadratseite mit der Formulierung von **E/R** beschrieben hat.

– Satz II,3 in der ersten *Risāla* der *Rasā'il iḫwān aṣ-ṣafā'* benutzt zwar die *ḍarb*-Terminologie, folgt aber **P** hinsichtlich der Reihenfolge von *Quadrat* und *Rechteck* und bewahrt darüber hinaus das in dieser Beziehung wesentliche *ḏālika l-qism*:[27]

[25] Ibn Sīnā 1976, S. 71, 1-3.
[26] Ibn Sīnā 1976, S. 71.
[27] *Rasā'il iḫwān aṣ-ṣafā'*, 1306 H., S. 29, 22f.

(مسألة) كلّ عدد قسم بقسمين فنقول انّ ضرب ذلك العدد في احد قسميه مساو لضرب ذلك لقسم في نفسه وفي القسم الآخر.

(Problem) Eine (beliebige) Zahl werde in zwei Teile geteilt. Dann sagen wir, daß das Produkt jener Zahl mit einem ihrer Teile gleich dem Produkt jenes Teils mit sich selbst und mit dem anderen Teil ist.

Es scheint folglich eine *ḍarb*-Version gegeben zu haben, die die charakteristischen Eigenheiten der *talbīn*-Version in II,3 besessen hat. Diese Erkenntnis wird auch durch die Form von II,3 in der lateinischen Übersetzung von Hermann von Carinthia und in der Version I von Adelard von Bath bestätigt.[28] Da diese beiden lateinischen Versionen Übersetzungen einer *ḍarb*-Version und offenkundig unabhängig voneinander sind, ist davon auszugehen, daß es wenigstens zwei Manuskripte dieser mit **P** im Hinblick auf die Struktur von II,3 übereinstimmenden *ḍarb*-Version gegeben hat. Dieser Befund besagt zunächst, daß es zwei verschiedene *ḍarb*-Versionen gegeben hat.

Bei der Analyse von **P** ist zugunsten der Annahme argumentiert worden, daß II,3 in **P** eine Bearbeitungsstufe verkörpert, die später als die Version II von al-Ḥağğāğ entstanden ist.[29] Einer der wichtigsten Gründe für diese Hypothese ist die Tatsache, daß **L** zum einen in II,3 prinzipiell mit **G** und darüber hinaus spezifisch mit **E/R** in der Verwendung und Stellung von *ḏālika l-qism* übereinstimmt und zum anderen in der Struktur von II,12 gegen **P** mit **G** korrespondiert, in der Sprache aber deutlich mit **P** gegen die Isḥāq/Ṯābit-Redaktion verwandt ist, so daß eine zu **G** zurückkehrende Bearbeitung durch an-Nairīzī aufgrund der Isḥāq/Ṯābit-Redaktion ausgeschlossen sein dürfte. Damit wäre für **E/R** zu schlußfolgern, daß es im Hinblick auf die Struktur von II,3 die der Version II von al-Ḥağğāğ näherstehende Variante verkörpert.

Da **L** andererseits keine präzisierende Ergänzung zu *ḏālika l-qism* enthält und somit in der Plazierung von b^2 nach $(a-b)b$ ambivalent bleibt, während die erste *Risāla*, Hermann und Adelard die Ambivalenz wie **P** durch die Umsortierung der beiden Satzteile überwunden haben, dürfte die präzisierende Formulierung von **E/R** weder zur originalen Fassung der *ḍarb*-Version noch zur Version II von al-Ḥağğāğ gehört haben. **E/R** wäre folglich ebenfalls als eine spätere Bearbeitungsstufe aufzufassen.

In II,5 handelt es sich, wie oben bereits diskutiert, um die Beschreibung des Rechtecks im ersten Teil der Behauptung. In **P** steht sie dem τὸ ὑπὸ τῶν ἀνίσων τῆς ὅλης τμημάτων περιεχόμενον ὀρθογώνιον [das die ungleichen Teile der ganzen (Strecke) umfassende Rechteck] in **G** deutlich näher als die in **E/R**, die erneut keine Übersetzung des griechischen Ausdrucks darstellt. Es soll jedoch angemerkt werden, daß **E/R** in diesem Fall der Formulierung

[28] Busard 1968, S. 41; Busard 1983, S. 73.

[29] Brentjes 1991.

in I,47 bei al-Ḫwārizmī verwandt ist, ein Umstand, für den ich z. Z. noch keine Erklärung habe. Bemerkenswerterweise stimmen in diesem Punkt Hermann von Carinthia und Adelard von Bath mit **E/R** überein.[30]

In II,6 bestehen die auffälligen Differenzen zwischen **E/R** und **P** in folgendem:

a. **E/R** hat *ḫaṭṭ āḫar* [eine andere Strecke] als Zusatz zur gegebenen Strecke gegenüber *baʿḍ az-ziyāda* in **P**;

b. **E/R** hat *al-ḫaṭṭ al-mazīd* [die hinzugefügte Strecke] als kleinerer Faktor bei der Bildung des Rechtecks im ersten Teil der Behauptung gegenüber *az-ziyāda* [der Zusatz/die Hinzufügung] in **P**;

c. **E/R** hat *iḏā uḍīfa ilaihi l-ḫaṭṭ al-mazīd* [wenn (mit) ihr die hinzugefügte Strecke verbunden wird] bei der zu quadrierenden Größe im zweiten Teil der Behauptung gegenüber *maʿa z-ziyāda* [mit dem Zusatz] in **P**.

Es ist sehr aufschlußreich für die Beurteilung des Verhältnisses zwischen den beiden Versionen, diese Stelle mit den ihnen entsprechenden in Satz II,6 von **L**, Haupttext und MS Teheran Malik 3586 sowie **G** zu vergleichen. Dabei zeigt sich folgendes:

a. **G** : τις . . . εὐθεῖα ἐπ' εὐθείας [irgendeine ... Gerade in gerader (Fortsetzung)]
L, Haupttext: *ḫaṭṭ āḫar mustaqīm* [eine andere gerade Strecke]
MS Teheran Malik 3586: *ḫaṭṭ mustaqīm ʿalā istiqāmatihī* [eine gerade Strecke in gerader (Fortsetzung)]

b. **G** : καὶ τῆς προσκειμένης [und des Zusatze]
L, Haupttext: *wa z-ziyāda* [und des Zusatzes]
MS Teheran Malik 3586: *wa z-ziyāda* [und des Zusatzes];

c. **G** : τῆς συγκειμένης ἔκ τε τῆς ἡμισείας καὶ τῆς προσκειμένης [(der aus der Hälfte) und dem Zusatz (Zusammengesetzten)]
L, Haupttext: *(niṣf al-ḫaṭṭ) maʿa z-ziyāda* [(der Hälfte der Strecke) mit dem Zusatz]
MS Teheran Malik 3586: *(al-ḫaṭṭ al-murakkab min niṣf al-ḫaṭṭ) wa z-ziyāda* [(der Strecke, die aus der Hälfte der Strecke) und dem Zusatz (zusammengesetzt ist)].[31]

L, Haupttext hat wie **E/R** *ḫaṭṭ āḫar* plus das aus der Isḥāq/Ṯābit-Redaktion übernommene *mustaqīm*, während das *baʿḍ az-ziyāda* von **P G** deutlich nähersteht. Im zweiten Fall ist in MS Leiden 399.1, Haupttext jeglicher früherer Ḥaǧǧāǧ-Wortlaut durch den der Isḥāq/Ṯābit-Redaktion ersetzt worden, aber

[30] Busard 1968, S. 43; Busard 1983, S. 75.

[31] *Euclidis Elementa* 1969, S. 75, 8f u. 10 u. 11f.; *Codex Leidensis 399,1.* 1897, S. 28, 20 u. 21; 30, 1; MS Teheran Malik 3586, f 274^{b}, 15 u. 16 u. 17.

sowohl **G** (und die Isḥāq/Ṯābit-Redaktion) als auch **P** stimmen inbezug auf die Wortwahl zur Beschreibung der kleineren Rechteckseite überein. Im dritten Fall sind **P** und **L**, Haupttext gleich und unterscheiden sich vom griechischen Wortlaut, wenn man die gesamte Beschreibung des Quadrates betrachtet, die von der Isḥāq/Ṯābit-Redaktion wörtlich wiedergegeben wird. Inbezug auf die Differenz zu **E/R** stehen sie jedoch durch die Verwendung von *az-ziyāda* **G** deutlich näher als das Ḥağğāğ-Zitat in **E/R**.

Diese Abweichungen in **E/R** stellen wohl das Ergebnis einer Bearbeitung dar. Darauf weist u.a. das *uḍīfa ilā* hin, das weder in **P** noch in **L**, Haupttext noch in MS Teheran Malik 3586 in Buch II zur Bezeichnung einer additiven Operation benutzt worden ist. Alle in diesen Manuskripten in Buch II verwendeten Ausdrücke für eine additive Operation sind von *ğmʿ* sowie im Fall der Isḥāq/Ṯābit-Redaktion noch von *rkb* abgeleitet.

Vergleicht man II,6 mit der in der ersten *Risāla* der *Rasāʾil iḫwān aṣ-ṣafāʾ* vorhandenen Variante der *ḍarb*-Version, so zeigt sich, daß diese an den kritischen Stellen **P** erheblich nähersteht als **E/R**:[32]

(مسألة) كلّ عدد قسم بنصفين ثمّ يزاد فيه زيادة ما . فاقول انّ ضرب ذلك العدد في تلك الزيادة ونصف العدد في نفسه مجموعا يكون مساويا لضرب نصف ذلك العدد مع الزيادة في نفسه .

(Problem) Eine (beliebige) Zahl werde in zwei Hälften geteilt. Danach werde ihr irgendein Zusatz hinzugefügt. Dann sage ich, daß das Produkt jener Zahl mit jenem Zusatz und die Hälfte der Zahl mit sich selbst zusammengenommen gleich dem Produkt der Hälfte jener Zahl (plus) den Zusatz mit sich selbst.

Hermann und Adelard stimmen weitgehend mit **P** überein. Aus diesem Befund kann man erneut folgern, daß entweder die Abweichungen in **E/R** das Ergebnis späterer Eingriffe sind oder daß es zwei verschiedene *ḍarb*-Versionen gegeben hat, wovon eine **P** bzw. dessen Archetypus eng verwandt gewesen ist. Dieses Ergebnis stimmt mit den bei II,3 gezogen Schlüssen überein.

In II,7 betrifft die substantielle Differenz den zweiten Teil der Behauptung. Zur besseren Verdeutlichung sei auch hier wieder die kritische Stelle mit dem **G**, **L**, Haupttext und MS Teheran Malik 3586 verglichen.

G: (τῆς ὅλης καὶ) τοῦ εἰρημένου τμήματος [(der Ganzen und) des genannten Teils]

E: *(al-ḫaṭṭ fī) l-qism allaḏī ḍuriba fī nafsihī* [(der Strecke mit) dem Teil, der mit sich selbst multipliziert wurde]

P: *(al-ḫaṭṭ fī) ḏālika l-qism* [(der Strecke mit) jenem Teil]

---

[32] *Rasāʾil iḫwān aṣ-ṣafāʾ*, 1306 H., S. 30, 9-11.

L, Haupttext: *(al-ḫaṭṭ kullihī wa-) ḏālika l-qism* [(der Strecke als Ganzes und) jenes Teils]

MS Teheran Malik 3586: *(al-ḫaṭṭ kullihī wa-) l-qism allaḏī ḏakarnā* [(der Strecke als Ganzes und) des Teils, den wir erwähnten]

Hermann von Carinthia stimmt mit **P** überein und Adelard mit **L**.[33] Die Ausdrucksweise von **P** und die bearbeitete Form von **L**, Haupttext stehen **G** deutlich näher als die von **E/R** gebrauchte Beschreibung. Letztere ist keinesfalls als Übersetzung der betreffenden Passage des edierten griechischen Textes charakterisierbar. Die Form des Nachsatzes von **E/R** ist analog zu der des Nachsatzes in II,3 konstruiert. Deshalb läßt sich die bei II,3 gezogene Schlußfolgerung auf II,7 übertragen, womit sich **E/R** erneut als Bearbeitung einer *ḍarb*-Grundversion präsentiert.

Neben diesen recht substantiellen Differenzen gibt es noch zwei kleinere Abweichungen zwischen **E/R** und **P**, die die in diesem Abschnitt erzielten Ergebnisse zu bestätigen scheinen. Sie finden sich in den Sätzen II,7 und II,9. In II,7 unterscheiden sich **P** und **E/R** an einer Stelle inbezug auf die benutzte Kopula. **P** benutzt wie MS Teheran Malik 3586 *wa* und stimmt darin mit καί in **G** überein, während **E/R** und MS Leiden 399.1, Haupttext *maʿa* verwenden. Die Verwendung von *maʿa* erweist sich beim Vergleich zwischen **P**, **L**, Haupttext und MS Teheran Malik 3586 als ein relativ gutes Indiz für Textveränderungen in der arabischen Überlieferung von Buch II der *Elemente*, da **P** es nur zweimal (für σύν und einen damit zusammengesetzten Ausdruck) verwendet und auch MS Teheran Malik 3586 es nie für die Wiedergabe von καί gebraucht. Der Mischtext MS Leiden 399.1 dagegen benutzt es vergleichweise oft, u.a. auch für καί. Allerdings könnte es in **E/R** auch durch einen Abschreiber eingedrungen sein.

In II,9 hat **E/R** *bi-qismaini mutasāwiyaini* [in zwei gleiche Teile] anstelle von *bi-niṣfaini* [in zwei Hälften] in **P**. Die Verwendung von *bi-qismaini mutasāwiyaini* ist ein Charakteristikum der Isḥāq/Ṯābit-Redaktion, das in **P** nie vorkommt. Ein Vergleich mit II,9 in der ersten *Risāla* der *Rasāʾil iḫwān aṣ-ṣafāʾ* zeigt, daß das *bi-qismaini mutasāwiyaini* offenkundig erst nach der Entstehung der mit *ḍarb* operierenden Ḥaǧǧāǧ-Version eingeflossen ist, denn die erste *Risāla* verwendet noch das *bi-niṣfaini* von **P**.[34]

2.3 Einflüsse in E/R aus der Isḥāq/Ṯābit-Redaktion

Eine dritte Klasse von Abweichungen zwischen **E/R** und **P** erweist sich nach ihrer Untersuchung als Übernahme von Formulierungen aus der Isḥāq/Ṯābit-Redaktion. Die Kürze und die Art dieser Übernahmen legen nahe, diese als Ergebnisse des Abschreibprozesses zu interpretieren. Sie treten in den Sätzen II,1, 2, 6, 7 und II,9 auf.

[33] Busard 1968, S. 45; Busard 1983, S. 77.

[34] *Rasāʾil iḫwān aṣ-ṣafāʾ*, 1306 H., S. 30, 24.

In II,1 fügt **E/R** ein *kam kānat* [wie es (gerade) war] ein, das für ὡς ἔτυχεν [wie es (gerade) geschah] in **G** steht. In **P** fehlt diese Wendung grundsätzlich. Da sie in den anderen Sätzen, in denen sie in **G** auftritt, in **E/R** ebenfalls fehlt, ist anzunehmen, daß sie sekundär, vermutlich durch einen Abschreiber, eingebracht worden ist. Da sie in der Isḥāq/Ṯābit-Redaktion regelmäßig vorhanden ist und im Unterschied zu den späteren Sätzen im Fall von I,1 wie in **E/R** wiedergegeben wird und der jeweilige Haupttext von **E/R** anscheinend dieser Redaktion angehört, ist diese eine mögliche Quelle.

In II,2 und II,6 betrifft es die Verwendung von *kulluhū* [als Ganzes] nach *al-ḫaṭṭ* [die Strecke] (II,2) bzw. *ǧamīʿ ḏālika* [der Gesamtheit dessen] (II,6) durch **E/R**. **P** hat nur *al-ḫaṭṭ* (II,2) bzw. *ǧamīʿ ḏālika*, **G** nur ὅλη und die Isḥāq/Ṯābit-Redaktion verwendet recht konsequent *al-ḫaṭṭ kulluhū* [die Strecke als Ganzes]. In II,6 doppelt dabei **E/R** durch diese Einfügung. II,6 zeigt damit deutlicher als II,2, daß diese Einfügung sekundär eingedrungen ist. Das wird zusätzlich dadurch bestätigt, daß in II,7 **R** diese Einfügung abermals besitzt, während sie in **E** fehlt. Erneut liegt es nahe, an einen Abschreiber und die Isḥāq/Ṯābit-Redaktion als Quelle zu denken.

Im Unterschied zu dieser Einfügung verwendet **E/R** in II,3 und II,4 *ǧamīʿ al-ḫaṭṭ* [die Gesamtheit der Strecke/die ganze Strecke] im ersten Teil der Behauptung, während **P** hier wieder nur *al-ḫaṭṭ* hat. Dieser Unterschied zwischen den beiden Versionen könnte wegen der Regelmäßigkeit in **P** aus einer gezielten Kürzung eines seiner Vorläufer resultieren. Als Übertragung von ὅλη durch al- Ḥaǧǧāǧ wäre dann *ǧamīʿ al-ḫaṭṭ* anzunehmen.

In II,7 tritt die folgende Wendung auf:

G: τοῦ λοιποῦ τμήματος [des restlichen Teils]

E: *al-qism al- bāqī* [des übrigen Teils]

P: *al-qism al- āḫar* [des anderen Teils]

L, Haupttext: *al-qism al-āḫar* [des anderen Teils]

MS Teheran Malik 3586: *al-qism al-bāqī* [des übrigen Teils][35]

P überträgt das τὸ λοιπὸν τμῆμα in der Regel mit *al-qism al-āḫar*, während MS Teheran Malik 3586 dafür stets *al-qism al-bāqī* hat. MS Leiden 399.1, Haupttext stimmt in dieser Beziehung i.a. mit **P** überein. Vergleicht man II,7 mit II,8 bezüglich der Verwendung von *al-qism al-āḫar* versus *al-qism al-bāqī*, so zeigt sich, daß **E/R** in II,8 **P** mit *al-qism al-āḫar* folgt. Daraus kann wiederum geschlußfolgert werden, daß das *al-qism al-bāqī* in II,7 im Ḥaǧǧāǧ-Zitat von **E/R** sekundär, vermutlich durch einen Abschreiber und aus der Isḥāq/Ṯābit-Redaktion, eingedrungen ist.

---

[35] *Euclidis Elementa* 1969, S. 77, 4f; *Codex Leidensis 399,1*, 1897, S. 32, 17; MS Teheran Malik 3586, f 275^2.

In II,9 hat **E/R**, wie bereits im vorhergehenden Abschnitt vermerkt worden ist, das *bi-niṣfaini* [in zwei Hälften] von **P** durch das *bi-qismaini mutasāwiyaini* [in zwei gleiche Teile] der Isḥāq/Ṯābit-Redaktion ersetzt.

2.4 Kleinere Differenzen zwischen E/R und P

Kleinere Differenzen treten in den Sätzen II,2, 4, 6, 7, 8 und II,9 auf. Sie sind z.T. auf Fehler beim Abschreiben zurückzuführen. Gelegentlich sind sie aber auch für die Beurteilung des Verhältnisses zwischen **E/R** und **P** von Nutzen.

In II,2 wechselt **E/R** in der Rektion von *qasama* [teilen] zu *qasama* + Akk. über, während **P** an *qasama bi-* festhält.

In II,4 weicht **E/R** in der Beschreibung des Rechtecks im zweiten Teil der Behauptung durch *(ḍarb) aḥad al-qismaini fī l-āḫar* [(Multiplikation) eines der beiden Teile mit dem anderen] von **P** ab, das nur *aḥadihimā fī l-āḫar* [einer der beiden mit dem anderen] hat. Die Verwendung von *ḍarb* vor jeder „Fläche" in **E/R** scheint anzuzeigen, daß in einem der Vorläufer von **P** ebenfalls vor jedem „Flächenstück" *talbīn* gestanden hat. Dieser Befund deutet darauf hin, daß im Prozeß des Kopierens der Vorläufer von **P** Teile ausgefallen bzw. zusammengefaßt worden sind.

In II,6 weicht **E/R** durch die Verwendung von *huwa* [er/es] vor *miṯl* ab, dessen Ausfall in **P** offensichtlich durch das wiederholte Kopieren des Textes verursacht worden ist, denn an anderen Stellen tritt es auch in dieser Version auf.

In II,7 fehlt in **E/R** im zweiten Teil der Behauptung die explizite Angabe einer Addition. **P** hat *ǧamīʿan* wie im Teil 1 der Behauptung. Vergleicht man diesen Punkt beispielsweise mit II,8 und II,9, so zeigt sich, daß **E/R** dort in den Ḥaǧǧāǧ-Zitaten in der Behauptung explizit Summationen verlangt, so bald es sich um zwei Glieder handelt, d. h. im Prinzip mit **P** übereinstimmt.

Im ersten Teil der Behauptung verwendet **E/R** *maǧmūʿaini* [(die beiden) zusammengenommen/summiert] statt *ǧamīʿan* [zusammen/insgesamt], das es in anderen Sätzen einsetzt und das in **P** mit Ausnahme der Sätze II,12 u. 13 die Regel ist. Dasselbe wiederholt sich in II,9. Außerdem ist das sonst übliche *inna* nach *fa-* ausgefallen und im zweiten Teil der Behauptung fehlt das sonst übliche *yakūnu min* zwischen *allaḏī* und *ḍarb*, wovon **R** noch das *min* erhalten hat.

In II,8 weicht **E/R** mit *ṭūl al-ḫaṭṭ* [Länge der Strecke] von **P** mit *ṭūlihī* [(ihre) Länge] ab. Nun ist nach meiner Auffassung diese Stelle als Ganzes das Result einer Bearbeitung, d.h. hier kann aus einem Vergleich mit **G** kein Aufschluß über die etwaige bessere Lesart gewonnen werden. Da die Formulierung der Umstrukturierung aber deutlich mit der Ausdrucksweise in II,6 korrespondiert, ist anzunehmen, daß sie nach deren Vorbild gestaltet worden ist. Folglich dürfte die bessere Lesart die mit II,6 in diesem Punkt übereinstimmende, d. h. **P**, sein, falls diese nicht selbst einer Veränderung unterworfen war. Hermann folgt an dieser Stelle **P** und Adelard **E/R**.

In II,9 fehlt **P** *al-qismaini* [zwei Teile] vor *al-muḫtalifaini* [verschieden] im ersten Teil der Behauptung, was **E/R** bewahrt hat. **E/R** dagegen doppelt *miṯlāni* [zweimal das Gleiche], was **P** an dieser Stelle nicht hat. Allerdings tritt diese Dopplung in **P** und die ihr entsprechende Formulierung in **L**, Haupttext an verschiedenen Stellen, wenn auch unregelmäßig auf. Sie dürfte von Abschreibern eingebracht worden sein, denn **G** und MS Teheran Malik 3586 haben stets nur einmal ein entsprechendes Wort. Die uneinheitliche Bezeichnung des kleineren Teils in **E** durch *aṣġar/aqṣar* [kleiner/-kürzer] gegenüber der regelmäßigen Benutzung von *aqṣar* in **P** und **R** zeigt an, daß *aqṣar* die ursprüngliche Form ist und *aṣġar* durch Lese- bzw. Abschreibfehler eingedrungen ist.

2.5 Das Porisma zu II,4

E/R überliefert interessanterweise neben den Sätzen auch das einzige in Buch II enthaltene Porisma, das in **P** fehlt, da es traditionell ja erst nach dem Beweis angegeben wird und die Beweise in **P** mit Ausnahme der Sätze II,11 und II,14 fehlen. Das bestätigt die aus den lateinischen Übersetzungen von Hermann von Carinthia und Adelard von Bath (Version I) gezogene Schlußfolgerung, daß es eine vollständige *ḍarb*-Version der *Elemente* gegeben zu haben scheint.

In der Formulierung des Porisma stimmt die Lesung von R mit *as-saṭḥaini l-mutawāziyaini l-aḍlā'* [die zwei Flächen mit parallelen Seiten] mit **G**, die von Parallelogrammen spricht,[36] überein, während E nur *as-saṭḥaini* [die zwei Flächen] hat. Diese Differenz könnte entweder durch einen durch Abschreiber verursachten Ausfall in **E** entstanden sein oder als Anleihe in **R**, möglicherweise erneut durch einen Abschreiber, aus der Isḥāq/Ṯābit-Redaktion eingebracht worden sein, die *as-suṭūḥ al-mutawāziyat al-aḍlā'* [die Flächen mit parallelen Seiten] hat, aber ansonsten sprachlich deutlich verschieden von der Formulierung in **E/R** ist:[37]

فقد تبيّن من هذا الشكل انّ السطوح المتوازية الاضلاع التي يكون على سطوح مربّعة هي ايضا مربّعة .

So ist durch diesen Satz erklärt worden, daß die Flächen mit parallelen Seiten, die an quadratischen Flächen sind, ebenfalls quadratisch sind.

[36] *Euclidis Elementa* 1969, S. 73.

[37] MS Teheran Malik 3586, f. 273^b, 8f.

3. Zusammenfassung

Im vorliegenden Artikel konnte gezeigt werden, daß **E/R** ungeachtet der charakteristischen Unterschiede zu **P** grundlegende Gemeinsamkeiten mit **P** aufweist und folglich letztendlich auf eine gemeinsame Urquelle zurückgeht, die nach der Überlieferung sowie nach inneren Merkmalen eine der Versionen von al-Ḥaǧǧāǧ b. Yūsuf b. Maṭar war. Es konnte des weiteren nachgewiesen werden, daß **E/R** wie **P** eine Bearbeitung dieser Urquelle zu verkörpern scheint. In einem zentralen Punkt (II,3) unterscheiden sich jedoch beide Versionen, wobei auf der Grundlage der bei der Analyse von **P** aufgestellten Hypothesen und Argumente **E/R** in dieser Hinsicht als die ursprünglichere Version identifiziert wurde. Daraus scheint zu folgen, daß nicht beide Versionen von al-Ḥaǧǧāǧ selbst stammen können, da aus sprachlichen Gründen nach meiner Ansicht keine von ihnen direkt auf Version I, d.h. die Übersetzung von al-Ḥaǧǧāǧ zurückzuführen ist.

Aufgrund des archaischeren Ausdrucks bei der Beschreibung von Quadraten und Rechtecken in **P**, der erheblich größeren Verbreitung der entsprechenden Ausdrucksweise in **E/R** und weiteren, über die Bearbeitungsstufe von **P** hinausgehenden Veränderungen in **E/R** wurde geschlußfolgert, daß **P** diejenige der beiden vorliegenden Versionen ist, die letztendlich direkt von al-Ḥaǧǧāǧ herrührt. Das bedeutet, daß **E/R** von einem anderen Bearbeiter stammen dürfte. Schließlich konnten Argumente dafür erbacht werden, daß zwischen **P** und **E/R** eine andere, noch näher mit **P** verbundene *ḍarb*-Version existiert und daß es eine komplette *ḍarb*-Version der *Elemente* gegeben hat.

Bibliographie

Abu l-Faḍl Ğamāl ad-Dīn M. b. Mukarram b. Manṣūr al-Ifrīqī al-Miṣrī 1375 H./1956: *Lisān al-ʿarab*. Bairūt, ğ 13.

Brentjes, S. 1991: *Textzeugen und Hypothesen zum arabischen Euklid in der Überlieferung von al-Ḥağğāğ b. Yūsuf b. Maṭar (zwischen 786 und 833)*, im Druck.

Brentjes, S. 1992: On the Arabic Transmissions of Euclid's *Elements*. in: *Tradition, Transmission, Transformation. Ancient Mathematics in Islamic and Occidental Cultures.* Department of the History of Science, The University of Oklahoma, im Druck.

Busard, H.L.L. 1968: *The Translation of the* Elements *of Euclid from the Arabic into Latin by Hermann of Carinthia (?).* Leiden.

Busard, H.L.L. 1983: *The First Translation of Euclid's* Elements *Commonly Ascribed to Adelard of Bath.* Toronto.

Codex Leidensis 399,1. 1897: Euclidis Elementa ex interpretatione al-Hadschdschadschii cum commentariis al-Nairizii. Arabice et latine ediderunt notisque instruxerunt R.O. Besthorn et J.L. Heiberg. p. I, f. II. Kopenhagen.

Djebbar, A. 1990: *Quelques commentaires sur les versions arabes des Éléments d'Euclide et sur leur transmission à l'occident musulman*, im Druck.

Euclidis Elementa. 1969: vol I, libri I-IV cum appendicibus post I.L. Heiberg edidit E.S. Stamatis. Leipzig.

Freytag, G.W. 1837: Lexicon Arabico-Latinum. Halle, t. IV.

al-Ḫwārizmī, M. b. Mūsā 1968: *Kitāb al-ğabr wa l-muqābala*, eds. ʿA.M. Mašrafa, M.M. Aḥmad. al-Qāhira.

Ibn an-Nadīm 1871/72: *Kitāb al-Fihrist von Abūʾ l-Farağ Muḥammad b. Isḥāq, bekannt unter dem Namen Ibn Abī Jaʿqūb el-Nadīm.* Hrsg. von G. Flügel, J. Roediger und A. Müller, Leipzig.

Ibn Sīnā 1976: *Kitāb aš-šifāʾ. al-Fann al-awwal min ğumlat al-ʿilm ar-riyāḍī. Uṣūl al-handasa.* eds. ʿA. Sabrā, ʿA.L. Mazhar. al-Qāhira.

Levy, J. 1879: *Neuhebräisches und chaldäisches Wörterbuch.* Leipzig, Bd. 2, 3.

MS Escorial ar. 907, Madrid.

MS Paris, BN, Persan 169.

MS Paris, BN, ar. 4946.

MS Princeton, Yahuda Collection, 3770.

MS Rabāṭ, Ḥasanīya 53.

MS Rabāṭ, Ḥasanīya 1101.

MS Teheran, Malik 3586.

Payne-Smith, R. 1883: *Thesaurus Syriacus.* Oxford, t. II, f. IV, L-M.

Rasāʾil iḫwān aṣ-ṣafāʾ. 1306 H., o. O.

Sayılı, A. 1962: Abdülhamid ibn Türkün katışık denklemlerde mantikî zaruretler adli yazısı ve zamanın cebri. Logical necessities in mixed equations by ʿAbd al-Hamîd ibn Turk and the algebra of his time. *Türk tarih kurumu yayınlarından* 7th series, 41, S. 145-153.

Ullmann, M. 1983: *Wörterbuch der klassischen arabischen Sprache.* Wiesbaden, Bd. II,1

Ocreatus

Charles Burnett

The name 'Ocreatus' is familiar to anyone who has studied the 'Adelard of Bath' versions of Euclid's *Elements*. A 'Joan. Ocreatus' is credited in one twelfth-century manuscript with having made the translation of Euclid's *Elements* from Arabic into Latin commonly known as the 'Adelard II version'.[1] A certain 'Ocreatus' – probably 'H. Ocreatus' – explained the Saracen way of calculation to his teacher Adelard of Bath.[2] In a previous article I advanced several explanations for the by-name Ocreatus.[3] The most convincing seemed to be that Ocreatus was a nickname attached to geometers who had 'ocreae'. These 'ocreae' appear in the Adelard II version of Euclid's *Elements* as the source for additional proofs in the earliest manuscripts which contain proofs. In *Elements*, V def. 16 and *Elements*, X.9, these additions are introduced by the words 'Ex ocrea Iohannis'.[4] However, in a passage added to *Elements*, IX.15, where we read 'Ex ocrea Regeneri' in six manuscripts, the

---

[1]The heading 'Euclidis Elementa ex Arab. in Lat. vers. per Joan. Ocreatum' was recorded in E. Bernard, *Catalogi librorum manuscriptorum Angliae et Hiberniae in unum Collecti* (Oxford, 1697), n. 8639, which has been identified as British Library, Royal 15 A.XXVII. Unfortunately the heading is no longer present, and was probably on the first folio which is now lost. For this manuscript see *Robert of Chester's (?) Redaction of Euclid's* Elements, *the So-Called Adelard II Version*, ed. H.L.L. Busard and M. Folkerts, Basel etc. 1992, pp. 54-5. I do not intend to discuss at this point the attribution of the Adelard II version of Euclid's *Elements* or the context of its composition. I would only say that I find convincing the arguments of its editors that it postdates and is dependent on the Adelard I version and that the latter has every likelihood of being by Adelard himself.

[2]Ocreatus's *Helcep Sarracenicum* was first edited (from Paris, Bibliothèque Nationale, MS lat. 6626) by C. Henry, 'Prologus N. Ocreati in Helceph ad Adelardum Batensem magistrum suum', *Abhandlungen zur Geschichte der Mathematik,* **3**, 1880, pp. 129-39; a new edition and English translation is included in C. Burnett, '*Algorismi vel Helcep decentior est diligentia*: The Arithmetic of Adelard of Bath and his Circle', to be published in the proceedings of the conference on *Mathematical Problems in the Middle Ages – Arabic and Latin*, held at Wolfenbüttel in June, 1990. A second manuscript of this rare work is MS Cashel (Tipperary), G.P.A. Bolton Library (formerly Cathedral Library), 1, which includes a calendar mentioning saints venerated in the diocese of Worcester. The *Helcep Sarracenicum* is found on pp. 110-20, apparently without ascription. (I am grateful to Jennifer Moreton for sending me information concerning this manuscript, including R. Wyse Jackson, 'The Ancient Library of Cashel', *Journal of the Cork Historical and Archaeological Society*, second series, **52**, 1947, pp. 128-(31), and a detailed typescript description by Professor Marvin Colker who first recognized the existence of the *Helcep* in this manuscript). There are no grounds for Henry's reading of the author's name as *N.* Ocreatus. The initial is clearly an 'H.', and the two diagonal strokes through the letter are more likely to be a decorative feature than an erasure.

[3]See '*Algorismi vel Helcep ...*' (n. 2 above), nn. 46 and 55.

[4]See *Robert of Chester's (?) Redaction of Euclid's* Elements (n. 1 above), pp. 881 and 931.

other manuscripts give 'Ex valle Regeneri' or 'Ex walle Regenere'.[5] I suggested that '*vallis/*wallis' was a synonym of 'ocrea'. '*Vallis' bears some resemblance to the Medieval Latin 'valisium/valesium' (= 'valise, a kind of long leather bag in which one places one's personal belongings').[6] On the other hand '*wallis' with a 'w' reflects the English word 'wallet'.[7] The word 'ocrea' can refer to any capacious leather container.[8] Thus the common meaning of 'ocrea' and '*vallis/*wallis' would seem to be 'leather satchel', of the kind in which students kept their books and their wax-tablets.[9] 'Ocreatus' would refer to a person equipped with such a satchel, and the 'Iohannes' from whose 'ocrea' additional proofs were taken could well be nicknamed 'Iohannes Ocreatus'. Moreover the same epithet could be applied to Reginerus and other scholars in the entourage of the redactor of the Adelard II version of the *Elements.*

This explanation might still be possible, but only as a pun on the name Ocreatus, which I now believe is none other than the Latin translation of the surname of a family who were particularly important in the region of Bath from the time of Domesday (1086 A.D.) onwards. This is the Hussey family, whose name in Anglo-Saxon and Norman form was 'Hosed' or 'Hosethe' (= 'wearing hose, booted') which was Latinized to 'Hosatus'. To demonstrate the likelihood of this one has to show (a) the connections of Adelard and Ocreatus with Bath and its region, (b) the extent of the involvement of the Hussey family in Bath affairs and (c) the plausibility of taking Ocreatus as an equivalent to Hosatus.

The manuscripts are undivided in calling Adelard the scientist and translator of works from Arabic into Latin, 'Adelard of Bath'. Adelard himself is

---

[5]Ibid. pp. 920-1. Nine manuscripts give (in a more or less garbled form) 'ex valle' and four give 'ex walle'. This same phrase might have originally stood at the end of a set of definitions added at the beginning of *Elements*, X, where we now read (in almost all the MSS), 'Vale Reginere': see ibid., pp. 925 and 929.

[6]Du Cange, *Glossarium mediae et infimae latinitatis*, new edition, Niort, 1883, s.v. valisia/valixia/valesia; R.E. Latham, *Revised Medieval Latin Word-List From British and Irish Sources*, Oxford, 1965, s.v. valisium/valesium. Cf. É. Littré, *Dictionnaire de la langue française*, Paris, 1881, s.v. valise: 'espèce de long sac de cuir, dans lequel on met des hardes'.

[7]According to the *Oxford English Dictionary* this word is 'of obscure origin ... Possibly from Old French or Anglo-French *walet, which might be derived from Teut. *wall- , to go on pilgrimage'.

[8]Thus 'ocrea' is used to describe the pouch into which a certain quantity of gold was sewn: see the 13th-century text – *Die Regensburger Schottenlegende – Libellus de fundacione ecclesie Consecrati Petri*, ed. P.A. Breatnach, Munich, 1977, p. 239: 'aperierunt archam regiam et duas ocreas latas et longas fecit impleri de electo auro' ('They opened the royal chest and made two broad and long 'ocreae' to be filled with choice gold'; later these bags were sewn up). 'Ocrea' was also a term for a container for wine made from a leather skin: see DuCange, *Glossarium*, s.v. ocrea (1), and H. Brunner, *Grundzüge der deutschen Rechtsgeschichte*, 2nd edn, Berlin, 1928, II, p. 530 n.11: 'ocrea id est hosa vini'.

[9]While 'ocrea' as a gloss written in yellow or brown ink ('ocrea' being a common spelling for 'ochra' = 'yellow ochre') might be another interpretation, this would fail to account for the synonym '*vallis/*wallis'.

proud to call himself 'Bathoniensis' in the work in which he was addressing the widest audience, his *Quaestiones naturales*:

> Neque enim ego iuxta obscuram portam, sed medium forum habito. Etiam – si id erras – non Stoicum me sed Bathoniensem dico. Quare non Stoicorum errata, sed meum intellectum docere debeo.

> Nor do *I* hide in some dark doorway, but I pass my life in the middle of the square. And, even if you are wrong in this, I do not call myself a Stoic but 'the Bathonian'. Therefore I ought to teach not the errors of the Stoics but what I think myself.[10]

A local by-name such as 'Bathoniensis' should indicate that its holder had moved away from the place from which he had taken the name, or at least that he had acquired a reputation outside his own town or region. The latter seems to apply to Adelard. That he was *in* Bath is suggested by his use of Bath as an example in both his treatise on the astrolabe (*De opere astrolapsus*) and his translation of an Arabic text on making magical talismans (*Liber prestigiorum Thebidis secundum Ptolomeum et Hermetem*). In the *De opere astrolapsus* he writes:

> Verbi gratia ad natale solum: Quia enim Bathonia quinquaginta duobus gradibus ab equinoctiali circulo et terra Arin distare cognoscitur, ideo et latitudo climatis eius totidem graduum esse perhibetur.

> For example, take our native soil. For because Bath is known to be distant from the equinoctial circle and the land of Arin by 52 degrees, the latitude of the climate is regarded to be of the same number of degrees.[11]

In the talisman text we read, in a section on getting rid of scorpions:

> Inter humandum hec oratio dicatur: hec species <et> omnis sui generis species a loco presenti, verbi gratia a Batonia, fuget<ur>, ut nulla eam

[10] Adelard of Bath, *Quaestiones naturales*, xxviii, ed. M. Müller, Beiträge zur Geschichte der Philosophie des Mittelalters, **31**, 2, Münster i. W, 1934, p. 34; see M. Gibson, 'Adelard of Bath' in *Adelard of Bath: An English Scientist and Arabist of the Early Twelfth Century*, ed. C. Burnett, London, 1987, pp. 7-16 (7).

[11] See C.H. Haskins, *Studies in the History of Mediaeval Science*, 2nd edn, Cambridge, Mass., 1927, p. 28. Haskins' transcript of the text from Cambridge, Fitzwilliam McClean 165, fol. 82v has been corrected with the aid of MS Salzburg, St Peter, Stiftsbibliothek, a.V.2, fol. 87r and the edition in B.G. Dickey, 'Adelard of Bath: An Examination Based on Heretofore Unexamined Manuscripts', unpublished Ph.D dissertation, University of Toronto, 1982, p. 166. There are two further references to Bath ('Bathonia') in the *De opere astrolapsus*, on McClean 165, fols. 84v (ed. Dickey, p. 193) and 85r (ed. Dickey, p. 195).

> vel intrare vel inhabitare queat.
>
> While burying [the talisman of a scorpion] this prayer should be recited: 'May this species [of insect] and every species belonging to the same genus be put to flight from this present place (e.g., from Bath), so that none of them should be able to enter or inhabit the place'.[12]

In the Pipe Roll of 1130, auditing the sheriffs' accounts at the Exchequer of Henry I, 'Adelardus de Bada' is mentioned among those holding tenancies in Wiltshire, the county immediately east of Somerset in which Bath is situated. In Bath itself one would not expect Adelard to be called 'of Bath'. It is, therefore, quite plausible that Adelard of Bath is the 'Aðelardus' mentioned in four charters of Bath Priory.[13] He is in the retinue of the bishop of Bath, John de Villula of Tours, described twice as 'dapifer' (i.e., 'steward'),[14] and twice as being among the 'milites et ministri' of Bishop John.[15] The fullest reference (and one to which we shall return) is in a charter of 1106 in which the Bishop is restoring to the Priory the land he took from the monks on his accession. After naming the witnesses on the Priory's side, the Bishop's men are given:

> ... Walterus vicecomes Hosatus,[16] Hildebertus dapifer, Turoldus, Willelmus, Aðelardus filius Fastradi et omnes alii milites et ministri mei qui affuerunt.
>
> ... Sheriff Walter Hosatus, Hildebert the steward, Turoldus, Willelmus, Aðelard son of Fastrad and all the other knights and officials (thegns) of mine who were present.

A certain Fastrad was a tenant of Giso, John de Villula's predecessor, and held land at Wells, Yatton and Banwell (all in Somerset).[17] Aðelard, son of Fastrad, could have been sent abroad to John de Villula's home town of Tours to continue his education. We know that Adelard of Bath was a student at Tours and taught at Laon before making a seven-year 'study tour' of the Mediterranean which included a spell in Southern Italy and Sicily.[18]

---

[12]MS Lyon, Bibliothèque muncipale, 328, fol. 73v.

[13]Gibson, 'Adelard of Bath' (n. 10 above), pp. 7-8; see W. Hunt, *Two Chartularies of the Prior of St Peter at Bath*, Somerset Record Society, 7, 1893, I, nos 34 (no date), 41 (1100), 53 (1106) and 54 (1122). Strictly speaking, the abbey of St Peter at Bath should be called the Cathedral Priory of St Peter only from ca. 1090-ca. 1190, when the abbey church served as the Cathedral for the bishops of Bath and Wells, but, to avoid confusion, the monastic house will be called 'Bath Priory' throughout this article.

[14]Hunt, *Two Chartularies* (n. 13 above), I, nos 34 and 41.

[15]Ibid., I, nos 54 and 55.

[16]'Hosatus' is written above 'vicecomes' both in the cartulary edited by Hunt (MS Cambridge, Corpus Christi College, 111) and in MS London, British Library, Harley, 358, fol. 40r.

[17]Gibson, 'Adelard of Bath' (n. 10 above), p. 7.

[18]See Haskins (n. 11 above), pp. 20-42, and M. Gibson, 'Adelard of Bath' (n. 10 above).

It was probably during this period that he became known as 'Bathoniensis'. He returned to England, however, some time during the reign of Henry I (1100-35),[19] and appears not only in the 1130 Pipe Roll cited above, but also as a witness to a charter of King Stephen drawn up between 1135 and 1139.[20] Bath was a stronghold of King Stephen and lies roughly midway between the region where Fastrad had held land and Wiltshire where 'Adelardus de Bada' had a tenancy. It is, therefore, possible that Adelard settled there after his wanderings.

Of Ocreatus, up to now, we know nothing; not even whether we are dealing with a single scholar, or several people with the same by-name. The rubric to *Helcep Sarracenicum* in the Paris manuscript reads:

> Prologus H. Ocreati in Helceph ad Aelardum Batensem magistrum suum.
>
> The Prologue of H. Ocreatus to the *Helcep*: to Aelard of Bath, his teacher.[21]

We know of no other student of Adelard. Ocreatus's debt to Adelard's teaching in the *Helcep* is apparent in his reference to a theorem in Euclid's *Elements* which does not occur in the pre-Adelardian Latin Euclid.[22] The language of *Helcep Sarracenicum* is reminiscent of that of the Adelard II version of the *Elements*. Ocreatus uses 'Age ergo' ('Well then ...') to introduce a proof, and 'Ecce' ('Behold!') to highlight a result, in a way that is characteristic of this version.[23] Moreover, the Cashel manuscript of *Helcep Sarracenicum* appears to come from the Worcester region,[24] and it is at Worcester Cathedral Priory that we find the earliest manuscript of a work

[19]Adelard of Bath, *Quaestiones naturales* (n. 10 above), p. 1: 'Cum in Angliam nuper redierim Henrico Wilhelmi Anglis imperante, quoniam a patria causa studii diu me exceperam ...' ('When I recently returned to England during the reign of Henry, son of William, king of the English, since I had absented myself from my native land for a long time for the sake of study ...').

[20]*Regesta regum anglo-normannorum 1066-1154*, III, ed. H.A. Cronne and R.H.C. Davis, Oxford, 1968, p. 282, no. 764; see my Introduction in *Adelard of Bath*, ed. C. Burnett (n. 10 above), pp. 1-6 (4).

[21]C. Burnett, '*Algorismi vel Helcep* ...' (n. 2 above). The form 'Aelardus' for Adelard occurs elsewhere (see ibid.). The alternation 'æðel/ æl' is common in both Old English and Norman: see F. Colman, 'The Name-Element *Æðel*- and Related Problems', *Notes and Queries*, new series, 28, 1981, pp. 295-301. The spelling 'helcep' (= Arabic *Ḥisāb*, 'calculation') occurs in the prologue itself.

[22]*Helcep Sarracenicum*, sentence **40**, from Euclid, *Elements*, VII.19. This is an enunciation that had been taken over from the Adelard I version into the Adelard II version of the *Elements*; see Burnett, '*Algorismi vel Helcep* ...' (n. 2 above).

[23]See *Helcep Sarracenicum*, sentences **76** ('Age ergo'), and **10** and **21** ('Ecce'); Busard and Folkerts pick out 'Age ergo' as one of the three most characteristic features of the Adelard II version (*Robert of Chester's (?) Redaction* [n. 1 above], p. 29), and give sixteen examples of 'Ecce' in the Index (ibid., p. 367).

[24]See n. 2 above.

by Adelard.[25]

Now, Hussey – or 'Hosatus' to give the form which occurs most frequently in Latin charters – was an important family in the Bath area. We have already met 'Walterus Hosatus' as the first witness among the Bishop's men in the charter of 1106 which is also witnessed by 'Aðelardus filius Fastradi'. In fact, there are more Hosati in the cartulary of Bath Priory than members of any other family. Willelmus Hosatus was given the manor of Charlecombe by the abbots of Bath Priory already before the Domesday Survey (1086 A.D.).[26] The Hosati were also tenants of Bath Easton and Shockerwick in Bathford. Charlecombe, Bath Easton and Bathford are all now suburbs of the City of Bath. The Walter Hosatus of the charter of 1106 was sheriff ('vicecomes') of Wiltshire some time between 1100 and 1110, and possibly (on the evidence of this charter) sheriff of Somerset as well.[27] A 'Henricus Hosatus' occurs in the Pipe Roll for 1156 as holding lands not only in Somerset and Wiltshire, but also in Berkshire, Nottingham and Derbyshire, and Sussex; he witnessed several royal charters.[28]

While other members of the family appear in Sussex and Berkshire,[29] the association with Bath Priory is particularly striking. Bishop John de Villula granted some land in Cold Ashton to Hubertus Hosatus,[30] and in 1123, after the death of Bishop John, the monks of the Priory confirm Willelmus Hosatus in the possession of these lands of his father (presumably Hubertus).[31] The witnesses of this latter charter include 'Henrici Hosati, Rodberti fratris ejus, Atselini Hosati, et fratrum ejus'. Atselinus (or Atscelinus) Hosatus appears in the cartulary between 1121 and 1123 three times amongst the chaplains and clerks, once as the 'dapifer' ('steward') of the Priory, and once more

---

[25]MS Oxford, Bodleian Library, Auct. F.1.9, which includes the hand of the historian, John of Worcester (d. ca. 1140). This manuscript contains Adelard's version of the *Astronomical Tables* of al-Khwārizmī, alongside astronomical writings by Walcher of Malvern and Petrus Alfonsi.

[26]See R.W. Eyton, *Domesday Studies: An Analysis and Digest of the Somerset Survey (according to the Exon codex) and of the Somerset Gheld Inquest of A.D. 1084*, 2 vols, London, 1880, I, p. 58 and (for the relevant charter) Hunt, *Two Chartularies* (n. 13 above), I, no. 33. At the time of the donation (probably between 1084 and 1086) there were two abbots. See also the entries in *Domesday Book, VIII: Somerset*, ed. C. and F. Thorn, Chichester, 1980, index, s.v. William Hussey.

[27]See J.A. Green, *English Sheriffs to 1154*, London, 1990, pp. 73 and 85.

[28]B.A. Lees, *Records of the Templars in England in the Twelfth Century: The Inquest of 1185*, London, 1935, p. 200 n. 3 and *Regesta regum anglo-normannorum 1066-1154*, II, Oxford, 1956, ed. H.A. Cronne and H.W.C. Davis, index, s.v. Hosatus.

[29]The Husseys were lords of Harting, Sussex and of Standen Hussey in Berkshire: a 'Henricus' founded Dureford Priory in Sussex in 1165 and annexed Standen chapel to it (Lees, *Records* [n. 28 above], p. 200, n. 3.). Another (?) Henry gave land at Sparsholt to the Templars in ca. 1145, and the charter confirming this includes the names 'Henrico Hosato avunculo meo et Hugo Hosato et Willelmo fratre suo' (ibid.).

[30]Hunt, *Two Chartularies* (n. 12 above), I, no. 51.

[31]Ibid., I, no. 52.

as the tenant of Charlecombe.[32] Willelmus Hosatus was a monk in 1100.[33] Sheriff Walter Hosatus, on the other hand, was one of the bishop's men. A 'Godefridus Hosatus' appears in a charter of 1155[34] and is presumably the same as the 'Geoffrey Hoset' in the 1168 Pipe Roll, who held land in Wiltshire.[35] Members of the family recur in the Bath cartulary later in the twelfth century, and throughout the thirteenth century.[36] They were knighted in the mid-thirteenth century and continued to be important landowners in the Bath region.[37] Even now the surnames Hussey, Hussy, Husey and Hosey are strongly represented in South-West England.[38] The family seal of Henry Hose of Cold Ashton, who granted some land to St John's Hospital in Bath between 1200 and 1225, was a triangular shield bearing a leg wearing a leather boot and a spur.[39]

The close connection of the Hosatus family with the Bath region, and in particular with Bath Priory, cannot be doubted. The question remains as to whether a Hosatus could call himself 'Ocreatus'. 'Hosatus' is not a Classical Latin word.[40] It is from a Latin spelling of a Germanic root 'hosa', meaning 'leggings' or 'boots'. 'Ocrea', on the other hand, is a good Classical Latin term (usually in the plural) meaning (a) 'a greave worn by a soldier' or (b) 'a legging worn for protection in farm work'.[41] Throughout the Middle Ages 'ocreae' were glossed in Anglo-Norman by some form of 'hosae'.[42] The semantic range of 'ocrea' and 'hosa' seems to have been the same, for both words were also used for containers of wine,[43] and by the fourteenth century at least 'ocrearius' and 'hosarius' interchangeably

---

[32]Ibid., I, nos 49, 52, 57 and 66.

[33]Ibid., I, no. 57.

[34]Ibid., I, no. 75.

[35]A Henry Hose of Cold Ashton obtained in 1219 certain lands in Wiltshire late of Geoffrey Hose: see *Medieval Deeds of Bath and District: I Deeds of St John's Hospital, Bath*, ed. B.R. Kemp, Somerset Record Society **73**, 1974, p. 21.

[36]Hunt, *Two Chartularies*, II, nos 74, 81 and 107 (Henry Hosat de Cherlecumb, in the time of Prior Robert, 1198-1223); II, no. 67 (Henry Hosat de Taddewicha, during the same period); II, no. 140 (John Hose and his heirs in the time of Prior Thomas, 1223-61), etc. A tentative family tree of some thirteenth-century Hosati is given in *Medieval Deeds of Bath* (n. 35 above), p. 21.

[37]*Medieval Deeds of Bath and District: II Walker-Heneage Deeds*, ed. D.M.M. Shorrocks, Somerset Record Society 73, 1974, pp. 95-6 and 100 – a detailed family tree showing the descendants of Henry Husee (Hose?) of Tadwick (alive 1236) up to the late 15th century.

[38]See C.W. Bardsley, *A Dictionary of English and Welsh Surnames*, London, 1901, s.v. and P.H. Reaney, *A Dictionary of British Surnames*, revised edn, London, 1976, s.v.

[39]See *Medieval Deeds of Bath* (n. 35 above), p. 21.

[40]Its earliest appearance in Latin is possibly in Isidore of Seville's *Etymologiae*, XIX.34.9 (in a chapter on footwear): 'Osas puto ab os primum factas, et quamvis nunc ex alio genere, nomen tamen pristinum retinent'.

[41]See *Oxford Latin Dictionary*, s.v. ocrea.

[42]For examples see T. Hunt, *Teaching and Learning Latin in Thirteenth-Century England*, 3 vols, Woodbridge, 1991, III, p. 112: 14 examples of 'ocree/ocreas' glossed in Anglo-Norman as 'hose/hese/heses/hesez/huese' etc.

[43]See n. 8 above.

designated a 'cellarman'.[44] Moreover, the words were substituted for each other in proper names. Henry I's brother Robert Curthose, duke of Normandy from 1087-1106, was also known as 'Robelinus curta ocrea'.[45] It would not be surprising, therefore, if a man called 'Hosatus' was referred to as 'Ocreatus'. This would be especially appropriate in the context of Latin learning. A scholar might well replace the vulgar form 'hosatus' with the Classical Latin equivalent 'ocreatus', which is a word used by Horace, Pliny and Vegetius.[46] Thus the twelfth-century Parisian magister Peter of Troyes was known both as Manducator and Comestor; 'comestor' being the correct Classical form; 'manducator' a Latin back-formation from the French 'mangeur'.

My conclusion, then, is that the 'H. Ocreatus' of *Helcep Sarracenicum* and the 'Joan. Ocreatus' associated with the Adelard II version of Euclid's *Elements* were, in all likelihood, members of the Hosatus family of Bath. To identify them any more closely with one or more of the Hosati mentioned in the charters would be rash. 'Henry' seems to have been a particularly common name in the family, but 'Hugo' and 'Hubert' also occur. All one can say is that there was at least one 'H. Ocreatus' living at the right time and in the right place. If I am right, this adds further evidence of Adelard of Bath's residence in Bath and his involvement in the teaching of the local nobility.[47]

I would like to end with a tantalizing reference which may also be to a scholarly 'Hosatus'. MS Cambridge, Trinity College, R.15.16 is a mid-twelfth-century manuscript of arithmetical works whose earliest known provenance is the Grey Friars of Coventry.[48] The manuscript begins with an introduction to

---

[44] *Dictionary of Medieval Latin from British Sources*, ed. R.E. Latham, D.R. Howlett et al., s.v. 'hosarius'.

[45] William of Malmesbury, *Gesta regum*, ed. T.D. Hardy (London, 1840), II, p. 607. He owed this name to his short, fat figure.

[46] See *Oxford Latin Dictionary* and *Thesaurus Linguae Latinae*, s.v. ocreatus.

[47] Ocreatus could, in that case, have been one of the 'scholars of Bath' who, in 1113, added some verses to the mortuary roll of Matilda, daughter of William the Conqueror. These verses (headed 'vox scholarium ejusdem urbis') were published by L. Delisle, *Rouleaux des morts*, Paris, 1886, pp. 177 and 192 and reprinted in Hunt, *Two Chartularies* (n. 13 above), p. xviii. For further evidence of scholarly activity at Bath note that, at the time of Domesday, a certain 'Hugo the translator' ('Hugo interpres') owned a house in Bath, and held Bathampton as tenant of the Priory of St Peter (see *Domesday Book, VIII: Somerset* [n. 26 above], 1.31 and 7.11); whilst the significance of the by-name 'interpres' is not clear, it is curious that this is the only mention of a 'translator' in the whole of Domesday. A century later, a 'magister Rog(er)us Hereford' witnessed a charter at Bath: *Medieval Deeds of Bath* (n. 35 above), no. 2 (pp. 15 and 80-1), between ca. 1187 and ca. 1189; cf. 'magister Rog(er)us medicus' in the immediately following charter: ibid. no. 3, pp. 15 and 81, between ca. 1189 and 1191. This could be the famous computist and astronomer *magister* Roger (Infans) of Hereford (the Bath references are not mentioned in J.C. Russell, 'Hereford and Arabic Science in England about 1175-1200', *Isis*, **18**, 1932, pp. 14-25).

[48] See M.R. James, *The Western Manuscripts in the Library of Trinity College, Cambridge*, 3 vols, Cambridge, 1901, II, pp. 354-5 and my '*Algorismi vel Helcep ...*' (n. 2

arithmetic which mentions amongst the 'modern geometers', 'Alardus [i.e., Adelard], Johannes' and 'Willelmus'. The introduction also refers to the algorism as 'helcep' and on the back fly-leaves are some arithmetical notes which correspond to sections of the *Helcep Sarracenicum.* On fol. 3v there are marks of ownership. Of the earliest of these, which has been partially erased and can only be read with difficulty under ultra-violet light, the following letters are discernable:

cass fratris radulfi de lychfeld et ysay ... osato de cabi ...[49]

above).

[49]I am grateful to Thomas Peiss of the *Mittellateinisches Wörterbuch*, David Howlett and Richard Sharpe of the *Dictionary of Medieval Latin from British Sources* and Kieran Devine of *Dictionary of Medieval Latin from Celtic Sources* for providing me with examples of the use of 'ocrea', and to Silke Ackermann, Julia Barrow, Jennifer Moreton and Richard Sharpe for providing further help and correcting several egregious errors.

Deux mathématiciens peu connus de l'Espagne du XIe siècle: al-Mu'taman et Ibn Sayyid

A. Djebbar

Une lecture attentive des sources traditionelles relatives à l'histoire de l'occident musulman, telles que les chroniques, les biographies de savants ou les traités de synthèse, ne laisse subsister aucun doute quant à l'importance qu'ont connue les activités mathématiques dans l'Espagne médiévale et plus particulièrement entre le Xe et le XIIe siècle. Mais, si l'on excepte l'astronomie et la trigonométrie, relativement mieux connues,[1] il s'avère difficile à l'historien des sciences de dépasser cette certitude pour se faire une idée des différents aspects de cette activité, c'est à dire, ses débuts, ses rapports avec les foyers originels de cette tradition et, enfin, les rapports complexes qu'elle a entretenus avec son environnement social et politique qui pésera, parfois plus qu'ailleurs, sur ses orientations et sur son devenir. En effet, contrairement aux autres foyers scientifiques, celui de l'Espagne ne nous a offert, jusqu'à maintenant, comme témoignage de sa période féconde, que peu d'écrits de première main, encore que ceux-ci ne concernent pas toujours les disciplines de base telles que la géométrie, l'arithmétique, la théorie des nombres et l'algèbre. Il faut, de plus, préciser que, pour le moment, très peu de ces sources constituent des ouvrages indépendants et qu'elles sont, la plupart du temps, dispersées dans des livres d'enseignement tardifs comme ceux des mathématiciens maghrébins des XIIIe-XIVe siècles et dans les traductions latines ou hébraïques.

En attendant de pouvoir disposer, un jour, de certains de ces travaux, s'il en existe des manuscrits, je me contenterai de présenter, à l'aide de témoignages indirects et épars, mais que j'espère convaincants, deux mathématiciens peu connues et n'ayant fait l'objet, jusqu'à ce jour, d'aucune étude particulière.

Les raisons qui m'ont poussé à les réunir, dans un même exposé, sont de plusieurs ordres: En premier lieu, ils illustrent à leur manière deux aspects de l'activité mathématique de cette époque, l'enseignement et la recherche, en révélant certaines de leurs particularités et de leurs difficultés. En second lieu, par leur appartenance à deux couches sociales différentes, ils représentent deux types de savants dont le statut social n'a peut-être pas été sans effet

[1] J. Vernet Ginès et J. Samsó Moya, Panorama de la ciencia andalusí en el siglo XI, in *Actas de las Jornadas de Cultura Arabe et Islámica (1978)*, Madrid 1981, pp. 142-56; M.V. Villuendas, *La trigonometría europea en el siglo XI, estudio de la obra de Ibn Muʿāḏ El Kitāb Maŷhūlāt*, Barcelone 1979, pp. IX-XXXVI.

sur leurs activités.[2] En troisième lieu, certains moments de leur vie symbolisent, dans le cadre espagnol et méditerranéen, les rapports contradictoires de deux mondes différents mais étroitement liés, pour une longue période encore, par des relations complexes faites de violence et d'échanges pacifiques. En effet, nos deux mathématiciens connaîtront, directement pour l'un, indirectement pour l'autre, une figure représentative de la chevalerie espagnole et européenne et qui fut, à la fois, la plus maudite et la plus glorifiée par les poètes. Il s'agit, bien sûr, de Rodrigo Diaz,[3] le fameux «as-Sayyid» dont les faits d'armes traverseront les siècles, faisant dire à Corneille

> «Sois désormais le Cid: Qu'à ce grand nom tout cède,
> Qu'il comble d'épouvante et Grenade et Tolède.»[4]

Al-Muʾtaman et son Kitāb al-Istikmāl

Al-Muʾtaman est le surnom royal de Abū ʿĀmir Yūsuf Ibn Aḥmad,[5] troisième prétendant de la dynastie des Banū Hūd.[6] Cette famille régna de 1039 à 1118 sur la province de Saragosse, une des vingt trois principautés aux frontières mouvantes que bâtirent les rois des Taïfa sur les débris du khalifat omeyyade d'Espagne.[7]

Al-Muʾtaman monta sur le trône de Saragosse en 1081 et régna jusqu'à sa mort en 1085. Mais, malgré cette fonction privilégiée, les biographes nous ont transmis peu d'informations sur sa vie publique de chef civil et militaire de la Marche supérieure d'al-Andalus comme disent les historiens, c'est à dire, de la région la plus exposée et donc la plus vulnérable du territoire sous contrôle musulman. Elle était en effet exposée directement aux offensives des castillans qui se concrétiseront d'ailleurs par la prise de Tolède, l'année même de la mort d'al-Muʾtaman.

---

[2] Al-Muʾtaman appartenait à l'aristocratie (la *Khāṣṣa*). Quant à Ibn Sayyid, le silence des biographes nous incite à le situer dans les couches moyennes ou dans la *ʿĀmma*. Pour ces différentes catégories sociales de l'Espagne du XIe siècle, voir: M. Benaboud, ʿAṣabiyya and social relations in al-Andalus during the period of the Taifa states, *Hesperis-Tamuda* Vol. XIX, 1980-81, pp. 5-45.

[3] Pour les louanges du Cid, on peut se référer à la célèbre épopée castillane (Cf. Colin Smith, *Poema de Mio Cid*, Oxford 1972) qui aurait été écrite vers 1140. Pour les malédictions, voir l'anthologie poétique d'Ibn Bassām, adh-Dhakhīra fī Maḥāsin Ahl al-Jazīra, édit. Iḥsān ʿAbbās, ad-Dār al-ʿArabiyya li-l-Kitāb, Libye-Tunisie 1978.

[4] P. Corneille, *Le Cid*, Classiques Larousse, p. 104, vers 1224-25.

[5] Suter, dans son *Die Mathematiker und Astronomen der Araber und ihre Werke* (Leipzig 1900, p. 108, n° 249), donne l'orthographie incorrectement: al-Muʾtamin. Sarton fait de même dans son *Introduction to the History of Science*, t. I, p. 759.

[6] Sur cette dynastie, voir les informations données par D.M. Dunlop dans *Encyclopédie de l'Islam II*, article Hūdides, pp. 560-562.

[7] Sur la période des rois des Taifa, on peut consulter: M.ʿA. ʿInān, *Duwal aṭ-Ṭawāʾif mundhu qiyāmihā ḥattā al-fatḥ al-Murābiṭī*, Le Caire 1960. Sur le royaume le plus important de l'époque, celui de Séville, voir la récente thèse, très documentée, de M. Benaboud, *at-Taʾrīkh as-siyāsī wa-l-ijtimāʿī li-Ishbīlya fī ʿahd duwal aṭ-ṭawāʾif*, Tétouan 1983.

Le peu d'éléments de source arabe dont nous disposons nous présente un royaume de Saragosse contrôlant un arrière pays étendu et disposant de ressources financières suffisantes pour lui permettre de payer tribut au pouvoir castillan. En outre, sa situation stratégique d'état-tampon semble lui avoir évité l'instabilité que connaîtront presque toutes les autres principautés musulmanes. Cette stabilité sera dûe d'ailleurs, en partie, aux interventions décisives du Cid et de ses troupes contre les armées chrétiennes du Nord.[8]

Ces facteurs et d'autres liés à la personnalité et au long règne d'al-Muqtadir le père d'al-Muʾtaman, considéré par les biographes non seulement comme un mécène mais également comme un savant, ont crée des conditions favorables à l'apparition d'un important foyer scientifique à Saragosse et ayant comme pôle important le palais royal lui-même. C'est dans cet environnement qu'al-Muʾtaman s'initiera aux différentes sciences enseignées à son époque et se spécialisera dans certaines d'entre-elles tout en exerçant jusqu'à 1081, les fonctions de Ḥājib.

En 1068, c'est à dire au moment où il écrivait dans ses *Ṭabaqāt al-Umam*, la brève notice sur al-Muʾtaman, Ṣāʿid al-Andalusī classait notre auteur dans la catégorie des jeunes savants, ce qui laisse supposer qu'il avait probablement, à l'époque, moins de quarante ans. En tout cas, à cette date, il était déjà connu et reconnu, dans le milieu scientifique de Tolède, comme étant un mathématicien très versé dans la philosophie. En le comparant aux savants de sa génération, Ṣāʿid précise en effet que, «quoique savant en mathématique, il se distingue d'eux parce qu'il sait en outre la logique et cultive les sciences naturelles ainsi que la métaphysique».[9] Malheureusement, Ṣāʿid n'ajoute aucune information au sujet d'éventuels travaux que le mathématicien aurait déjà publiés à son époque. On ne sait donc pas si al-Muʾtaman s'était fait connaître par des ouvrages scientifiques ou par ses seules interventions dans les cénacles de spécialistes.

En tout cas, une chose est sûre: Son ouvrage de mathématique, le *Kitāb al-Istikmāl*.† – le livre du perfectionnement – qui le fera connaître, non seule-

[8]Pour une bibliographie concernant le royaume de Saragosse, cf. D.M. Dunlop, *E.I. II*, op. cit. p. 562.

[9]Ṣāʿid al-Andalusī, *Kitāb Ṭabaqāt al-Umam*, trad. R. Blachère, Paris 1935, p. 139.

†Note des éditeurs: Le présent article fut rédigé en 1984 et publié la même année comme Pré-publication du Département des Mathématiques, Orsay, Université de Paris-Sud. L'année suivante, quatre fragments anonymes appartenant à l'*Istikmāl* furent identifiés par A. Djebbar and J.P. Hogendijk. Toutefois, la découverte de ces fragments n'apporte aucun renseignement nouveau sur la vie d'al-Muʾtaman, et rien n'est à changer dans les conclusions de l'article de M. Djebbar. Voir A. Djebbar, "La contribution mathématique d'al-Muʾtaman et son influence au Maghreb", (Colloque Maghrébin sur les sciences Arabes. Bayt al-Ḥikma, 14-15 Février 1986), publié dans les *Actes* de ce colloque, Tunis 1990; A. Djebbar, "Las matematicás en el-Andalus a través de las actividades de tres sabios del siglo XI", dans: *El legado científico Andalusí. Museo Arqueológico Nacional.* Madrid 1992, pp. 23-26; A. Djebbar, "L'arithmétique euclidienne et ses prolongements dans le Kitāb al-Istikmāl d'al-Muʾtaman," à paraître; J.P. Hogendijk, "Discovery of an 11-th century geometrical compilation, the Kitāb al-Istikmāl of al-Muʾtaman ibn Hūd", *Historia Mathematica* 13 (1986), pp. 43-52; J.P. Hogendijk, "Le roi-géométre al-Muʾtaman

ment en Espagne mais, à partir du XIIe siècle, au Maghreb et en Egypte, était encore inachevé à cette époque et il semble même que son auteur n'ait pas eu le temps de le publier intégralement puisque, selon trois témoignages au moins, la version qui a circulé était imcomplète. On peut même supposer, en nous fondant sur les brèves informations rapportées par Ibn Khaldūn dans son *Kitāb al-ʿIbār*,[10] et par al-Maqqarī dans ses *Analectes*,[11] qu'avant le *Kitāb al-Istikmāl*, l'auteur aurait publié d'autres ouvrages, en particulier un traité d'optique. Mais les sources consultés ne fournissent aucune information supplémentaire à ce sujet.

La version du *Kitāb al-Istikmāl* qui a servi ultérieurement dans l'enseignement et que Ibn al-Qifṭī qualifie de belle somme,[12] n'est, en fait, qu'une partie d'un vaste projet qui devait combler, semble-t-il, une lacune dans l'enseignement supérieur des mathématiques. Il s'agissait de mettre à la disposition des étudiants un traité de la catégorie des ouvrages intermédiaires – les Mutawassiṭāt – susceptibles de prolonger les ouvrages de base grecs en géométrie, astronomie et théorie des nombres. Plusieurs éléments confortent en effet cette hypothèse: dans son livre «la médicine des âmes», le philosophe et astronome Joseph ben Judah as-Sabtī, communément appelé Ibn ʿAqnīn, conseille de l'étudier en même temps que les *Eléments* d'Euclide, l'*Almageste* de Ptolémée et les autres livres intermédiaires.[13] C'est d'ailleurs en arrivant au Caire, après avoir quitté Fès, c'est à dire après avoir acquis de solides connaissances mathématiques, qu'Ibn ʿAqnīn lui-même aurait abordé d'étude du livre d'al-Mu'taman, sous la direction de Maïmonide.[14]

Quant au contenu de l'ouvrage, aucun auteur connu n'en donne le détail. Mais, en comparant les informations fournies indirectement par le physicien et encyclopédiste Ibn al-Akfānī à ce qu'en disent les mathématiciens maghrébins Ibn Munʿim et Ibn al-Bannā', on peut reconstituer une partie de l'ouvrage et se convaincre déjà de son importance.

Empruntant son vocabulaire à la logique, al-Mu'taman divise son livre d'abord en genres puis en espèces puis en sections et enfin en propositions, ce qui n'est pas la tradition dans l'agencement des ouvrages mathématiques arabes.[15] Le premier genre englobe au moins trois espèces qui sont autant

---

ibn Hūd et son livre de la perfection (Kitāb al-Istikmāl), *Premier Colloque International d'Alger sur l'Histoire des Mathématiques Arabes, Actes du Colloque*, Alger 1988, pp. 51-66; J.P. Hogendijk, "The geometrical parts of the *Istikmāl* of Yūsuf al-Mu'taman ibn Hūd (11th century), an analytical table of contents," *Archives Internationales d'Histoire des Sciences* 41 (1991), pp. 207-281.

[10]Ibn Khaldūn, *Kitāb al-ʿIbār* ..., Dār al-Kutub, Beyrouth 1968, t. IV, p. 351.

[11]Al-Maqqarī, *Nafḥ aṭ-Ṭīb* ..., édit. Dozy, Leyde 1860, t. I: p. 288.

[12]Ibn al-Qifṭī, *Akhbār al-ʿUlamā bi-Akhbār al-Ḥukamā*, Beyrouth, Dār al-Āthār, non datée, p. 210.

[13]G. Sarton, *Introduction to the history of science*, op. cit.

[14]Ibn al-Qifṭī, op. cit.

[15]Il y a peut-être un lien entre cet agencement et la classification des sciences établie par les Ikhwān aṣ-Ṣafā. Cf. *Rasā'il*, Dār Ṣādir Beyrouth, non datée, pp. 266-72.

de livres indépendants. En effet, d'après le *Fiqh al-Ḥisāb* d'Ibn Munʿim,[16] la première espèce est un traité de théorie des nombres qui contient, entre autres, une importante section, la quatrième, sur les nombres amiables. Une note non datée, écrite par un certain Muḥammad al-Marāghī, à la suite du traité des nombres amiables de Thābit ibn Qurra d'un manuscrit d'Istanbul,[17] précise même que la section d'al-Mu'taman est cette même épitre d'Ibn Qurra mais en «plus long». Comme le nombre des propositions de cette quatrième section est, d'après les références d'Ibn Munʿim, au moins de quatorze et qu'il n'y en a que dix dans le traité d'Ibn Qurra, il y a là une interrogation sur les compléments introduits par al-Mu'taman et qui pourraient renseigner sur d'éventuels travaux poursuivis, dans ce domaine, en occident musulman, à partir de ceux de l'école de Baghdad. Mais, Ibn Munʿim est silencieux sur ce point, comme il l'est d'ailleurs sur les autres espèces du premier genre qui ne concernent peut-être pas l'arithmétique, la théorie des nombres et l'analyse combinatoire puisque ce sont là les trois sujets essentiels du *Fiqh al-Ḥisāb*.[18]

Nous savons malgré tout, grâce à une petite épitre d'Ibn al-Bannā', sa *Risāla fī-t-Taksīr*, que le troisième espèce était consacrée à la géométrie.[19] Ce texte montre, qu'en plus des propositions des *Eléments*, cette espèce en établissait d'autres comme la proposition I de la *Mesure du cercle* d'Archimède,[20] ou le théorème de Héron sur l'aire du triangle en fonction du périmètre et qui correspond chez al-Mu'taman à la proposition 28 de la deuxième section de la première espèce du premier genre.[21]

Ces sections qui traitent des figures rectilignes et curvilignes étaient probablement suivies par d'autres relatives aux sections coniques et à la géométrie dans l'espace. C'est du moins ce que l'on peut déduire d'une remarque importante faite par Ibn al-Akfānī dans sa petite encyclopédie *Irshād al-Qāṣid*:[22] Dans le paragraphe sur la géométrie, il définit cette science et donne le contenu de ses dix parties. Puis, avant d'énumérer les titres des ouvrages qui traitent de tel ou tel aspect de cette discipline, il fait la remarque suivante: «Et, jusqu'à maintenant, je n'ai pas vu un livre englobant ces dix parties. Mais si le *Kitāb al-Istikmāl* d'al-Mu'taman ibn Hūd, que Dieu lui soit miséricordieux, avait été achevé, il aurait été suffisant.»

[16]Ms. Rabat n° 416 Q, pp. 315-25.

[17]Ms. Aya Sofya n° 4830, f. 121b. Il faut signaler que la numérotation d'Ibn Haydūr, telle qu'on la lit dans une copie de son *at-Tamḥīs fī Sharḥ at-Talkhīṣ*, ne correspond pas à celle de nos auteurs. Cf. R. Rashed, Nuṣūṣ li-ta'rīkh al-aʿdād al-mutaḥābba wa-ḥisāb al-tawāfuqāt, *Journal for the History of Arabic Sciences* Vol. 6, n° 1-2, 1982, p. 68.

[18]A. Djebbar, *L'analyse combinatoire dans l'enseignement d'Ibn Munʿim (XIIe-XIIIe siècles)*, pré-publication mathématique d'Orsay n° 83 T 03, 1983, p. 85.

[19]Ms. Tunis n° 9002, ff. 131a-132b. La numérotation de cette espèce est confirmée par al-Marāghī qui renvoie, par exemple, à la 5e proposition de la 1e section de la 3e partie de la 3e espèce du 1e genre. Cf. Ms. Aya Sofya, op. cit. f. 178b, marge supérieure.

[20]Archimède, *La mesure du cercle,* trad. Ch. Mugler, Paris 1970, pp. 138-39.

[21]Pour différents aspects mathématiques et historiques liés à ce problème voir: E.S. Kennedy et M. Mawaldi, Abū al-Wafā' and the Heron theorems, *Journal for the History of Arabic Sciences* Vol. 3, n° 1, 1979, pp. 19-30.

[22]Ms. Escurial, n° 949, ff. 56b-58b.

C'était là l'essentiel de ce qu'on pouvait déduire des renseignements épars et parfois allusifs concernant le contenu de cet ouvrage. Il faut ajouter, pour conclure sur ce point, que ces mêmes sources nous permettent d'affirmer que ce livre a joué un rôle non négligable dans l'enseignement des mathématiques à partir du XIIe siècle: Nous avons déjà signalé la place que lui attribuait Ibn ʿAqnīn dans le cursus d'un étudiant en mathématique. A la même époque, au Maghreb, Ibn Munʿim dira à plusieures reprises dans son *Fiqh al-Ḥisāb* que les livres VII, VIII et IX des *Eléments* d'Euclide ne sont pas suffisants pour comprendre tous les aspects de son propre ouvrage et qu'il est indispensable à l'étudiant de posséder le contenu de la première espèce du premier genre du livre d'al-Mu'taman. De plus, il utilisera systématiquement l'analyse et la synthèse comme outil de démonstration après avoir rappelé textuellement la définition qu'en donne al-Mu'taman au début de son ouvrage.[23] A la fin du XIIIe siècle, c'est à ce même livre que renverra Ibn al-Bannā' pour compléter son résumé sur les figures planes et sur les équations qui peuvent en découler.[24] Son commentateur Ibn Haydūr s'y réfèrera également, dans la seconde moitié du XIVe siècle ou au début du XVe, à propos de la recherche des nombres amiables.[25] On ne sait pas si l'ouvrage a bénéficié d'un commentaire au Maghreb mais, à l'Est, il semble qu'après les révisions et corrections de Maïmonide, le livre a donnee lieu à une nouvelle rédaction, exécutée vers la fin du XIIIe siècle ou au début du XIVe, par le même Muḥammad al-Marāghī précédemment évoqué.[26]

Ces quelques éléments sont loin de restituer les aspects essentiels de l'ouvrage d'al-Mu'taman. mais, malgré leur caractère disparate, ils auront permis, je l'espère, d'en suggérer la richesse et l'importance dans l'enseignement des mathématiques, en attendant que d'autres aspects de son oeuvre soient révélés par une conaissance plus complète de la tradition d'enseignement en Espagne puis, après le XIIe siècle, au Maghreb.

Les travaux d'Ibn Sayyid

Pour illustrer certains aspects de la recherche mathématique dans l'Espagne du XIe siècle, j'ai choisi un auteur encore moins connu qu'al-Mu'taman et qui, de surcroît, n'a même pas publié les travaux qui lui vaudront le titre de

---

[23] Ibn Munʿim, *Fiqh al-Ḥisāb*, op. cit. pp. 215, 216, 256.

[24] Ibn al-Bannā', *Risāla fī-t-Taksīr*, op. cit.

[25] R. Rashed, Nuṣūṣ li-ta'rīkh al-aʿdād al-mutaḥābba wa-ḥisāb al-tawāfuqāt, op. cit. p. 68.

[26] Ms. Aya Sofya, op. cit. Dans des gloses en marge des *Rasā'il d'al-Kūhī* (relatives au volume du paraboloïde – ff. 161b-165a – à l'inscription d'un pentagone régulier dans un carré – ff. 168b-171a – aux compléments aux *Données* d'Euclide – ff. 173b-180b –), al-Marāghī se réfère à sa rédaction du *Kitāb al-Istikmāl* qu'il a intitulé *Kitāb al-Ikmāl*. Cette rédaction est antérieure à l'année 728 H. (1327) au cours de laquelle il dit avoir écrit ses gloses (f. 165a et 171a, marges supérieures). Ces gloses montrent également qu'il a respecté l'agencement d'al-Mu'taman mais qu'il a subdivisé les espèces en parties (ff. 162b marge droite, 169a marge gauche, 178b marge supérieure).

géomètre qui accompagne régulièrement son nom. Il s'avère être pourtant, à travers les témoignages qui nous sont parvenus, un exemple significatif de ce que pouvait être, à une époque donnée, les possibilités de la recherche mathématique arabe et les contraintes de son environnement.

Le peu de choses que l'on connaît d'Ibn Sayyid provient, à l'heure actuelle, de quatre sources: Deux biographies, les *Ṭabaqāt* de Ṣā'id al-Andalusī qui était son contemporain et la *Takmila* d'Ibn al-Abbār, plus tardive. Les deux autres sources sont mathématiques: Il s'agit tout d'abord du *Fiqh al-Ḥisāb* d'Ibn Mun'im, déjà mentionné et, surtout, de trois pages d'un manuscrit de l'Escurial qui ont été classées anonymes dans le catalogue de Dérenbourg[27] et que j'ai identifiées comme étant un écrit du philosophe Ibn Bājja, dont la copie d'Oxford était la seule connue jusqu'à maintenant.[28]

On sait peu de choses sur la formation mathématique d'Ibn Sayyid et sur le milieu scientifique de Valence où il acquit sa notoriété: On apprend, par Ibn al-Abbār, qu'en 1063 il étudiait encore la science des héritages à Jativa, ville toute proche de Valence.[29] En 1068, Ṣā'id le cite dans sa catégorie des jeunes savants de son temps qui, selon sa propre expression «étudient avec zèle la philosophie dont l'entendement est sûr, les idées nobles et qui ont déjà acquis la connaissance des diverses parties de cette science».[30] Ṣā'id ne semble d'ailleurs pas l'avoir connu comme grand géomètre puisque, comme il le précise lui-même, c'étaient Ibn al-Aḥmar et Ibn Jawshān qui étaient alors considérés comme les meilleurs spécialistes de cette discipline.

Aucune information ne nous est parvenue concernant ses activités pendant les vingt années qui suivirent et qui furent pour Valence et pour son foyer scientifique les dernières années de calme. En effet, dès 1085, la ville est occupée par les troupes du Cid avec, semble-t-il, la bénediction d'al-Musta'īn, le fils d'al-Mu'taman, puis reprise par les Almoravides appelés par les rois des Taïfa pour contenir la poussée chrétienne vers le Sud.[31]

La ville fut à nouveau conquise par le Cid qui la gouvernera jusqu'à sa mort en 1099, date à laquelle les Almoravides en reprennent le contrôle. C'est durant ces périodes de troubles, au cours desquels les habitants ont subi de dures épreuves dues aux multiples affrontements et aux sièges répétés de la

---

[27]Ms. Escurial n° 972/6, ff. 33a-34b. Dérenbourg écrit ceci: «Fragment géométrique anonyme, d'une lecture difficile, le haut des feuillets étant mouillé.» Cf. *Les manuscrits arabes de l'Escurial*, t. II, fascicule 3: *Sciences exactes et sciences occultes, revues et complétées par H. P.-J. Renaud.*

[28]Ms. Oxford Pocock n° 206, ff. 119b-120a. Je tiens à remercier le Professeur J. Alaoui, de l'université de Fès, de m'avoir communiqué une copie de ce manuscrit et de m'avoir fait bénéficier, avant sa parution, de son édition critique: *Rasā'il Falsafiyya li Abī Bakr Ibn Bājja*, Dār ath-Thaqāfa, Beyrouth – Dār al-Nashr al-Maghribiyya, Casablanca, 1983. Une traduction française du texte d'Ibn Bājja, basée sur les deux manuscrits et accompagnée d'une analyse, est en préparation.

[29]Ibn al-Abbār, *at-Takmila li Kitāb aṣ-Ṣila*, édit. Codera et Zaydin, Madrid 1886, Vol. II, p. 550, n° 1553.

[30]Ṣā'id, *Ṭabaqāt al-Umam*, op. cit. p. 138.

[31]M. Benaboud, Ma'rakat az-Zallāqa wa-l-Wāqi'al-Andalusī, in *al-Baḥth al-'Ilmī* (revue de l'I.U.R.S.), Rabat 1982, n° 33, pp. 49-67.

ville successivement par l'un et l'autre protagoniste du conflit, qu'Ibn Sayyid semble avoir élaboré ses travaux en géométrie. Ibn Bājja les situe, en effet, dans les années 80 du Ve siècle de l'hégire, c'est à dire entre 1087 et 1096. Il précise d'ailleurs qu'Ibn Sayyid ne les a pas consignés dans un livre et qu'il ne les a pas achevés à cause des «vicissitudes de son temps».[32] Cette dégradation de la situation générale influera bien évidemment sur l'enseignement puisqu'Ibn sayyid ne trouvera que deux étudiants pour les initier à ses travaux. Le premier de ces deux étudiants disparaîtra très tôt dans la tourmente des guerres locales et le second, qui n'est autre que le futur philosophe Ibn Bājja, s'efforcera de poursuivre et de transmettre cette tradition, malgré ses multiples activités et les péripéties de sa vie publique, en Espagne puis au Maghreb.[33] C'est d'ailleurs son unique témoignage qui nous permetra d'esquisser les grandes lignes des travaux essentiels d'Ibn Sayyid.

Mais, avant de les aborder, il me faut dire quelques mots d'un autre travail de ce savant, dont la tradition mathématique maghrébine nous a conservé le contenu et qui concerne un chapitre célèbre de la théorie des nombres, celui des nombres figurés: D'après les informations d'Ibn Munʿim qui en parle dans son *Fiqh al-Ḥisāb*,[34] il s'agirait d'une épitre très condensée qui étudie les nombres figurés selon le point de vue des suites arithmétiques, en établissant des résultats concernant la sommation de ces suites et de toute une famille de sous-suites, extraites ou non des lignes et des colonnes du tableau de Nicomaque.

Il serait long d'aborder le détail de ce qui, au regard de la tradition arabe connue dans ce domaine, est spécifique à cette épitre. Mais, son analyse que j'ai faite ailleurs[35] permet néanmoins de dégager quelques points essentiels:

En premier lieu, cet écrit ainsi que certains chapitres du livre d'al-Mu'taman confirment l'existence, en Espagne et avant le XIe siècle, d'une solide tradition de recherche en arithmétique dont le point de départ semble avoir été la traduction faite par Thābit Ibn Qurra de l'*Introduction arithmétique* de Nicomaque. On a d'ailleurs confirmation de la présence de cet ouvrage en Espagne grâce à la traduction faite au XIIIe siècle par Kalonymos, de l'arabe à l'hebreu, d'un résumé de cette introduction qui, si l'on en croit Steinschneider, est accompagné de commentaires rédigés au Xe siècle par Rabīʿibn Yaḥyā, évêque d'Elvira.[36]

En second lieu, le travail d'Ibn Sayyid a contribué, probablement à côté d'autres que nous ignorons, à maintenir une tradition de recherche dans ce domaine, entre le XIIe et le XIVe siècle, d'abord en Espagne puis au Maghreb.

---

[32] J. Alaoui, *Rasāʾil Falsafiyya li Abī Bakr Ibn Bājja*, op. cit. p. 86.

[33] J. Alaoui, *Muʾallafāt Ibn Bājja*, Dār ath-Thaqāfa, Beyrouth – Dār al-Bashr al-Maghribiyya, Casablanca, 1983, pp. 10-13; qui contient des éléments importantes relatifs à la vie publique d'Ibn Bājja ainsi qu'une bibliographie complète.

[34] Ibn Munʿim, *Fiqh al-Ḥisāb*, op. cit. p. 298.

[35] *Nombres figurés et séries finies dans les ouvrages maghrébins des XIIe-XIIIe siècles*, à paraître.

[36] M. Steinschneider, *Mathematik bei den Juden*, 1964, p. 125.

C'est ce que montre l'étude comparative des chapitres aritmétiques du ***Fiqh al-Ḥisāb*** d'Ibn Mun'im ainsi que ceux du *Raf' al-Ḥijāb* d'Ibn al-Bannā'. Le premier complètera les écrits d'Ibn Sayyid et d'Ibn Ṭāhir, en établissant certaines preuves et en corrigeant d'autres.[37] Sa contribution dans ce chapitre concernera en particulier la sommation de sous-suites de nombres figurés, compte tenu de la parité de leurs valeurs, de celles de la longueur de leurs côtés et de celle de leur nombre. Son analyse aboutira d'ailleurs à la mise en évidence et à l'utilisation de suites arithmétiques polynomiales de variable n et de suites arithmétiques de raisons négatives.

Le second mathématicien reprendra, sans le dire, une partie de ce travail, puis, en changeant de point de vue, il introduira, sans s'en attribuer la paternité, un début d'interprétation combinatoire de certains nombres figurés.[38]

L'autre volet des travaux mathématiques d'Ibn Sayyid, c'est à dire celui qui concerne la géométrie, est plus important pour l'histoire des sciences en Espagne, mais il est plus difficile à appréhender, faute de textes mathématiques détaillés qui auraient permis, par une analyse interne des méthodes et des techniques, de le situer par rapport à la tradition grecque dans ce domaine et de le relier aux travaux des géomètres arabes qui l'ont précédé et dont il s'est peut-être nourri. Il est possible, toutefois, sur la base du témoignage de son élève, de se faire une idée des lignes directrices de son projet et de se convaincre de son importance pour l'histoire de la géométrie:

La première partie de ces travaux concernerait la matière même des *Coniques* d'Apollonius, l'agencement et le nombre de leurs propositions, ainsi que les outils ayant servi à leur établissement. On comprend en effet, du texte d'Ibn Bājja, que la nouvelle étude repose sur une série de définitions concernant les coniques qui sont équivalentes main non identiques à celles d'Apollonius et qu'elle utilise une notion de diamètre plus génerale que celle des *Coniques* puisque ces lignes perdent désormais leur propriété de diviser les segments ordonnés en deux parties égales. On ne sait rien de plus sur la nature et les propriétés de ces nouvelles lignes. Mais, on comprend que leur introduction aurait permis de se passer d'un grand nombre de propositions aux démonstrations longues et difficiles et qu'elle aurait ouvert des voies nouvelles pour l'établissement d'autres propositions «dont l'intérêt serait plus grand et les utilisations plus nombreuses», selon les propres termes d'Ibn Bājja.[39]

Comme on le voit, la pauvreté de l'information ne permet même pas d'avancer des conjectures. Tout au plus, peut-on faire deux constatations: la première concerne l'importance de ce travail et son niveau au yeux de l'excellent mathématicien qu'a été Ibn Bājja. Ce dernier n'hésite pas d'ailleurs à considérer que, dans ce domaine, Ibn Sayyid a surpassé les Anciens

[37]Ibn Mun'im, *Fiqh al-Ḥisāb*, op. cit. pp. 298-315.

[38]A. Djebbar, *Enseignement et recherche mathématiques dans le Maghreb des XIIIe-XIVe siècles* (étude partielle), Publication mathématique d'Orsay n° 81-02, 1981, pp. 76-87.

[39]Ms. Escurial, op. cit. f. 33b, lignes 1-2. Ms Oxford, op. cit. f. 119b, lignes 25-27.

eux-mêmes qui se sont occupés de problèmes analogues. C'est un type de jugement peu fréquent, sous la plume d'un savant arabe, pour être signalé. La seconde constatation se rapporte à la tradition arabe des coniques: Le projet d'Ibn Sayyid n'est ni une rédaction partielle ou totale des *Coniques*, ni un commentaire. Il se rattacherait plutôt à la tradition de recherche sur les sections planes des surfaces de révolution, tradition qui a été entretenue par les savants des IXe et Xe siècles, comme Thābit Ibn Qurra et as-Sijzī.[40]

La seconde partie des travaux d'Ibn Sayyid, celle dont il ne semble pas avoir achevé la rédaction, concerne, selon la terminologie moderne, certaines courbes gauches et certaines courbes planes de degré supérieur à deux. On sait, par le même témoignage, que dans un premier temps il étudia les courbes issues des intersections de surfaces coniques et non coniques, et, nous dit Ibn Bājja, «il trouva que leurs intersections communes n'étaient pas tout entières dans un même plan».[41] On ne sait pas s'il a été amené à caractériser ces courbes ou à en établir une quelconque classification, à partir des différents types de surfaces qu'il a utilisées pour les engendrer. Le silence de son élève et la suite de son propos laisse à penser que les courbes gauches n'auraient été pour lui qu'un instrument auxiliaire pour l'étude des courbes planes. Ces dernières seront en effet obtenues par Ibn Sayyid comme projection de certaines de ces courbes gauches, selon le procédé suivant: Partant de deux section coniques C_1 et C_2, tracées dans un plan $\mathcal{P}$ et se coupant dans ce plan, il construit deux cônes de sommets O_1 et O_2 situés d'un même côté de $\mathcal{P}$ et passant respectivement par les courbes C_1 et C_2. Grâce à une analyse préalable qui lui aura permis de choisir les hypothèses convenables, concernant les positions des courbes de base et celles des sommets O_1 et O_2, il obtient, comme intersection des deux cônes, une courbe gauche qu'il projette selon une direction donnée, déterminée également par analyse, sur un plan $\mathcal{Q}$ orthogonal à $\mathcal{P}$. Il établit ainsi l'existence et les propriétés de courbes planes autres que les sections coniques.

A ce sujet, la dernière partie du résumé d'Ibn Bājja est assez explicite pour nous permetter d'affirmer que ce ne sont pas les particularités purement géométriques de ces nouvelles courbes qui sont mises en évidence, pour les distinguer des anciennes, mais plutôt une propriété de nature plus algébrique dirions-nous aujourd'hui et qui s'apparenterait à la notion de degré. Ibn Bājja utilise pour cela deux termes distincts, celui de *quwwa* et celui du *martaba* que je traduirai provisoirement par les termes de *puissance* et d'*ordre* respectivement.[42] Pour lui, la puissance de la courbe C_3 obtenue

[40]Thābit Ibn Qurra, *Kitāb Quṭūʿ al-Usṭuwāna wa Basīṭuhā*, dont un résumé a été publié par L. Karpova et B. Rosenfeld dans *Archives internationales d'histoire des sciences*, n° 94, 1974, pp. 66-72, sous le titre: *The treatise of Thābit Ibn Qurra on section of cylinder and on its surface.* Pour l'étude d'al-Sijzī portant sur les sections de paraboloïde, d'ellipsoïdes et d'hyperboloïdes, cf.: Ms. Paris n° 2457/40.

[41]Ms. Escurial, op. cit. f. 33b, ligne 8. Ms. Oxford, op. cit. f. 119b, ligne 30 et f. 120a, ligne 1.

[42]Ms. Escurial, op. cit. f. 34a. Ms. Oxford, op. cit. f. 120a, lignes 8-19.

par projection de la courbe gauche est égale à la puissance des deux courbes C_1 et C_2 réunies et son ordre est distinct de celui des coniques, comme le confirme la suite de la construction. Ibn Sayyid considère en effet, dans une seconde étape, deux courbes de même ordre ou d'ordres différents, se coupant dans le plan $\mathcal{P}$ et constituant les bases de deux nouveaux cônes dont l'intersection, projetée sur le plan $\mathcal{Q}$, fournira encore une nouvelle courbe qui, selon les termes d'Ibn Bājja, «sera en puissance de la puissance» des deux courbes tracées sur le plan.[43] D'autre part, il est explicitement dit dans le texte que ce procédé de génération de courbes s'étend indéfiniment et une analogie est même établie entre cette démarche et celle de la proposition 116 du livre X des *Eléments* d'Euclide qui dit qu' «il résulte d'une médiale une infinité d'irrationelles dont aucune n'est la même qu'aucune de celles qui la précèdent».[44]

La dernière partie du texte, la plus concise mais dont le contenu est le plus facile à délimiter, concerne le troisième et dernier volet des travaux d'Ibn Sayyid, celui de l'application des recherches précédentes à la généralisation de deux des plus célèbres problèmes hérités de la tradition grecque et qui ont donné lieu, entre le IXe et le XIe siècle, à de multiples publications de la part de mathématiciens arabes. Il s'agit du problème de la trisection de l'angle et celui de la détermination des deux moyennes proportionelles. Se reférant à la construction précédente des courbes de degré supérieur à deux, Ibn Bājja nous apprend en effet que «par cette méthode [Ibn Sayyid] a extrait n'importe quel nombre de segments entre deux segments [donnés] de sorte qu'ils se succèdent tous selon un même rapport et, par cette voie, il a également divisé l'angle selon un rapport numérique quelconque».[45]

C'était là l'essentiel du résumé que nous a laissé Ibn Bājja. Il n'est pas aisé de faire l'analyse de ce véritable programme de recherche attribué à Ibn Sayyid et dont il ne nous est parvenu qu'un simple canevas. Il est toutefois possible, même dans l'état où se trouve notre information, d'avancer quelques remarques concernant le contenu de ces travaux et la tradition mathématique dans laquelle il s'insère et qu'il prolonge.

Tout d'abord, rien dans les textes publiés ou connus ne contredit l'affirmation d'Ibn Bājja concernant l'originalité du second volet des travaux d'Ibn Sayyid relatif aux courbes de degré supérieur à deux. Mais, cette originalité se situe-t-elle au niveau des résultats seulement ou bien s'étend-elle aux méthodes qui ont permis de les établir? Ibn Bājja évoque bien l'analyse comme méthode de démonstration, à différents stades de son investigation, en particulier dans le choix de la direction de projection des courbes gauches.

---

[43]Op. cit.

[44]P. Peyrard, *Les oeuvres d'Euclide*, Paris (Blanchard) 1966, p. 391. Cette proposition est numérotée par Heath 115, à la suite de Heiberg qui considère comme douteuse son attribution à Euclide. Cf. T.L. Heath, *Euclid, the thirteen Books of the Elements*, second edition, Dover New York 1956, t. III, p. 225.

[45]Ms. Escurial, op. cit. f. 34a, lignes 8-10. Ms. Oxford, op. cit. f. 120a, lignes 16-18.

Mais, cette analyse est-elle restée purement géométrique ou a-t-elle abouti à la mise en évidence puis à l'utilisation des propriétés algébriques des nouvelles courbes, à partir de celles des coniques? Sa manière de distinguer les différentes courbes qu'il obtint par intersection de cônes puis par projection (qui apparaît comme un embryon de classification de ces courbes), ainsi que leur utilisation dans le dernier chapitre, rendent cette hypothèse tout à fait plausible.

La seconde remarque concerne la résolution des deux problèmes déjà évoqués et leurs liens avec d'éventuels recherches antérieures. Là aussi, les textes d'Ibn Bājja sont silencieux et permettent donc d'avancer diverses interprétations, en particulier celle qui suppose une transmission d'Est en Ouest de recherches réalisées par les savants du centre et de l'est de l'empire mais, également de certains problèmes non résolus. On sait, par exemple, que la détermination des n grandeurs proportionelles a préoccupé un mathématicien comme Ibn al-Haytham. Ibn al-Qifṭī et Ibn Abī Uṣaybiʿa citent, à ce propos, son épitre sur la détermination de quatre grandeurs proportionelles entre deux grandeurs données.[46] On sait également que ce problème s'est posé naturellement à des algébristes du XIe siècle, à l'occasion de la résolution géométrique d'équations de la forme: $x^{2n+1} = a$, avec $n \geq 2$, ou celles qui s'y ramènent. C'est d'ailleurs à propos de la résolution de ces équations que ʿUmar al-Khayyām évoque, dans son *Traité d'algèbre*, le travail d'Ibn al-Haytham en ces termes: «Si on dit: *Un carré quelconque est égal à un nombre de parties du cube de son côté*, la solution de ceci n'est pas possible par les méthodes que nous avons exposées, car il est besoin de faire apparaître quatres droites entre deux droites pour que les six soient en proportion continue. Ceci a été démontré par Abū ʿAlī Ibn al-Haytham: seulement, c'est très difficile, aussi ne pouvons-nous le joindre à notre livre.»[47]

Cela ne fait que confirmer l'importance du travail d'Ibn Sayyid, son successeur, non seulement parcequ'il en est une généralisation mais, surtout, parcequ'il pourrait signifier l'introduction d'un point de vue ou d'instruments nouveaux liés à son étude des courbes algébriques de degré supérieur à deux. Cette conjecture est renforcée par la formulation d'Ibn Abī Uṣaybiʿa qui parle de «la géométrie d'Ibn Sayyid» et non de simples travaux géométriques[48] et par le jugement d'Ibn Bājja relatif à ces problèmes «dans l'étude desquels», dit-il, «Ibn Sayyid a été le seul, parmi les géomètres qui ont précédé, à se spécialiser».[49]

Pour conclure, il reste à dire quelques mots des prolongements connus de cette tradition mathématique: Le témoignage d'Ibn Bājja est, là encore, très précieux puisqu'après avoir étudié ces travaux sous la direction d'Ibn

[46]Ibn al-Qifṭī, *Akhbār al-ʿUlamā* ... op. cit. p. 116. Ibn Abī Uṣaybiʿa, *ʿUyūn al-Anbāʾ fī Ṭabaqāt al-Aṭibbāʾ*, Dār ath-Thaqāfa, Beyrouth 1979, t. III, p. 161.

[47]R. Rashed et A. Djebbar, *L'oeuvre algébrique d'al-Khayyām*, Alep, I.H.A.S., 1981, trad. p. 65.

[48]*ʿUyūn al-Anbāʾ* ..., op. cit. t. III, p. 103.

[49]Ms. Oxford, op. cit. f, 120a, ligne 8 (illisible dans le manuscrit de l'Escurial).

Sayyid lui-même, il dit les avoir poursuivis en les complétant et en résolvant des problèmes nouveaux. On sait également, par le contenu d'une de ses lettres adressée à son ami Ibn al-Imām,[50] qu'il avait projeté de rédiger sur le sujet un livre qui devait renferner également sa contribution propre. On ne sait pas si les activités politiques d'Ibn Bājja, à Saragosse puis à Fès au service des Almoravides, lui ont permis de réaliser ce projet ou tout au moins d'enseigner le contenu de ces travaux. Les ouvrages de mathématiciens tels qu'al-Ḥasan al-Marrākushī ou Ibn al-Bannā', pour ne citer que ceux qui, au Maghreb, sont connus pour avoir étudié la géométrie des coniques, n'y font jamais allusion dans leur ouvrages connus. On n'en trouve également aucune trace dans la liste des ouvrages mathématiques qui ont bénéficié, à partir du XIIe siècle, de traductions latines ou hébraïques.

Cela ne permet pas, malgré tout, de conclure à une rupture rapide et totale de la tradition en ce domaine. Il faut donc espérer que la découverte de nouveaux textes médiévaux permettra, d'une part, de compléter nos connaissances sur ces activités mathématiques et, d'autre part, d'amorcer des études comparatives, à partir de l'analyse des travaux européens antérieurs à ceux de Clairaut, c'est à dire en particulier ceux de Léonard de Vinci,[51] de Guarini,[52] du Père Courcier[53] et de Frézier[54] pour les courbes gauches et ceux de Thomas Céva[55] pour l'utilisation des courbes de degré supérieur à deux dans la résolution du problème de la multisection d'un angle.[56]

En attendant que les recherches à venir jettent de nouvelles lumières sur ce sujet, il est d'ore et déjà permis, me semble-t-il, d'accorder aux travaux d'Ibn Sayyid ou simplement à son projet, une place dans l'histoire de certains chapitres des mathématiques comme la géométrie des coniques et l'étude des courbes de degré supérieur à deux, ainsi que dans l'histoire des outils géométriques comme l'analyse et la synthèse ou les méthodes projectives.

---

[50] J. Alaoui, *Rasā'il Falsafiyya li-Abī Bakr Ibn Bājja*, op. cit. p. 88.

[51] G. Libri, *Histoire des sciences mathématiques en Italie*, 1865, t. III p. 46 note n° 2: «Les douze premiers feuillets du volume I des manuscrits de Léonard sont remplis de recherches géométriques dans lesquelles il cite plusieurs fois son traité de géométrie (...). Il a cherché une méthode géométrique pour étendre sur le plan les surfaces courbes (V.B. f.27). Il a distingué les lignes à double courbure des lignes simples (Ms. L.V. Vol.N, f.65).»

[52] C. Guarini, *Euclides adauctus et methodicus mathematicaque universalis*, Turin 1671, Tractatus XXXII: De superficiebus corporum in planum redigentis, pp. 572-96.

[53] P. Coursier, *Opusculum de sectione superficiei sphaericae, per superficiem sphericam, cylindricam, conicam ...*, Paris 1663.

[54] Frezier, *Traité de stéréotomie*, Paris 1754, t. I. Cf. également Hachette, *Traité de géométrie descriptive*, Paris 1822, pp. XI, 87, 88, 91-97, 112-115.

[55] T. Ceva, *Opuscula mathematica*, Med. 1699, Cycloidum anomalarum descriptio.

[56] Pour le recherche de n moyennes proportionnelles, voir les éléments fournis par E. Knobloch sur les méthodes de Nunes et surtout de Clavius, dans «La vie et l'oeuvre de Clavius», qui utilise sa récente publication de l'*Opera mathematica* de ce mathématicien (collection Gerstenberg-reprints, Berlin 1984).

The Volume of Domes in Arabic Mathematics

Yvonne Dold-Samplonius

1. Introduction

The Arabic word for dome or cupola is *qubba* pl. *qibāb* or *qubab*. By extension *qubba* also means: cupolaed structure, dome-shaped edifice; domed shrine, memorial shrine, kubba (esp. of a saint). In pre-Islamic times the *qubba* was a small domed leather tent, carried by a camel, in which certain tribes kept sacred stones. Also the dome located in front of the miḥrāb[1] – as exemplified for instance in the Great Mosques of Damascus, Qayrawān, Cairo, and Cordoba – might have had a special meaning. From the late 9th and the 10th centuries A.D. the building of commemorative structures over certain burial places, especially those of Shīʿī saints, occurred. Throughout the entire Muslim world, all the special names for sepulchral buildings, which vary with country and language as well as with the person interred, come under the generic name of *qubba*.[2]

Fig. 1: The Qubba of the Sāmānids in Bukhārā (Lewis).

[1] A recess in a mosque wall indicating the direction of prayer.

[2] cf. Diez [1838/1986] and Ettinghausen [1976] p. 65 and 68.

There are basically two types of monuments: the circular, tower-like form, and the often more grandiose square or polygonal type. Both can be covered either by a circular dome or a conical or pyramidal roof. Its original, and later stereotyped, form is a square building covered by a dome. The oldest preserved example is the *Qubba of the Sāmānids* in Bukhārā (Fig. 1), constructed around 907 but certainly before 943. It consists of a square structure with a large central dome and four small corner ones set over a gallery.

As early as the Seljuq period (11th century) the construction of domes with double shells was tried, which came to a successful development in Timurid times (first half 15th century). The aim of a drum and a double-shelled dome is to give a towering effect to the exterior. A striking example is the *Qubba of Tamerlane and Ulugh Beg* in Samarkand (Fig. 2).

As long as domes consisted of cones or sphere-segments, their mensuration was automatically included when measuring solids, and the qubba *per se* did not have to be mentioned. It is therefore not astonishing that only Ghiyāth al-Dīn al-Kāshī explains the measuring of complicated qubbas in his Book "On Measurements", one of the five parts of his *Key of Arithmetic*. In this paper I first discuss the computation of the hemisphere qubba and secondly the short section[3] by al-Kāshī in which he calculates the volume of a more complicated type of qubba. In both cases the results are checked by modern methods.

Fig. 2: The Qubba of Tamerlane with a double-shelled dome (Pougatchenkova).

[3]Fols. 83r/83v; Nader [1977] pp. 375-380.

2. Practical Mathematics.

Under the reign of Caliph al-Ma'mūn (813-33) the mathematician and astronomer Abū Ja'far Muḥ. b. Mūsā al-Khwārizmī[4] wrote in Baghdad his well-known *Algebra*, a work of elementary practical mathematics. Its purpose is explained by the author as providing "what is easiest and most useful in arithmetic, such as men constantly require in cases of inheritance, legacies, partition, lawsuits, and trade, and in all their dealings with one another, or where the measuring of lands, the digging of canals, geometrical computations, and other objects of various sorts and kinds are concerned."[5] It appears that men needed advice in settling problems arising out of legacies most frequently, as this section, the third and last part of the treatise, is by far the longest. These problems involve only arithmetic or simple linear equations but require considerable knowledge of the complicated Islamic law of inheritance. The small second part of al-Khwārizmī's treatise concerns practical mensuration.[6] He gives rules for finding the area of various plane figures, including the circle, and for finding the volume of a number of solids, including cone, pillar with circular base, pyramid, and truncated pyramid. This section is really concerned with the practical application of mensuration, as the first few lines already demonstrate: "Know that the meaning of the expression *one by one* is mensuration: one cubit [*dhirā'*] (in length) by one cubit (in width) understood."[7]

More than 600 years later, in 1427 in Samarkand during the reign of Ulugh Beg (1394-1449), the Timurid astronomer and mathematician Ghiyāth al-Dīn Jamshīd al-Kāshī[8] completed his *Key of Arithmetic*. The Arabic title "Miftāḥ al-ḥisāb" can also be read as "Miftāḥ al-ḥussāb", *The Calculators' Key*, both versions indicating the practical purpose of the treatise: arithmetic is seen as the key to solving all problems which can be reduced to calculations. In the introduction al-Kāshī explains: "I redacted this book and collected in it all that is needed for him, who calculates carefully, avoiding tedious satiation and disgraceful brevity." This work is a kind of encyclopaedia of elementary, and not so elementary, mathematics. Inheritance problems do not occur here. The work is divided into five books, the most extensive being book IV *On Measurements*. Rules are given here for measuring areas of all kinds, including such plane figures as ring-like and star-shaped surfaces as well as curved areas, and for finding volumes of solids, from cylinder, cone, truncated cone, intersection of cones, sphere, etc. to real buildings. Its last chapter nine *Measuring Structures and Buildings*, is indeed practical mathematics, as al-Kāshī explains: "The specialists merely spoke about this (i.e. measuring)

[4]cf. Toomer in DSB VII, pp. 358-365.

[5]Rosen [1831] trans., p. 3.

[6]Rosen l.c. trans., pp. 70-86.

[7]In Gandz' translation this passage reads: "Know that the meaning of *one upon one* is the area, and its meaning is: one cubit upon one cubit". Gandz [1932] p. 67.

[8]cf. Dold-Samplonius [1992/3].

for the arch and the vault, and besides that it was not thought necessary. But I present it among the necessities together with the rest, because it is more often required in measuring buildings than in the rest." Chapter nine is subdivided into three sections:

1. Measuring the Arch [*ṭāq*] and the Vault [*āzaj*].
2. Measuring the Qubba.
3. Measuring the Muqarnas.[9].

Al-Kāshī distinguishes the following categories of qubba: "They occur either in the form of a hollow hemisphere, or in the form of a segment of a hollow sphere, or in the form of a polygonal cone, or in the form, which arises by imagining the rotation of the façade of the arch, i.e. of an arch as mentioned in section 1, around the line of its elevation, that is the line, which connects its upper limit with the middle of the line between its fundaments." Al-Kāshī only indicates how to calculate the last category of qubba. The properties of the first two categories, he remarks, have been pointed out earlier in book IV, in connection with measuring the sphere and its segments, i.e. the surfaces in chapter 6, sections 4 and 5, and their volumes in chapter 7, sections 5 and 6. Category three has been mentioned in connection with measuring the cone, i.e. in chapter 6, section 3 and chapter 7, section 2 for surface and volume respectively.

The first mention of measuring the qubba is found in a textbook by Abū'l-Wafā al-Būzjānī,[10] written between 961 and 976. This work on practical arithmetic, *Book on What is Necessary from the Science of Arithmetic for Scribes and Businessmen* [K. fīmā yaḥtāju ilayhi al-kuttāb wa'l-'ummāl min 'ilm al-ḥisāb][11] consists of seven parts [*manāzil*]. The first three parts, the only ones contained in the Leiden MS., are purely mathematical, i.e. on ratio, multiplication and division, and mensuration. The other four parts contain the solutions of practical problems, i.e. taxes, exchange and shares related to harvest, problems concerning payment for work, and construction estimates.[12] In the introduction to the third part it is announced that chapter 6, *On the Measuring of Solids*, will be about the measuring of "...(and) other things falling under the shapes of spheres, as cylinders [*asāṭīn*], cones [*makhrūṭāt*], and **qubbas**." The solids considered by Abū'l-Wafā in the said chapter 6 are the cube, the cuboid, the prism, the cylinder, the cone, the frustrum of the cylinder, the frustrum of the cone, the sphere, as well as spherical sections. The qubba is not mentioned anymore. However, a cone or the sections of a sphere include all the qubbas existing in Abū'l-Wafā's time (cf. above).

From then on the calculation of volume and surface areas of qubbas forms

---

[9]cf. Dold-Samplonius l.c.

[10]cf. Youschkevitch in DSB I, pp. 39-43.

[11]edited by Saidan [1971], based on the Leiden MS. Or. 103 and the Cairo MS. Riyāḍa 42 M.

[12]cf. Saidan [1974] pp. 369-375.

a regular part of many practical arithmetical textbooks. At present it seems that only qubbas in the form of hollow hemispheres have been considered, as found in the treatise *Ghunyat al-Ḥussāb* (The Indispensable for Calculators) (MS. Ayasofya Kutuphanesi 2728 fol. 26r-125v) by Aḥmad b. I(a)bāt (d. 1234).[13] The complicated type of qubba al-Kāshī calculated was already in consideration by that time but attained its full development at a much later date.

Fig. 3: 11th Century Qubba in Monsaraz (Portugal), Christianized in the 14th Century.

3. Measuring a Hemisphere Qubba.

A hemisphere qubba (cf. Fig. 3) is assumed to consist of the solid shell between two concentric hemispheres. In practice (cf. Fig. 2) the inner and outer surfaces of the shell are never really parallel, because in the lower part, up to an angle of 61°, pressure occurs and in the upper part pull.[14] When the inner and outer diameter of a hemisphere qubba are known, its volume and the inner and outer surface areas can be calculated as follows: We know how to compute the surface area of a sphere with diameter equal to the outer diameter. Half of this amount is the outer surface area of the qubba. In the same way, the inner surface area of the qubba can be computed.

To calculate the volume of the qubba, we compute the volumes of the outer and the inner sphere and take each time its half. The difference between these two amounts is the volume of the qubba.

The formulas for computing the area and the volume of a sphere are:

$$\text{Area(sphere)} = (2r)^2 \times \pi \text{ and } \text{Vol(sphere)} = \text{Area(sphere)} \times \frac{r}{3},$$

where $2r$ is the diameter.

[13] I thank Dr Ulrich Rebstock [Tübingen] for giving me this information and sending me photocopies of the relevant parts of as well Aḥmad b. I(a)bāt's treatise as of the commentary on it by ibn al-Ḥanbālī.

[14] Prof. Erich Rossmann, Architect in Karlsruhe, kindly supplied this information.

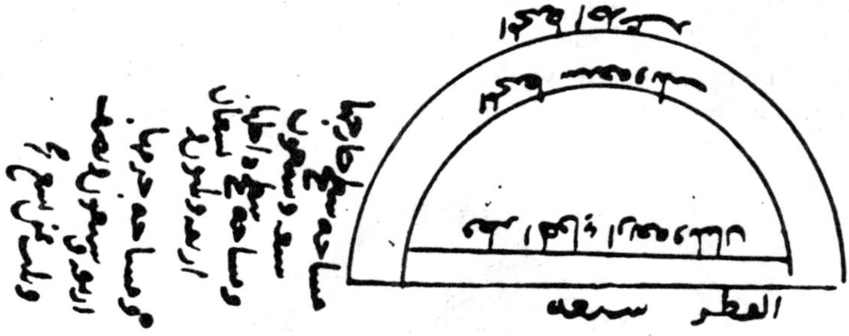

Figure 4: Figure for Calculating the Hemisphere Qubba.[15] The two diameters have their values attached to them and the amounts calculated for volume and areas are written at the left.

In the above mentioned treatise by Aḥmad b. I(a)bāt an example is given (cf. Fig. 4), in which the outer diameter equals seven and the inner diameter four and two thirds, $(2r)_{\text{outer}} = 7; (2r)_{\text{inner}} = 4\frac{2}{3}$. π is expressed in the usual way: discard one seventh and half of one seventh and multiply with four: $\pi = (1 - \frac{1}{7} - \frac{1}{2} \times \frac{1}{7}) \times 4 = \frac{11}{14} \times 4$, which equals $3\frac{1}{7}$.

With these values the surface areas and the volume of the qubba are computed:

$$\text{Area(sphere)}_{\text{outer}} = 7^2 \times \frac{11}{14} \times 4 = 154 \Rightarrow \text{Area(qubba)}_{\text{outer}} = \frac{154}{2} = 77.$$

$$\text{Area(sphere)}_{\text{inner}} = (4\frac{2}{3})^2 \times \frac{11}{14} \times 4 = 68\frac{4}{9} \Rightarrow \text{Area(qubba)}_{\text{inner}} = 34\frac{2}{9}.$$

To compute the volume of the qubba, Aḥmad b. I(a)bāt proceeds as follows, using the wrong, but widespread, formula Vol(sphere)$= (2r)^3 \times \frac{11}{14} \times \frac{11}{14}$:

$$\text{Vol(sphere)}_{\text{outer}} = (7)^3 \times \frac{11}{14} \times \frac{11}{14} = 211\frac{3}{4} \Rightarrow \frac{1}{2}\text{Vol(sphere)}_{\text{outer}} = 105\frac{7}{8}.$$

$$\text{Vol(sphere)}_{\text{inner}} = (4\frac{2}{3})^3 \times \frac{11}{14} \times \frac{11}{14} = 62 + \frac{2}{3} + \frac{2}{3 \times 9} \Rightarrow$$

$$\Rightarrow \frac{1}{2}\text{Vol(sphere)}_{\text{inner}} = 31 + \frac{1}{3} + \frac{1}{3 \times 9}.$$

[15]Fol. 116v.

$$\text{Vol(qubba)} = \frac{1}{2}\text{Vol(sphere)}_{\text{outer}} - \frac{1}{2}\text{Vol(sphere)}_{\text{inner}} = 74 + \frac{1}{2} + \frac{1}{3} \times \frac{1}{8} \times \frac{1}{9}.$$

When we calculate the volume of the qubba by means of the correct formula, while assuming for π the value $3\frac{1}{7}$, we obtain:

$$\frac{1}{2}\text{Vol(sphere)}_{\text{outer}} = \frac{1}{2}\text{Area(sphere)}_{\text{outer}} \times \frac{r}{3} = 77 \times \frac{7}{2 \times 3} = 89\frac{5}{6}.$$

$$\frac{1}{2}\text{Vol(sphere)}_{\text{inner}} = \frac{1}{2}\text{Area(sphere)}_{\text{inner}} \times \frac{r}{3} =$$

$$= 34\frac{2}{9} \times \frac{4\frac{2}{3}}{2 \times 3} = 26 + \frac{1}{3} + \frac{2}{9} + \frac{1}{3 \times 6} + \frac{1}{3 \times 6 \times 9}.$$

$$\text{Vol(qubba)} = \frac{1}{2}\text{Vol(sphere)}_{\text{outer}} - \frac{1}{2}\text{Vol(sphere)}_{\text{inner}} = 63 + \frac{1}{6} + \frac{1}{3 \times 9} + \frac{1}{9 \times 9}.$$

The relative error is independent of the diameter. Comparing the results calculated with the correct formula and with the formula as found in Aḥmad b. I(a)bāt's treatise we find the difference (D):

$$D = (2r)^3 \times \frac{11}{14} \times \frac{11}{14} - \pi(2r)^2 \times \frac{r}{3} = (2r)^3 \left\{ (\frac{11}{14})^2 - \frac{\pi}{6} \right\}, \text{with } \pi = 3\frac{1}{7},$$

$$D = (2r)^3 \left(\frac{121}{196} - \frac{22}{42} \right) = (2r)^3(0.6173 - 0.5238) = 0.0935(2r)^3.$$

This means that the calculated volume is 17.86 % (!) more than the correct volume, independent of the size of the diameter.

Al-Kāshī does not carry out the calculation of the hemisphere qubba, but refers to his calculation of the sphere. There he uses, as expected, the right formulas for area and volume expressing π as the ratio between the circumference and the diameter of a circle. Also Abū'l-Wafā gives the right definitions or formulas, mentioning Archimedes.[16] However, al-Karajī,[17] who lived a century after Abū'l-Wafā, employs in his *The Sufficient on Arithmetic* [Al-Kāfī fī'l ḥisāb] exactly the same formulas as Aḥmad b. I(a)bāt.[18] Here he also explains to measure a hollow hemisphere by subtracting the hollow inside from the hemisphere.

[16] Saidan [1971] l.c., p. 268.
[17] cf. Rashed in DSB VII, pp. 240-246.
[18] Hochheim [1877-80] part 2, p. 28.

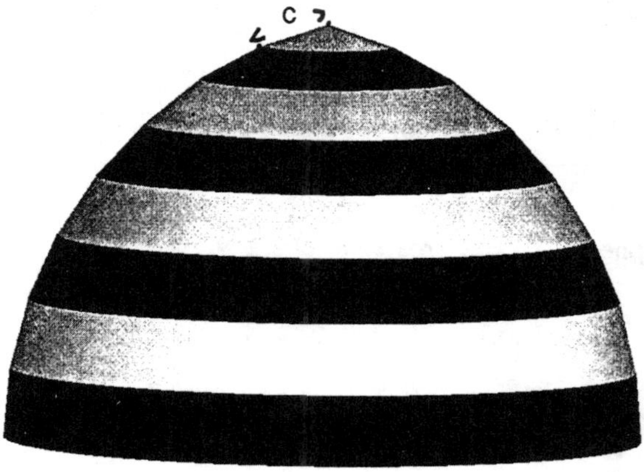

Fig. 5: Qubba, sliced in eight slices (Computer-simulation Kindl).

4. Measurement of the Qubba by al-Kāshī.

As mentioned above, al-Kāshī only indicates how to calculate the complicated type of qubba, i.e. the dome originating by rotating an arch around its vertical axis. The method is the following (Fig. 5): The dome is divided into parallel slices by drawing circles from the axis on its surface. These circles have to be so close that the curves between two of them equal the corresponding chords. Seven or eight of these circles should normally suffice, according to al-Kāshī. In this way the dome is cut up in a cone and several frusta. We first measure all the circles on the surface of the dome. The next step is to measure the distance from the apex of the dome to the nearest circle, i.e. the chord (Fig. 5: *c*) equalling the curve on the circle. By multiplying half the circumference of the nearest circle by this amount we obtain the surface area of the cone. Thereupon we multiply half the sum of every two neighboring circles by their distance to obtain the surface area of all frusta. The sum of these products yields the surface area of the qubba.

To obtain the volume of the qubba, which is a hollow solid, we first measure the volumes of the cone and the frusta, which fill the outer surface of the shell, and add these. From this sum we then subtract the sum of the volumes of the cone and the frusta filling the inside of the shell. The difference between these two sums is the volume of the qubba, as we have seen before in the case of the hemisphere qubba.

In the section *Measuring the Arch and the Vault* al-Kāshī describes five methodes for drawing the façade of an arch, all five types being based on circle-segments. To illustrate the general method he applies it to a qubba of the fourth type, which is constructed as follows[19] (Fig. 6):

[19]Fol. 79v; Nader l.c., p. 359.

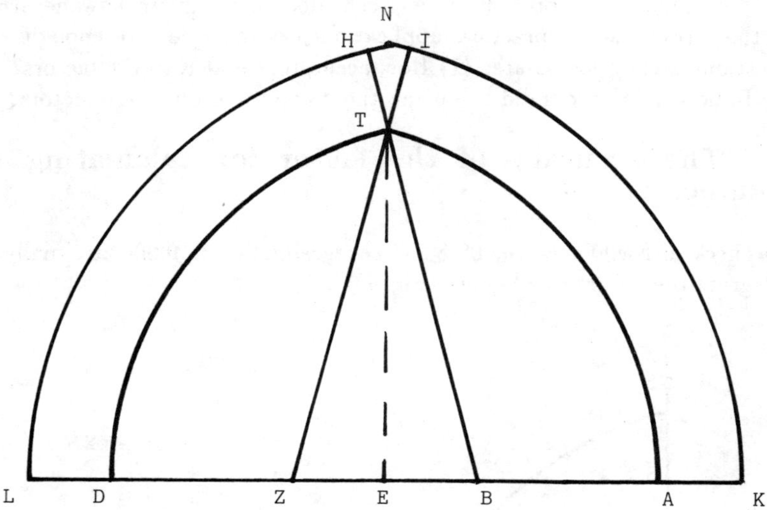

Fig. 6: Construction of an arch of the fourth type.

We divide AD, i.e. the inner width of the arch, through the points B and Z into three equal parts. From point B we draw arc DT with BD as radius and from point Z arc AT with AZ as radius. We connect BT and ZT and produce them by the size of the thickness of the arch to the points H and I. In the same way AD is produced in both directions to the points K and L. Then we draw from point B with radius BL arc LH and from point Z with radius ZK arc KI. We construct in the points H and I the perpendiculars HN and IN on TH and TI. The complex of the three sections TK, TN, and TL constitutes thus the façade of the arch. For the calculations it suffices to say that the façade of the arch consists of the two sections between the two parallel arcs, i.e. the sections ND and NA. I suppose that al-Kāshī mentions the closing wedge TN because of its importance in practice.

To obtain the surface area of the interior of the dome we have to multiply the square of the diameter of the base of the hollow (= inner) dome by 1°46′32″, if we compute sexagesimally, or by 1.775,[20] computing in the decimal system. When we multiply the square of the diameter of the base of the (outer shell of the) dome by the same number, we obtain the exterior curved surface area of the dome, as the inner and outer surfaces are supposed to be parallel to each other. When we multiply the cube of the diameter of the base of the hollow dome as well as the cube of the diameter of the base of the dome by 0°18′23″ in the sexagesimal system, or by 0.306 in the decimal system, and take the difference of these two products, we obtain the volume of the hollow qubba.

[20]On the calculation of this factor and the checking of its value by modern methods cf. Dold-Samplonius [1992].

"To simplify the procedure" al-Kāshī does not explain how he arrived at these results. For practical application just the rules are enough. The questions arising for us are: (1) How accurate are al-Kāshī' s factors? and (2) In how many slices did he cut up the dome to obtain these factors?

5. The accuracy of the factor for calculating the volume.

To check al-Kāshī's value, 0°18′23″ sexagesimally, or 0.306 decimally, we integrate the volume, as follows (Fig. 7):

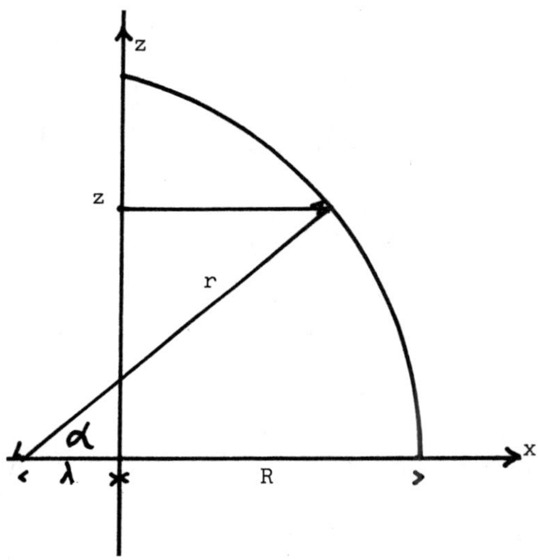

diameter $2R$

$r = R + \lambda$

$0 \leq \alpha \leq \mu$

$x = r\cos\alpha - \lambda$

$z = r\sin\alpha$

$0 \leq z \leq r\sin\mu$

$dz = r\cos\alpha\, d\alpha$

Figure 7.

The area $F(z)$ of the circular disc at the level z is $F(z) = x^2\pi$, hence the volume of the qubba is

$$\text{Vol(Qubba)} = \int_0^{r\sin\mu} F(z)dz, \text{where } \cos\mu = \frac{\lambda}{r},$$

$$\int_0^{r\sin\mu} F(z)dz = \pi\int_0^{r\sin\mu}(r\cos\alpha - \lambda)^2 dz = \pi r\int_0^{\mu}(r\cos\alpha - \lambda)^2 cos\alpha\, d\alpha =$$

$$= \pi r\int_0^{\mu}(r^2 cos^3\alpha - 2\lambda r\cos^2\alpha + \lambda^2 cos\alpha)d\alpha =$$

$$= \pi r^3\int_0^{\mu}\cos^3\alpha\, d\alpha - 2\lambda\pi r^2\int_0^{\mu}\cos^2\alpha\, d\alpha + \lambda^2\pi r\int_0^{\mu}\cos\alpha\, d\alpha =$$

$$= \frac{\pi r^3}{3}(\sin\mu\,\cos^2\mu + 2\sin\mu) - \lambda\pi r^2(\sin\mu\,\cos\mu + \mu) + \lambda^2\pi r\sin\mu.$$

Since $\cos\mu = \frac{\lambda}{r}$, whence $\sin\mu = \frac{\sqrt{r^2-\lambda^2}}{r}$, we obtain

$$\mathrm{Vol(Qubba)} = \pi\left\{\sqrt{r^2-\lambda^2}\times\frac{\lambda^2+2r^2}{3} - \lambda r^2 \arccos(\frac{\lambda}{r})\right\}.$$

In al-Kāshī's practical example the interior diameter of the qubba, $2R$, equals 6, which means that $r = 4$ and $\lambda = 1$:

$$\mathrm{Vol} = \pi\left\{\sqrt{15}\times\frac{33}{3} - 4^2\arccos(\frac{1}{4})\right\} = 67.584955.$$

$$\frac{\mathrm{Vol}}{(2R)^3} = 0.3128933 \text{ (decimally)} = 0°18'46'' \text{ (sexagesimally)}.$$

Here again, for the same reason as in the case of the surface area of the qubba, our computed value is larger than al-Kāshī's computed value. However, in the case of the surface area the error was only 0.48 %, whereas in the case of the volume it is an error of 2.2 %. This is a very moderate error in view of the almost 18 % occurring in the calculation of the hemisphere qubba above and acceptable for modern architects. But, as al-Kāshī calculated a factor with only 0.48 % deviation for the surface area, why didn't he do the same in the case of the volume?

There is another obvious error, as al-Kāshī uses the same factor for the outer and the inner, hollow, volume. What happens when the diameter increases? Let us assume that the outer diameter ($2R$) of the dome equals 6.5 (cf. Fig. 6). We calculate its volume by the formula developed above:

$$\mathrm{Vol(Qubba)} = \pi\left\{\sqrt{r^2-\lambda^2}\times\frac{\lambda^2+2r^2}{3} - \lambda r^2 \arccos(\frac{\lambda}{r})\right\}.$$

$2R = 6.5, \lambda = 1$, so $r = \frac{6.5}{2} + 1 = 4.25$. Hence,

$$\mathrm{Vol} = \pi\left\{\sqrt{\frac{17^2}{4} - 1}\times\frac{2\times(\frac{17}{4})^2+1}{3} - (\frac{17}{4})^2\arccos\frac{4}{17}\right\} = 84.932507.$$

$$\frac{\mathrm{Vol}}{(2R)^3} = \frac{84.932507}{(6.5)^3} = 0.3092672.$$

In this case al-Kāshī's factor of 0.306 for calculating the volume is too small by only 1.0564 %. This means that in taking the difference of the outer volume, too small by 1.0564 %, and the inner volume, too small by 2.2 %, the resulting volume of the qubba is more or less correct. Thus in a practical example, where the inner diameter of the qubba measures 6 m and where the shell of the qubba has a width of 25 cm, whence the outer diameter equals 6.5 m, the volume of the qubba-shell calculated by al-Kāshī's method is correct for all practical purposes.

Let us now look again at the calculation of the surface area and see what happens in this case with the outer surface area. Applying our formula, developed in a former paper,[21] we have:

$$\mathbf{F} = 2\pi r \left\{ \sqrt{r^2 - 1} - \arccos \frac{1}{r} \right\}$$

$2R = 6.5, r = R + 1.$ so $r = 4.25$, hence

$$\mathbf{F} = 8.5\pi \left\{ \sqrt{(\frac{17}{4})^2 - 1} - \arccos \frac{4}{17} \right\} = 74.700548.$$

When we divide the surface area by the square of the diameter, we obtain

$$\frac{\mathbf{F}}{(2R)^2} = \frac{74.700548}{6.5^2} = 1.768.$$

Here al-Kāshī's value of 1.775 is larger than our computed value by 0.39 %, which does not pose a problem for practical purposes. If the builders have to calculate the surface of the inner as well as the outer area, for example to order the necessary amount of plaster, the errors even counterbalance.

6. How to slice the dome.

How many circles did al-Kāshī draw on the surface of the dome? Or, to put it differently, how many slices did he use to obtain his factors? In Figure 8 we see a computer-simulation of three qubbas, sliced in seven, thirty-six, and six slices from left to right. Rather amazingly, already six slices, which means only five circles drawn on the surface, approximate the desired curve nicely. As I have mentioned in connection with the calculations on the muqarnas[22] al-Kāshī had a lot of practical knowledge.

To establish the number of slices, we calculated the volume and surface areas for several cases. The computations were made on the computer according to the methods indicated by al-Kāshī. From the results we deduced the respective factors:

Surface-area: As al-Kāshī's factor for the surface area is more exact than his factor for the volume we first calculated the surface area, adding the surface area of the cone at the top and all the surface areas of the frustra. Seven or eight circles, eight or nine slices, normally suffice according to al-Kāshī.

[21] Dold-Samplonius [1992] l.c.

[22] Dold-Samplonius [1992/3], l.c.

Fig. 8: Qubbas, cut into seven, thirty-six, and six slices resp. (computer-simulation Kindl).

Calculating with eight slices we obtain the following: The surface area of the cone at the top plus the surface area of the seven frusta amounts to 63.918, in which case the factor would be 1.77551. Calculating with nine slices the result for the dome is a surface area of 67.1268, or a factor equal to 1.77721. With six slices we obtain a total surface area of 63.6908, which means the factor equals 1.76919.

Al-Kāshī proposes as factor 1.775, hence he calculated with eight slices, otherwise expressed with seven circles drawn on the surface of the dome. Such a qubba is shown in Figure 5.

Volume: Calculating the volume of the dome in the case of eight slices, we have to add the volume of the cone at the top and the volumes of the seven frusta, which amounts to 67.0055 or a factor of 0.310211.

With six slices the total volume would be 66.557, which means a factor of 0.308134. And with five slices we obtain a volume of 66.1078, or a factor equalling 0.306055. The last factor is the value indicated by al-Kāshī, but I do not think that this factor resulted from calculating with only five slices. As he procured an excellent value in case of the surface area, he would have been perfectly able to do the same in case of the volume.

My conclusion is, that the deviation in the factor for the volume is on purpose, for practical reasons as explained above.

7. Conclusion

In all examples the inside and outside surface of the qubba-shell are supposed to be parallel. In practice the lower part of a dome has to be reinforced and the top can be relatively thin. However, this effect affects only the outside surface, which becomes somewhat deformed. It will especially matter if a dome is very large, as in the case of the Hagia Sophia. For smaller qubbas the effect should not be significant.

An acceptable result for the volume can probably still be computed with the factor ascertained by al-Kāshī. In the case of the hemisphere qubba the error of 18 % is much too large, so I assume that this so-called practical problem was more of a theoretical nature.

Acknowledgements

I thank Prof. Dr W. End (Heidelberg) for his help in calculating on the computer, Christoph Kindl (IWR, Heidelberg), who made beautiful computer-simulations, and Dr U. Rebstock (Tübingen) for sending me the MS. with the computation of the hemisphere qubba. Furthermore I am grateful to all those, who were willing to listen and comment, esp. Dr Sonja Brentjes (Leipzig) and Dr J.P. Hogendijk (Utrecht). Dr Xenia Young (EMBL, Heidelberg) was so kind to correct the English language. Mistakes and errors are entirely my responsability.

Bibliography.

Bretanizkij, L. S. and Rosenfeld, Boris A., 1956: "Architecture in the treatise 'Key of Arithmetic' by Ghiyāth al-Dīn al-Kāshī", (Russian) *Iskusstwo Azerbaidjana* **5**, pp. 87-130.

DSB, 1970-80: *Dictionary of Scientific Biography*, 16 vols. New York.

Diez, E., 1938/1986: "Ḳubba", *Encyclopaedia of Islam*, Ergänzungsband, pp. 139-146, repr. in the 2nd ed. Vol. V, pp. 289-296.

Dold-Samplonius, Yvonne 1992: "The XVth Century Timurid Mathematician Ghiyāth al-Dīn Jamshīd al-Kāshī and his Computation of the Qubba". "*Amphora*", *Festschrift for Hans Wussing on the Occasion of his 65th Birthday*. Ed. S.S. Demidov, M. Folkerts, D.E. Rowe, Ch.J. Scriba. Basel et al., pp. 171-181.

Dold-Samplonius, Yvonne 1992/3: "Practical Arabic Mathematics: Measuring the Muqarnas by al-Kāshī". *Centaurus* **35**, pp. 193-242.

Ettinghausen, Richard, 1976: "*The Man-made Setting*". The World of Islam. Ed. B. Lewis. London, pp. 57-88.

Gandz, Solomon, 1932: "The Mishnat ha Middot." *Quellen und Studien zur Geschichte der Mathematik, Astronomie und Physik*. Abt. A Quellen, Vol. 2. Berlin.

Hochheim, Adolf, 1877-80: *Kāfī fī'l Ḥisāb* (Genügendes über Arithmetik). Halle.

Al-Kāshī, Ghiyāth al-Dīn, 1427/1558: *Miftāḥ al-Ḥisāb* (Key of Arithmetic). MS. Or. 185, Leiden.

Lewis, Bernard (ed.), 1976: *The World of Islam*. London.

Luckey, Paul, 1951: *Die Rechenkunst bei Jamshīd b. Masʿūd al-Kāshī mit Rückblicken auf die ältere Geschichte des Rechnens*. Wiesbaden.

Nader, Nabulsi, 1977: *Al-Kāshī, Ghiyāth al-Dīn: Miftāḥ al-Ḥisāb* (Key of Arithmetic). Edition, Notes, and Introduction. Damascus.

Pougatchenkova, Galina A., 1981: *Chefs-d'oeuvre d'architecture de l'Asie Centrale, XIVe-XVe siècle*. Paris.

Rosen, Frederic (ed. and transl.), 1831: *The Algebra of Mohammed Ben Musa*. London.

Rosenfeld, Boris A. and Youschkevitch, Adolf P., 1954: *Al-Kāshī, Ghiyāth al-Dīn: Miftāḥ al-Ḥisāb* (Key of Arithmetic). Russian Translation and Commentary. Matematicheskije Traktaty, *Istoriko-Matematicheskije Issledovaniya* **7**, pp. 13-326.

Saidan, Ahmad S., 1971: *The Arithmetic of Abū al-Wafāʾ al-Būzajānī*, (Arabic) Edition, Introduction, Commentaries and Reference to *The Arithmetic of al-Karajī*. Amman.

Saidan, Ahmad S., 1974: "The Arithmetic of Abū'l-Wafāʾ", *Isis* **65**, pp. 367-375.

Die *Rithmachia* des Werinher von Tegernsee

Menso Folkerts

1. Einleitung

Die Rithmimachie ist ein Zahlenkampfspiel, das vom 11. bis zum 17. Jahrhundert in Westeuropa stark verbreitet war; danach geriet sie in Vergessenheit und wurde erst vor gut 100 Jahren von Mathematikhistorikern wiederentdeckt. Mit den Arbeiten [Peiper 1880] und [Wappler 1892] begann die wissenschaftliche Aufarbeitung ihrer Geschichte, die durch die grundlegende Darstellung [Borst 1986] zu einem vorläufigen Abschluß kam.[1] Borst gelang es, die Entstehung des Spiels aufzuhellen, er edierte alle Texte, die bis zur Mitte des 12. Jahrhunderts entstanden, und zeigte die Verbindungen zwischen ihnen auf; darüber hinaus gab er im Kapitel 11 seiner Arbeit[2] einen Überblick über die weitere Entwicklung des Spiels.

Ein wichtiges Bindeglied zwischen den älteren Rithmimachie-Texten und den Bearbeitungen des 13.-15. Jahrhunderts bildet die Abhandlung, die Werinher von Tegernsee kurz vor 1180 verfaßte. Obwohl die Bedeutung dieses Textes schon von Peiper und Wappler erkannt wurde, konnte die Schrift bislang nicht näher untersucht werden, weil sie bis vor kurzem als verschollen galt. Ich stieß bei meinen Studien der Rithmimachie-Texte auch auf die Handschrift London, BL, Addit. 22790, die mit großer Wahrscheinlichkeit den Text des Werinher enthält, und machte A. Borst auf diesen Codex aufmerksam, so daß dieser die Schrift noch für sein Buch über das mittelalterliche Zahlenkampfspiel verwerten und begründete Informationen über ihren Inhalt und ihre Stellung innerhalb der Rithmimachie-Texte geben konnte.[3] Im vorliegenden Aufsatz wird der Text erstmals ediert und der Inhalt näher beschrieben.

2. Rithmimachie-Texte vor Werinher

Zum besseren Verständnis des Werinher-Textes sollen im folgenden einige Bemerkungen über diejenigen Rithmimachie-Texte gemacht werden, die Werinhers Bearbeitung vorausgehen.[4]

---

[1]Siehe die Bibliographie am Ende des Aufsatzes. Das gegenwärtige Wissen ist zusammengefaßt bei M. Folkerts: *Rithmimachia*, in: [Maß 1989], S. 331-344, und: *Rithmimachia*, in: [Verfasserlexikon 1990], Sp. 86-94.

[2]S. 212-244.

[3][Borst 1986], S. 216-219.

[4]Sie fassen die Ergebnisse von [Borst 1986] zusammen.

In der in Würzburg entstandenen Abhandlung *De aggregatione naturalium numerorum* wurde die auf die Pythagoreer zurückgehende Proportionenlehre, wie man sie etwa in der *Arithmetik* des Boethius findet, mit der Bruchrechnung verbunden; auf diese Weise entstanden aus den Proportionen Zahlenketten, die auch für das Spielen geeignet waren. Im Zusammenhang mit dem sog. Wormser Schulstreit erfand ein Würzburger Geistlicher namens Asilo um 1030 das später „Rithmimachie" genannte Zahlenkampfspiel, in dem die in *De aggregatione naturalium numerorum* vorkommenden Zahlen auf Spielsteinen aufgetragen werden, die sich nach bestimmten Regeln bewegen, Steine des Gegners schlagen und schließlich im gegnerischen Feld eine Reihe von Steinen aufstellen, deren Zahlen eines der pythagoreischen Mittel (arithmetisches, geometrisches, harmonisches Mittel) repräsentieren. Asilos Arbeit war ein Entwurf; Aufbau des Spielbretts und Spielregeln waren noch nicht genau festgelegt.

Auch der Reichenauer Mönch Hermannus Contractus gab um 1040 noch keine eigene Spielanleitung, sondern erläuterte und verbesserte Asilos Entwurf und klärte einige strittige Fragen.

Um 1070 verfaßte ein unbekannter Lothringer Geistlicher („Lütticher Anonymus") einen weiteren Rithmimachie-Text. Er versuchte, verbindliche Regeln zu geben, legte die Größe des Spielfelds (12×8 Felder) und der Steine fest und gab Beispiele für die Anwendung der Regeln beim Schlagen. Aufgrund dieses Textes war es erstmals möglich, die Aufstellung der Steine zu rekonstruieren.

Der verbreitetste Traktat zur Rithmimachie vor Aufkommen der Druckkunst dürfte von Odo von Tournai (um 1090) stammen. Er kannte die Arbeiten von Asilo, Hermann und vom Lütticher Anonymus und bemühte sich, im Sinne der Frühscholastik Ordnung ins Spiel zu bringen. Seine Darstellung verbindet Zahlentheorie und Rechenpraxis und zeichnet sich durch Geschlossenheit aus.

In den zwei Jahrzehnten nach Odo entstanden drei Bearbeitungen, deren Autoren Borst „Regensburger Anonymus", „Fränkischer Kompilator" und „Bayerischer Kompilator" nennt. In St. Emmeram schuf um 1090 der „Regensburger Anonymus" eine Darstellung, die sich in ihren theoretischen Teilen an Asilo und Hermann orientiert, und eine Tabelle, die dem Lütticher Anonymus folgt. Das Brett ist jetzt 14×8 Felder groß. Erstmals wird eine Liste aller Spielwerte gegeben. Die Fassung des „Fränkischen Kompilators" entstand um 1100 in Moselfranken (Trier?); sie beruht auf den Darstellungen Odos und des Regensburger Anonymus und ist ebenso wenig ausgereift wie die „Bayerische Kompilation", die fast gleichzeitig (um 1105) wohl in der Nähe von Regensburg, Ilmmünster oder Tegernsee abgefaßt wurde und auf der Fränkischen Kompilation beruht. In ihr wird erstmals ein Spielbrett von 16 Feldern Länge vorgeschlagen.

Um 1130 verfaßte ein nicht eindeutig zu identifizierender „Fortolfus" in Mainfranken (Würzburg?) die erste ausführliche Darstellung des Spiels. Die Schrift ist ein Lehrbuch mit literarischen Ansprüchen, das sich auch an Laien wendet. Unter Rückgriff auf die „Fränkische Kompilation“ gab der Autor verbindliche Anweisungen zur Felderzahl, zum Spielplan und zu den Formen der Steine, stellte ausführlich die Möglichkeiten, eine Siegesposition zu errichten, dar und ging auf den Zusammenhang mit der Musiktheorie ein. Zu Fortolfs Darstellung gibt es einen ausführlichen scholastischen Kommentar, der in der 2. Hälfte des 12. Jahrhunderts in Frankreich (vielleicht in Paris) verfaßt und von Borst mit dem Namen „Pseudo-Abaelard“ versehen wurde.

3. Werinher als Verfasser einer Rithmimachie

Schon lange ist bekannt, daß ein Tegernseer Mönch im 12. Jahrhundert eine Rithmimachie verfaßte. In der Tegernseer Briefsammlung, die B. Pez 1729 veröffentlichte, heißt es im Brief eines O. an den Mönch W.:[5]

> Amico dilectissimo W. O. amicitiae foedus inscissum.
> Quamvis locorum nos disparent spatia, ea tamen, quae scindi nescit, conglutinat amicitia: quia, ut quidam ait, numquam vera amicitia fuit, quae aliquando cessare potuit. Hac ergo fiducia roboratus peto, ut Mappam, quam etiam pridem mihi promiseras, facias, et regulas Rhythmimachiae a te factas, mihi transmittas. Dominum nostrum utique liberalem, et benignum Rutpertum Abbatem super his allocutus sum, qui ut vere pius adiutorium tibi ad haec perficienda promptissimum spopondit. Si enim accepero, in te redundabit, quia et meliora, imo quaeque optima de te confido. Secundum fidem ergo meam, quod desidero, accipiam.

Es wurde lange gerätselt, wer sich hinter den Buchstaben *O.* und *W.* versteckt. Pez vermutete *Odo (?)* und *Wilhelmus.* Inzwischen ist klar, daß nicht Odo der Absender war, sondern Propst Otto von Rottenbuch; der Empfänger, den frühere Forscher als „Wilhelm von Tegernsee“ bezeichneten, war Werinher von Tegernsee.[6] Der Abt Rupert von Tegernsee (1155-1186), der im Brief genannt ist und der versprochen hatte, Werinhers Arbeit zur Rithmimachie zu fördern, war Ottos leiblicher Bruder. Daß in Tegernsee die Fächer des Quadriviums im 10. bis 12. Jahrhundert einen hohen Stellenwert besaßen, zeigen die erhaltenen Handschriften aus dieser Zeit, die sich jetzt in der Bayerischen Staatsbibliothek befinden; drei von ihnen enthalten die *Arithmetik* des Boethius.[7] Borst weist darauf hin, daß auch die *Quirinalia* des Metellus von Tegernsee, die um 1165 entstanden, mit der boethisch-nikomachischen Zahlentheorie verbunden sind.[8] So verwundert es nicht, daß ein Mönch dieses Klosters sich für die Rithmimachie interessierte.

Ottos Brief an Werinher muß zwischen 1155 und 1180, dem Todesjahr Ottos, geschrieben worden sein; die Münchner Abschrift, nach der Pez den

---

[5][Pez 1729], S. 55; veröffentlicht nach der Tegernseer Handschrift München, Clm 19411.
[6][Zimmermann 1946], vor allem S. 203f.
[7]Clm 18208, 18480, 18764.
[8][Borst 1986], S. 216.

Brief edierte, geschah frühestens 1178.[9] Wir wissen nicht, auf welche Weise Werinher, über dessen literarische Aktivität sonst nichts bekannt ist, seiner Aufgabe nachkam; es ist auch ungewiß, ob er überhaupt die Arbeit zu einem Abschluß brachte.

Einige ältere Erforscher der Rithmimachie haben versucht, Werinhers Schrift unter den zahlreichen anonym überlieferten Texten zu finden. Peiper konnte ihm ebenso wenig eine erhaltene Schrift zuweisen[10] wie Chicco, der sich relativ ausführlich mit dieser Frage beschäftigte.[11] Bubnov suchte, vermutlich durch Pez beeinflußt, die Rithmimachie in der Odo zugeschriebenen Handschrift Wien 2503, wobei er also den mutmaßlichen Schreiber des Briefs als Autor ansah.[12] Wappler kam der Wahrheit am nächsten, indem er annahm, die Handschrift Dresden C 19 enthalte den Text Werinhers:[13] tatsächlich beruht die in Dresden vorhandene Version auf der Fassung des Werinher, die sie teilweise wörtlich zitiert (siehe Abschnitt 7). In Wirklichkeit dürfte der Werinher-Text in einer aus der Bibliothek von Guglielmo Libri stammenden Handschrift stehen, die seit mehr als hundert Jahren unbeachtet in London in der British Library geschlummert hat (Additional 22790).[14] Auch wenn in der Handschrift kein Verfasser genannt ist, gibt es gewichtige Gründe, die Arbeit Werinher zuzuschreiben: Die Schrift benutzte die Bayerische Kompilation, die um 1105 im Umkreis von Tegernsee entstand, und ergänzte sie durch Elemente aus drei anderen deutschen Fassungen (Regensburger Sammlung, Fränkische Kompilation, Fortolf). Die anspruchsvolle Arbeit wurde offenbar nicht zu Ende geführt, da einige Teile, die zu erwarten gewesen wären, fehlen. Schließlich muß die Schrift vor 1200 verfaßt worden sein, weil die einzige bekannte Handschrift um diese Zeit oder kurz vorher geschrieben wurde.

4. Die Handschrift London, BL, Addit. 22790

Die Handschrift London, BL, Addit. 22790 (im folgenden **L** genannt), ist bisher in der Wissenschaftsgeschichte kaum zur Kenntnis genommen worden. Dies hängt sicher damit zusammen, daß sie sich zunächst in Privatbesitz (G. Libri) befand und man nicht bemerkte, daß sie nach dem Verkauf der Sammlung (1859) in das British Museum gelangt war. Sie ist im Versteigerungskatalog der Sammlung Libri relativ ausführlich, aber in den Einzelheiten ungenau beschrieben;[15] demgegenüber widmet der gedruckte

---

[9] Siehe [Plechl 1962], S. 438, 452, 485-488.

[10] [Peiper 1880], S. 220.

[11] [Chicco 1977], S. 82.

[12] [Bubnov 1899], S. XCVIII.

[13] [Wappler 1892], S. 6-8.

[14] Es ist A. Borsts Verdienst, die Zuschreibung des Textes an Werinher auf eine sichere Grundlage gestellt zu haben ([Borst 1986], S. 216-219).

[15] [Libri 1859], S. 103f. (Nr. 483) mit Schriftproben auf S. XXIX.

Katalog der Additional-Handschriften ihr nur wenige Zeilen.[16] Peiper und Wappler erwähnen die Handschrift nach der Beschreibung im Katalog der Libri-Sammlung.[17] Bubnov kannte die Londoner Handschrift wenigstens beiläufig, benutzte sie aber nicht für seine Ausgabe und bemerkte auch nicht ihre Identität mit dem Libri-Codex.[18] Abgesehen von van de Vyver, blieb sie den Bearbeitern der frühen Astrolab-Traktate ebenfalls verborgen.[19]

Die Handschrift **L** enthält auf den ersten vier Blättern drei Traktate des Hermannus Contractus (*De utilitatibus astrolabii*, *De mensura astrolabii* und *De horologio viatorum*), die auch sonst oft zusammen überliefert werden. Es folgen die *Rithmachia* des Werinher und verschiedene Texte zum Quadrivium und Trivium. Im einzelnen sind vorhanden:

f.1r: Hermannus Contractus, *De utilitatibus astrolabii*, Kap. I,1, II,1-6 in der Zählung von Bubnov (S. 114,5-120,16). Inc.: *Quicumque astronomice peritiam discipline et celestium sperarum geometricaliumque mensurarum altioremque scientiam* ... Expl.: ... *quibus ad radium solis sublevatur.* Der Text weicht z.T. von Bubnovs Edition ab. Zur Frage des Verfassers siehe [Bergmann 1985], Kap. 2, vor allem 2.2.1 und 2.2.2, und Kap. 2.4; Zusammenstellung der Handschriften auf S. 227f.[20] Nach Auskunft von Arno Borst gehört der Text dieser Handschrift zum 21-Kapitel-Traktat, der, wie Bergmann gezeigt hat, eine Bearbeitung des Gerbertschen Werks ist, die von Hermann angefertigt wurde.

f.1r Ende: Zahlentafel, offenbar zum Rithmimachie-Traktat f.4vff. gehörig.

f.1v-3v: Hermannus Contractus, *De mensura astrolabii.* Ed. [Drecker 1931], S. 203-212. Überschrift: *De compositione astrolabii.* Inc.: *Hermannus christi pauperum peripsima et philosophie tyronum asello* ... Expl.: ... *Et hec de astrolabii compositione dicta sufficiant.* Der Text weicht in den letzten Zeilen von Dreckers Edition ab. Zur Autorschaft siehe [Bergmann 1985], Kap. 2.2.3; Zusammenstellung der Handschriften auf S. 227 (h), ohne Erwähnung dieser Handschrift.

f.3v-4r: Hermannus Contractus, *De horologio viatorum.* Ed. [Migne 1853], S. 405-412. Überschrift: *De compositione hologici* (!) *instrumenti.* Inc.: *Componitur quoddam simplex et parvum viatoribus horologicum instrumentum* ... Expl.: ... *per singula signa et menses metiri valeamus. Finis propositorum huius negotii.* Siehe [Bergmann 1985], Kap. 2.5; Zusammenstellung der Handschriften auf S. 228 (hv), ohne Erwähnung dieser Handschrift.

f.4v-6r, 11v: Werinher von Tegernsee, *Rithmachia* (siehe unten).

---

[16][Catalogue 1875], S. 733.

[17][Peiper 1880], S. 213f.; [Wappler 1892], S. 6.

[18][Bubnov 1899], S. XXXVIII, nennt Add. 22790 und Libri 483 nacheinander.

[19]Sie fehlt auch bei [Bergmann 1985]. Van de Vyver bemerkte beiläufig die Identität der Londoner Handschrift mit dem Codex Libris ([van de Vyver 1931], S. 273, Anm. 26). Arno Borst wird die Handschrift in seiner Ausgabe der Werke von Hermannus Contractus angemessen berücksichtigen.

[20]J(19) bzw. J(21), ohne Erwähnung dieser Handschrift.

f.6r unten: Notiz: *Macometus vel albateni<us> in li<bri> LVI. cap<itulo> dicit secundum quorundam opinionem periferiam triplam diametro cum adiectione VII. partis. secundum tolomeum et al<ios> astrol<ogos> triplam cum adiectione decima tercie et quarta parte sexte periferie.*

f.6v: Kapitelüberschriften von Ptolemaeus, *Almagest*. Inc.: *Cap<itula> libri almagesti ptolomei philudensis* ... Die Überschriften unterscheiden sich nur geringfügig von den beiden Fassungen der Übersetzung von Gerhard von Cremona.

f.7r-v: Grammatischer Text. Inc.: *Circa appositionem tria consideranda sunt* ... Expl.: ... *quod primum est adverbiorum.*

f.8r: (Beginn des 2. Teils der Handschrift) Ausradierter Text. Erkennbar nur die Figur eines rechtwinkligen Dreiecks mit der Beschriftung *basis katetus hipotenusa.*

f.8v: Synonyme griechischer und lateinischer Worte. Inc.: *Gonia dicitur corpus* ... Expl.: ... *decanos super x.*

f.9r-11r: Text zur Logik. Inc.: *Impedimenta reductionis* ... Expl.: ... *et esse essentiali et esse compositis.*

f.12r-v: Weitere Texte, überwiegend zur Logik, am Ende unvollständig. Inc.: *Ex utrisque ... s vera conclusio* ... Expl.: ... *vocem omnis discipl<ine> stud. nulla medie. stud.*

Über die Geschichte der Handschrift ist wenig bekannt. Sie dürfte im letzten Viertel des 12. Jahrhunderts von einem nordfranzösischen Urkundenschreiber geschrieben worden sein.[21] Offenbar wurde der Text auf f.1r-6r und 8r-12v von nur einer Hand geschrieben; von einer zweiten Hand stammen die Texte auf f.6v-7v. Die Blätter 1-7 und 8-12 bilden jeweils eine Lage, bei der das letzte Blatt herausgeschnitten wurde. Beide Lagen sind unterschiedlich groß. Sie wurden, wie die Bearbeitung des Pergaments für die Bindung zeigt, schon bald nach dem Beschreiben zusammengebunden. Wo sich die Handschrift befand, bevor Libri sie erwarb, ist unbekannt. Am 31. März 1859 ging sie in den Besitz des British Museum über.

5. Inhalt von Werinhers *Rithmachia*

Der Rithmimachie-Text, den die Handschrift **L** bietet, ist kein abgeschlossenes Ganzes: Vor dem Incipit gibt es auf zwei Seiten (f.1r, 4v) Zahlenreihen, die mit dem Haupttext zusammenhängen; ein weiterer Abschnitt ist zwischen Texte zur Logik eingefügt (f.11v). Es entsteht der Eindruck, daß es sich um einen Entwurf handelt, der nicht zu einem Ende gebracht wurde.

Der erste Abschnitt (Kap. 1) folgt dem Auszug aus Hermanns *De utilitate astrolabii*. Es handelt sich um Zahlenreihen, die mit der Erzeugung der in

[21][Borst 1986], S. 217. Nach einer Information der British Library entstand sie in Südengland.

den Proportionen vertretenen Zahlen aus der Einheit zu tun haben (siehe Kap. 13 Ende).

Die zweite Tabelle, die dem Incipit unmittelbar vorausgeht (Kap. 2), enthält die Kopfzahlen der 48 Steine, die für das Spiel benötigt werden. Die *multiplices* werden als *pedites* bezeichnet, die *superparticulares* als *comites* und die *superpartientes* als *duces*. Die Steine der ersten Gruppe werden – kurioserweise – *unigoni* genannt, die zweiten *trigoni* und die dritten *tetragoni*. Besonders herausgehoben sind die beiden Pyramiden als *reges*: die *pyramis perfecta* (91) der geraden und die *pyramis tercurta* (190) der ungeraden Partei.

Der eigentliche Text beginnt mit Erörterungen über *nomen*, *materia*, *intentio*, *finis* und den Erfinder des Spiels (Kap. 3). Der Name *ritmachia* wird von *rith* = *numerus* und *chaia* = *pugna* abgeleitet; es handle sich um einen „Kampf der Zahlen"(*pugna numerorum*), weil hier auf vortrefflichste Weise die *virtus numeri* im Kampf stehe. Die *materia* des Spiels ist die gerade und ungerade Zahl, die beide gegeneinander gleichsam ein Lager aufbauen, mit den *pedites*, *comites* und *duces* vorrücken, wie in einer wirklichen Schlacht taktische Mittel gegeneinander anwenden, bis eine der Pyramiden gefallen ist, und welche dann, nachdem das gegnerische Lager erobert ist (*quasi expugnatis castris*), mit dem harmonischen oder einem anderen Mittel die Siegeszeichen im Herrschaftsbereich des Feindes aufrichten und daraufhin den Kampf nicht weiterführen. Die *intentio* des Spiels ist zweifach: Der Angreifer muß sich einerseits in acht nehmen, andrerseits dem Gegner schaden; beides ist miteinander verbunden. Das Ziel (*finis*) des Spiels besteht darin, die gegnerische Pyramide zu erobern, *medietates* zu errichten und dadurch den Sieg zu erringen. Bei der Beantwortung der anfangs gestellten Frage, *cui parti philosophie supponitur*, stellt Werinher fest, das Spiel sei dem Gebiet der *practica* zuzurechnen, der als *theorica* die Arithmetik zuzuordnen sei. Die *rithmachia* habe zur Arithmetik eine ähnliche Beziehung wie Zither und Orgel zur Musiktheorie, das Rechenbrett zur Geometrie oder Sonnenuhr und Astrolab zur Astronomie. Es geht also darum, das theoretisch Gelernte in die Praxis umzusetzen. Als Erfinder des Spiels wird der Römer Boethius genannt; parallel dazu nennt Werinher die Griechen Pythagoras, Nikomachos und andere als Begründer der Arithmetik.

Nach diesem theoretischen Abschnitt kommt Werinher zu seiner eigentlichen Aufgabe (*negotium*), der Beschreibung des Spiels (Kap. 4). Das Spielbrett hat 16 $\times$ 8 Felder. Auf ihm werden die Steine der geraden und ungeraden Partei aufgestellt. Von den denkbaren Arten werden nur die *multiplices*, *superparticulares* und *superpartientes* berücksichtigt; zwei weitere Spezies, der *multiplex superparticularis* und *multiplex superpartiens*, werden für das Spiel *propter multiplicitatem et inexplicabilitatem monadum et dyadum* beiseite gelassen. In der ersten Reihe stehen die *multiplices*; an sie schließen sich die *superparticulares* und an diese die *superpartientes* an. Auch hier werden die verschiedenen Spezies mit *pedites*, *comites* und *duces* verglichen. Werinher gibt nicht genau an, welchen Platz jeder einzelne Stein auf dem Brett einnehmen soll, aber seinen Ausführungen kann man entnehmen, daß wohl die Aufstellung des Fortolf gemeint ist, da die *multiplices* in der ersten Reihe stehen, die *superparticulares* sie umgeben (*his adhereant ... ipsos ambientes*) und die *superpartientes* hinter ihnen stehen (*retro hos et in postrema acie*). Es folgt eine ausführliche Auflistung, welche *multiplices*, *superparticulares* und *superpartientes* bei der Partei der Geraden bzw. Ungeraden vorkommen (*dupli*, *quadrupli* usw. bis *superseptipartientes*, *supernovempartientes*). Das Kapitel endet mit der Angabe der Züge: Die *multiplices* können gerade oder im Winkel (*angulariter*) ins nächste Feld rücken, die *superparticulares* entsprechend in das dritte und die *superpartientes* in das vierte Feld.

Nach diesem ersten praktischen Abschnitt folgt ein langer theoretischer Teil, der sich mit der Einteilung und den Eigenschaften der Zahlen beschäftigt (Kap. 5-13). Zunächst (Kap. 5) wird gesagt, daß es drei Arten von (geraden) Zahlen gibt: *pariter par*, *pariter impar* und *impariter par*. *Pariter pares* können bis zur Einheit halbiert werden (sie sind also von der Form 2^m). *Pariter impares* lassen nur eine Halbierung zu (also $2 \times (2n+1)$); bei Zahlen *impariter pares* sind mehrere Halbierungen möglich, aber das Verfahren kommt vor der 1 zum Ende (also $2^m \times (2n+1)$, $m > 1$). Als Beispiele für die drei Arten nennt Werinher die Zahlen 64, 42 und 72, die bei den *pedites* der Geraden (64) bzw. bei den *comites* der Geraden (42, 72) vorkommen.

Im folgenden Kap. 6 wird die Erzeugung der genannten drei Spezies dargestellt: Wenn man die Eins fortlaufend verdoppelt, erhält man die Folge der *pariter pares*: 2, 4, 8, 16, ... Nimmt man die Folge der ungeraden Zahlen und multipliziert eine jede von ihnen mit 2, so ergeben sich die *pariter impares* 6, 10, 14, 18, ... Um *impariter pares* zu bekommen, schreibe man die *pariter pares* in eine Reihe und darunter die ungeraden Zahlen und multipliziere eine beliebige Zahl aus der oberen mit einer aus der unteren Reihe.

Während sich die Kapitel 5 und 6 mit der Einteilung der geraden Zahlen beschäftigen, liefern die drei folgenden Kapitel dasselbe für ungerade Zahlen. Hier wird zwischen Primzahlen (*numerus primus et incompositus*), zusammengesetzten Zahlen (*numerus secundus et compositus*) und relativ primen Zahlen (*ad se quidem secundus et compositus, ad alium relatus primus et incompositus*) unterschieden. Eine Primzahl (Kap. 7) wird nur durch die Einheit gemessen. Um herauszufinden, ob eine Zahl prim ist, prüft man, ob sie durch irgendeine Zahl teilbar ist. Unnötigerweise sagt Werinher, man solle untersuchen, ob sie 2, 3, 4 usw. als Teiler besitzt. – Im Falle der zusammengesetzten Zahlen (Kap. 8) findet man die Vielfachen von n, indem man die ungeraden Zahlen der Reihe nach aufschreibt und, ausgehend von n, jeweils $(n-1)$ ungerade Zahlen überspringt. Diese Regel wird für jeden der Fälle $n = 3, 5, 7, 9$ einzeln formuliert. – Um Zahlenpaare zu finden, die jede für sich zusammengesetzt, relativ zueinander aber prim sind (Kap. 9), wird angegeben, wie mit Hilfe der Wechselwegnahme, also mit dem Euklidischen Algorithmus, der größte gemeinsame Teiler gefunden wird. Der letzte Rest ist der g.g.T. Die in der Handschrift beigefügten Zahlen lassen vermuten, daß das Verfahren am Beispiel (9, 25) erläutert werden soll, jedoch stimmen die in **L** stehenden Zahlen nicht mit den Koeffizienten der Wechselwegnahme überein.

Zu Beginn des Kapitels 10 teilt Werinher überraschenderweise mit, daß Primzahlen im Spiel in jedes Feld ziehen können und ebenfalls die vollkommenen (*perfecti*) Zahlen, *quia ex eis nascuntur.* Es folgt die übliche Definition der vollkommenen Zahl, die gleich der Summe ihrer Teiler ist. Die ersten vier *numeri perfecti* werden richtig genannt (6, 28, 496, 8128). Unrichtig dagegen ist die auch sonst verbreitete Ansicht, es gebe genau eine vollkommene Zahl zwischen zwei aufeinanderfolgenden Zehnerpotenzen (*in unoquoque limite*); gerade dadurch gelten diese Zahlen als *nobiles* (*quia non in modis vulgi passim habundant, sed in unoquoque limite unus tantum tamquam princeps presidet*). Bevor Werinher in Kap. 12 die auf Euklid IX.36 beruhende bekannte Regel zur Erzeugung vollkommener Zahlen bringt, behandelt er in Kap. 11 die nicht vollkommenen Zahlen, die, je nachdem, ob die Summe ihrer Teiler kleiner oder größer als die Zahl selbst ist, als *numeri diminuti* bzw. *superflui* bezeichnet werden. Zu den ersten gehört die 8, zu den zweiten die 12.

Nachdem sich die bisherigen Kapitel mit der Einteilung und den Eigenarten der natürlichen Zahlen beschäftigt haben, behandelt Kap. 13 Zahlenverhältnisse (*de relatione numerorum ad se*). Es werden die verschiedenen Arten der *multiplices* (*duplum*, *triplum*, *quadruplum* usw.), *superparticulares* (*sesqualtrum*, *sesquitertium* usw.) und *superpartientes* (*superbipartiens*, *supertripartiens* usw.) genannt; auf die übrigen geht Werinher, wie schon in Kap. 3 gesagt, nicht ein (*reliquum ob predictam causam omittimus*). Er erwähnt auch nur kurz, daß aus drei Einheiten die *dupli*, aus den *dupli* die *tripli*, aus diesen die *quadrupli* usw. erzeugt werden können. Aus den *multiplices* können, wenn man ihre Reihenfolge umkehrt (*ex conversis multiplicibus*), die *superparticulares*, aus diesen in ähnlicher Weise die *superpartientes* gebildet werden; kehrt man die Reihenfolge nicht um, so entstehen aus den *superparticulares* die *multiplices superparticulares* und aus den *superpartientes* die *multiplices superpartientes.* Die kurzen Angaben lassen kaum die Gesetzmäßigkeit erkennen: Wenn die drei Ausgangsgrößen mit a_1, a_2, a_3 und die sich daraus ergebenden gesuchten Größen mit b_1, b_2, b_3 bezeichnet werden, so gilt: $b_1 = a_1$, $b_2 = a_1 + a_2$, $b_3 = a_1 + 2a_2 + a_3$. Dies alles kann man bei Boethius nachlesen, den Werinher wohl meint, wenn er schreibt: *alibi plenius et perfectius determinatum habetur*; hat man einmal die Gesetzmäßigkeit erkannt, so lassen sich die Zahlenreihen am Ende dieses Kapitels und die ähnlichen Listen in Kap. 1 leicht deuten.

Nach den theoretischen Erörterungen über die (ebenen) Zahlen und ihre Verhältnisse greift Werinher zu Beginn von Kap. 14 die Spielanweisungen von Kap. 4 Ende wieder auf, indem er wiederholt, daß die *multiplices* ins nächste, die *superparticulares* ins 3. und die *superpartientes* ins 4. Feld ziehen. Dies folge aus ihrer Erzeugung aus den Einheiten.[22] Auch die nächste Feststellung, daß Primzahlen und vollkommene Zahlen in jedes beliebige Feld ziehen können, wurde schon vorher getroffen (Kap. 10). Die darauf folgende Formulierung, daß Drei-, Vier-, Fünfecke und die weiteren Polygone dem Gesetz jener Art von Zahlen folgen sollen, in denen sie selbst entstehen,[23] verwirrt, weil Fünfecks- und höhere Polygonalzahlen im Spiel nicht vorkommen; vielleicht wollte Werinher nur andeuten, daß jede Figur in der Weise ziehen darf, die ihr aufgrund ihrer Eigenart und Entstehung zukommt. Der nächste Satz, daß jede Figur in dem Feld, in das sie ziehen kann, so lange sicher ist, bis sie, von gegnerischen Figuren bedrängt, nicht entfliehen kann, gehört zum Allgemeingut der Rithmimachie-Traktate seit Asilo; die Formulierung überrascht an dieser Stelle, weil es sonst in diesem Kapitel nur um die Zugmöglichkeiten der einzelnen Figuren geht. Die folgende Feststellung, daß die rote Pyramide in alle roten, die weiße in alle weißen Felder ziehen kann, kommt genauso unerwartet, da von verschiedenfarbigen Pyramiden bisher ebenso wenig die Rede war wie von roten oder weißen Spielfeldern; im übrigen stehen beide Pyramiden auf gleichartigen Feldern, sofern man den Spielplan von Fortolf zugrunde legt. Die Pyramide kann alle Pyramidenseiten des Gegners schlagen (gemeint ist vermutlich, wenn sie ihr als Einzelsteine in den Weg treten), nur im Winkel kann sie auf diese Weise den Basisstein nicht schlagen.[24]

Am Ende des Kapitels über das Bewegen der Steine kündigt Werinher an, jetzt über Dreieckszahlen reden zu wollen. Dies geschieht allerdings erst im übernächsten Kapitel 16. Das dazwischen liegende Kap. 15 beschäftigt sich mit der Einteilung der Zahlen nach ihrer Dimension in *numeri lineares*, *plani* und *solidi vel cubici*. Zahlen, die nur eine Dimension besitzen, heißen *lineares*; solche, die 2 bzw. 3 Seiten haben, sind die *plani* bzw. *solidi vel cubici*. Ähnlich wie in der Geometrie, wo das Prinzip der Linie ein Punkt ist, das Prinzip der Fläche eine Linie und das Prinzip des Körpers eine Fläche, entstehen aus den *lineares numeri* die *plani* und aus diesen die *solidi vel cubici*. Der Rest des Kapitels behandelt die verschiedenen Arten der dreidimensionalen Zahlen. Zu ihnen gehören die *cubici*, bei denen alle drei Seiten gleich lang sind (also n^3); eine Teilmenge von ihnen sind die *numeri sperici vel cyclici*, die in sich zurückkehren:[25] gemeint ist, daß die letzte Ziffer von n und n^3 gleich ist, was bei 5 oder 6 der Fall ist. Räumliche Zahlen, die nicht von der Form n^3 sind, heißen *solidi numeri*. Zu ihnen gehören die *laterculi* bzw. *asseres vel tignuli*, bei denen zwei Seiten gleich und die dritte kleiner bzw. größer als die beiden übrigen ist, und die *paralellepipedes*, bei denen die mittlere Seite gleich viel von den beiden anderen abweicht. Sind alle Seiten verschieden, so handelt es

---

[22] *Quoniam multiplices ab unitatibus proximi sunt in nascendo ...; superparticulares ..., quia tertii ab unitatibus prodeunt; superpartientes ... quia quarti sunt ab unitatibus.*

[23] *... sequantur legem illius speciei numerorum, in quibus ipsi fiunt.*

[24] *Pyramis latera omnia pyramidis aufert, tantum angulariter basim non tollit, nisi caracterum aliquis.*

[25] *qui in idem punctum unde creverunt reflectuntur.*

sich um einen *numerus scalenos vel cuneus vel speniscus.* All diese Zahlen kann man auch als *parte altera longiores* bezeichnen.

In Kap. 16 wird eine andere Unterteilung der *numeri solidi* angeboten: in die *circulares*, die offenbar mit den *cyclici* in Kap. 15 identisch sind, und die *pyramides*. Pyramidalzahlen entstehen durch Aufeinanderschichten von gleichartigen Flächenzahlen; diese können Vierecke, Dreiecke, Fünfecke oder Sechsecke sein. Erst jetzt löst Werinher sein Versprechen vom Ende des Kap. 14 ein und behandelt die *numeri trigoni.* Auch sie können in jedes beliebige Feld ziehen, weil sie in sich selbst und weil alle anderen Polygonalzahlen in sie aufgelöst werden können.[26] Die Dreieckszahlen entstehen aus den eindimensionalen *numeri lineares*, die man in der Folge der natürlichen Zahlen anordnen kann. Indem man zu einer Zahl die nächste addiert, erhält man zunächst die 3, die *actu* eine Dreieckszahl ist, während eine aus drei Punkten bestehende *linea* dies nur *potestate* ist. Auf diese Weise entsteht die Folge der Dreieckszahlen 1, 3, 6, 10, ... Wenn man in der Folge der natürlichen Zahlen jeweils eine Zahl überspringt, erhält man durch Addition die Viereckszahlen $1 + 3 = 4$, $1 + 3 + 5 = 9$, ...; läßt man zwei Zahlen aus, so kommt man zu den Fünfeckszahlen, usw. Aus dem Gesagten ergibt sich die Richtigkeit der obigen Aussage, daß sich nämlich beliebige *multiangule figure* in *trigoni* auflösen lassen: Aus zwei Dreieckszahlen entsteht eine Viereckszahl (genauer: $a_n^3 + a_{n-1}^3 = a_n^4$), aus Vierecks- und Dreieckszahl ergibt sich eine Fünfeckszahl ($a_n^4 + a_{n-1}^3 = a_n^5$), aus Fünfecks- und Dreickszahl eine Sechseckszahl ($a_n^5 + a_{n-1}^3 = a_n^6$), usw. Das Ganze wird – etwas dilettantisch – durch drei Figuren illustriert: *tetragonus* mit der Fläche 9, *pentagonus* mit der Fläche 12 und *exagonus* mit der Fläche 15.

Nachdem die Polygonal- und Pyramidalzahlen eingeführt sind, kann jetzt die Einteilung der Pyramiden behandelt werden (Kap. 17). Es gibt gerade und ungerade Pyramiden, die sich offenbar darin unterscheiden, daß eine gerade Pyramide aus einer geraden Anzahl, eine ungerade aus einer ungeraden Zahl von Schichten gebildet wird. Warum aber eine drei- oder viereckige Pyramide zu den *pares*, eine fünf- oder sechseckige jedoch zu den *impares pyramides* gehört, ist unklar. Bei den Figuren, die zur Erläuterung beigefügt, jedoch teilweise fehlerhaft gezeichnet sind, handelt es sich um folgende Pyramiden mit 6 bzw. 5 Schichten:

(Dreiecke)	$1 + 3 + 6 + 10 + 15 + 21$	$= 56$
(Vierecke)	$1 + 4 + 9 + 16 + 25 + 36$	$= 91$
(Fünfecke)	$1 + 5 + 12 + 22 + 35$	$= 75$
(Sechsecke)	$1 + 6 + 15 + 28 + 45$	$= 95.$

Klar ist demgegenüber die Unterteilung in vollkommene (*perfecta*) und verkürzte (*curta*) Pyramide: Wenn sie sich bis zur Einheit erhebt, ist sie vollkommen; fehlen oben einige Schichten, so heißt sie – entsprechend der Zahl der fehlenden Schichten – *biscurta*, *tercurta* usw.

[26] *cum ipsi in se et in eos omnes alii quorumlibet angulorum numeri resolvantur.*

Nach Beendigung dieses theoretischen Abschnittes geht Werinher wieder auf die Spielpraxis ein. Das Kapitel 18 enthält die Regeln über das Schlagen der Steine. Es werden fünf Möglichkeiten angeboten:

1) Wenn ein Stein im nächsten Zug einen gegnerischen Stein mit derselben Kopfzahl erreichen kann, wird dieser fortgenommen.

2) Wenn man gegnerische Steine mit eigenen Steinen umgeben (*vallare*) kann, die denselben Betrag (? *summa*) darstellen und die die gegnerischen Felder erreichen können (*respitiant*), so kann man diese fortnehmen. (Der Sinn dieser Regel ist nicht klar.)

3) Wenn man eine Zahl mit Zahlen umgibt, die multipliziert (*multiplicati*) oder addiert (*iuncti*) diese Zahl ergeben, so wird sie genommen. (Es wird nicht ausdrücklich gesagt, daß die angreifenden Steine auf Feldern stehen müssen, von denen aus die gegnerische Figur erreicht werden kann.)

4) Wenn das Produkt aus der zwischen einer eigenen und einer gegnerischen Figur liegenden Felderzahl und der eigenen Zahl die Kopfzahl des Gegners ergibt, so fällt diese Figur.[27]

5) Wenn eine Figur durch gegnerische Steine so eingemauert ist, daß sie keinen rechtmäßigen Zug machen kann, so wird sie genommen.

Es folgt die Schlagregel für die Pyramide: Wenn ihre Basis nach den Regeln 1) – 5) genommen werden kann, so fallen auch ihre anderen Seiten.

Nach der Einnahme der Pyramide muß man versuchen, in der gegnerischen Hälfte eine perfekte Harmonie zu errichten. Um ein harmonisches Mittel zu präsentieren, muß man einen dem Gegner geraubten Stein verwenden; ihn kann man dann *potestative* für seine Harmonie wie einen eigenen gebrauchen.

Der Begriff *medietas* erinnert Werinher daran, daß er erklären muß, was man darunter versteht. Dies geschieht in Kap. 19. Nachdem zunächst *medietas* definiert wird,[28] folgen Erläuterungen der drei Mittel zwischen zwei Zahlen a und b:[29] Beim arithmetischen Mittel A ist die Differenz zwischen der größten und der mittleren Größe so groß wie zwischen der mittleren und der kleinsten (also $b - A = A - a$); beim geometrischen Mittel G gilt Entsprechendes für die Proportionen (nämlich $b : G = G : a$); für das harmonische Mittel H gilt: $b : a = (b - H) : (H - a)$. Für jedes der Mittel gibt Werinher Beispiele an, die korrekt sind.[30] Demgegenüber befriedigen seine Angaben, wie man die Mittel berechnen kann, nur teilweise: Seine Vorschriften zur Bestimmung von A treffen zu.[31] Korrekt ist auch die Bemerkung, daß das Verhältnis zwischen den Differenzen gleich dem zwischen benachbarten Gliedern ist,[32] während

[27] Aus unserer Sicht unnötig, werden zwei Möglichkeiten unterschieden: Felderzahl mal Kopfzahl bzw. Kopfzahl mal Felderzahl.

[28] *Medietas in numero vocatur habitudo quedam vel differentiarum vel proportionum.*

[29] Im folgenden sei $b > a > 0$.

[30] A: (10, 25, 40); G: (1, 2, 4) und (10, 20, 40); H: (3, 4, 6).

[31] $A = \frac{a+b}{2}$ oder $A = a + \frac{b-a}{2}$.

[32] $(G - a) : (b - G) = a : G$. *Huius proprietas est quod quecumque inter terminos est proportio, eadem est inter differentias terminorum.*

man aus der Formulierung, wie man das geometrische Mittel G finden soll, nur mit Mühe die Aussage herauslesen kann, daß G gleich der mittleren Proportionalen zwischen a und b ist.[33] Bei der Ermittlung des harmonischen Mittels begeht Werinher einen Fehler: Das Produkt ab müßte durch das arithmetische Mittel $\frac{a+b}{2}$ *dividiert* werden, während Werinher *triplica* schreibt.

In Kap. 20 werden die verschiedenen Mittel aufgelistet, die auf beiden Seiten möglich sind. Genannt werden:

Partei der *pares*:

A: (4, 6, 8), (2, 4, 6), (9, 45, 81), (8, 25, 42)

G: (2, 4, 8), (4, 16, 64), (4, 6, 9).

Partei der *impares*:

A: (5, 7, 9), (3, 5, 7), (7, 28, 49), (12, 56, 100), (7, 64, 121)

G: (5, 25, 125[34]), (49, 28, 16).

Werinher betont zu Recht, daß es mehr arithmetische als geometrische Mittel gibt. – Für die harmonischen Mittel H benötigt man geraubte Steine. Folgende Möglichkeiten sind angegeben (die geraubten Steine sind mit * gekennzeichnet):

Partei der *pares*:

(15, 20, *30), (4, 6, *12), (4, 6, *3), (10, 25, *225).[35]

Partei der *impares*:

(3, 5, *15[36]), (5, 9, *45), (9, 17, *42),[37] (7, 28, *4).

Der folgende Hinweis, daß bei allen Mitteln auf der geraden Seite der mittlere Wert gerade, auf der ungeraden Seite ungerade ist, trifft ebenso wenig zu[38] wie die Bemerkung, bei den Harmonien der geraden Partei bestehe zwischen den äußeren Zahlen ein ungerades Verhältnis und umgekehrt.[39] In den Harmonien (3, 4, 6) und (2, 3, 6) liegen nach Werinher die musikalischen Konsonanzen verborgen; tatsächlich lassen sich in ihnen die Verhältnisse 2:1 (Oktave), 3:2 (Quinte) und 4:3 (Quarte) finden.

Das folgende Kap. 21 behandelt die einfachen Siegmöglichkeiten. Falls man kein harmonisches Mittel bilden kann, weil man keine dafür geeigneten gegnerischen Steine erworben oder seine eigenen durch Sorglosigkeit verloren hat, so genügt es zum Sieg, nach dem Raub der gegnerischen Pyramide ein geometrisches oder arithmetisches Mittel aufzubauen. Diese Mittel können von der eigenen Partei, aber auch unter Zuhilfenahme erbeuteter Steine hergestellt werden, und zwar in gerader oder abgeknickter (*per angulos*) Aufstellung. Man soll sich vom ersten Zug an bemühen, eine *medietas* im

[33] $b : G = G : a$. *Proportionem maximi ad minimum divide et mediam pone* [$\overline{\text{ii}}$.] *in quadrupla proportione, ut X et XL medius intercidit XX duplam utrinque custodiens proportionem, que est subdupla quadruple.*

[34] In **L** fälschlich: 200.

[35] Fehlerhaft. Vermutlich ist gemeint: (25, 45, *225).

[36] Die 15 fehlt in **L**.

[37] Fehlerhaft. Vermutlich ist gemeint: (7, 12, *42) oder (9, 16, *72).

[38] Gegenbeispiel: (9, 45, 81).

[39] Gegenbeispiele: (3, 4, 6), (5, 9, 45) und andere.

gegnerischen Lager zu erzeugen, so daß kein fremder Stein diese Aufstellung unterbrechen kann.

Im Kap. 22, mit dem der Haupttext der *Rithmachia* endet, werden die Schlagmöglichkeiten für die Figuren der geraden Partei (diese sind im folgenden fettgedruckt) angegeben:

2: 3 Möglichkeiten (12 im 6. Feld, 16 im 8., 28 im 14.)
4: 5 Möglichkeiten (12 im 3., 16 im 4., 28 im 7., 36 im 9., 56 im 14.)
6: 4 Möglichkeiten (12 im 2., 30 im 5., 36 im 6., 66 im 11.)
8: 4 Möglichkeiten (16 im 2., 56 im 7., 64 im 8., Pyramide mit Basis 64 im 8.)
9: 5 Möglichkeiten (9 im 3. [2 Fälle],[40] 36 im 4., 81 im 9., 90 im 10.)
15: 3 Möglichkeiten (30 im 2., 90 im 6., 120 im 8.)
16: 3 Möglichkeiten (16 im 2.,[41] 64 im 4., Pyramide mit Basis 64 im 4.)
20: 2 Möglichkeiten (100 im 5., 120 im 6.)
25: 2 Möglichkeiten (100 im 4., 225 im 9.)
45: 2 Möglichkeiten (90 im 2., 225 im 5.)

Es überrascht, daß die Schlagmöglichkeiten der ungeraden Partei nicht aufgeführt sind. Dies ist ein weiteres Zeichen dafür, daß der vorliegende Text unvollständig ist.

Auf f.11v sind weitere Teile eingetragen, die offenbar zur *Rithmachia* gehören, ohne daß ihre Funktion an dieser Stelle erkennbar wäre; es sind Zusätze, die auch hätten fehlen können. Zunächst (Kap. 23) werden noch einmal (wie in Kap. 2) die Spielsteine aufgelistet, unterteilt in die beiden Parteien und dort jeweils in die *pedites*, *comites* und *duces*; es fällt auf, daß die *pedites* und *comites* der geraden Partei zweimal vorhanden sind. Im abschließenden Kap. 24 wird zunächst eine Etymologie des Begriffs *armonia* gegeben. Dann folgt ein Satz über die Rechensteine (*calculi*) mit dem Hinweis, daß die Alten mit Steinen gerechnet haben. Das Ganze endet mit weitschweifigen Bemerkungen vor allem über Pyramiden, die die Aussagen in Kap. 17 z.T. wiederaufnehmen, z.T. erweitern. Dieses Kapitel macht einen unausgereiften Eindruck.

6. Quellen der Schrift

Wie viele andere Rithmimachie-Schriften beschränkt sich Werinher in seinem in der Handschrift **L** überlieferten Text nicht darauf, Spielregeln zu formulieren, sondern geht außerdem auch auf die Grundlagen des Spiels ein. Die Theorie wird ausführlicher als die Praxis behandelt: die Kapitel 1, 3,

[40]Da die „gerade“ 9 zu den *superparticulares* gehört, rückt sie in das 3. Feld vor und kann dort gemäß der Schlagregel 1) jeden der beiden gegnerischen Steine mit der Kopfzahl 9 bedrohen.

[41]Da die „gerade“16 zu den *multiplices* gehört, bedroht sie die gegnerische 16 nach der Regel 1) im nächsten Feld.

5-13, 15-17, 19, 24 sind im wesentlichen theoretischen, die Kapitel 2, 4, 14, 18, 20-23 praktischen Fragen gewidmet.

Werinher bezeichnet das Spiel, das er behandelt, mit dem Wort *rithmachia* und wählt damit den ältesten spezifischen Namen für das Zahlenkampfspiel; er wurde seit etwa 1080 gebraucht und ist zum erstenmal in einer Abhandlung über den Abakus nachweisbar, der von einem unbekannten Autor *G.* stammt.[42]

Der eigentlichen Behandlung des Themas ist eine Einleitung vorangestellt, die nach dem im 12. Jahrhundert üblichen Schema des *accessus ad auctores* gegliedert ist (Kap. 3). Fortolf wählte, offenbar beeinflußt durch Konrad von Hirsaus *Dialogus super auctores*,[43] zu Beginn seiner Schrift eine ähnliche Einteilung.[44] Es ist gut möglich, daß Werinher hier Fortolf folgte,[45] jedoch konnte er eine derartige Gliederung auch anderen Texten entnehmen, die in der Tradition von Boethius' Übersetzung der *Kategorien* von Aristoteles stehen.[46]

Da alle Elemente der Rithmimachie aus der antiken Zahlentheorie entstammen, verwundert es nicht, daß es für fast alle theoretischen Erörterungen Entsprechungen in der *Arithmetik* des Boethius gibt.[47] Von dieser Schrift, die zu den verbreitetsten mathematischen Texten im westlichen Mittelalter gehörte,[48] standen auch in Tegernsee Codices zur Verfügung.[49] Daher wird man annehmen können, daß die *Arithmetik* des Boethius die Hauptquelle für die theoretischen Abschnitte war. Bemerkenswert ist jedoch die Tatsache, daß der Autor der *Rithmachia* nur selten wörtlich aus der *Arithmetik* des Boethius zitiert.

Neben Boethius' Schrift selbst könnte dem Autor der Spielanleitung auch einer der Kommentare zu Boethius' *Arithmetik* vorgelegen haben. A. Borst hat in diesem Zusammenhang auf die Handschrift Cod. math. qu. 33 in der Landesbibliothek Stuttgart hingewiesen, die einen dieser Kommentare enthält (in Zukunft zitiert mit: **S**).[50] Er nennt gleich im Anfang die

[42] Siehe [Borst 1986], S. 113, und [Borst 1990], S. 277.

[43] [Borst 1986], S. 186, 191.

[44] Buch I, Kap. I-V: [Borst 1986], S. 429-432.

[45] Werinher behandelt wie Fortolf folgende Begriffe: *nomen* (bei Fortolf: *inscriptio*); *materia*; *intentio*; *finis* (bei Fortolf: *utilitas*); *cui parti philosophiae supponatur*.

[46] So schreibt etwa der anonyme Kommentator von Boethius' *Arithmetik* in der Handschrift **S** (hierzu siehe unten) auf f.1r: ... *quippe cum in principio cuiuslibet artis decem inquirere iubeamur auctoritate boetii in IIII^{to} topicorum. Sunt autem haec decem: Quae sit ars ista. Quod genus artis. Quae materia. Quod officium. Quis finis. Quae partes. Quae species. Quod instrumentum. Quis artifex. Quae nominis interpretatio.*

[47] Die betreffenden Stellen sind im Similienapparat der Edition (Kapitel 9) verzeichnet.

[48] Über 200 Handschriften sind erhalten, von denen mindestens 50 vor dem Jahr 1000 geschrieben wurden.

[49] Siehe Anm. 7.

[50] [Borst 1990], S. 256, Anm. 3. Die Handschrift stammt aus dem 12. Jahrhundert. Der Kommentar zur *Arithmetik* des Boethius trägt die Überschrift *Glossae super arithmeticam* und steht auf f. 1r-34r.

Rithmimachie:[51]

> Instrumentum eius (= arithmeticae) est, ut quidam volunt, abacus vel, ut aliis placet, mensa pythagorea, quae dicitur rithmachia, quia in istis maximum huius artis habetur exercicium, vel ipsae figurae quae subponuntur sunt eius instrumentum.

An einer späteren Stelle äußert sich der unbekannte Kommentator zu den Zielen der Rithmimachie und gibt eine Etymologie des Namens, die derjenigen des Werinher ähnelt:[52]

> Inde est, quod in rithmachia, quae pugna numeri dicitur, ubi summa est victoria maximam armoniam in campis alterius constituere, ex utraque parte ponitur pyramis, quae vim habet maximam ad constituendam maximam armoniam in rithmachia.

Auch an anderen Stellen kann man Ähnlichkeiten zwischen dem Text des Werinher und dem des anonymen Kommentators feststellen;[53] allerdings sind sie nicht so stringent, daß man einen direkten Einfluß annehmen muß.

Werinher dürfte zwar die *Arithmetik* des Boethius studiert haben, aber die theoretischen Teile in seiner *Rithmachia* erwecken den Anschein, als ob er den mathematischen Inhalt nicht immer voll erfaßte.[54] So verwundert es, daß er bei der Frage, ob eine vorgegebene Zahl prim ist, zunächst ihre Teilbarkeit durch 2 und danach durch 4 untersucht (Kap. 7). Bei der Bestimmung des größten gemeinsamen Teilers mit Hilfe der Wechselwegnahme wird ein Zahlenbeispiel gebracht, bei dem die im Codex vorhandenen Zahlen nicht mit dem Verfahren in Einklang gebracht werden können (Kap. 9).[55] Die Einteilung der räumlichen Zahlen in Kap. 15 und 16 läßt keine klare Linie erkennen; im übrigen werden die Kreiszahlen zunächst als *numeri cyclici* und später als *circulares* bezeichnet.[56] Auch der Begriff *parte altera longiores* für die *numeri solidi*, die am Ende von Kap. 15 genannt werden, ist wenig glücklich, da in der üblichen Terminologie dies der Fachausdruck für die Rechteckszahlen ist. Die Unterscheidung der Pyramiden in gerade und ungerade nach der Anzahl ihrer Schichten ist ungewöhnlich; die Behauptung, drei- und viereckige Pyramiden seien gerade, fünf- und sechseckige dagegen ungerade, ist geradezu unsinnig (Kap. 17). Die am Ende dieses Kapitels gezeichneten Pyramiden enthalten nicht nur einzelne Fehler; vielmehr gehen sie offenbar von der unzutreffenden Annahme aus, man könne die Pyramidalzahlen

---

[51]**S**, f.1v.

[52]**S**, f.25r; vgl. Werinher, Z. 39-40, in der Edition in Kapitel 9.

[53]Einige davon sind in Kapitel 9 im Similienapparat angegeben.

[54]Dabei wird unterstellt, daß die in der Handschrift **L** überlieferte Fassung im großen und ganzen den ursprünglichen Text des Werinher wiedergibt und die im folgenden genannten Fehler nicht dem Schreiber dieser Handschrift anzulasten sind.

[55]In der Dresdner Handschrift C 19, die auf Werinhers Text beruht (hierzu siehe Abschnitt 7), ist das Zahlenschema zur Bestimmung von g.g.T. (9, 25) korrekt. Somit sind die falschen Zahlen in der Handschrift **L** wohl dem Kopisten und nicht Werinher zuzuschreiben.

[56]Bei Boethius heißen sie stets *cyclici*.

ähnlich wie die Dreieckszahlen zweidimensional in Dreiecksform darstellen.[57] Auch im Text über die *medietates* gibt es inhaltliche Fehler: Während ein Leser, der Boethius' Anweisung zur Berechnung des geometrischen Mittels kennt, in den Text des Werinher vielleicht noch einen Sinn hineinlesen kann,[58] ist dies im Fall des harmonischen Mittels[59] unmöglich, da Werinhers Anweisung falsch ist, während Boethius eine zwar komplizierte, aber korrekte Methode angibt.[60]

Die wichtigste Quelle, die Werinher für die praktischen Teile seiner *Rithmachia* benutzte, war die „Bayerische Kompilation". Dies beweisen Übereinstimmungen, von denen nur die wichtigsten hier genannt seien: In Kap. 2 und Kap. 4 bezeichnet Werinher die Steine der *multiplices* als *pedites*, die *superparticulares* als *comites* und die *superpartientes* als *duces*. In den bekannten Rithmimachie-Texten vor Werinher finden wir diese Terminologie, die aus der Tradition der Schachtraktate stammt,[61] nur in der Bayerischen Kompilation bzw. in dazugehörigen Anhängen. Auch der Name *rex* für die Pyramide dürfte auf diese Schrift zurückgehen.[62] – Die Größe des Spielfelds (16 × 8 Felder; Kap. 4) kann Werinher nur von der Bayerischen Kompilation oder von Fortolf übernommen haben, da in den früheren Texten das Brett entweder kürzer ist oder über die Länge keine explizite Angabe gemacht wird. – Die Zugregeln für die verschiedenen Figuren (Kap. 4) werden in allen Rithmimachie-Schriften vor Werinher ähnlich formuliert, jedoch gebraucht Werinher die Begriffe *proximum* (statt *secundum*) und *sursum*, die vor ihm nur in der Bayerischen Kompilation benutzt werden. – Auch die Formulierung *ut non possit ... ultra rebellare* findet sich in den Rithmimachie-Texten vor Werinher wohl nur in der Bayerischen Kompilation.[63]

Die „Bayerische Kompilation" war nicht die einzige Quelle, aus der Werinher schöpfte. Daneben muß er auch die sogenannte „Regensburger Sammlung" gekannt haben, da eine Passage in Kap. 21 (Z. 315-320) der *Rithmachia* fast wörtlich mit einem Abschnitt der „Regensburger Sammlung" übereinstimmt, während parallele Formulierungen in anderen Texten davon abweichen. Auch die Angabe, daß die Spielfelder abwechselnd rot und weiß sind (Kap. 14), könnte auf diesen Text zurückgehen, da in Thiemos Spielplan, der auf der Regensburger Sammlung beruht, dieselben Farben verwendet werden.[64]

[57]Die Figuren der Pyramiden mit dreieckiger und fünfeckiger Grundfläche sind völlig falsch.

[58]Z. 280-282 der Edition.

[59]Z. 286-287.

[60]*Arithmetik* II 50 (S. 163,29-170,13): Wenn $b : H = H : a$ sein soll, ergibt sich H aus a und b mit Hilfe von $H = [(b-a) \cdot a] : (b+a) + a$. Ebenso Fortolf II 4 (S. 455,636-642). (Im folgenden werden alle Rithmimachie-Texte vor Werinher stets nach [Borst 1986] und Boethius' *Arithmetik* nach [Friedlein 1867] zitiert.)

[61][Borst 1986], S. 46.

[62]In der Bayerischen Kompilation heißt die Pyramide *regina*.

[63]*non possit diutius rebellare*: S. 417,52-53.

[64]Eine andere denkbare Quelle dafür könnten Anweisungen über das Schachspiel sein.

Sprachliche Übereinstimmungen zeigen, daß Werinher auch aus der „Fränkischen Kompilation“ Anleihen nahm. Besonders gravierend ist die Formulierung am Anfang von Kap. 21 (Z. 311-313), die fast wörtlich mit der „Fränkischen Kompilation“ (und nur mit dieser) identisch ist. Auch die 5. Regel über das Schlagen von Figuren, die Werinher in Kap. 18 bringt, muß aus der „Fränkischen Kompilation“ stammen, da sie später nicht mehr in derselben Weise formuliert wird.[65] – Es gibt noch ein weiteres schwerwiegendes Indiz dafür, daß Werinher die „Fränkische Kompilation“ kannte und für seine Schrift heranzog: die Liste der Schlagmöglichkeiten für die Steine der geraden Partei (Kap. 22). Zwar gibt es analoge Listen auch in der Arbeit des Regensburger Anonymus und in der Bayerischen Kompilation. In der zuletzt genannten Schrift wird aber auch eine Schlagmöglichkeit angegeben, die erst durch die Verlängerung der Felderzahl von 14 auf 16 möglich war[66] und daher in den früheren Texten noch nicht vorkommt. Obwohl Werinhers Brett 16 Felder lang ist und daher diese Möglichkeit bestünde, kommt sie in seiner Liste nicht vor. Er hat also eine frühere Tabelle benutzt. Der Regensburger Anonymus kommt ebenfalls nicht in Frage, weil Werinher zwei Schlagmöglichkeiten der 4 erwähnt, die nicht in der Schrift aus Regensburg, wohl aber in der Fränkischen Kompilation aufgeführt sind.[67]

Schließlich dürfte Werinher auch die ausführliche Arbeit von Fortolf gekannt haben. Aus ihr könnte er das einleitende Kap. 3[68] und insbesondere die Etymologie des Begriffs *rithmachia* entnommen haben; es ist auch möglich, daß Werinher theoretische Teile seiner Schrift nicht direkt aus Boethius' *Arithmetik* oder aus einem Kommentar dazu, sondern aus Fortolfs Werk übernahm.

Obwohl für große Teile der *Rithmachia*, wie sie uns in der Handschrift **L** vorliegt, die Quellen angegeben werden können, gibt es zahlreiche Passagen, für die dies nicht möglich ist. Man wird annehmen dürfen, daß diese zumindest teilweise vom Autor der *Rithmachia* stammen. Auch hier seien die wichtigsten Stellen genannt:

Der Begriff *unigoni*, der – analog zu *trigoni* und *tetragoni* – für „runde Steine“ steht (Kap. 2), begegnet in den erhaltenen Rithmimachie-Traktaten vor Werinher nicht. Runde Steine werden wohl erstmals in einem Spielplan abgebildet, der zum Text des Odo gehört ([Borst 1986], 131); bei Fortolf heißen sie *rotundae* (S. 439, 264). Auch die dreieckige Form für die *superparticulares* ist in schriftlichen Texten vor Werinher nicht belegt. – Die Bemerkungen über die Erfinder des Spiels (Kap. 3) lassen sich ebenfalls in dieser Form vor Werinher nicht nachweisen. – Der Autor begründet, warum die *multiplices superparticulares* und *multiplices superpartientes* nicht verwendet werden (Kap. 4) und warum die drei Zahlenarten in das 2., 3. bzw. 4. Feld ziehen (Kap. 14). – Einige Regeln, die wir erstmals bei Werinher finden,

[65] Diese Schlagregel findet man erstmals im Anhang d) zu Odos Text (S. 371).
[66] Die 6 schlägt die 90 im 15. Feld: Bayer. Komp. 12 (419, 85).
[67] 16 im 4. Feld, 56 im 14. Feld.
[68] Hierzu siehe oben.

sind unnötig und stiften für die Spielpraxis eher Verwirrung: Primzahlen und vollkommene Zahlen dürfen in jedes Feld ziehen (Kap. 10, 14). Dreiecke, Vierecke, Fünfecke (die im Spiel überhaupt nicht vorkommen) sollen sich in einer Weise bewegen, die ihrer Art entspricht (Kap. 14). Die rote Pyramide kann in alle roten, die weiße in alle weißen Felder gehen (Kap. 14).[69] – Der Sinn der 2. Schlagregel in Kap. 18 wird nicht klar, da durch die Begriffe *multiplicati aut iuncti* in der 3. Regel nach dem Verständnis der Vorgänger sowohl das Multiplizieren als auch das Addieren abgedeckt ist. – Ein Zusatz in der 4. Schlagregel (Kap. 18) zeigt, daß das kommutative Gesetz der Multiplikation noch nicht selbstverständlich ist.

Auch die Zusammenstellungen der arithmetischen, geometrischen und harmonischen Mittel, die in der Rithmimachie möglich sind und für Siegpositionen ausgenutzt werden können (Kap. 20), haben keine direkten Parallelen in früheren Texten. Naturgemäß lassen sich einige Tripel auch in anderen Listen von arithmetischen[70] und harmonischen Mitteln[71] finden, jedoch unterscheidet sich die Auswahl, die Werinher getroffen hat, von diesen Zusammenstellungen, so daß man annehmen muß, daß er die Tripel aus verschiedenen Quellen ausgesucht und vielleicht auch teilweise selbst gefunden hat.[72] Ob die zahlreichen Fehler bzw. welche von ihnen Werinher oder dem Schreiber der Handschrift **L** anzulasten sind, ist nicht zu klären. – Auch die (unzutreffenden) Aussagen am Ende dieses Kapitels 20 darüber, wann die Mittelbildung zu geraden und wann zu ungeraden Zahlen führt, haben keine Entsprechung in anderen Rithmimachie-Texten vor Werinher.

Dem Verfasser der *Rithmachia* in der Handschrift **L** lagen also mehrere Texte vor, die er zu einem eigenen Entwurf verarbeitete: für die theoretischen Teile zunächst die *Arithmetik* des Boethius, vielleicht ein Kommentar dazu und Abschnitte aus Fortolfs zusammenfassender Darstellung; für die praktischen Regeln stützte er sich vorwiegend auf die „Bayerische Kompilation", daneben aber auch auf die „Regensburger Sammlung" und die „Fränkische Kompilation". Der Kompilator suchte, so gut es ging, die Vorlagen zu einem Ganzen zu verarbeiten. Daß ihm dies nur teilweise gelingen konnte, hängt damit zusammen, daß die Quellen von unterschiedlichen Voraussetzungen ausgingen. So ist das Ergebnis zum Teil widersprüchlich und eignet sich nicht als Anweisung zum Spielen. Der Kompilator versuchte nach besten Kräften, die heterogenen Vorlagen zu vereinheitlichen, und fügte Ergänzungen hinzu,

---

[69] Rote und weiße Pyramiden sind sonst nicht belegt.

[70] Odo, Anhang f (373, Kap. 3); Anhang i (377,4-6. 12-14); Regensburger Sammlung, Anhang b (398, Kap. 1-2).

[71] Für die gerade Partei: Odo 21 (354,113-118); Odo, Anhang f (373, Kap. 1); Anhang i (377,7-11); Fränk. Komp. 17 (409,105-113); Fortolf II 11 (461,790-808). Für die ungerade Partei: Odo 22 (354,119-124); Odo, Anhang f (373, Kap. 2); Anhang i (377,15-18); Fränk. Komp. 21 (410,138-144); Fortolf II 11 (461,809-811).

[72] Bemerkenswerte Übereinstimmungen gibt es nur im Fall der harmonischen Mittel der ungeraden Partei zwischen Werinher einerseits und den Anhängen f und i zu Odo bzw. der Fränkischen Kompilation andrerseits.

die sein Bemühen erkennen lassen, ohne daß man den Eindruck erhält, daß er alle Einzelheiten voll erfaßte.

All dies erhöht die Wahrscheinlichkeit, daß Werinher der Autor dieser Kompilation war. Zwar wissen wir nichts über seine mathematischen Fähigkeiten. Aber die wenigen Informationen, die wir durch die Tegernseer Briefsammlung über Werinhers Bemühungen um die Rithmimachie besitzen, passen gut zum Text der *Rithmachia*, der in der Handschrift **L** überliefert ist: Wenn Werinher dem Wunsch des Propstes Otto von Rottenbuch nachkam, die Regeln der Rithmimachie ihm mitzuteilen, so wird er sich zunächst auf die „Bayerische Kompilation" gestützt haben, die im Umkreis von Tegernsee entstand, und sie durch weitere ihm zugängliche Entwürfe ergänzt haben; daß er dabei auch die *Arithmetik* des Boethius heranzog, war für einen Mönch, der im 12. Jahrhundert in Tegernsee lebte, selbstverständlich. Der Entwurf, der im Codex **L** vorliegt, erfüllt all diese Bedingungen. Daß er offenbar nicht vollendet wurde, weil einige Abschnitte, die man erwarten würde, fehlen,[73] und zum Teil Widersprüche aufweist, erhöht noch die Wahrscheinlichkeit, daß wir es wirklich mit dem Text des Werinher zu tun haben. Viele der Fehler, die in der Handschrift **L** vorhanden sind, deuten aber darauf hin, daß es sich nicht um das Autograph handelt, sondern um eine Abschrift, die relativ bald nach der Entstehung von Werinhers Kompilation von einem Kopisten gemacht wurde, der nicht besonders sachverständig war.

7. Fortwirken

Obwohl Werinhers Darstellung nicht ausgereift und vermutlich nicht sehr weit verbreitet war – darauf deutet die Tatsache hin, daß nur eine Handschrift erhalten ist –, war seine Kompilation für die folgenden Rithmimachie-Texte recht bedeutend. A. Borst hat wahrscheinlich gemacht,[74] daß die *Rithmachia* des Werinher für eine anonyme Kompilation, die zu Beginn des 13. Jahrhunderts in England entstand,[75] und für den Abschnitt über den Zahlenkampf in der Versdichtung *De vetula*, die um 1240 in Frankreich vielleicht von Richard de Fournival geschrieben wurde, benutzbar gemacht wurde.

Am deutlichsten zeigt sich das Fortwirken von Werinhers Text aber an einer anderen Fassung, die in einer bisher unedierten Handschrift des frühen 15. Jahrhunderts vorliegt.[76] Der unbekannte Autor dieses Textes, höchstwahrscheinlich ein italienischer Geistlicher, übernahm ganze Passagen

[73]In Kap. 22 sind zwar die Schlagmöglichkeiten der „geraden" Figuren aufgelistet, eine entsprechende Zusammenstellung für die Figuren der „ungeraden" Partei fehlt jedoch. Man vermißt ebenfalls Informationen über die verschiedenen Siegmöglichkeiten, vor allem über mehrfache Harmonien, die durch vier Steine erreicht werden können.

[74][Borst 1986], S. 220. 224.

[75]Überliefert in Oxford, BL, Auct. F. 5. 28, f.15v-16r; unediert.

[76]Dresden, Sächsische Landesbibliothek, C 19, 3. Teil, 7 Blätter. Hierzu siehe [Wappler 1892], S. 6-8, und [Borst 1986], S. 239-241.

aus Werinhers Bearbeitung, so daß vor dem Auftauchen der Handschrift **L** die Meinung vertreten wurde, es handle sich um ein zweites Exemplar von Werinhers Text.[77] In Wirklichkeit hat der Kompilator Werinhers Fassung zugrunde gelegt und diese durch Zusätze erweitert. Die meisten davon betreffen die theoretischen Grundlagen der Rithmimachie, die der Autor der *Arithmetik* des Boethius entnehmen konnte.[78] Am Anfang ist ein Vorwort über die Bedeutung der Rithmimachie, am Ende ein Kapitel über die Harmonien in der Musik hinzugefügt. Bisweilen hat der Autor der italienischen Kompilation Werinhers Anordnung verändert.[79] Aus der Tatsache, daß sein Text fast alle inhaltlichen Fehler aufweist, die auch in der Handschrift **L** vorkommen,[80] muß man schließen, daß dem italienischen Autor ein Text vorlag, der nicht wesentlich besser als die Fassung in **L** war. Da auch die einzige bekannte Handschrift dieser italienischen Bearbeitung nicht vollständig zu sein scheint,[81] können keine Rückschlüsse gezogen werden, ob dem Schreiber eine Textfassung von Werinhers *Rithmachia* vorlag, die umfangreicher als diejenige in **L** war.

8. Bemerkungen zur Edition

Da von der *Rithmachia* des Werinher nur eine Handschrift bekannt ist, besteht nicht die Möglichkeit, mit den üblichen textkritischen Methoden den ursprünglichen Text wiederherzustellen. Daher gibt die Edition im Prinzip die Lesart der einzigen Handschrift **L** wieder. Es wurde darauf verzichtet, Schreibweisen zu vereinheitlichen, so daß man – entsprechend der Handschrift – Formen wie *impares* oder *inpares* nebeneinander findet. Bei der Auflösung von Abkürzungen wurde die Schreibweise gewählt, die der Schreiber üblicherweise benutzte. Von der Lesart der Handschrift wurde lediglich in folgenden Fällen abgewichen: bei der Groß- und Kleinschreibung, der *e caudata*, den Zahlen und der Interpunktion. Zur besseren Lesbarkeit wurde der Text in der Edition durch Punkte, Kommata, Semikola und Doppelpunkte gegliedert. Neue Sätze beginnen stets mit Majuskeln. Die

[77] [Wappler 1892], S. 6.

[78] Größere Einschübe behandeln: Definition der Zahl, Einteilung in gerade und ungerade (zwischen Kap. 4 und 5 von Werinhers Text); Gründe dafür, daß die „gerade" Partei auch ungerade Zahlen enthält und umgekehrt (zwischen Kap. 9 und 10); Erklärung der *multiplices*, *superparticulares* und *superpartientes* und die gegenseitige Erzeugung (in Kap. 13); Erzeugung der Kubikzahlen (vor Kap. 18).

[79] Z.B. bei den Abschnitten über die räumlichen Zahlen (Kap. 15 und 16); der Aufbau wird dadurch klarer.

[80] Ausnahmen: Bestimmung des g.g.T. (Werinher, Kap. 9) und des harmonischen Mittels (Kap. 19).

[81] Der Text endet mitten auf f.6r. Die nächste Seite (f.6v) ist leer geblieben, während auf f.7v nur der Anfang eines neuen Kapitels mit der Überschrift *Quid sit proporcio Capitulum primum* steht. Von den in der Handschrift **L** vorhandenen Abschnitten fehlt im Dresdner Codex das Ende von Kap. 21 Mitte an.

(seltene) *e caudata* wurde mit gewöhnlichem *e* wiedergegeben. Die römischen Zahlen, die der Schreiber, wie auch sonst üblich, im allgemeinen mit Punkten einrahmte, wurden einheitlich mit großen lateinischen Buchstaben geschrieben. Dabei wurde darauf geachtet, daß die in der Handschrift vorliegende Form möglichst bewahrt wurde, so daß z.B. *.iii.*^{us} mit *III*^{us} und nicht mit *tertius* wiedergegeben wurde. Nur offenkundig unrichtige Worte wurden verbessert, wobei die Lesart der Handschrift selbstverständlich im Variantenapparat angegeben ist. Spitze Klammern ⟨ ⟩ bezeichnen Buchstaben oder Worte, die vom Herausgeber eingefügt wurden, eckige Klammern [ ] zu tilgende Buchstaben. Aus Gründen der Lesbarkeit wurde eine Kapitelzählung eingeführt. Sie lehnt sich an die Untergliederung der Handschrift mit Hilfe des Paragraphen-Zeichens (§) an; da dieses Zeichen vom Schreiber nicht konsequent benutzt wurde, ist es in der Edition fortgelassen worden. – Neben dem Variantenapparat wurde ein Similienapparat beigegeben. Er verzeichnet ähnliche Texte in den vor Werinher entstandenen Rithmimachie-Traktaten sowie in der *Arithmetik* des Boethius und einige Stellen im Boethius-Kommentar der Handschrift Landesbibliothek Stuttgart, Cod. math. qu. 33 (zitiert als: **S**; hierzu siehe oben). Die Seitenzahlen bei Boethius' *Arithmetik* beziehen sich auf die Ausgabe [Friedlein 1867], die bei den Rithmimachie-Texten auf [Borst 1986]. – Die Edition wurde mit Hilfe von EDMAC (John Lavagnino / Dominik Wujastyk) erstellt.

9. Edition

⟨1⟩ *1r*

I	I	I	unitas
I	II	IIII	dupli
IIII	II	I	dupli conversi
IIII	VI	IX	sesqualtri
VIIII	VI	IIII	sesq⟨ualtri⟩ conversi
VIIII	XV	XXV	superbipartiens
IIII	VI	VIIII	sesqualter non con⟨versus⟩
IIII	X	XXV	duplex sesqui⟨alter⟩
VIIII	XV	XXV	[duplex] superbi⟨partiens⟩
IX	XXIIII	LXIIII	duplex superbi⟨partiens⟩

Hee sunt nativitates numerorum

⟨2⟩ | Pedites parium numerorum *4rb*

II	IIII	VI	VIII	pedites
IIII	XVI	XXXVI	LXIIII	parium

Unigoni

Pedites inparium qui sequuntur

III	V	VII	VIIII
IX	XXV	XLIX	LXXXI

Comites ex parium proportione

VI	XX	XLII	LXXII
VIIII	XXV	XLIX	LXXXI

Trigoni

Comites imparium

XII	XXX	LVI	XC
XVI	XXXVI	LXIIII	C

---

21 LXXII] lxxiii **L** *22* LXXXI] lxxxiii **L**

2–11: Boeth. Arithm. I 32 (66,18-72,19); Fortolf I 9 (435,169-438,241) *13* Pedites: Bayer. Komp. 3 (415,11f.); Bayer. Komp., Anh. 7 (424,48) *20* Comites: Bayer. Komp. 4 (415,17); Bayer. Komp., Anh. 8 (424,49)

Duces parium numerorum

XV	XLV	XCI	CLIII
XXV	LXXXI	CLXVIIII	CCLXXXVIIII

Tetragoni

Duces imparium numerorum

XXVIII	LXVI	CXX	CXC
XLVIIII	CXXI	CCXXV	CCCLXI

Pyramis perfecta XCI

Pyramis tercurta rex inparium CXC

| Incipit Rithmachia. 4v

⟨3⟩ Nomen, materia, intentio, finis, cui parti philosophie supponitur, quis inventor, hic in principio requiritur.

Nomen est ritmachia, et sic dicitur a rith, quod est numerus, et a chaia, quod est pugna, quasi pugna numerorum. Pugnat enim pulcherrime hic virtus numeri. Materia eius par et impar numerus, qui quodammodo contra se quasi castra ponunt et alterni vel pedites vel comites vel duces procedunt eminus tam diu, donec per vicarios tractus auferendo, nunc prodeundo, nunc cedendo, nunc sibi eminendo, nunc retro vel per devia fugiendo, quousque piramidum altera capiatur, ut tunc quasi expugnatis castris illata armonica vel quavis alia medietate quasi fixa in hostili regno victricia signa universitatem numeri deinceps non sinant rebellare. Intentio duplex est, quarum neutra sine altera esse potest. Nam bellator quivis tum sibi cavere, tum nocere adversario debet. Finis est victoria per expugnationem piramidum et inductionem medietatum. Practice ludus iste supponitur, que sic diffinitur: Practica est perfecta manuum operatio in propatulo existens secundum preeuntis theorice intellectum, que quidem est arismetica. Non enim aliter arismetice opus rithmachia representat, quam musica in cytharis et organis, et geometria in abaci opere, et astronomia in ho⟨ro⟩scopis et astrolabii sollertia consistit. Inventor ludi huius apud romanos boetius fuit, quemadmodum arismetice apud grecos pytagoras et nicomacus et alii quam plures. His premissis ad negotium transeamus.

29 CLXVIIII] clxviii **L** *29* CCLXXXVIIII] cclxxxviii **L** *39* a rith] arith **L** *39* a chaia] achaia **L** *50* que] qui **L** *51–52* theorice] teorice *(?)* **L** *et corr. in* thorice *53* quam] quod **L** *56* quam] quod **L**

27 Duces: Bayer. Komp., Anh. 9 (425,52) *35* rex: Bayer. Komp.8 (416,34) (*regina*) *37–38* Nomen ... requiritur: Fortolf 429,38-42 *39–40* Nomen ... numerorum: **S**, f.25r; Fortolf I 1 (430,57-60) *41* Materia ... numerus: Fortolf I 2 (430,70) *48–49* Nam ... debet: Fortolf I 14 (443,343-345) *54–56* Inventor ... plures: **S**, f.1v: *Arithmeticae disciplinae apud graecos pythagoras primus fuit auctor, post nichomachus, apud latinos vero apuleius, deinde boetius*

⟨4⟩ Sit tabula XVI in longitudine, VIII in latitudine campis distincta, et super eam disponantur ex una parte pares numeri, ex alia inpares. Omnium specierum multiplices dico, superparticulares et superpartientes duobus, scilicet multiplici superparticulari et multiplici superpartienti, propter multiplicitatem et inexplicabilitatem monadum et dyadum sepositis ad presentem ludum. Et erunt in prima fronte multiplices tamquam pedites in acie prima. His adhereant superparticulares tamquam comites ipsos ambientes. Retro hos et in postrema acie superpartientes tamquam duces locantur. Et sunt multiplices ex parte parium dupli, quadrupli, sescupli, octupli; superparticulares vero sesqualtri, se⟨s⟩quiquarti, sesquisexti, sesquioctavi; superpartientes vero superbipartientes, superquadripartientes, supersesquipartientes, superoctopartientes. Ex parte | imparium multiplices tripli, quincupli, septu- *4vb*
pli, novemtupli; superparticulares se⟨s⟩quitertii, sesquiquinti, sesquiseptimi, sesquinoni; superpartientes supertripartien⟨te⟩s, superquinquepartientes, super⟨septi⟩partientes, supernovempartientes. Et multiplices quidem in proximum campum ante et retro, sursum, deorsum et angulariter trahuntur, superparticulares in tertium eodem modo et superpartientes in quartum campum eodem modo. Nunc de divisione et nativitate istorum nu⟨mer⟩orum videamus ipsorum diffinitionem commutantes, quippe que presentis non est negotii.

⟨5⟩ De divisione

Numerorum alius pariter par, alius pariter impar, alius impariter par. Et hec est paris numeri prima divisio. Pariter par numerus est qui in duo equa dividitur, eius medietas in duo equa, et medietas medietatis, usque ad insecabilem monadem, veluti in hoc parium pedite vides, scilicet LXIIII, in cuius divisione statur in unitate. Pariter impar numerus est qui in duo equa dividitur, sed neutra medietatum in duo equa dividi potest, ut in hoc comite parium: XLII. Inpariter par est qui in duo equa dividitur et eius medietas et medietas medietatis, ita tamen, ut priusquam perveniat ad unitatem, incidat numerum insecabilem, ut in hoc parium comite: LXXII.

⟨6⟩ Si vis scire qui sint pariter pares, pone ab uno omnes qui sese duplo vincunt, et habebis pariter parem, ut I, II, IIII, VIII, XVI, XXXII, LXIIII. Si vis scire, unde nascantur pariter inpares, naturalem pone imparium summam, et eam multiplica per binarium, et habebis pariter impares.

64 comites] duces **L** *67* sesquioctavi] sexquioctavi **L** *76* ipsorum *corr.* **L** *ex* ipsius *(?)*

58 Sit ... distincta: Bayer. Komp. 1 (415,1-2); Fortolf I 6 (432,113-115) *59–60* super ... superpartientes: Bayer. Komp. 1 (415,2-3); Fortolf I 7 (433,124-125) *63–65* Et ... locantur: Bayer. Komp. 2-4 (415,6-416,20) und öfter; zu *pedites*, *comites*, *duces* siehe Kap. 2 *65–72* Et ... supernovempartientes: Fortolf I 7 (433,127-135) und öfter *72–75* Et ... modo: Bayer. Komp. 5 (416,22-25); Fortolf I 13 (442,338-341) und öfter *72–73* proximum: Bayer. Komp. 5 (416,22) *73* sursum: Bayer. Komp. 5 (416,22) *78* Numerorum ... par (*2.*): Boeth. Arithm. I 8 (17,2-4) *79–82* Pariter ... unitate: Boeth. Arithm. I 9 (17,9-17) *82–84* Pariter ... XLII: Boeth. Arithm. I 10 (21,24-22,5) *84–86* Inpariter ... LXXII: Boeth. Arithm. I 11 (25,5-12) *87–88* Si ... LXIIII: Boeth. Arithm. I 9 (18,4-12) *88–90* Si ... impares: Boeth. Arithm. I 10 (22,23-23,13)

II	II	II	II
VI	X	XIIII	XVIII
III	V	VII	IX

Impariter pares sic creantur: Pariter pares loca superius, subtus naturali numero disponens inpares, et multiplica inferiores per superiores in hunc modum:

	IIII	VIII	XVI	XXXII
⟨III⟩	V	VII	VIIII	XI

⟨7⟩ Numeri inparis hec fit divisio: Inpar numerus alius primus et incompositus, alius secundus et compositus; alius ad se quidem secundus et compositus, ad alium relatus primus et incompositus. Primus et incompositus est quem nullus numerus preter unitatem metitur, ut sunt III, V, VII, XI, et tales hoc modo invenientur: Vide si propositum numerum nec in II nec III nec IIII aut plura equa potes dividere. Quod sic, erit primus et incompositus.

⟨8⟩ Secundus et compositus ita creatur: Posito naturali numero inparium ab unitate quantum vis primus ternarius duobus intermissis imparibus numeris, scilicet V et VII, tertium a se creat et metitur. Duobus post hunc intermissis quintum a se quinquies metitur et creat, scilicet XV. Duobus item post hunc imparibus omissis XXI, qui in tercio loco creat, septenaria dimensione metitur, et sic usque | [usque] in infinitum. Quemadmodum ternarius II imparibus interiectis numeros a se creat, sic quinarius IIII transilitis imparibus quintum a se creat et metitur, scilicet XV; septenarius VI interiectis a se creat; novenarius VIII interiectis; et ita semper secundum ordinem legitima omissione a se numeros distantes permetiuntur.

⟨9⟩ Eos porro, qui per se secundi et compositi, ad alios autem primi sunt et incompositi, sic invenies: Duobus numeris propositis minorem de maiori tam diu aufer, donec minus minore relinquatur, et id ipsum de minore detrahatur, donec vel unitas vel alius numerus, a quo nil detrahi possit, remaneat. Et ille erit communis illorum inparium numerorum mensura.

<table>
<tr><td>IX</td><td colspan="2">XXV</td></tr>
<tr><td rowspan="2"></td><td>I</td><td rowspan="2"></td></tr>
<tr><td>III</td></tr>
<tr><td>VI</td><td colspan="2">XXI</td></tr>
</table>

104 sic] sit **L** 106 *post* intermissis *add. et del.* **L** tertium a se 119 *post* illorum *add. et del.* **L** mensura

94–96 Impariter ... modum: Boeth. Arithm. I 11 (26,19-27,20) *99–101* Numeri ... incompositus *(1.)*: Boeth. Arithm. I 13 (30,5-14) *101–102* Primus ... XI: Boeth. Arithm. I 14 (30,16-31,21) *105–114* Secundus ... permetiuntur: Boeth. Arithm. I 17 (33,24-35,27) *115–119* Eos ... mensura: Boeth. Arithm. I 18 (37,21-38,11)

⟨10⟩ Item sciendum quod primi et incompositi in quemvis campum trahuntur et etiam hii, qui perfecti dicuntur, sequentes legem primorum et incompositorum, quia ex eis nascuntur. Dicuntur autem numeri perfecti, qui constant ex denominationibus suis, ut VI in limite denarii, XXVIII in limite centenarii, in limite millenarii CCCCXCVI[I], in limite X milium $\overline{\text{VIII}}$ CXXVIII. Perfecti ergo et nobiles dicuntur, quia non in modis vulgi passim habundant, sed in unoquoque limite unus tantum tamquam princeps presidet, quod non accidit in imperfectis.

⟨11⟩ Imperfectorum autem alii diminuti, alii superflui. Diminutus est ille, qui partes suas vincit, ut VIII. Partes dico multiplicativas ipsius, scilicet I, II, IIII. Superfluus est ille, quem partes proprie excedunt – non dico constitutive – ut XII, partes sunt I, II, IIII, VI.

⟨12⟩ Sic autem creantur perfecti: Pone duplos omnes ab uno et sequentem iunge priori, dum ostendas quemvis primum et incompositum. Per hunc multiplica numerum in aggregatione ultimum, ut uni duos iungens habebis III, qui est primus et incompositus. Dic: ter duo sex erunt, qui est primus perfectus. Alium quoque similiter produces uni duos iungens et his IIII^{or}: VII habebis, qui primus est et incompositus. Dic: septies IIII, et sunt XXVIII, qui in centenarii limite perfectus presidet. Nam in quibusque limitibus ad generandum perfectos illorum principes idem facies.

⟨13⟩ ⟨N⟩unc de relatione numerorum ad se videamus. Nam alius ad alium vel est multiplex vel superparticularis vel superpartiens vel multiplex superparticularis vel multiplex superpartiens. Et multiplicis quidem species sunt duplum, triplum, quadruplum etc.; superparticularis sesqualtrum, sesquitertium etc.; superpartientis superbipartiens, supertripartiens etc. Reliquum ob predictam causam omittimus. Qualiter autem ex unitatibus III dupli, ex duplis tripli, ex triplis quadru|pli et ita in ordine nascantur; rursum qualiter ex *5rb* conversis multiplicibus su⟨per⟩particulares ita, ut ex conversis duplis sesqualteri, ex conversis triplis sesquitertii, ex quadruplis sesquiquarti oriantur; et item qualiter ex conversis sup⟨er⟩particularibus superpartientes ita, ut conversi sesqualteri forment superbipartientes, conversi se⟨s⟩quitercii supertripartientes, superquadripartientes a sesquiquartis formentur; et qualiter ex non conversis sup⟨er⟩particularibus multiplices superparticulares, ex non con-

128 post $\overline{\text{VIII}}$ *add. et del.* **L** x *132* diminuti] dimiduti **L** *136* ab uno **L** *et corr. in* a duobus *141* et *(1.) superscr.* **L** *152* ex *(2.) corr.* **L** *ex* de *(?)* *153–154* conversi] a **L** *155* sesquiquartis] sesquiquitis **L** *156 post* conversis *add. et del.* **L** parti *156* superparticulares *corr.* **L** *ex* superpartientes

126–127 Dicuntur ... suis: Boeth. Arithm. I 19 (41,6-12) *127–129* ut ... $\overline{\text{VIII}}$ CXXVIII: Boeth. Arithm. I 20 (42,3-5) *132–135* Imperfectorum ... VI: Boeth. Arithm. I 19 (39,20-40,23) *136–143* Sic ... facies: Boeth. Arithm. I 20 (42,11-44,16) *144–146* Nam ... superpartiens: Boeth. Arithm. I 22 (46,6-9) *146–147* Et ... etc.: Boeth. Arithm. I 23 (47,3-6) *147–148* superparticularis ... etc. *(1.)*: Boeth. Arithm. I 24 (49,17-19) *148* superpartientis ... etc. *(2.)*: Boeth. Arithm. I 28 (58,14-18) *148–149* ob predictam causam: Z. 61-63 *149–158* Qualiter ... habetur: Boeth. Arithm. I 32 (66,5-72,19); Fortolf I 8 (434,151-435,168)

versis sup⟨er⟩partientibus multiplices superpartientes producantur, alibi plenius et perfectius determinatum habetur.

I	I	I	
I	II	IIII	dupli
I	III	IX	tripli
I	IIII	XVI	quadrupli

dupli	IIII	II	I
sesqualteri	IIII	VI	IX
tripli	IX	III	I
sesquitertii	IX	XII	XVI

IX	VI	IIII	sesqualter
IX	XV	XXV	superbipartiens
XVI	XII	IX	sesquitercius
XVI	XXVIII	XLIX	supertripartiens

⟨14⟩ Habemus ecce omnes numerorum species quomodo nascantur et quomodo trahantur. Nam multiplices in pr⟨ox⟩imum, superparticulares in tertium, superpartientes in IIII. Quorum tractuum hec est ratio: Quoniam multiplices ab unitatibus proximi sunt in nascendo, in proximum campum trahuntur: superparticulares in tertium, quia tertii ab unitatibus prodeunt; superpartientes in quartum, quia quarti sunt ab unitatibus. Primi et incompositi et perfecti, cuiuscumque sint speciei, in quemvis campum trahuntur. Et trianguli, quicumque fuerint, et tetragoni et pentagoni etc. ultra sequantur legem illius speciei numerorum, in quibus ipsi fiunt. Et quicumque numerus eorum, quos dixi, potest ire in quemvis campum tutus, nisi ita sit adversariis septus, ut per legitimos tractus nequeat evadere. Pyramis rubea in omnes rubeos campos, alba in omnes albos vadit. Pyramis latera omnia pyramidis aufert, tantum angulariter basim non tollit, nisi caracterum aliquis. Nunc de trigonis dicemus.

⟨15⟩ ⟨N⟩umerorum alii lineares, alii plani, alii solidi vel cubici. Lineares unam habent dimensionem, ut bis vel ter. Plani duo latera habent, ut bis bini, ter terni. Solidi III lateribus continentur. Sicut autem principium constitutionis linee est punctum et superficiei linea et corporis superficies, ita a linearibus numeris plani, a planis solidi fiunt. Et nota quod cubici numeri sunt illi, qui omnia latera habent equalia, ut bis bini bis, ter terni ter. Et

175 post in *add. et del.* **L** secundum *177* trahuntur] thrauntur **L** *178* et *(1.)*] cetera *(?)* **L**

172–173 Nam ... IIII: siehe Kap. 4 (Z. 72-75) *176–177* Primi ... trahuntur: siehe Kap. 10 (Z. 124-126) *179–181* Et ... evadere: Bayer. Komp. 7 (416,31f.); Fortolf I 15 (445,409f.) *182* rubeos campos ... albos: Spielplan Thiemos (403f.) *185–186* Lineares ... ter: Boeth. Arithm. II 5 (90,6-16) *187–188* Sicut ... superficies: Boeth. Arithm. II 4 (87,15f.); Fortolf I 16 (446,428-433) *189–190* Et ... ter *(2.)*: Boeth. Arithm. II 25 (111,23-112,7); Fortolf II 13 (463,863f.) *190–192* Et ... XXV: Boeth. Arithm. II 30 (121,8-21)

cuborum quidam a privo nomine sperici vel cyclici dicuntur, et sunt sperici qui in idem punctum unde creverunt | reflectuntur, ut quinquies V XXV. So- *5va*
lidi numeri sunt, qui III habent dimensiones, licet imparium laterum, ut bis bini ter. Solidorum alii laterculi, alii asseres, alii paralellipedi, alii scalenos. Laterculus est, qui tertium latus brevius habet duobus reliquis, ut bis bini semel. Asser vel tignulus est, cuius tertium latus duobus est altius, ut bis bini ter. Paralellepipes est numerus, cuius medium latus a duobus reliquis equidistanter excrescit, ut bis terni bis. Scalenos vel cuneus vel speniscus est, cuius omnia latera sunt inequalia, ut bis terni quater. Qui omnes uno nomine parte altera longiores dici possunt.

⟨16⟩ Item solidorum alii circulares, alii pyramides. Circulares ut quinquies V, XXV. Pyramides dicuntur numeri, qui a qualibet basi profecti a lato ad summam colliguntur unitatem per adiectionem minorum ad maiores. Pyramis alia est ex tetragonis, alia ex trigono, alia ex pentagono, alia ex exagono. Trigoni autem sunt numeri, qui in quemvis campum trahuntur, cum ipsi in se et in eos omnes alii quorumlibet angulorum numeri resolvantur. Surgunt autem trigoni a linearibus numeris. Sunt autem numeri lineares quivis in ordine naturali dispositorum. Et hoc modo inde nascuntur: Sequentem antecedenti semper copula, ut uni duo, et habebis III, qui numerus actu trigonus est, linea vero tripunctalis potestate. Rursum uni et duobus iunge III, et habebis VI, qui etiam est trigonus. Ad formandum tetragonum sequentem semper copula proximo sui proximi, qui est ei proximus uno interposito, ut III uni adde, et erunt IIII; V tribus et uni, et erunt IX. In formando pentagono duobus intermissis sequentem copula precedenti, et in formando exagono tribus interpositis idem fac, ut in hac figura:

trigonus	I	III	VI	X	XV
tetragonus	I	IIII	IX	XVI	XXV
pentagonus	I	V	XII	XXII	XXXV
exagonus	I	VI	XV	XXVIII	XLV

191 quidam] quidem **L** *191* privo *corr.* **L** *ex* privato *194* ter] t'er **L** *194* paralellipedi] parallepipedi paralellipedi **L** *et del.* parallepipedi *203* minorum *corr.* **L** *ex* minorem *206 post* et *add. et del.* **L** omnes *211 post* habebis *add. et del.* **L** ii *(?)*

194 Solidorum ... scalenos: Boeth. Arithm. II 25 (111,12-17) *195–196* Laterculus ... semel: Boeth. Arithm. II 29 (120,12-20) *196–197* Asser ... ter: Boeth. Arithm. II 29 (120,20-121,2) *196* tignulus: Boeth. Arithm. II 4 (86,17) *197–198* Paralellepipes ... bis *(2.)*: Boeth. Arithm. II 25 (114,21-115,1); Fortolf II 13 (464,868-873) *198–199* Scalenos ... quater: Boeth. Arithm. II 25 (114,8-14); Fortolf II 13 (463,864-464,868) *201–202* Circulares ... XXV: **S**, f.21v: *Sperici sunt numeri circulares, qui sicut circulus a quo puncto inchoatur in eundem revertitur, ut cum dico: quinquies quinque quinquies, sicut a quinario fit multiplicatio, ita summa multiplicationis in quinarium desinit. Sunt enim CXXV.* *202–204* Pyramides ... exagono: Boeth. Arithm. II 21 (105,4-7) *206–207* Surgunt ... numeris: Boeth. Arithm. II 4 (86,11-16) *207–208* Sunt ... dispositorum: Boeth. Arithm. II 5 (90,8-16) *208–211* Sequentem ... trigonus: Boeth. Arithm. II 9 (94,4-18) *210* tripunctalis: **S**, f.22v *211–213* Ad ... IX: Boeth. Arithm. II 12 (96,3-17) *213–214* In ... precedenti: Boeth. Arithm. II 14 (98,2-6) *214–215* et ... fac: Boeth. Arithm. II 15 (99,4-13)

His consideratis quomodo in trigonos multiangule figure solvantur et ab ipsis formentur, patet. Nam ex II^bus trigonis, veluti ex III et VI, fit unus tetragonus, scilicet IX, ut in preiacenti formula patet; et ex VI et X alius tetragonus oritur, scilicet XVI; ex X et XV, qui sunt II trigoni, III^us tetragonus, scilicet XXV, oritur. Ex tetragono, qui est IX, et trigono III facies pentagonum, et ex pentagono, scilicet XII, et predicto triangulo facies exagonum, ut similitudo formationis habeatur.

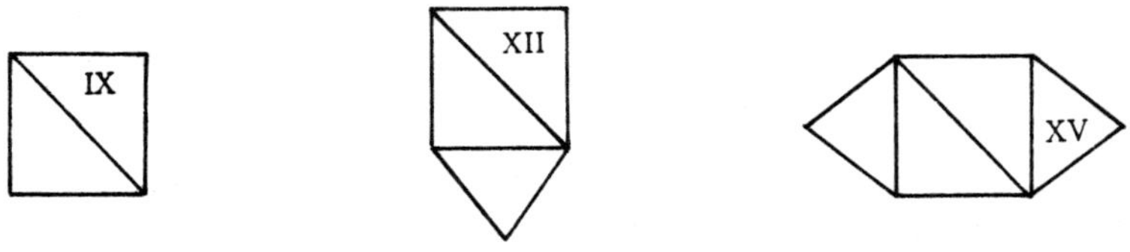

tetragonus pentagonus exagonus

⟨17⟩ Pyramis alia par, alia impar. Par est que | paris numeri aggregatione 5vb
constat, ut in trigona vel tetragona patet pyramide. Impar que imparium est laterum, ut in pentagona et exagona patet pi⟨ramide⟩.

Item pyramis alia perfecta, alia curta. Perfecta est que a⟨d⟩ summam vi basium qualibet erigitur unitatem. Curta est que non pervenit ad unitatem, sed desunt ei II vel III vel plures gradus, unde ipsa totiens et totiens curta dicetur. Nam si II gradus desunt, dicitur biscurta; si III, tercurta dicetur.

Cum ergo IIII^or sint py⟨ramides⟩, erit trigona LVI, tetragona vero XCI, pentagona vero LXXV, exagona XCV.

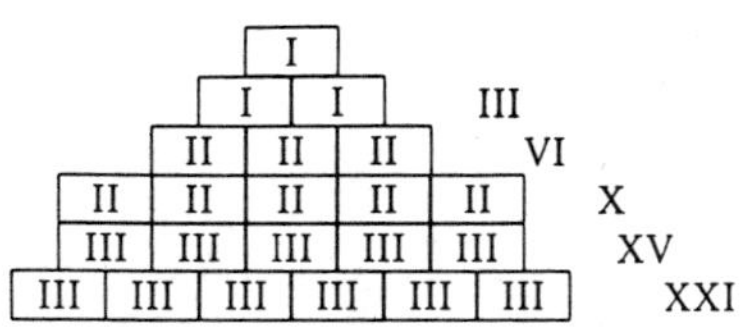

basis py<ramidis> trigone XXI

piramis tetragona

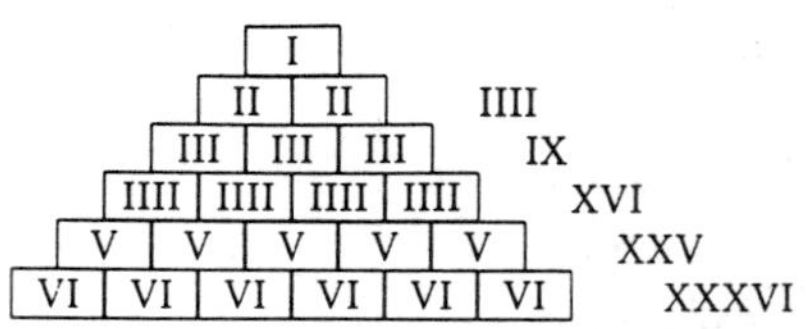

Basis py<ramidis> tetragone XCI

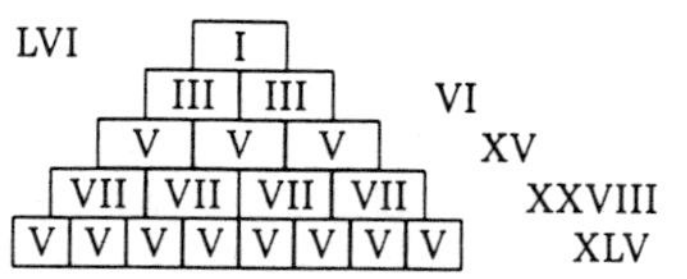

piramis exagona

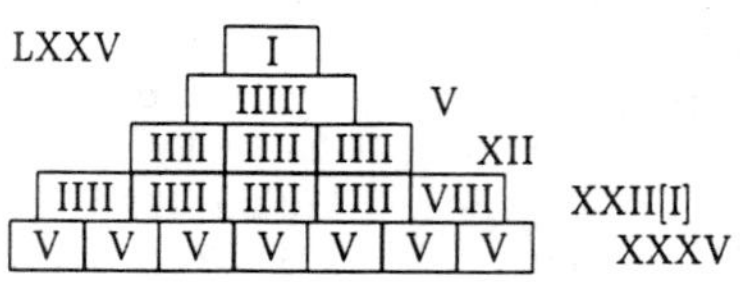

pyramis <p>entagona

---

220 ab *corr.* **L** *ex* ap *232–233* vi basium] tibasium *(?)* **L** *241 post* II *(3.) add. et del.* **L** II *244* XXI *corr.* **L** *ex* XXV

220–226 His ... habeatur: Boeth. Arithm. II 18 (102,4-103,9) *232–235* Item ... dicetur *(2.)*: Boeth. Arithm. II 24 (109,24-111,5); Fortolf I 16 (447,452-458)

⟨18⟩ Post hec dicendum est quomodo numerus numerum auferat. Ad quod istas regulas observa: Quicumque numerus contrarie partis numerum eiusdem quantitatis in suo legitimo tractu offendit, auferat eum. Item si numeros aliquos adverse partis aliquibus tuis eandem summam presentantibus vallare potes, qui illorum respitiant campos, aufer eos. Item si numeros numero circumponas, qui multiplicati aut iuncti compositum efficiant, auferatur. Item si quantitas interiacentium camporum cum numero vel numerus cum interiacentibus campis multiplicatus numerum quemvis adversum in legitima via offendit, auferatur. Item si quis numerus ita septus fuerit hostibus, ut omnis ei rectus tractus denegetur, auferatur absque ratione. Ablata basi pyramidis cum quavis istarum regularum latera quoque omnia tolle, et tunc maximam et perfectam armoniam in regno adversarii statuere conare, ut non possit hostilis exercitus, quantuscumque fuerit, ultra rebellare. Itaque victor, qui cepit piramidem, positurus armo|nicam medietatem numquam potest eam *6ra* ⟨ponere⟩ totam ex suis, sed per rapinam adverse partis numerum capit; quo tam potestative in armonia sua postmodum utitur, ac si suus esset. Nunc dicendum est, quod sit medietas.

⟨19⟩ Medietas in numero vocatur habitudo quedam vel differentiarum vel proportionum. Sunt autem III: arismetica, geometrica, armonica. A⟨rismetica⟩ est cum eadem differentia est minimi ad medium que medii ad maximum, ut X, XXV, XL, nam XV utrobique est differentia. Sic autem repperies eam: Inter quoscumque numeros maiorem minori adde et postea medietatem illius summe mediam loca. Aut sic: Differentiam minoris ad maiorem divide et eam minori appone. Sed in hac medietate consideratio proportionum negligitur et speculatio differentiarum custoditur. G⟨eometrica⟩ m⟨edietas⟩ est dum inter III terminos eadem que est minoris ad medium et medii ad maximum est proportio, ut I, II, IIII. Huius proprietas est quod quecumque inter terminos est proportio, eadem est inter differentias terminorum. Sic autem invenies: Proportionem maximi ad minimum divide et mediam pone [ii.] in quadrupla

---

256 qui] quod **L** *258* numerus] numeris **L** *259 post* adversum *add. et del.* **L** adversum *269 post* vel *(1.) add. et del.* **L** in *273* minori *corr.* **L** *ex* minorem *275 post* appone *add. et del.* **L** G. med'. est c *280 post* ad *add. et del.* **L** medi

---

253–254 Quicumque ... eum: Bayer. Komp. 6 (416,26f.); Fortolf I 14 (443,345f.) und öfter *256–257* Item ... auferatur: Bayer. Komp. 6 (416,27f.); Fortolf I 14 (443,355f.) und öfter *257–260* Item ... auferatur: Bayer. Komp. 6 (416,28-30); Fortolf I 14 (444,368-370) und öfter *260–261* Item ... ratione: Odo, Anhang d (371); Fränk. Komp. 7 (406,33f.) *262–263* et ... conare: Odo 10 (348,48f.); Fränk. Komp. 10 (407,50f.) und öfter *263* in regno adversarii: Bayer. Komp. 10 (417,54) *263–264* ut ... rebellare: Bayer. Komp. 10 (417,52f.) *264–266* Itaque ... suis: Regensb. Anon. 12 (388,68f.) *266–267* sed ... esset: Bayer. Komp. 11 (418,67f. 71) *270* Sunt ... armonica: Boeth. Arithm. II 41 (139,2f.) *270–272* A⟨rismetica⟩ ... differentia: Boeth. Arithm. II 43 (140,19-29); Fortolf II 5 (455,646f.) *273–275* Inter ... appone: Boeth. Arithm. II 50 (163,1-15); Fortolf II 4 (454,622-626) *275–276* Sed ... custoditur: Boeth. Arithm. II 42 (139,24-28; 140,15-17) *276–278* G⟨eometrica⟩ ... IIII: Boeth. Arithm. II 44 (145,9-14); Fortolf II 5 (456,666f.) *278–279* Huius ... terminorum: Boeth. Arithm. II 44 (146,2-19); Fortolf II 5 (456,665f.) *279–282* Sic ... quadruple: Boeth. Arithm. II 50 (163,15-28); Fortolf II 4 (454,630-635)

proportione, ut X et XL medius intercidit XX duplam utrinque custodiens proportionem, que est subdupla quadruple. Armonica med⟨ietas⟩ est quando eadem, que est maioris extremitatis ad minorem proportio, est et differentie maioris et medii ad differentiam minoris et medii, ut III, IIII, VI. In his numeris eadem [est], que est extremorum proportio, est et differentiarum terminorum. Hanc sic invenies: Maximum per minimum multiplica et maiorem minori adde et medium triplica, et habebis ar⟨monicam⟩ medietatem.

⟨20⟩ Videamus post hec, qui ex utraque parte arismetici sint vel geometrici vel armonici. Ex parte parium ar⟨ismetici⟩ sunt IIII, VI, VIII, vel sic: II, IIII, VI. Item IX, XLV, LXX⟨X⟩I. Item VIII, XXV, XLII. Geometrici vero termini sunt II, IIII, VIII. Item IIII, XVI, LXIIII. Item IIII, VI, VIIII. Ex parte imparium ar⟨ismetici⟩ sunt V, VII et VIIII. Item III, V, VII. Item VII, XXVIII, XLVIIII. Item XII, LVI, C. Item VII, LXIIII, CXXI. Geometrica medietas est ibi V, XXV, CXXV. Item XLVIIII, XXVIII, XVI. Et sciendum quod maxime ex utraque parte habundant termini arismeti⟨ci⟩, minus geometrici. Sed armonici nequaquam habentur per se, nisi per rapinam prede. Que si acquisita fuerit, multas ultra statuere potes. Verbi gratia ex parte parium armonici termini sunt XV, XX et XXX per predam acquirendus. Item par armonia IIII, VI, XII acquirendus. | Item par armonia IIII et VI, 6r
III acquirendus. Item arm⟨onia⟩ X, XXV, CCXXV acquirendus. Ex parte imparium impar armonia est III, V, ⟨XV⟩ acquirendus. Item ar⟨monici⟩ termini imparium V, VIIII, XLV acqu⟨irendus⟩. Item ar⟨monia⟩ VIIII, XVII, XLII acqu⟨irendus⟩. Item VII, XXVIII, ar⟨monia⟩ IIII a⟨c⟩quirendus. Et sciendum quod in ar⟨ismetica⟩, g⟨eometrica⟩ et ar⟨monica⟩ medietate ex parte parium medium tantum parem numerum esse oportet, ex parte imparium inparem numerum. Item notandum quod inter paris armonie terminos impar proportio et inter inparis par proportio invenitur et preterea omnes musice consonantie, ut in his terminis:

III	IIII	VI	par ar⟨monia⟩
II	III	VI	impar

⟨21⟩ Rursum sciendum quod si armonicam non potes ponere medietatem terminis alienis non acquisitis vel tuis per incuriam perditis, sufficere tibi debet ad victoriam rapta piramide aris⟨metica⟩ vel ge⟨ometrica⟩ medietas. Arismetici enim et g⟨eometrici⟩ termini per se sunt; utrobique etiam rapi possunt. Sed semper unus armonicorum per predam acquirendus est. Tali altercatione alternorum tractuum omnis numerus a prime positionis campis studeat in campis adverse partis medietates efficere aris⟨meticam⟩ et geome-

294 CXXV] cc **L** *298* sunt] sint **L** *306* impar *corr.* **L** *ex* par *316* omnis] omnib **L** *317–318* geometricam] armonicam **L**

282–286 Armonica ... terminorum: Boeth. Arithm. II 47 (152,2-22); Fortolf II 5 (456,672-678) *294–296* Et ... prede: Regensb. Anon. 12 (388,67-69); Fränk. Komp. 11 (407,57f.) und öfter *311–313* Rursum ... medietas: Odo 10 (348,51-349,53); Fränk. Komp. 10 (407,53-55) *314–315* Arismetici ... est: siehe Kap. 18, Z. 264-267 *315–318* Tali ... geometricam: Regensb. Anon. 10 (387,58-60)

tricam, quarum utraque in III terminis constans: minimo, medio, maximo, si fiat per angulos sive in directum, tali diligentia ponenda est, ut nullus ex alienis terminos eorum possit interrepere. Aliter non titulantur victorie.

⟨22⟩ Ex pari propositione binarius in adversa parte III lapides tollit: XII in VI^{to} campo, XVI in VIII°, XXVIII in X^{o}IIII. Quaternarius tollit V: XII in tertio, XVI in IIII^{to}, X⟨X⟩VIII in VII°, XXXVI in nono, LVI in XIIII°. Senarius IIII^{or} tollit: XII in proximo, XXX in V^{o}, XXXVI in VI^{to}, LXVI in XI°. Octonarius IIII: XVI in secundo, LVI in VII°, LXIIII et piramidem cum suis in VIII°. VIIII. tollit V: duos novenarios adverse partis in III° secundum cursum, XXXVI in IIII^{to}, LXXXI in nono, XC in X^{o}. Quindecim III: XXX in secundo, XC in VI°, C et XX in VIII°. Sedecim III: in adversa parte XVI in secundo, LXIIII et py⟨ramidem⟩ cum suis in IIII°. Viginti tollit II: C in V^{to}, CXX in VI°. Viginti quinque tollit II: C in IIII^{to}, CCXXV in nono. XLV tollit II: XC in secundo, CCXXV in quinto.

⟨23⟩ | Acies parium numerorum et pedites: *11va*

II	IIII	VI	VIII
IIII	XVI	XXXVI	LXIIII

Com⟨i⟩tes parium numerorum sunt hi:

VI	XX	XLII	LXXII
VIIII	XXV	XLVIIII	⟨LXXXI⟩

Exercitus parium pedites:

II	IIII	VI	VIII
IIII	XVI	XXXVI	LXIIII

comites:

VI	XX	XLII	LXXII
IX	XXV	XLVIIII	LXXXI

duces:

		rex	
XV	XLV	XCI	CLIII
XXV	LXXXI	CLXVIII⟨I⟩	CCLXXXIX

Exercitus imparium pedites:

III	V	VII	VIIII
IX	XXV	XLVIIII	LXXXI

322 XXVIII] xxiiii **L** *324* Senarius] Senerius **L** *325* LXIIII *corr.* **L** *ex* lxxxiiii *(?)* *329 post* IIII° *add. et del.* **L** lxxx in v^{to} *336 ante* XX *scr. et del.* **L** viiii *342* XLII] xiii **L** *et corr. in* xiiii

318–320 quarum ... victorie: Regensb. Anon. 11 (388,61-64) *321–331* Ex ... quinto: Regensb. Anon. 21-23 (392,146-394,172); Fränk. Komp. 14f. (407,69-408,81); Bayer. Komp. 12 (419,73-106) *332–357*: Kap. 23: siehe Liste in Kap. 2

⟨comites:⟩

XII	XXX	LVI	XC
XVI	XXXVI	LXIIII	C

⟨duces:⟩

		rex	
XXVIII	LXVI	CXX	CXC
XLIX	CXXI	CCXXV	CCCLXI

⟨24⟩ | Armonia interpretatur adunatio ab ar, quod est ad, et monos, quod *11v* est unum. Est enim armonia adunatio plurium vocum et differentium a se invicem natura.

Calculus dicitur lapis et ponitur pro numero. Antiqui enim lapillis numerare solebant.

Quemadmodum triangulus principium est planorum numerorum, ita pyramis primus est numerus cubicorum. Py⟨ramis⟩ autem alia perfecta, alia imperfecta sive curta; curtarum alia monocurta, alia bicurta, tricurta, tetracurta. Sive ergo py⟨ramis⟩ sit trigona sive tetragona, si aufertur primum latus, dicitur monocurta, si II, dicitur bi⟨curta⟩, si auferuntur tria latera, tricurta. Et quotcumque fuerint trigoni vel tetragoni vel pentagoni in py⟨ramide⟩ coniugati, semper ultimus et latissimus coniugatorum erit basis, in qua residet pyramis. Et sicut ex duobus numeris linearibus superficies trigone vel tetragone vel plurium angulorum fiunt, ita ex II trigonis vel tetragonis vel pentago⟨nis⟩ vel plurium angulorum pyramis. Et sicut superficialium numerorum quidam latiores, quidam strictiores aliis pro numeri quantitate, ita piramidum alia maior, alia minor. Nam que VIII triangulis constat, maior est ea, que IIII vel V. Et cum unitas quilibet numerus sit potestate, erit primus trigonus potentia et primus quadratorum potentia, alii autem omnes ab ipso actu trigoni vel tetragoni sunt. Et quotcumque trigoni sunt in py⟨ramide⟩ vel tetragoni vel p⟨entagoni⟩ vel ex⟨agoni⟩, tot costas vel latera pyramis habet, et secundum decrementum numerorum extenuatur py⟨ramis⟩ usque ad unitatem, et tot unitates habebit per latera ipsa pyramis in quovis latere, quot numeros habet in sui aggregatione. Latera autem dico seriem unitatum in extremitatibus usque ad summum. Et in singulis costis ordinatim ponuntur unitates secundum quantitatem pyramidis. Est autem monadica prima

368 quotcumque] quodcumque **L** *376* trigonus] trignus *(?)* **L** *377* quotcumque] quodcumque **L** *381* quot] quod **L**

358–360 Armonia ... natura: Boeth. Arithm. II 33 (126,16f.); **S**, f.29r: *Armonia est proprie multarum adunatio vocum in unam consonantiam* *361* Calculus ... numero: Boeth. Arithm. II 33 (127,27); Fortolf I 12 (440,289) *361–362* Antiqui ... solebant: Isid. Etymol. X 43 *363–364* Quemadmodum ... cubicorum: Boeth. Arithm. II 21 (104,25-105,2) *364–367* Py⟨ramis⟩ ... tricurta: Boeth. Arithm. II 24 (109,24-111,6) *368–370* Et ... pyramis: Boeth. Arithm. II 22 (106,3-11) *370–383* Et ... pyramidis: Boeth. Arithm. II 23 (106,14-109,22) *383–384* Est ... ternarius: Asilo 13 (334,60f.); Hermann 11 (339,59f.); Regensb. Anon. 15 (389,86f.)

superficies, scilicet trigona ternarius. Cubus autem primus VIII habet [viii.] angulos, XII latera, VI superficies, ut in tessara perlucide videri potest.

384–385 Cubus ... superficies: Boeth. Arithm. II 49 (158,24f.); Fortolf II 12 (462,831f.)
385 tessara: Fortolf I 11 (439,265)

Literatur

[Bergmann 1985] Bergmann, Werner: *Innovationen im Quadrivium des 10. und 11. Jahrhunderts. Studien zur Einführung von Astrolab und Abakus im lateinischen Mittelalter.* Stuttgart 1985 (Sudhoffs Archiv, Beiheft 26).

[Borst 1986] Borst, Arno: *Das mittelalterliche Zahlenkampfspiel.* Heidelberg 1986 (Supplemente zu den Sitzungsberichten der Heidelberger Akademie der Wissenschaften, Philosophisch-historische Klasse, Band 5, Jahrgang 1986).

[Borst 1990] Borst, Arno: *Rithmimachie und Musiktheorie*, in: Geschichte der Musiktheorie, Bd. 3: Rezeption des antiken Fachs im Mittelalter, Darmstadt 1990, S. 253-286.

[Bubnov 1899] Bubnov, Nicolaus: *Gerberti postea Silvestri II papae Opera Mathematica (972-1003).* Berlin 1899.

[Catalogue 1875] *Catalogue of Additions to the Manuscripts in the British Museum in the Years MDCCCLIV - MDCCCLX.* London 1875.

[Chicco 1977] Chicco, Adriano: *La rithmomachia*, in: Bonus Socius. Bijdragen tot de cultuurgeschiedenis van het schaakspel en andere bordspelen. Jubileumuitgave voor Meindert Niemeijer ter gelegenheid van zijn 75ste verjaardag, 's-Gravenhage 1977, S. 81-101.

[Drecker 1931] Drecker, J.: *Hermannus Contractus Über das Astrolab*, in: Isis 16 (1931), 200-219.

[Friedlein 1867] *Anicii Manlii Torquati Severini Boetii de institutione arithmetica libri duo, de institutione musica libri quinque. Accedit geometria quae fertur Boetii*, ed. Gottfried Friedlein. Leipzig 1867.

[Libri 1859] *Catalogue of the extraordinary collection of splendid manuscripts ... formed by M. Guglielmo Libri.* London 1859.

[Maß 1989] *Maß, Zahl und Gewicht. Mathematik als Schlüssel zu Weltverständnis und Weltbeherrschung*, hg. von Menso Folkerts, Eberhard Knobloch und Karin Reich. Weinheim 1989.

[Migne 1853] Migne: *Patrologia Latina* 143. 1853.

[Peiper 1880] Peiper, R.: *Fortolfi Rythmimachia*, in: Abhandlungen zur Geschichte der Mathematik 3 (1880), 169-227.

[Pez 1729] Pez, Bernhard: *Thesaurus anecdotorum novissimus*, tom. VI, pars II. Augsburg / Graz 1729.

[Plechl 1962] Plechl, Helmut: *Die Tegernseer Handschrift Clm 19411, Beschreibung und Inhalt*, in: Deutsches Archiv für Erforschung des Mittelalters 18 (1962), 418-501.

[Verfasserlexikon 1990] *Die deutsche Literatur des Mittelalters. Verfasserlexikon*, 2. Auflage, Bd. 8. Berlin / New York 1990.

[van de Vyver 1931] van de Vyver, André: *Les premières traductions latines (X^e^ - XIe s.) de traités arabes sur l'astrolabe*, in: 1er Congrès International de Géographie Historique, 2, Brüssel 1931, S. 266-290.

[Wappler 1892] Wappler, E.: *Bemerkungen zur Rythmomachie*, in: Zeitschrift für Mathematik und Physik, historisch-litterarische Abtheilung, 37 (1892), 1-17.

[Zimmermann 1946] Zimmermann, Alfons M.: *Die Familia s. Quirini im Mittelalter*, in: Studien und Mitteilungen zur Geschichte des Benediktiner-Ordens 60 (1946), 190-217.

The Arabic version of Euclid's *On Divisions*

Jan P. Hogendijk

1. Introduction

The Greek text of Euclid's work *On Divisions* (περὶ διαιρέσεων) is lost, and the only Greek references to this work are to be found in Proclus' commentary on the *Elements* of Euclid. Proclus tells us little more than that *On Divisions* is a work by Euclid which included the division of figures into "like and unlike" parts.[1] In the *Fihrist*, the 10th-century Arabic bibliographer Ibn al-Nadīm mentions Euclid's *Book of the division, corrected by Thābit* (= Thābit ibn Qurra, 836-901),[2] but as far as is known, no manuscripts of the Arabic translation or the revision by Thābit have survived. In 1851 F. Woepcke published[3] a French translation of an abstract of *On Divisions* which had been composed by the 10th-century Persian geometer Aḥmad ibn Muḥammad ibn ʿAbdaljalīl al-Sijzī.[4] Al-Sijzī rendered the first sentences of all propositions in *On Divisions* but the proofs of only four of them. He stated that he omitted the other proofs because he found them too easy. In an anonymous 10-th century Arabic text called *Various geometrical problems*, C. Schoy discovered a proposition of Euclid's *On Divisions* together with a geometrical construction by the anonymous translator of the work. Schoy published German translations of these passages in 1926.[5]

The modern studies of *On Divisions* (such as Archibald's reconstruction of 1915)[6] are based on Woepcke's French translation and on possible traces of

---

[1] Cf. J.L. Heiberg, *Litteraturgeschichtliche Studien über Euklid.* Leipzig (Teubner) 1882, pp. 36-37.

[2] Ibn al-Nadīm, *Kitāb al-Fihrist.* Mit Anmerkungen herausgegeben von Gustav Flügel, Leipzig 1871, vol. 1, p. 266 lines 15-16.

[3] F. Woepcke, Notice sur des traductions arabes de deux ouvrages perdus d'Euclide, *Journal Asiatique*, 4ᵉ série, 18 (1851), pp. 217-247. Reprinted in F. Woepcke. *Études sur les mathématiques Arabo-Islamiques*, ed. F. Sezgin, Frankfurt: Institut für Geschichte der arabisch-islamischen Wissenschaften, 1986, vol.1.

[4] See F. Sezgin, *Geschichte des arabischen Schrifttums*, Band 5, Mathematik bis ca. 430 H, Leiden: Brill, 1974, pp. 329-334. This work will henceforth be abbreviated as GAS V.

[5] C. Schoy, Graeco-Arabische Studien nach mathematischen Handschriften der Vicekö-niglichen Bibliothek zu Kairo. *Isis* 8 (1926), 21-40, see especially pp. 32-34. Reprinted in C. Schoy, *Beiträge zur arabisch-islamischen Mathematik und astronomie*, ed. F. Sezgin, Frankfurt: Institut für Geschichte der arabisch-islamischen Wissenschaften, 1988, vol. 2, pp. 589-591.

[6] R.C. Archibald, *Euclid's Book on division of Figures. With a restoration based on Woepcke's text and on the Practica Geometriae of Leonardo Pisano.* Cambridge (at the University Press) 1915.

On Divisions in medieval Latin texts (which do not contain explicit references to Euclid's work). No further attention has been paid to the Arabic texts themselves. Thus in Volume 8 of Euclid's *Opera Omnia*,[7] Heiberg and Menge printed Woepcke's translation of Euclid's *On Divisions* but not the Arabic text on which the translation is based.

The main purpose of this paper is to make available the Arabic texts translated by Woepcke and Schoy as well as literal English translations. Text **A** below is the abstract of Euclid's *On Divisions* by al-Sijzī, which was translated by Woepcke. Text **A** exists in a unique manuscript: Paris, Bibliothèque Nationale, Fonds Arabe, 2457, 53b-55b.[8] The manuscript was believed (by Woepcke and others) to be an autograph by al-Sijzī, written in 358 H./A.D. 968-9. However, on f. 216 there is a table of contents in the same hand, dated 657 H./A.D.1259, so the manuscript must be a copy of the manuscript written by al-Sijzī.[9]

Text **B** is part of the anonymous compilation entitled *Various geometrical problems*, which probably dates from the late 10th century.[10] The treatise exists in a unique manuscript in Cairo, Dār al-Kutub Muṣṭafā Fāḍil Riyāḍa 41m, 165b-170b,[11] copied in 1153 H./ 1740 A.D. Text **B** consists of props. 5 and 6 of the treatise. Proposition 5, which is said to be "from the Book of Euclid on division", deals with the construction of a segment of a circle contained between two parallel lines and equal in area to a given sector. The proposition is closely related to but not identical to al-Sijzī's proposition 28, on the construction of a segment of a circle contained between two parallel lines and equal in area to a given part of the circle. Prop. 6 is a trisection of the angle, which the text attributes to "the translator of it" (i.e. of *On Divisions*).

Euclid's *On Divisions* had a considerable influence on the Arabic geometrical tradition. For example, in Chapter 8, *On the division of triangles*, and Chapter 9, *On the division of quadrilaterals*, of the *Book on the geometrical constructions necessary for the craftsman* (GAS V, 324 no. 2), the 10-th century geometer Abu l-Wafā' al-Būzajānī included many solutions of

---

[7]Euclides, *Opera Omnia*, ed. J.L. Heiberg et H. Menge. Leipzig (Teubner) 1883-1916. 8 vols.

[8]See M. De Slane, *Catalogue des manuscrits arabes (de la Bibliothèque Nationale)*, Paris 1893-1895, p. 431.

[9]This was first noticed by F. Sezgin in *Geschichte des arabischen Schrifttums*, Band VI, Leiden 1978, p. 188 no. 1.

[10]See GAS V, op.cit., p. 396.

[11]On the manuscript see H. Suter, Der V. Band des Katalogs der arabischen Bücher der vicekōniglichen Bibliothek in Kairo, *Zeitschrift für Mathematik und Physik, Historisch-litterarische Abteilung* 38 (1893), p. 24; reprinted in H. Suter, *Beiträge zur Geschichte der Mathematik und Astronomie im Islam*, Frankfurt: Institut für Geschichte der arabisch-islamischen Wissenschaften, 1986, vol. 1; D.A. King, *A catalogue of the scientific manuscripts in the Egyptian National Library* (in Arabic), Part 1. Cairo 1981 (published by the General Egyptian Book Organization), p. 447; D.A. King. *A survey of the scientific manuscripts in the Egyptian National Library*, Winona Lake: Eisenbrauns, 1986, p. 52 no. B 98.

problems in *On Divisions* (corresponding to *On Divisions* props. 1-9, 11-16 and 27-28). Text **C** below is one (unnumbered) proposition from Chapter 9. In this proposition Abu l-Wafāʾ constructs a segment of a circle contained between two parallel lines and equal to one-third of the circle. The wording of the solution is extremely close to that of al-Sijzī's abstract, prop. 28.[12] I have included text **C** here because it sheds light on the relationship between texts **A** and **B**.

In my edition of text **C** I have used the manuscripts: **F** = Istanbul, Aya Sofya 2753, f. 44:2-7, and **L** = Uppsala Tornberg 324, Vet. 27, f. 39b:6-16. Both manuscripts are undated. In the Uppsala manuscript the text is incorrectly attributed to al-Fārābī. Only the first and last page of the manuscript (f. 1b, 60a) belong to a text by al-Fārābī,[13] and the rest is a large continuous fragment of the treatise by Abu l-Wafāʾ, which begins abruptly on f. 2a in the middle of a sentence. Text **C** can also be consulted in the edition by Ṣ. A. al-ʿAlī,[14], in the Russian translation by Krasnova[15] (based on **F**), and in al-Fārābī's *Mathematical works* [16] (based on **L**). The relationship with Euclid's *On Divisions* is not mentioned in these three publications.

I now comment on the relationship between the three texts **A**, **B** and **C**.

The following terminological argument shows that al-Sijzī's abstract (text **A**) was probably based on the revision of *On Divisions* by Thābit ibn Qurra. The rectangle contained by two segments AB and GD, which for Euclid is a geometrical concept, is rendered in al-Sijzī's abstract in an arithmetical way as *alladhī yakūnu min AB fī GD*, meaning "that which is (the result) of (the multiplication) AB times GD". This expression is unusual in Arabic geometrical texts, but Thābit ibn Qurra used similar expressions in his translations of Book V-VII of the Conics, namely *alladhī yakūnu min ḍarb AB fī GD* "that which is (the result) of the multiplication (of) AB times GD".[17]

The close resemblance between text **C** and al-Sijzī's abstract prop. 28 shows that Abu'l-Wafāʾ must have had a version of *On Divisions* in front of him when he wrote Chapters 8 and 9 of his own work. Because

[12] Cf. F. Woepcke. Analyse et extrait d'un recueil de constructions géométriques par Aboul Wafā. Journal Asiatique, 5ᵉ série, 5 (1855), pp. 340-341. Reprinted in Woepcke, Études, op. cit. vol. 1.

[13] See GAS V, op. cit. p. 296 no. 1.

[14] See Ṣ.A. al-ʿAlī, *Ma yaḥtāju ilayhi al-ṣāniʿ min ʿilm al-handasa li-Abi l-Wafāʾ... al-Būzjānī*, Bagdad 1979, p. 126.

[15] S.A. Krasnova: Abu-l-Vafa al-Buzdzhani, Kniga o tom, chto neobchodimo remesleniku iz geometricheskikh postroennii (translation with introduction and commentary) in: *Fiziko-Matematicheskie Nauki v stranakh vostoka, sbornik statei i publikatsii*, vypusk I (IV), Moscow (Nauka) 1966, pp. 42-159, see esp. p. 105.

[16] Al-Farabi, *Matematicheskie Traktaty*, Per. ... A. Kubesov, B.A. Rozenfeld, Alma Ata 1971, pp. 168-169.

[17] Cf. G.J. Toomer, *Apollonius' Conics Books V to VII. The Arabic translation of the lost Greek original in the version of the Banū Mūsā*, New York: Springer 1990. Sources in the History of Mathematics and Physical Sciences 9, e.g. p. 5 line 17, p. 11 line 1, index p. 870, ḍarb.

Abu'l-Wafā' presents problems with solutions, he probably had the whole text of *On Divisions* and not the abstract of al-Sijzī, in which most constructions and proofs are omitted. Because al-Sijzī used the revision of *On Divisions* by Thābit ibn Qurra, this must also be the version which Abu'l-Wafā' used and copied. The coincidences between his Text **C** and al-Sijzī's text **A** suggest that al-Sijzī did not change the wording of the parts of *On Divisions* which he put together in his abstract.

There are considerable differences between prop. 28 of Text **A** and Text **C** on one hand and prop. 5 of Text **B** on the other hand. One can explain these differences by the assumption that Text **B** is based not on Thābit's revision but on the original translation of *On Divisions.* If this is true, Prop. 5 in text **B** was probably deleted by Thābit ibn Qurra in the course of his revision of the text. The trisection in prop. 6 in text **B** may have undergone the same treatment. This trisection is based on a *neusis*, but Thābit ibn Qurra preferred to trisect the angle by conic sections.[18]

By studying the Arabic tradition, and in particular the work of Abu'l-Wafā', one can improve the reconstruction of *On Divisions* published by Archibald in 1915. In his reconstruction Archibald took the problems from Al-Sijzī's abstract and the solutions (constructions and proofs) from Leonardo Fibonacci's *Practica Geometria*, written in 1220. Most constructions can also be taken from Abu l-Wafā'; these constructions are probably closer to the original *On Divisions* than those of Leonardo.

In the following editions and translations, passages in angular brackets are additions made by me in my attempt to restore the text, and my own explanatory additions are in parentheses or square brackets. The critical apparatus for the text is printed at the bottom of each page.

[18]Thābit's trisection can be consulted in W.R. Knorr, Textual studies in ancient and medieval geometry, Boston 1989, pp. 277-281, 361-366.

Manuscripts:

Text A: MS. Paris, Bibliothèque Nationale, Fonds Arabe 2457, 53b-55b.

Text B: MS. Cairo, Dār al-Kutub Muṣṭafā Fāḍil Riyāḍa 41m, 167b-168a.

Text C: **F** = MS. Aya Sofya 2753, f. 44;
L = MS. Uppsala, Tornberg 324, f. 39.

Translation of text A

In the name of God, the Merciful, the Compassionate.

9[19] The Book of Euclid on the division.

1. We wish to demonstrate how we bisect a known triangle by a line parallel to its base.

2. We wish to demonstrate how we divide a known triangle into three equal parts by two lines parallel to its base.

3. We wish to bisect a known triangle by a straight line drawn from a known point which is on one of its sides.

4. We wish to bisect a known trapezium by a line parallel to its base. And[20] the known trapezium can be divided into three equal parts just as we divided the triangle, by a construction corresponding to that construction.

< 5 > We wish to bisect a parallelogram by a straight line drawn from a known point which is on one of its sides.

< 6 > We wish to demonstrate how we cut off from a known parallelogram an assumed part, any part we wish, by a straight line drawn from a known point which is on one of its sides.

< 7 > We wish to demonstrate how we bisect a known trapezium by a straight line drawn from a known point which is on the line of the highest[21] (part) of the trapezium.

8. We wish to cut off from a known trapezium an assumed part by a straight line drawn from a known point which is on the < line of the > highest (part) of the trapezium.

9. We wish to demonstrate how we bisect a parallelogram by a straight line drawn from a known point outside it.

10. We wish to demonstrate how we cut off from a parallelogram an assumed part by a straight line drawn from a known point outside it.

11. We wish to demonstrate how we bisect a known trapezium by a straight line drawn from a point beyond the highest (part) of that trapezium. It is necessary that the point is not beyond the place of the meeting of the two sides.[22]

12. We wish to demonstrate how we cut off from a known trapezium an assumed part by a straight line drawn from a known point beyond the line

[19] This is the ninth text in the manuscript Paris, Bibliothèque Nationale 2457.

[20] In the margin of the manuscript, this corollary is numbered "5", the next two propositions (my propositions < 5 > and < 6 >) are numbered "6" and "7", and the subsequent proposition (my proposition < 7 >) is not numbered. I assume that the corollary was numbered "5" by mistake, and that the scribe made up for his mistake by not assigning a number to my proposition < 7 >.

[21] Meaning: on the longer side of the trapezium.

[22] I do not understand this condition. The problem is solvable for all points outside the trapezium. Woepcke's interpretation (pp. 34-35) does not convince me.

Text A

بسم الله الرحمن الرحيم . ط كتاب اقليدس فى القسمة .

ا نريد ان نبيّن كيف نقسم مثلّثا معلوما بنصفين بخطّ يوازي قاعدته .

ب نريد ان نبيّن كيف نقسم مثلّثا معلوما بثلاثة اقسام متساوية بخطّين موازيين [1] لقاعدته .

ج نريد ان نقسم مثلّثا معلوما [2] بقسمين متساويين بخطّ مستقيم يخرج من نقطة معلومة تكون على احد اضلاعه .

د نريد ان نقسم منحرفا معلوما بقسمين متساويين بخطّ يوازي قاعدته

[3] وقد يقسم المنحرف المعلوم بثلاثة اقسام متساوية كما قسمنا المثلّث بعمل نظير ذلك العمل .

ه [4] نريد ان نقسم سطحا متوازى الاضلاع بقسمين متساويين بخطّ مستقيم يخرج من نقطة معلومة تكون على ضلع من اضلاعه .

و [5] نريد ان نبيّن كيف نفصل من سطح متوازي الاضلاع معلوم جزءا مفروضا اي جزء اردنا بخطّ مستقيم يخرج من نقطة معلومة تكون على ضلع من اضلاعه .

< ز > نريد ان نبيّن كيف نقسم منحرفا معلوما بقسمين متساويين بخطّ مستقيم يخرج من نقطة معلومة تكون على خطّ اعلى [6] المنحرف .

ح نريد ان نفصل من منحرف معلوم جزءا مفروضا بخطّ مستقيم يخرج من نقطة معلومة تكون على < خطّ > اعلى [7] المنحرف .

ط نريد ان نبيّن كيف نقسم سطحا متوازي الاضلاع بقسمين متساويين بخطّ مستقيم يخرج من نقطة معلومة خارجة عنه .

ي نريد ان نبيّن كيف نفصل من سطح متوازي الاضلاع جزءا مفروضا [54a] بخطّ مستقيم يخرج من نقطة معلومة خارجة عنه .

يا نريد ان نبيّن كيف نقسم منحرفا معلوما بقسمين متساويين بخطّ مستقيم يخرج من نقطة خارجة عن خطّ اعلى [7] ذلك المنحرف ويحتاج ان لا يكون النقطة خارجة [8] عن موضع ملتقى الضلعين .

يب نريد ان نبيّن كيف نفصل من منحرف معلوم جزءا مفروضا بخطّ مستقيم يخرج

---

1 موازيين : متوازيين ، موازيان (crossed out) . 2 مثلّثا معلوما : مثلّث معلوم . 3 ه (in margin)
4 و . 5 ز . 6 خطّ اعلى : حطا على . 7 اعلا . 8 الخارجة .

of the highest (part) of that trapezium. It is necessary that the point is not beyond the place of the meeting of the two sides of the trapezium.[23]

13. We wish to demonstrate how we bisect a known quadrilateral by a straight line drawn from a known angle of it.

14. We wish to demonstrate how we cut off from a known quadrilateral some assumed part by a line drawn from a known angle of it.

15. We wish to demonstrate how we bisect a known quadrilateral by a straight line drawn from a known point which is on one of its sides.

16. We wish to demonstrate how we cut off from a known quadrilateral an assumed part by a straight line drawn from a known point which is on one of its sides.

17. (Figure 1) We wish to apply to a straight line a rectangle equal to the area contained by lines AB, AG, and deficient from the completion of the line by a square area.[24] Since we have proved what we said, if someone says : why was it so that it was not possible for us to apply to AB an area equal to $AB \cdot AG$,[25] deficient from its completion by a square area, and equal to the area $AE \cdot EB$, we say: that is not possible because AB is greater than BE and AG is greater than AE, so $BA \cdot AG$ is greater than $AE \cdot EB$. So if there is applied to AB a parallelogram equal to $AB \cdot AG$ it is (like) area $AZ \cdot ZB$.[26]

Figure 1

[23]See the preceding footnote.

[24]It would have been more correct to say: We wish to apply to ***the straight line AB*** a rectangle equal to the area contained by AB, AG and ***deficient from its completion*** by a square. The problem is mathematically equivalent to the following: to construct a point X on AB such that $AX \cdot XB = AB \cdot AG$. To explain the terminology, draw three perpendiculars AA', XX' and BB' to AB such that $AA' = XX' = BB' = XB$. Now rectangle $AXX'A'$ is equal to "the area contained by AB, AG" , $XBB'X'$ is a square, and $ABB'A'$ is a "complete" rectangle with side AB. The problem is of a well-known type in Greek geometry, the so-called ***application of areas***. See T.L. Heath, ***Euclid. The thirteen Books of the Elements***. Vol. I, pp. 342-345, Vol. II, pp. 257-259.

[25]The notation $AB \cdot AG$ will be used to translate Arabic expressions like: ***alladhī yakūnu min AB fī AG***, literally; "which is (the result) of AB times AG". The Arabic preposition ***fī*** is normally used in multiplications, so a multiplicative notation seems appropriate.

[26]Woepcke (p.36) assumed that the sentence is incomplete, but there is no good reason for this assumption. The text means that a point X such that $AX \cdot XB = AB \cdot AG$ cannot lie between A and G; hence it must lie between G and B. The text is not concerned with the necessary condition for the existence of point X.

من نقطة معلومة خارجة عن خطّ اعلى [9] ذلك المنحرف ويحتاج الّا يكون النقطة خارجة عن موضع ملتقى ضلعي المنحرف .

يجَ نريد ان نبيّن كيف نقسم شكلا ذا اربعة اضلاع معلوما بقسمين متساويين بخطّ مستقيم يخرج من زاوية معلومة منه .

يدَ نريد ان نبيّن كيف نفصل من ذى اربع اضلاع معلوم [10] جزءا ما مفروضا بخطّ يخرج من زاوية منه معلومة .

يهَ نريد ان نبيّن كيف نقسم شكلا ذا اربعة اضلاع معلوما [11] بقسمين متساويين بخطّ مستقيم يخرج من نقطة معلومة يكون على احد اضلاعه .

يوَ نريد ان نبيّن كيف نفصل من ذى اربع اضلاع معلوم جزءا مفروضا بخطّ مستقيم يخرج من نقطة معلومة تكون على احد اضلاعه .

يزَ نريد ان نضيف الى خطّ مستقيم سطحا قائم الزوايا مساويا للسطح الّذى يحيط به خطّا ابَ اجَ وينقص عن تمام الخطّ سطحا مربّعا . واذ قد بيّنّا ما قلنا فانّه ان قال قائل لم صار لا يمكننا ان نضيف الى ابَ سطحا مساويا للّذى يكون من ابَ فى اجَ ينقص عن تمامه سطحا مربّعا ويكون مثل سطح اهَ فى هبَ قلنا انّ ذلك لا يمكن لانّ ابَ اعظم من بهَ وانّ اجَ اعظم من اهَ فالّذى يكون من باَ فى اجَ اعظم من الّذى يكون من اهَ فى هبَ فاذا اضيف الى ابَ سطح متوازى الاضلاع مساو للّذى يكون من ابَ فى اجَ كان سطح ازَ فى زبَ .

< يحَ > نريد ان نبيّن كيف نقسم مثلّثا معلوما بقسمين متساويين بخطّ مستقيم يمرّ بنقطة معلومة داخل المثلّث . فليكن المثلّث المعلوم مثلّث ابجَ والنقطة الّتى فى داخله دَ ونريد ان نجيز على دَ خطّا مستقيما يقسم مثلّث ابجَ بقسمين متساويين . فنخرج من نقطة دَ خطّا موازيا لخطّ بجَ وهو دهَ . ونضيف الى دهَ سطحا مساويا لنصف الّذى يكون من ابَ فى بجَ وليكن طبَ فى هدَ . ونضيف [12] الى خطّ طبَ [13] سطحا متوازى الاضلاع مساويا [54b] للّذى يكون من بطَ فى بهَ ينقص عن تمامه سطحا مربّعا وليكن السطح المضاف سطح بحَ فى حطَ . ونصل خطّ دحَ ونخرجه الى زَ . فاقول انّه قد اخرج خطّ حدزَ [14] فقسم مثلّث ابجَ بقسمين متساويين وهما حبزَ حزجاَ .

برهان ذلك انّ الّذى يكون من طبَ فى بهَ مساو للّذي يكون من طحَ فى حبَ

---

9 اعلا . 10 معلوما . 11 معلوم . 12 الى خطّ لبَ سطحا مساويا لنصف الّذى يكون من ابَ فى بجَ وليكن سطح طبَ فى هدَ ونضيف (added in MS.) . 13 كبَ . 14 حبزَ .

< 18 > (Figure 2) We wish to demonstrate how we bisect a known triangle by a straight line which passes through a known point inside the triangle. Thus let the known triangle be triangle ABG, and (let) the point which is in its interior (be) D. We wish to let pass through D a straight line which bisects triangle ABG. Thus we draw from point D a line parallel to line BG, namely DE. We apply to DE an area equal to half $AB \cdot BG$, let it be $TB \cdot ED$. We apply to line TB a parallelogram equal to $BT \cdot BE$, deficient from its completion by a square area, let the applied area be area $BH \cdot HT$.[27] We join line DH and we extend it towards Z. I say that line DHZ has been drawn such as to divide triangle ABG into two equal parts, namely HBZ, $HZGA$.

Proof of this: $TB \cdot BE$ is equal to $TH \cdot HB$, so the ratio of BT to TH is equal to the ratio of HB to BE. Separando, the ratio of TB to BH is also equal to the ratio of BH to HE. But the ratio of BH to HE is equal to the ratio of BZ to ED. Thus the ratio of TB to BH is equal to the ratio of BZ to ED. So $TB \cdot ED$ is equal to $BH \cdot BZ$. But $TB \cdot ED$ is equal to half of $AB \cdot BG$, and the ratio of $BH \cdot BZ$ to $AB \cdot BG$ is equal to the ratio of triangle HBZ to triangle ABG, because the angles at point B are common. So triangle HBZ is half of triangle ABG. So triangle ABG has been divided into two equal parts, namely BHZ, $AHZG$. If we apply to TB a parallelogram equal to $TB \cdot BE$, deficient from its completion by a square, and if this area is $AB \cdot AT$,[28] then if we join AD and extend it towards K, we prove in the same way that triangle ABK is half of triangle ABD. That is what we wanted to demonstrate.

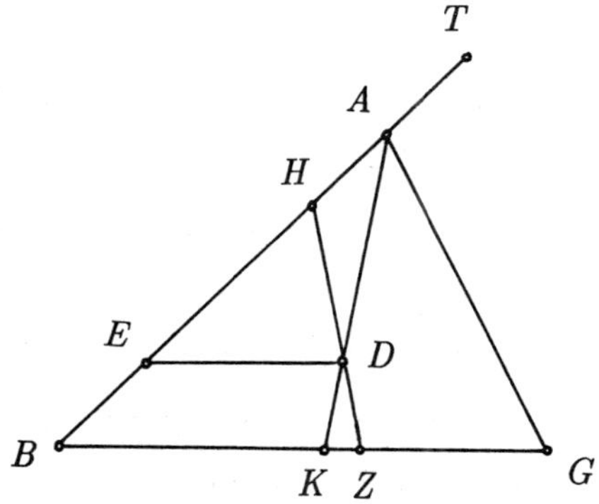

Figure 2

[27]The text does not discuss the possibility of this construction. Point H is between E and T as a consequence of the end of prop. 17.

[28]This means that point H coincides with A.

فيكون [15] نسبة بط الى طح كنسبة حب الى به [16] . واذا فصلنا ايضا يكون نسبة طب الى بح كنسبة بح الى حه . ولكن نسبة بح الى حه كنسبة بز الى هد . فنسبة طب الى بح كنسبة بز الى هد . فالّذى يكون من طب فى هد مساو للّذى يكون من بح فى بز [17] . ولكن الّذى يكون من طب فى هد هو مثل نصف الّذى يكون من اب فى بج ، ونسبة ما يكون من بح فى بز الى الّذى يكون من اب فى بج كنسبة مثلّث حبز الى مثلّث ابج ، وذلك انّ الزوايا الّتى [18] عند ب مشتركة ، فيكون مثلّث حبز نصف مثلّث ابج . فقد قسم مثلّث ابج بقسمين متساويين وهما بحز احزج . فان كنّا اذا اضفنا الى طب سطحا متوازى الاضلاع يساوى الّذى يكون من طب فى به [19] ينقص عن تمامه مربّعا كان ذلك السطح هو الكائن من اب فى اط ، فاذا وصلنا خطّ اد ونفذناه الى ك بيّنّا بمثل هذه السبيل انّ مثلّث ابك هو نصف مثلّث ابد وذلك ما اردنا ان نبيّن .

< يط > نريد ان نبيّن كيف نفصل من مثلّث معلوم جزءا مفروضا بخطّ مستقيم يجوز على نقطة معلومة داخل المثلّث . فليكن المثلّث المعلوم مثلّث ابج والنقطة المعلومة الّتى داخله د ونريد ان نجيز على نقطة د خطّا مستقيما يفصل من مثلّث ابج جزءا مفرودا فليكن الجزء المفرود الثلث . ونخرج من نقطة د خطّا موازيا لخطّ بج وهو ده . ونضيف الى ده سطحا قائم الزوايا مساويا لثلث ما يكون من اب فى بج وليكن ذلك السطح بز فى هد . ونضيف الى زب [20] سطحا مساويا للّذى يكون من زب فى به ينقص عن تمامه سطحا مربّعا وليكن السطح المضاف سطح بح فى حز . ونصل خطّ حد ونخرجه الى ط . وبمثل هذا المسلك نبرهن انّ مثلّث حطب ثلث مثلّث ابج وبمثل هذا العمل نقسمه بايّ جزء اردنا وذلك ما اردنا ان نبيّن .

< ك > وليكن اربعة خطوط ا ب ج د وليكن ا فى د اعظم من ب فى ج فاقول انّ نسبة ا الى ب اعظم من نسبة ج الى د .

< كا > وايضا ان كان ا فى د اقلّ من ب فى ج فاقول انّ نسبة ا الى ب اقلّ من نسبة ج الى د .

< كب > وان كان خطّان مستقيمان عليهما اب ده [55a] ويكون نسبة اب الى بج اعظم من نسبة ده الى هز فاقول انّه يكون على جهة التفصيل نسبة اج الى جب اعظم من نسبة دز الى زه .

< كج > وعلى هذه الصورة بعينها [21] ليكن نسبة اج الى جب اعظم من نسبة دز الى زه فاقول انّه يكون على جهة التركيب نسبة اب الى بج اعظم من نسبة ده الى هز .

---

15 ويكون . 16 بد . 17 بز : هز . 18 الذى . 19 به : بد . 20 زب : ده . 21 بعينه .

< 19 > (Figure 3) We wish to demonstrate how we cut off from a known triangle an assumed part by a straight line which passes through a known point inside the triangle. Thus let the known triangle be triangle ABG and (let) the known point which is inside it (be) D. We wish to let pass through point D a straight line which cuts off from triangle ABG an assumed part, and let the assumed part be one-third. We draw from point D a line parallel to line BG, namely DE. We apply to DE a rectangle equal to one-third of $AB \cdot BG$, let this area be $BZ \cdot ED$. We apply to ZB[29] an area equal to $ZB \cdot BE$, deficient from its completion by a square area, let the applied area be area $BH \cdot HZ$.[30] We join line HD and extend it towards T. In a similar way we prove that triangle HTB is one-third of triangle ABG. By a similar construction we divide it into any part we wish. That is what we wished to demonstrate.

< 20 > Let there be four lines A, B, G, D and let $A \cdot D$ be greater than $B \cdot G$. I say that the ratio of A to B is greater than the ratio of G to D.

< 21 > Again, if $A \cdot D$ is less than $B \cdot G$, I say that the ratio of A to B is less than the ratio of G to D.

< 22 > (Figure 4)[31] If there are two straight lines AB, DE and the ratio of AB to BG is greater than the ratio of DE to EZ, I say that, separando, the ratio of AG to GB is greater than the ratio of DZ to ZE.

< 23 > In exactly the same figure, let the ratio of AG to GB be greater than the ratio of DZ to ZE. I say that, componendo, the ratio of AB to BG is greater than the ratio of DE to EZ.

< 24 > (Figure 5) Again, let us make the ratio of AB to BG less than the ratio of DE to EZ. Then, separando, the ratio of AG to GB is less than the ratio of DZ to ZE.

< 25 > We wish to demonstrate how we bisect a known triangle by a straight line drawn through a known point outside the triangle.

< 26 > We wish to demonstrate how we cut off from a known triangle an assumed part by a straight line drawn from a point outside the triangle.

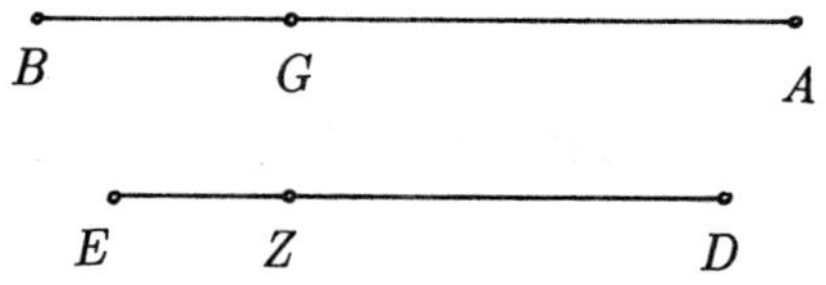

Figure 5

< 27 > (Figure 6) We wish to bisect a known figure contained by an arc and two straight lines which contain an angle. Thus let there be a known

---

29 Instead of ZB the manuscript has DE by scribal error.

30 The text does not say for which points D this construction is possible. Point H is between E and Z as a consequence of the remark at the end of proposition 17.

31 Figure 4 in the manuscript contains points T, H and K, which were probably used in the proofs of props. 22 and 23.

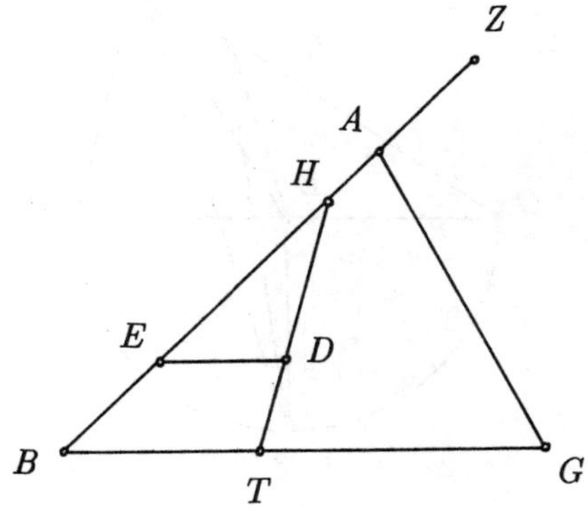

Figure 3

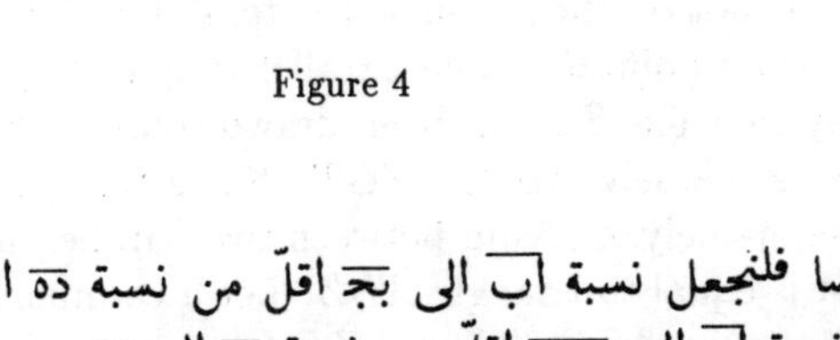

Figure 4

< كد > وايضا فلنجعل نسبة اب الى بج اقلّ من نسبة ده الى هز فانّه يكون على جهة التفصيل نسبة اج الى جب اقلّ من نسبة دز الى زه .

< كه > نريد ان نبيّن كيف نقسم مثلّثا معلوما بقسمين متساويين بخطّ مستقيم يخرج من نقطة معلومة خارجة عن المثلّث .

< كو > نريد ان نبيّن كيف نفصل من مثلّث معلوم جزءا مفروضا بخطّ مستقيم يخرج من نقطة خارجة عن المثلّث .

< كز > نريد ان نقسم شكلا معلوما يحيط به قوس وخطّان مستقيمان يحيطان بزاوية بقسمين متساويين .

فليكن شكل معلوم عليه ابج يحيط به قوس بج وخطّا با اج المستقيمان المحيطان بزاوية باج ونريد ان نخرج خطّا مستقيما يقسم شكل ابج بقسمين متساويين .

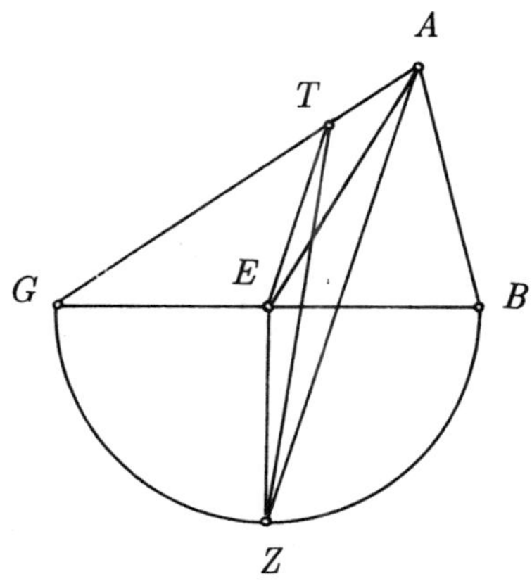

Figure 6

figure ABG, contained by arc BG and the straight lines BA, AG, containing angle BAG. We wish to draw a straight line which bisects figure ABG. Thus we join line BG, and we bisect it at point E. We draw from point E a line perpendicular to line BG, namely EZ. We join the straight line AE. Since line BE is equal to line EG, area BZE is equal to area EZG. But triangle ABE is equal to triangle AEG, so figure $ABZE$ turns out to be equal to figure $ZGAE$.

If line AE is on the rectilinear extension of line EZ, figure ABG has been divided into two equal parts, namely $ABZE$, $GAEZ$.

If line AE is not on the rectilinear extension of line ZE, we join line AZ and we draw from point E a line parallel to line AZ, namely ET. We join line TZ. I say that line TZ has been drawn so as to divide figure ABG into two equal parts, namely $ABZT$, ZGT. Since triangles TZA, EZA are on the same base, namely AZ, and between two parallel lines, namely AZ, TE, triangle ZTA is equal to triangle AEZ. Let a common addition be made to them, namely figure AZB. Then $TZBA$ turns out to be equal to $ABZE$, which is half of figure ABG. Thus the straight line TZ has been drawn so as to divide $BZGA$ into two equal parts, namely $ABZT$, TZG. That is what we wished to demonstrate.

< 28 > (Figure 7) We wish to draw in a known circle two parallel lines which cut off from the circle some assumed part. Thus let us make the part one-third and the circle ABG, and we wish (to do) what we have said. We make the centre of circle ABG <point D, and we make the> side of the triangle inscribed in this circle, namely AG, and we draw lines AD, DG. We let pass through point D a line parallel to line AG, namely DB. We join line GB. We bisect arc AG at point E. We draw from point E a line parallel to line BG, namely EZ. We draw line AB. I say that the parallel lines EZ, GB have been drawn so as to cut off from circle ABG one-third of it, namely figure $ZBGE$.

فنصل خطّ بج ونقسمه بنصفين على نقطة ه ونخرج من نقطة ه خطّا قائما على خطّ بج على زوايا قائمة وهو هز . ونصل خطّ اه المستقيم .

فلانّ خطّ به مساو لخطّ هج يكون سطح بزه مساويا لسطح هزج [22] . ومثلّث ابه مساو لمثلّث اهج فيصير شكل ابزه مساويا لشكل زجاه . فان كان خطّ اه على استقامة خطّ هز [23] فقد قسم شكل ابج [24] بقسمين متساويين وهما ابزه جاهز .

وان لم يكن خطّ اه على استقامة خطّ زه فانّا نصل خطّ از ونخرج من نقطة ه خطّا موازيا لخطّ از وهو هط فنصل خطّ طز فاقول انّه قد اخرج خطّ طز فقسم شكل ابج بقسمين متساويين وهما ابزط زجط . فلانّ مثلّثا طزا هزا على قاعدة واحدة وهي از وفيما بين خطّين متوازيين وهما از طه [25] يكون مثلّث زطا مساو لمثلّث اهز . وليزاد عليهما زيادة مشتركة وهي شكل ازب فيصير طزبا مثل ابزه وهو نصف شكل ابج . فقد اخرج خطّ طز المستقيم فقسم بزجا بقسمين متساويين وهما ابزط طزب [26] وذلك ما اردنا ان نبيّن .

< كح > نريد ان نخرج فى دائرة معلومة خطّين متوازيين يفصلان من الدائرة جزءا ما مفروضا .

فلنجعل الجزء الثلث و الدائرة ابج ونريد ما قلنا . فنجعل مركز دائرة ابج < نقطة د ونجعل > ضلع المثلّث الّذى يعمل فى هذه الدائرة وهو اج ونخرج خطتي اد دج ونجيز على نقطة د خطّا موازيا [55b] لخطّ اج وهو دب ونصل خطّ جب . ونقسم قوس اج بنصفين على نقطة ه ونخرج من نقطة ه خطّا موازيا لخطّ بج وهو هز

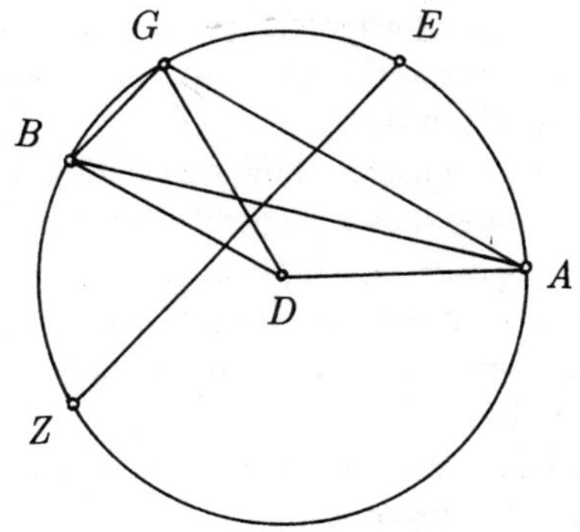

Figure 7

22 هزح . 23 هد . 24 + ابز مساويا (crossed out) . 25 + مثلّثان (crossed out) . 26 طزح .

Proof of this: Line AG is parallel to line DB, so triangle DAG is equal to triangle BAG. Let to them a common addition be made, namely the circular segment AEG. Then the whole figure $DAEG$ turns out to be equal to the whole figure $BAEG$. But $DAEG$ is one-third of circle ABG, so figure $BAEG$ is one-third of this circle.

Since EZ is parallel to GB, arc EG is equal to arc BZ. But EG is equal to EA, so EA turns out to be equal to ZB. We make arc EGB common. Then the whole arc AB is equal to the whole arc EZ, so the straight line AB is equal to the straight line EZ, and the circular segment $AEGB$ turns out to be equal to the circular segment $EGBZ$. We drop the common part, namely the circular segment GB, then by subtraction figure $EZBG$ is equal to figure $BAEG$.

But figure $BAEG$ is one-third of circle ABG. So figure $EZBG$ is one-third of circle ABG. That is what we wanted to demonstrate.

If we wish to cut off from the circle one-fourth of it or one-fifth of it or another assumed part by means of two parallel lines, we draw in this circle the side of the square or the pentagon which are inscribed in it, and we draw to it from the centre two lines as we have drawn them here, and we proceed as in this construction.[32]

< 29 > We wish to divide a known triangle into two parts such that the ratio between them is a known ratio, by means of a line parallel to its base.

< 30 > We wish to divide a known triangle by means of lines parallel to its base into parts in known ratios.

< 31 > We wish to divide a known trapezium into two parts such that the ratio between them is a known ratio, by a line parallel to its base.

< 32 > We wish to demonstrate how <we divide> a known trapezium into parts such that the ratios between them are equal to known ratios and such that the division is by lines parallel to the base of the trapezium.

< 33 > We wish to demonstrate how we divide a known quadrilateral into two parts such that the ratio between them is a known ratio, by a line drawn from a known angle of it.

< 34 > We wish to demonstrate how we divide a known quadrilateral by straight lines drawn from a known angle of it into parts in known ratios.

< 35 > If these things are known, it is possible for us to divide a known quadrilateral in a known ratio or known ratios by a straight line or straight lines drawn from a known point on one of its sides, if we bear in mind the conditions mentioned above.

The book is finished. We have restricted ourselves to the statements without proof, because the proof is easy.

---

[32]This is worked out in detail in the fragment of *On Divisions* preserved in the *Various geometrical problems*, see Text C below. Note that Thābit ibn Qurra wrote a short treatise on the problem of constructing a segment equal to one-sixth of the circle; this is published in J. Sesiano, Un complément de Ṯābit ibn Qurra au Περὶ διαιρέσεων d'Euclide, *Zeitschrift für Geschichte der arabisch-islamischen Wissenschaften* 4 (1988), 149-159.

ونخرج [27] خطّ اب فاقول انّه قد اخرج خطّا هز جب المتوازيين يفصلان [28] من دائرة ابج ثلثها وهو شكل زبجه . برهان ذلك انّ خطّ اج مواز لخطّ دب فمثلّث داج مساو لمثلّث باج وليزاد عليهما زيادة مشتركة وهي قطعة دائرة اهج فيصير جميع شكل داهج مساويا لجميع شكل باهج . ولكن داهج هو ثلث دائرة ابج فشكل باهج هو ثلث هذه الدائرة . ولانّ هز يوازي جب يكون قوس هج مساوية لقوس بز ولكن هج مثل ها فيصير ها مثل زب . ونجعل قوس هجب مشتركة فجميع قوس اب مساو لجميع قوس هز فخطّ اب المستقيم مساو لخطّ هز المستقيم ويصير قطعة دائرة اهجب مساويا لقطعة دائرة هجبز . ونسقط المشترك وهو قطعة دائرة جب فيبقى شكل هزبج مساويا لشكل باهج ولكن شكل باهج هو ثلث دائرة ابج فشكل هزبج هو ثلث دائرة ابج وذلك ما اردنا ان نبيّن . فاذا اردنا ان نفصل من الدائرة ربعها او خمسها او جزءا آخر مفروضا بخطّين متوازيين فانّا نخطّ فى هذه الدائرة ضلع المربّع او المخمّس الّذى يعمل فيها ونخرج اليه من المركز خطّين مستقيمين كما اخرجناها هنا ونعمل مثل هذا العمل .

< كط > نريد ان نقسم مثلّثا معلوما بقسمين يكون نسبة احدهما الى الآخر كنسبة معلومة بخطّ يوازى قاعدته . < ل > نريد ان نقسم مثلّثا معلوما بخطوط توازى قاعدته اقساما على نسب معلومة [29] .

< لا > نريد ان نقسم منحرفا معلوما بقسمين يكون نسبة احدهما الى الآخر كنسبة معلومة بخطّ يوازى قاعدته . < لب > نريد ان نبيّن كيف < نقسم > منحرفا معلوما باقسام يكون نسبها بعضها الى بعض كنسب معلومة ويكون القسمة بخطوط موازية لقاعدة المنحرف .

< لج > نريد ان نبيّن كيف نقسم شكلا ذا اربعة اضلاع معلوم بقسمين يكون نسبة احدهما الى الآخر كنسبة معلومة بخطّ مستقيم يخرج من زاوية منه معلومة .

< لد > نريد ان نبيّن كيف نقسم ذا اربعة اضلاع معلوم بخطوط مستقيمة تخرج من زاوية منه معلومة باقسام على نسب معلومة .

فان كانت هذه الاشياء معلومة فقد يمكننا [30] ان نقسم ذا [31] اربعة اضلاع معلوم بنسبة معلومة او نسب معلومة بخطّ مستقيم او خطوط مستقيمة تخرج من نقطة معلومة على ضلع من اضلاعه اذا نحن اتّبعنا الشرائط الّتى [32] تقدّم ذكرها .

تمّ الكتاب اقتصرنا بالدعاوي دون البرهان لانّ البرهان عليه سهل .

---

27 ونصل (crossed out) نخرج 28 يفصلا . 29 معلوم . 30 يمكنا . 31 اذا . 32 الّذى .

Text B

5.[33] (Figure 8) By Euclid, in the Book of the Division. Circle ADG[34] is known, and in it there is a known sector ABG. We wish to cut off from the circle a segment contained by two parallel lines and two arcs, such that the segment is equal to the sector. Thus we bisect arc AG at point E, and we join AG. We draw line BZ parallel to AG, and we join ZA. We draw line EH parallel to ZA. Then arc ZH is equal to arc AE, and this is clear. So arcs HZ (plus) AE are equal to arc AG. We make arc AZ common, then arc ZG is equal to arc EH. We join ZG. Then triangle AZG is equal to triangle ABG. So the segment contained by arc GA and lines AZ, ZG is equal to the sector. We make the segment contained by chord AZ common, then the segment of which the chord is GZ is equal to the sector plus the segment of which the chord is AZ. But the segment of which the chord is GZ is equal to the segment of which the chord is EH. We drop the common segment, of which the chord is AZ, then by subtraction the segment which is contained by arcs HZ, AE and the chords AZ, EH, which are parallel, is equal to the sector. That is what we wished to demonstrate.

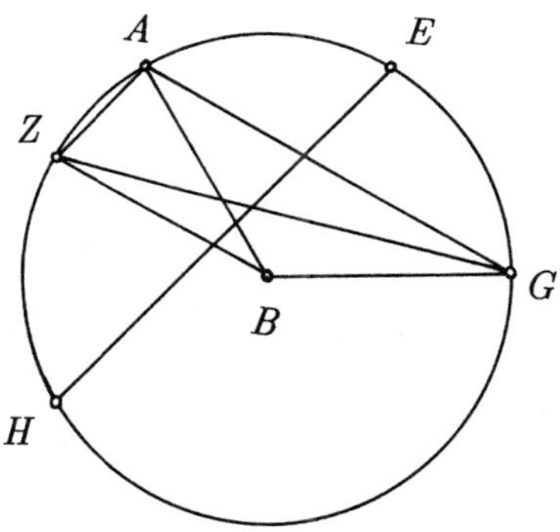

Figure 8

6.[35] (Figure 9) By the poor translator of it. We wish to trisect an acute rectilinear angle such as angle ABG. Thus we extend GB on the side of B rectilinearly and indefinitely. We draw with centre B and distance GB

[33]This is no. 5 of the *various geometrical problems*. The figure in the manuscript consists of a circle only, without letters. In Schoy's reconstruction, point B is on the circumference of the circle. It is more likely that B was the centre, because a "sector" (Arabic: qaṭṭāʿ) is usually part of a circle contained by an arc and two radii, and because the text is consistent with prop. 28 of Al-Sijzī's abstract if B is the centre.

[34]It is not clear where exactly point D is located.

[35]This is no. 6 of the *various geometrical problems*. A trisection of the angle is necessary if one wishes to apply prop. 5 of Text B and prop. 28 of Al-Sijzī's abstract to find the a segment equal to for example 1/9th of the circle.

Text **B**

ﻫ . لاقليدس فى كتاب القسمة . دائرة ادج معلومة فيها قطّاع ابج معلوم ونريد ان نفصل من الدائرة قطعة يحيط بها خطّان متوازيان وقوسان فيكون القطعة مساوية للقطّاع . فنقسم قوس اج بنصفين على نقطة ه ونصل اج ونخرج خطّ بز يوازى اج ونصل زا ونخرج خطّ هح يوازي زا فقوس زح مثل قوس اه وذلك ظاهر فقوسا حز اه مثل قوس اج ونجعل قوس از مشتركة فيكون قوس زج مساوية لقوس هح ونصل زج فمثلّث ازج مثل مثلّث ابج فيكون القطعة الّتى يحيط بها قوس جا وخطّا از زج مثل القطّاع . ونجعل القطعة الّتى يحيط بها وتر از مشتركة فيكون القطعة الّتى وترها جز مثل القطّاع والقطعة الّتى وترها از . لكن القطعة الّتى وترها جز مثل القطعة الّتى وترها هح فنسقط القطعة الّتى وترها از مشتركة تبقى القطعة الّتى يحيط بها قوسا حز [1] اه ووترا از هح وهما متوازيان مثل القطّاع وذلك ما اردنا ان نبيّن .

و . لمترجمه الفقير . نريد ان نثلّث زاوية حادة مستقيمة [2] الخطّين كزاوية ابج . فنخرج جب من جهة ب على استقامته الى غير النهاية وندير على مركز ب ببعد جب نصف دائرة جاد ولتقطع ضلع با على نقطة < ا ونضع > طرف المسطرة على نقطة ا وطرفه الآخر ... ... [3] ونحرّكها حتّى يكون الخطّ < الّذى بين > خطّ بز وبين النقطة الواقعة تحت المسطرة من محيط نصف الدائرة كنقطة ﻫ مساويا لنصف قطر الدائرة . فاذا وجدنا ذلك نصل خطّي اهز به . فلكون زاوية اهب الخارجة من مثلّث بهز مساوية لمجموع زاويتي هبز هزب المتساويتين يكون زاوية باه المساوية لزاوية بها ضعف زاوية بزه . ولكون زاوية ابج الخارجة من مثلّث ازب مساوية لمجموع زاويتي باز ازب يكون ثلاثة امثال زاوية ازب . فنعمل على نقطة ب من خطّ بج زاوية جبح مساوية لزاوية بزه فهي ثلث زاوية ابج وذلك ما اردنا ان نعمل .

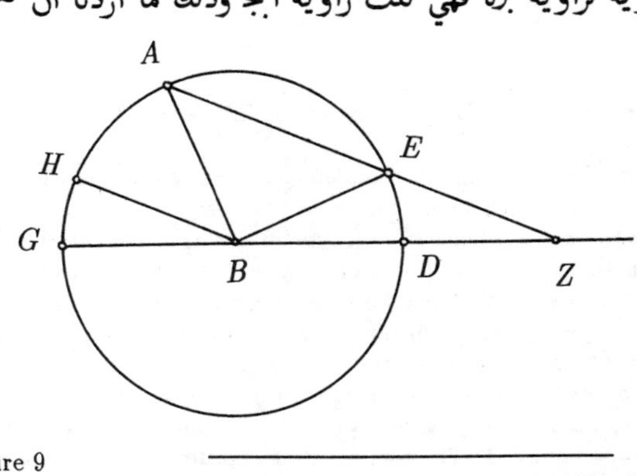

Figure 9

1 جز . 2 مستقيم . 3 illegible

semicircle GAD, let it intersect side BA at <point A. We put>[36] the end of the ruler on point A and its other end <......>[37] and we move it until the line segment <between> line BZ and the point under the ruler on the circumference of the semicircle, such as point E, is equal to the radius of the circle.[38] If we have found this, we join lines AEZ, BE. Because angle AEB, which is exterior to triangle BEZ, is equal to the sum of the equal angles EBZ, EZB, angle BAE, which is equal to angle BEA, is twice angle BZE. Because angle ABG, which is exterior to triangle AZB, is equal to the sum of angles BAZ, AZB, it is three times angle AZB. So we make on point B of line BG angle GBH equal to angle BZE. Then it is one-third of angle ABG. That is what we wished to do.

Text C

(Figure 7) If he said: how do we cut off from circle ABG one-third of it, or one-fourth of it, or any part we wish, by two parallel lines, then we make the centre of the circle point D, and we draw in the circle the chord of one-third of it, namely line AG, and we draw line BD parallel to AG. We join BG. We bisect arc AG at point E. We draw from point E line EZ parallel to BG. Then figure $ZBGE$, which is between the two parallel lines, is one-third of the circle. This is the figure for it.[39]

Arabic text:[40]

فان قال كيف نفصل من دائرة ابج ثلثها او ربعها [40a] او ايّ جزء شئنا بخطّين متوازيين فنجعل مركز الدائرة نقطة د ونخرج فى الدائرة وتر ثلثها وهو خطّ اج ونخرج خطّ [40b] بد يوازى اج ونصل بج ونقسم قوس اج بنصفين على نقطة ه ونخرج من نقطة ه خطّ هز يوازي بج فيكون شكل زبجه الّذى فيما بين الخطّين المتوازيين ثلث الدائرة وهذه صورته [40c] .

[36]The manuscript is damaged by water. The words that could be read only faintly are in brackets.

[37]Two words are wiped out completely. The meaning must have been: we put the other end of the ruler on GB extended.

[38]This construction belongs to a type of construction called *neusis* in Greek geometry. A *neusis* is the insertion of a straight segment of given length between two given (straight or curved) lines such that the segment verges (Greek: νευεῖν) to a given point, that is to say that the given point is on the rectilinear extension of the segment.

[39]The figure is the same as that in *On Divisions*, prop. 28 in the abstract of al-Sijzī (Figure 7 above). Line AB is also drawn in manuscripts **F** and **L**. Abu l-Wafā᾽ does not give proofs, because they were supposed to be unnecessary for the craftsmen for whom the book was intended. The margin of the Istanbul manuscript contains a proof which I do not render here.

[40]Apparatus: a ثُلُثُهُ او رُبُعُهُ **F**. b MS. **F** omits خَطّ. c صُورَتُهَا **L**.

Robert Recorde and The First Published Equation

Barnabas Hughes

Robert Recorde (1510-1558) was the most influential 16th century textbook writer in the English language. His *Castle of Knowledge* introduced the Copernican System to English readers. Besides being a medical doctor, he became Comptroller of Mines and Monies in Ireland. Due to some irregularity (apparently, failure to pay a judgment against him in a law suit), he was sent to prison where he died. On November 12 in the year before his death he published *The Whetstone of Witte*,[1] the second part of his *Arithmetike*, which contains *The Cossike practise, with the rules of Equation.* Recorde first of all writers on topics mathematical published a symbolic statement that can properly be named an equation. In fact he offers six examples of what he calls *equations* [Ff.j.v]:

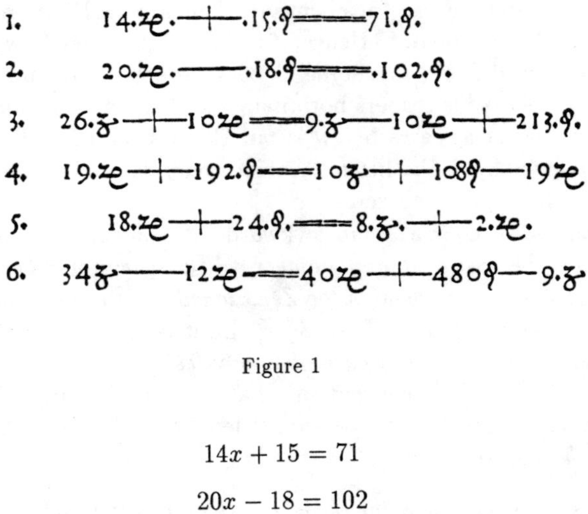

Figure 1

That is,

(1) $$14x + 15 = 71$$

(2) $$20x - 18 = 102$$

(3) $$26x^2 + 10x = 9x^2 - 10x + 213$$

(4) $$19x + 192 = 10x^2 + 108 - 19x$$

(5) $$8x + 24 = 8x^2 + 2x$$

(6) $$34x^2 - 12x = 40x + 480 - 9x^2.$$

[1]My copy is the Da Capo Press facsimile (New York 1969). I have put all direct quotations in italics followed by their locations, signature numbers in brackets.

Here we see all the proper symbols for given numbers, unknown numbers and the relationship of equality. (The numerous periods are traditional marks made to prevent adding digits to the numbers.) True, the symbols for the unknowns have changed and we no longer use a symbol to signal a known number. But those symbols had been understood and named by English and continental mathematicians as symbols for constant, root and square, beginning nearly a hundred years previously.[2]

In Recorde's tract on elementary algebra the reader finds all the information needed to lay a solid foundation for understanding and using the *Cossike Art*. Symbols, concepts, properties, methods and typical problems: all are discussed, defined and exemplified. We shall inspect each in turn and remark on his sources and originality.

Apart from the Arabic numerals, plus and minus signs, and the equal sign, the remaining symbols in his equations are strange to modern readers. Constants are accompanied by what appears to be the Greek letter φ but really is a stylized *d*, the initial letter of *dragma* (either the ancient coin or a unit of weight). The root or first degree unknown is represented by a stylized *r*, the first letter of *radix*. The squared term is designated by *z*, the initial letter of *zensus*, and so on. Recorde lists these and more, which according to Adam Riese "zeichen ader benennung Di in gemeinen brach teglich gehandelt werdenn."[3] Hence, it is difficult to judge how much Recorde actually borrowed from, say, some algebraic work of Johann Scheubel.[4] In fact, Recorde offers his readers both names and symbols for unknowns to the 80th power. What appears below is but the first and last lines of his table [U.j.*r*]. Noteworthy is the list of Arabic numerals used to identify what we call powers. It begins with zero.

The names, of course, seem awkward. A constant is *nomber absolute.* Then follow the *roote of any nomber, ... a square nomber, ... a Cubike nomber, ... a square of squares, or Zenzizensike, ... a Sursolide* (x^5), ... *zenzizenzicubike* (x^{12}), and so on [S.j.*v*]. Be it noted further that Recorde is wont to call all of these terms *cossike nombres.*

Recorde had a clear concept of what an equation is. We read in the introductory paragraph of the section titled *The rule of equation, commonly called Algebers Rule*:

> *This Rule is called the Rule of Algeber, after the name of the inuentoure, as some men thinke: or by a name of singular excellencie, as other judge. But of his use it is rightly called, the rule of equation: bicause that by equation of nombers, it doeth dissolue doubtfull*

[2]Dresden Codex C 80, much of which was written by Johannes Widmann von Eger in 1486, identifies the symbols by sign and name. See W. Kaunzner (1968). *Johannes Widmann von Eger*. (Munich: Deutsches Museum, Reihe C, Quellentexte und Übersetzung, Nr. 7), 19.

[3]Quoted by F. Cajori (1928). *A history of mathematical notations* I (La Salle IL: Open Court), 134 n.1.

[4]Recorde does not identify which text of Scheubel he used.

questions: And unfolde intricate ridles [Ee.v.*v*].

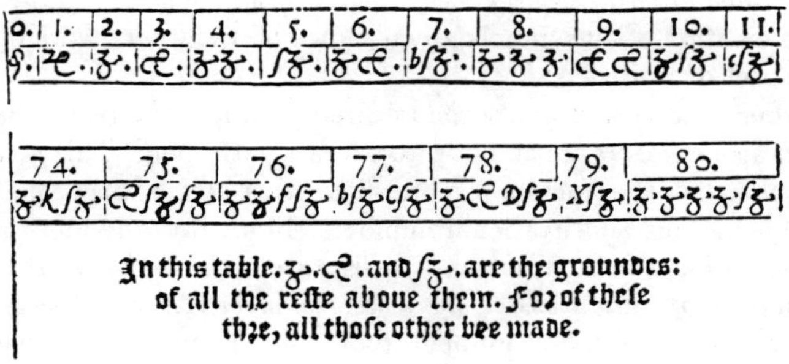

Figure 2

Farther on he explains what he means by *equation of nombers.* (The book is written in dialogue form with two actors.) The *Scholar* remarks that the rule sounds like the method of false position which finds the correct answer to a problem by starting with an incorrect guess. The *Master* replies,

> *This rule doeth farre excell that other. And dooeth not take a false nomber, but a true nomber for his position, as it shall be declared anon. Wherby it maie bee thoughte, to bee a rule of wonderfull inuention, That teacheth a manner at the firste worde, to name a true nomber, before he knoweth resolutely, what he hath named* [Ff.j.*r*].

Recorde is saying, equivalently, "Let us assume we know the value of the unknown and call it x." This is precisely what his first example shows, two numbers equal to one number; that is, $14x + 15 = 71$. Eighty years later Descartes would write in *La Geometrie* :

> ...we find it possible to express a single quantity in two ways. This will constitute an equation, since the terms of one of these two expressions are together equal to the terms of the other.[5]

Descartes honed the concept sharper. Recorde had the right idea, the right word and, first of all, the complete symbolization.

It is in this section that Recorde defines and shows his great creation, the equal sign. There had been other symbols for equality before him and subsequent writers would offer others. But his symbol took hold, probably due to the English mathematicians of the next century. His reason for selecting it has oft been quoted, *a paire of parallels, or Gemowe lines of one lengthe, thus: =, bicause noe .2. thyngs, can be more equalle.* What I have never seen

[5]D.E. Smith & M.L. Latham (Eds. 1954). *The Geometry of René Descartes.* (New York: Dover repr.), 9.

elsewhere is his reason for creating the symbol in the first place. He tells us, *And to auoide the tediouse repetition of these woordes: is equalle to: I will sette as I doe often in woorke use, a paire of paralleles ...* [Ff.j.*v*]. So it is with the creation of all symbols, to cut down on the weary work of repeating the same set of words.

In order to solve first and second degree equations the reader had to learn what *Cossike nombers* are and how to operate with them. This is the focus of nearly one half of the tract on algebra. The other half is fairly equally divided between equations and irrational numbers. First, Recorde looks at the new numbers. In his words, *Nombres Cossike, are soche as bee contracte unto a denomination of some Cossike signe as .1.nomber .1.roote .1.square .1.cube. etc* [S.j.*r*]. For all of these: number, root, square and so on, he has symbols, some of which are shown above. Further, he notes that the *signs or figures* can represent irrational numbers as well as rational; they are *commodiously* general. In this section, however, the reader's attention is drawn only to integral and rational numbers, both positive and negative, each of which is denominated by symbols such as we have seen above. Irrational numbers, *surds* he calls them, with their *Cossike denominations* are discussed in the last quarter of the tract.

The symbols are a substantive part of the operations. Recorde makes it clear that the rules are very much like those in arithmetic; indeed he uses denominate numbers from arithmetic to exemplify operating with the algebraic symbols. An overview of the section on operations shows a separation between working with integers and with fractions. Further, he considers operations on monomials *(nombers uncompounde)* before explaining operations on polynomials *(nombers compounde)*. The common operations are numeration (how many are there?), addition, subtraction, multiplication, division and extraction of roots. Restricted to integers are progressions (both arithmetic and geometric) and the *Golden Rule* (Rule of Three or proportion).

Regarding the foregoing, four remarks need be made. First, for extraction of roots of monomials, Recorde expects the reader to memorize the powers of 2, 3 and 4 to the tenth power and to understand the table of powers of the unknowns. In order to find the roots of polynomials, he offers the same set of rules found in many algebra texts today. Secondly, the dialog between *Scholar* and *Master* shows that Recorde had excellent insights into difficulties experienced by learners. Any modern teacher could profit from reading the give-and-take between the two. For instance, in one place after hearing the *Master* begin to discuss addition of like terms, the *Scholar* remarks, *If this be al, I can marke it well inough* [U.ij.*rv*]. The *Master* rejoins, *There is somewhat more to be considered ...* and adds considerably to what preceded. Whereupon the *Scholar* acknowledges, *By examples, I shall better conceiue those rules.* Third, the only difficulty today's reader may experience is keeping track of the symbols for the various powers of the unknown above the cube. Hence, keeping a list of symbols at hand is most helpful. Finally, the entire section on symbols and operations is prefatory to the part on methods for solving equations.

Recorde acknowledges only two varieties of equations. The first type is *When one nomber is equall to an other: meanyng that thei bothe must be simple nombers Cossike, and uncompounde* [Ff.iij.*v*]. In other words, one term of whatever degree equals a single term of whichever degree; such as (in modern symbols), $7x^5 = 567x$. The second type is *When one simple nomber Cossike, is compared as equalle to .2. other simple nombers Cossike, of seueralle denominations, and like distaunce* [Ff.iv.*v*]. The last phrase means that the entire equation is equivalent to a quadratic trinomial. This second type receives the lion's share of attention, since there are three kinds of these.

The first type of equation is solved very much the way it is done today. The unknown must be isolated on the left (!) side of the equal sign. So, by using the addition and subtraction laws (our additive inverse) one proceeds toward the isolation. Recorde justifies these laws by remarking, *For all springeth of those principles Geometricalle* [Ff.ij.*r*]. Then, if necessary, divide both sides by the coefficient of the unknown. Finally, be sure to check the answer. The example shown above is handled with dispatch: divide both sides by $7x$ to reach $x^4 = 81$; then take the fourth root to find the solution $x = 3$. A word problem, for example, says, *There is a floor Paued with Square Bricke, the lengthe of the floore beying longer than the breadth, by 1/7, and the whole Pauemente containeth .3584. brickes: I require to knowe the bredthe and lengthe* [Gg.ij.*v*]. (The answer is 56 by 64.)

For all practical purposes Recorde solves the quadratic types by the quadratic formula. His terminology, however, is different. In fact it suggests a geometric mindset, at least unconsciously so. First and in modern terminology: regardless of which type the equation is, the coefficient of its highest power must either be unity or be made such by the multiplicative inverse. At the same time, if the highest power is greater than two, then *his* equations are reducible to quadratics. All of this leads to $x^2 = (b/a)x + c/a$, with a, b and c integers. Secondly and again regardless of the type, he says, *the roote of those .2. nombers compounded . . . shall be extracted . . .* [Ff.iv.*v*]. In other words Recorde is looking on the right side of the equation as the area of the square x^2 on the left side. This is what I mean by "a geometric mindset."

Our author identifies three varieties of the second kind of equation, one cossike number equal to two cossike numbers [Dd.iiij.*r* to Ee.iij.*v*]:

$$x^2 = px + q \Longrightarrow x = \sqrt{(p/2)^2 + q} + (p/2) \tag{1}$$

$$x^2 = q - px \Longrightarrow x = \sqrt{(p/2)^2 + q} - (p/2) \tag{2}$$

$$x^2 = px - q \Longrightarrow x = (p/2) \pm \sqrt{(p/2)^2 - q}. \tag{3}$$

The three types he got from Scheubel, as will be remarked upon below. Recorde has one general rule for solving all of these equations, *extracte the roote of that compousite nomber* [Gg.j.*v*]. For the third case he offers a second method, namely to factor q and find which pair of factors can be summed to equal p and multiplied to equal q [Ee.ij.*v*]. Further, Recorde recognizes neither negative roots nor double identical roots [*idem*]. He is aware of negative numbers and operations on them; he calls such *an absurd nomber lesse than nothyng* ... [which] *you maie woorke, ... as with any other* [Bb.iv.*rv*]. Consequently, two roots emerge where both are positive. In every example and problem, a check of the root(s) is made by substitution into the original equation. Finally, Recorde provided for an anomaly. The rule for collecting like terms where subtraction is involved is always to take the smaller from the larger number. This could produce an equation equal to zero! The situation is saved, he says, by translating *the nombers with – from that side of the equation, and set them on the contrary side, with the sign of* + [Ii.iij.*r*]. He is careful with his examples to assure that this procedure always works.

Some indication of the kind of problems offered late sixteenth century students of *The Arte of Cossike Nombers* can be sifted from the following examples. Coins are mentioned: the *crown* worth 10 shillings and the *angell* worth about 7 shillings. The spelling and some terminology have been modernized.

> A poor man who died had four children. All his goods came to 72 crowns. These he would have distributed thus: the second and third child should have seven times as much as the first. And the portions of the third and fourth should be five times as much as the second. And the first and the fourth should have twice as much as the third. If you work the solution[6] well, you may seem worthy to be master of those wards [Hh.ij.*v*].

> A gentleman, willing to prove the cunning of a bragging Arithmetician, said thus: I have in both my hands 8 crowns. If I add to the amount in each hand the square and cube of what lies in each and then add all together, the sum is 196. Now tell me (quoth he) what is in each hand,[7] and I will give all for your labor [Ii.j.*v*].

> There were two several men, which had certain sums of angelles, in such rate, that the second man his sum was three and a fourth that of the first; and if their 2 sums were multiplied together, and to that total the 2 first sums added, there would amount $142\frac{1}{2}$. I demand of you, what was each[8] of their sums in angelles? [Ii.iij.*r*]

[6] *Answer*: A: $4\frac{1}{2}$; B: $11\frac{1}{4}$; C: $20\frac{1}{4}$; D: 36.
[7] *Answer*: 3 and 5.
[8] *Answer*: 6 and $19\frac{1}{2}$.

Recorde ends the section on algebra with a problem that leads to a cubic equation: There is a number whose square less 16 is multiplied by the original number augmented by 8. The product is 2560. The *Scholar* produces the correct equation: $x^3+8x^2-16x-128 = 2560$. After transposing the negative terms to the right side, he remarks, *All that is above my cunning. For hitherto I have learned no rule for any of them both.* He does not know how to solve an equation with two terms equal to two terms. The *Master* wisely gives the answer, *12*. And then, after several more similar examples, closes the section with, *But of these and many other very excellent and wonderful works of equation, at an other time I will introduce you farther, if I see your diligence applied well in this, that I have taught you. And therefore here will I make an end of Cossike nombes, for this time* [Ll.ij.*v*]. Recorde died before he could get far into the project.

In sharp contrast with his earlier *The Grounde of Artes* (1542), *The Whetstone of Witte* clearly identifies its sources both in general and in particular. At least twice Recorde acknowledges *some other bokes*[9] and some other men.[10] Clearly these general references suggest that he had several books at hand. Their titles remain unknown, with one exception. Some five times he cites Euclid's *Elements* by book and proposition.[11] As for the remainder of his references we can only make educated guesses.

According to the subtitle the text is tripartite: *thextraction* (sic) *of roots: The Cossike practise, with the rule of Equation: and the woorkes of Surde Nombres.* Using this division I have separated the references accordingly. First we consider the part on finding square and cube roots. Here Michael Stifel is acknowledged once,[12] in a discussion on the component factors of *diametralle nombers.* These are products of the smaller two numbers in a set of Pythagorean triplets. Recorde puts the two factors in fractional form, thus: 3/4, 5/12, 7/24. He remarks that Stifel prefers *for the more delectable contemplation, to behold their forme of progression* to show them as mixed numbers: $1\frac{1}{3}$, $2\frac{2}{5}$, $3\frac{3}{7}$, $4\frac{4}{9}$ and so on. (I have not noticed in any history of mathematics text that the pair of numbers $n(2n+1)+n$ and $2n+1$ from the mixed number $n+n/(2n+1)$, n a natural number, can always produce a Pythagorean triplet. Was Stifel the first to observe this relationship?) The first three references to Euclid's *Elements* are: VIII,18 regarding the mean proportional between two plane numbers; XII,18 for the ratios of spheres to their diameters; and VIII,19 about the mean proportional between two solid numbers. The third name cited is Girolamo Cardano. The pertinent discussion considers the steps one takes for finding cube roots. With respect to one step in the process the *Master* tells the *Scholar* that there are as many ways of going about it as there are writers. Then he quotes Cardano's rule,

---

[9]Bb.j.*v*.

[10]Ff.i.*r*.

[11]N.ii.*r*, P.iv.*v*; Aa.j.*r*, Q.i.*r*, Df *r*.

[12][F.iv.]*v*.

Multiplie the roote squarely, and againe by 3 and that nomber shall be the diuisor vnto the remainer [O.iv.*v*]. A few pages farther on he refers again to *Cardanes rule* which fails to produce the correct answer. Scheubel fares better in the discussion on finding cube roots. Where Recorde faults Cardano's rule, he notes that Scheubel's rule is better;[13] but on the next page he observes that it is not always so helpful.[14] The rule cited by Recorde is *Triple the roote, and the square of it also, and add bothe those nombers together, and .1. more: And so have you a denominator for your numeratour*. So much for clues to explicit references[15] in the first section.

Only two authors are noted in the part on algebra and equations. The first is Euclid where the discussion focuses upon simplification of algebraic fractions.[16] To establish his point the *Master* uses an example from *vulgare nombers* which he proves by a reference to *Elements* V,19. The second author is Johann Scheubel (1494-1580), Ordinary Professor of Euclid and Arithmetic at the University of Tübingen and widely read author of books in mathematics. It was his text, *Algebrae Compendiosa Facilisque Descriptio, qua depromuntur magna Arithmetices miracula* (Parisiis, apud Gulielmum Cauellat, 1551), upon which Recorde leaned heavily for this section of the tract.[17]

There are strong similarities between Recorde's and Scheubel's respective treatises on algebra. Both begin with the same presentation: the meaning of the symbols for what we call variables (although this concept is not present in either text), and they use the same symbols. Karpinski cites Stifel as the source of the symbols. Since Recorde uses so much of Scheubel's text farther on, I prefer to think that he referred to this text while preparing these pages. The similarities between the two tracts continue. Both authors discuss how to add and subtract, multiply and divide algebraic statements, both integral and rational, embodying variables, coefficients and constants. The examples, however, are not the same, and Recorde makes a better presentation. He continues to use the symbols he had explained, while Scheubel employs collateral terminology for the unknowns (e.g., x is *prima*, x^3 is *secunda* and so on).

Next follows the section on algebra and equations, in both texts. At the outset Scheubel defines equation: *vbi duae res vel quantitates inter se*

---

[13] P.iii.*rv*.

[14] P.iii.*v*.

[15] Potential sources for determining these references to Stifel and Cardano are not at my disposal. They are Stifel's *Arithmetica integra* (1539) and *Deutsche arithmetica* (1545) and Cardano's *Libellus qui dicitur computus minor* (1539) and *Practica arithmeticae generalis* (1539).

[16] Aa.j.*r*.

[17] This statement is worded strongly because I disagree with the late Louis Karpinski who claimed that Stifel was Recorde's major source. He did not quote or refer to Stifel once in the algebraic part of the tract. Karpinski's position from where he views mostly the arithmetic section of the text was published as "The Whetstone of Witte (1557)," *Bibliotheca Mathematica* 1912-1913 (Series 3), **13**, 223-228.

aequales esse proferuntur.[18] Recorde describes it obliquely. That is, he identifies just two kinds of equations: *when one nomber is equalle vnto one other; when one nomber is compared as equalle vnto .2. other nombers* [Ff.i.*v*]. He obviously judges these descriptors sufficient for understanding the concept of equation. Further, he selected those descriptors to simplify the explanation of solving equations. He makes a point of this because in his judgement Scheubel made the explanation more complicated than it warrants. Further, Recorde produced[19] Scheubel's *varieties of equations* so that the reader *maie perceive, how thei bee conteined in those .2. formes, named by me.* Then he adds, parenthetically, *As for the manyfolde varieties, that some other doe teache, I accoumpte it but an idle bablying, or (to speake moare fauourably of them) an vnnecessary distinction* [F.iij.*rv*]. Recorde may have had Cardano in mind because in his *Artis Magnae*, Ch. II, he lists many kinds of first and second degree equations.[20] Further, Scheubel had at copy of this book, although it is not known when he obtained it.[21]

The final section on *Surde nombres* focuses mostly upon arithmetical operations on simple and complex irrational numbers. Only Euclid is mentioned by name and in a negative sense. The context is compound surds, specifically *bimedialles* and *residualles.* He remarks, *Euclides definitions doe not very aptly agree to this place, as at an other tyme I will shewe you, and therfore I doe omitte them for thys tyme* [Df *r*]. Nonetheless, both Recorde and Scheubel start off with quite similar definitions of surds. Then both texts develop the arithmetic operations on them. Symbolism is the same but examples differ. Recorde, however, stops short of discussing the solution of equations involving surds which Scheubel develops. Too bad! But then, as the text tells us, a messenger came knocking at the door and called *The Master* away. A promise (or was it just a wish) to complete the tract was left unfulfilled. Yet our task is done. Scheubel's text was the one closest to Recorde's hand as he wrote.

---

[18] Johann Scheubel, *Algebrae compendiosa et facilis descriptio* (Paris,1551), 13*v*. I wish to express my appreciation to Walter Jones of the Marriott Library, The University of Utah, who obtained for me a xerox copy of its copy of Scheubel's text.

[19] These appear on Ff.iij.*v*, [Ff.iv.]*v*, and Gg.i.*rv*.

[20] T. R. Witmer (Ed. 1968). *The Great Art or The Rules of Algebra by Girolamo Cardano.* (Cambridge MA: MIT Press), 23-26.

[21] *See my* (1972) The private library of Johann Scheubel, sixteenth-century mathematician, Viator **3**, 422.

Über die beiden nachgelassenen mathematischen Handschriften von Adam Ries

Wolfgang Kaunzner

Einleitung

Seit dem Jahre 1860 weiß man aufgrund einer unvollständigen Edition der „Coß“ von Adam Ries (1492-1559) durch Bruno Berlet in groben Zügen über den fachlichen Inhalt dieses algebraischen Werkes Bescheid. Im Jahre 1959 wies Willi Saemann darauf hin, daß aus der Feder von Adam Ries ein noch früheres bislang nicht als solches erkanntes algebraisches Manuskript stammt, nämlich Kodex C 349 der Sächsischen Landesbibliothek Dresden. Beide Handschriften wurden in der letzten Zeit für die Benutzung nur schwer zugänglich aufbewahrt.

1992 war Jubiläumsjahr anläßlich des 500. Geburtstages von Adam Ries. Nach langen Vorbereitungen und durch die Mithilfe vieler Beteiligter wurde es möglich, daß diese beiden derzeit bekannten algebraischen Handschriften von Adam Ries, nämlich Kodex Dresden C 349 teilweise und die „Coß“ mit der Signatur $0^{M}_{1}0$ im Erzgebirgsmuseum in Annaberg in Sachsen vollständig durch eine Edition zugänglich gemacht wurden.

Hier im folgenden soll nicht die Tätigkeit von Adam Ries in Abhängigkeit von den Quellen gewertet werden, aus denen er schöpfte, weil er etwa Kodex C 349 zumindest großenteils von einer derzeit verschollenen Vorlage abschrieb, während er die „Coß“ als ein eigenes algebraisches Lehrbuch zusammenstellte, sondern es soll nur aufgezeigt werden, daß er sich auch mit Problemen der genannten Art abgab. Inwieweit er den Text in C 349 evtl. in den ersten Kapiteln selbst mitgestaltete, zu solchen Fragen soll hier nicht Stellung genommen werden. Es geht also nur darum, darauf aufmerksam zu machen, daß Adam Ries – und zwar offensichtlich vor seiner Zeit als Rechenmeister in Erfurt und als Rechenmeister und Bergbeamter in Annaberg – sich auch mit der damaligen theoretischen Mathematik abgab und hier gleichrangig neben den anderen Cossisten seiner Periode steht.

Ein herzlicher Dank sei gerichtet an die Leitung der Sächsischen Landesbibliothek Dresden und an die Leitung des Erzgebirgsmuseums Annaberg, weil die Genehmigung zur Reproduktion entsprechender Textstellen gegeben wurde, ferner an Herrn Professor Dr. Menso Folkerts, München, für die Möglichkeit, in seinem Filmarchiv zu arbeiten.

Zum Inhalt der beiden mathematischen Handschriften von Adam Ries

Wenn von Adam Ries die Rede ist, dann stellt man sich vermutlich einen untersetzten, grobgliedrigen Mann vor, mit großem Hut und langem Bart, einen, der typisch das Aussehen repräsentiert, wie man es von einem biederen Rechenmeister erwartet. Das einzige Bildnis, das wir von ihm haben, in seinem dritten Rechenbuch,[1] es kommt unseren Vorstellungen wohl ziemlich nahe (Abb. 1).

Wenn hier einiges von den wissenschaftlichen Leistungen des Mathematikers Adam Ries gesagt werden soll, dann muß zunächst deutlich hervorgehoben werden, daß dies nur im Rahmen jener Zeit, also des frühen bis mittleren 16. Jahrhunderts, gesehen werden kann. Die Mathematik hatte im deutschsprachigen Gebiet bis zur Mitte des 15. Jahrhunderts ein sehr bescheidenes Dasein gefristet. Auf den Latein- und den Stadtschulen des 13. bis 15. Jahrhunderts hatte sie im lateinischen Unterricht keine Bedeutung erlangt, und auch auf den Universitäten findet man in den Vorlesungsverzeichnissen nur die Anfangsgründe der Euklidischen Geometrie aufgeführt und ein bißchen praktische Zahlenlehre, den Algorismus. Dies erscheint uns unvorstellbar, wird aber durch die Tatsache verständlich, daß es damals im praktischen Leben keine Anwendungsmöglichkeit für Mathematik im heutigen Sinne gab. Rechnen, wie man es im deutschen Sprachbereich seinerzeit betrieb, fußte auf dem bescheidenen Wissen der Römer, und in der Geometrie bediente man sich zum Teil einfacher Näherungsformeln aus der Überlieferung heraus. Schulden und Guthaben wurden entweder auf einem Kerbholz eingeritzt oder bei Bäckern und Müllern mit Knoten in Schnüren angemerkt, und wenn schon gerechnet werden mußte, dann mit den Händen und mit den Fingern; der Tauschhandel herrschte vor; bei Geldgeschäften verwendete man ein Rechenbrett bzw. später einen Rechentisch oder ein Rechentuch, wo einzelne waagrechte Linien bzw. Felder feste Stellenwerte inne hatten. Wenn wirklich etwas aufgeschrieben werden mußte, dann geschah dies in römischen Zahlen.

Vieles hiervon änderte sich noch im Verlaufe des 15. Jahrhunderts. Der Buchdruck mit beweglichen Lettern, der Übergang von Pergament als Schreibmaterial zu Papier, die Entstehung einer Umgangssprache aus den einzelnen Dialekten heraus, die Abkehr vom Tauschhandel zum Geldhandel als Folge der verstärkten Edelmetallgewinnung, das Aufkommen einer internationalen Waren- und Geldwirtschaft, die Gründung von Rechenschulen, in denen berufene Rechenmeister in der Muttersprache unterrichteten – all dies sind Fakten, die teils aus der historischen Entwicklung heraus zu sehen sind, die sich teils aber auch gegenseitig bedingten. Zwar gab es noch keine

---

[1] Adam Ries: *Rechenung nach der lenge, auff den Linihen vnd Feder. Darzu forteil vnd behendigkeit durch die Proportiones, Practica genant, Mit grüntlichem vnterricht des visierens. Durch Adam Riesen. im 1550. Jar.* Leipzig 1550.

allgemeine Schulpflicht, aber so mancher, der lernen wollte, fand nun viel leichter als früher Zugang zu etwas Geschriebenem oder Gedrucktem. Es gab jedoch noch keine einheitliche deutsche Schriftsprache. So ist es leicht verständlich, daß man deutsche Texte aus jener Zeit heute ohne weiteres schon aufgrund ihrer Mundart in das Gebiet ihrer Entstehung einordnen kann.

Die von Italien ausgehende humanistische Bewegung erstreckte sich auch auf den österreichischen sowie den süd- und mitteldeutschen Raum und gab den Anstoß, daß man sich hier gezielter als vorher mit den lateinischen mathematischen Werken befaßte, die in immerhin beachtlicher Anzahl bereits in den Bibliotheken vorhanden waren. Nun drehte es sich aber nicht mehr nur darum, diese Abhandlungen weiterhin zu kopieren, wie dies vom 13. bis zum 15. Jahrhundert der Fall gewesen war, sondern man wollte sie auch erschließen und verstehen. Im 15. Jahrhundert findet man folglich – erstmals – eine eigene theoretische Mathematik auf deutschem Boden. Im einzelnen ging es hierbei vorwiegend um die griechische Geometrie und um die Arbeiten zur Trigonometrie, zur Zahlenlehre und zur Gleichungslehre der Muslime. Alle diese Texte waren in der Zeit seit ihrer Entstehung um 300 v. bzw. um 800 n. d. Z. auf verschiedenartige Weise ergänzt oder kommentiert worden. Bei uns lagen sie lateinisch vor.

Im deutschen Sprachraum setzte man sich hiermit nachweislich ab etwa 1460 kritisch auseinander. Zum einen wurde versucht, die überkommenen Arbeiten wieder fehlerfrei zu gestalten. Zum anderen wurde hier bei uns erkannt, daß die überlieferte mathematische Darstellung einer neuen Form bedurfte, denn es gab noch keine Symbole, die einen mathematischen Text als solchen klar erkenntlich und zumindest dem Fachmann leicht verständlich gemacht hätten. Dies wurde die große Stunde der deutschen Mathematiker des 15. und 16. Jahrhunderts. Jene Wissenschaftler bemühten sich nämlich erstmals erfolgreich darum, mathematische Fachausdrücke zu synkopieren, d.h. brauchbar abzukürzen, bzw. gar zu symbolisieren. Als deutlichstes Kennzeichen dieser Periode – man nennt sie heute die „deutsche Coß“ – verwendet man seither das Plus- und das Minuszeichen.

Damals, um das Jahr 1480, traten zwei neue Berufsgruppen auf, nämlich der Rechenmeister und der akademisch gebildete Mathematiker.

Der Rechenmeister widmete sich der elementaren Ausbildung in der Rechenschule, er unterrichtete deutsch. Sein Ausbildungsziel bestand darin, dem späteren Kaufmann oder Handwerker einfaches und praxisnahes Wissen zu vermitteln. Er führte seine Schüler gewöhnlich zuerst über die Linienrechnung in die elementaren Rechenkenntnisse ein, wo eine neue Form des traditionsreichen Abacusrechnens auf Rechentischen, Rechenbrettern oder Rechentüchern praktiziert wurde, ehe er zum Rechnen mit der Feder oder Kreide überging, das ist unsere heutige schriftliche Methode; dort wurden die indisch-arabischen Ziffern erläutert und die Grundrechenarten bis zum Dividieren, Wurzelziehen und arithmetischer und geometrischer Reihenlehre in sehr aufwendiger Weise nach mehreren Verfahren gelehrt. Ferner wurden in der Rechenschule viele Textbeispiele behandelt, die meistens mit dem Dreisatz zu lösen waren, der *regula de tri*.

Der wissenschaftlich gebildete Mathematiker schrieb und unterrichtete lateinisch, z.B. an den Universitäten in Erfurt, Leipzig und Wien. Ein neues Arbeitsgebiet dieser Fachleute war die Coß, die algebraische Gleichungslehre. Im deutschen Sprachraum ging es ihnen seinerzeit vorwiegend darum, eine sinnvolle algebraische Symbolik zu schaffen und hierauf sowohl einen Kanon von lösbaren Gleichungstypen aufzubauen als auch die allgemeine Lösung der quadratischen Gleichung zu finden. Durch die ganze Zeit hindurch war dieses im 9. Jahrhundert von Muḥammad ibn Mūsā al-Ḫwārizmī (vor 800 bis nach 847) schriftlich formulierte Wissensgebiet, von wenigen Ausnahmen abgesehen, ohne jede Symbolik behandelt worden. Wie stellten sich unsere damaligen Mathematiker dazu? Aus dem 15./16. Jahrhundert liegen erstmals deutsch geschriebene Algebratexte vor, ferner gelang es ihnen, eine Symbolik einzuführen, die noch bis ins 17. Jahrhundert hinein, bis zu Johannes Kepler (1571-1630) hin, verwendet wurde. Ein Grund für die langsame Entwicklung in der Mathematik lag u.a. auch darin, daß nur positive Zahlwerte vorkommen durften, sowohl im Gleichungsansatz als auch im Ergebnis, daß irrationale Zahlwerte nur bedingt als brauchbar anerkannt wurden, negative im allgemeinen überhaupt noch nicht.

Die damalige Gleichungslehre hieß die „Coß". Die Muslime hatten für die Unbekannte in einer Gleichung die Begriffe „Ding" oder „Wurzel" gewählt, in den lateinischen Übersetzungen ab etwa 1150 findet man hierfür wörtlich *res*, *causa* oder *radix*, die Italiener sagten *cosa*, so nannten die hiesigen Fachleute das ganze Fachgebiet die „Coß", die „Lehre von dem Ding". Ein erheblicher Teil des Aufwands der „deutschen Coß" galt dem Bemühen, eine eigene Potenzrechnung der algebraischen Unbekannten aufzubauen, die in gewisser Weise als Vorstufe zum Logarithmenrechnen angesehen werden kann.

Wie läßt sich Adam Ries hier einordnen? Seine praktischen Arbeiten, die seinen Ruhm begründeten, liegen in drei gedruckten Rechenbüchern vor,[2]

---

[2]Adam Ries: *Rechnung auff der linihen gemacht durch Adam Riesen vonn Staffelsteyn, in massen man es pflegt tzu lern in allen rechenschulen gruntlich begriffen anno 1518. vleysigklich vberlesen, vnd zum andern mall in trugk vorfertiget.* Erfurt 1525 – so lautet der Titel des ersten Rechenbuches gemäß der Auflage von 1525. Nach Fritz Deubner: Adam Ries, der Rechenmeister des deutschen Volkes. *Zeitschrift für Geschichte der Naturwissenschaften, Technik und Medizin (NTM)*, Jg. 1 (o. J.), Heft 3, S. 11-44, hier S. 23, kennt man von diesem Rechenbuch derzeit nur Auflagen von 1525, 1527 und 1530, nicht von 1518. Man sehe hierzu Stefan Deschauer: *Das 1. Rechenbuch von Adam Ries.* Algorismus, Studien zur Geschichte der Mathematik und der Naturwissenschaften, herausgegeben von Menso Folkerts, Heft 6. München 1992.

Adam Ries: *Rechenung auff der linihen vnd federn in zal, maß vnd gewicht auf allerley handierung, gemacht vnnd zu samen gelesen durch Adam Riesen von Staffelstein Rechenmeyster zu Erffurdt im 1522 Jar* – dies ist der Titel der Erstauflage des sogenannten zweiten Rechenbuches, von dem Deubner, S. 24-34, insgesamt 108 Auflagen nachweist. Hierzu auch Stefan Deschauer: *Das 2. Rechenbuch von Adam Ries.* Algorismus, Studien zur Geschichte der Mathematik und der Naturwissenschaften, herausgegeben von Menso Folkerts, Heft 5. München 1991.

ferner weiß man um seine Tätigkeit als Rechenmeister in Annaberg und als Bergbeamter und Münzfachmann in Sachsen Bescheid, weil von dort her Berg-, Münz- und Zehntrechnungen aus seiner Feder vorliegen, sowie Preistabellen und Brotordnungen.[3] Was Erfurt anlangt, da gibt es noch offene Fragen, obwohl sein Aufenthalt in Erfurt als die wissenschaftlich fruchtbarste Spanne seines Lebens gilt.

Bis zur Mitte des 19. Jahrhunderts war dies der Kenntnisstand um das berufliche Wirken von Adam Ries. Er hatte jedoch bereits in seinen Rechenbüchern auf ein algebraisches Werk aus seiner Feder verwiesen, und zwar auf die „Regeln Algebre“ bzw. auf die Regeln der Coß, die er zum Druck bringen wollte.[4] Um die Mitte des 19. Jahrhunderts z.B. war jedoch nur mehr vage bekannt, daß der Dresdener Rechenmeister Martin Kupffer im Jahre 1664 die damals noch vorhandenen algebraischen Aufzeichnungen von Adam Ries hatte zusammenbinden lassen. Als der Annaberger Bruno Berlet im Jahre 1855 einen Beitrag „Ueber Adam Riese“ verfaßte, da nahm er folglich an, daß man diese „Coß“ von Adam Ries für verloren halten müßte.[5] Das Werk wurde jedoch noch im nämlichen Jahr, also 1855, in Marienberg in Sachsen wiederentdeckt,[6] nachdem es an die 200 Jahre lang verschollen gewesen war.

Adam Ries schrieb den ersten Teil seiner „Coß“ – wir nennen diese ersten 325 Seiten Coß 1 – in den Jahren zwischen 1518 und 1524, und zwar in Erfurt und in Annaberg;[7] Coß 2 zeichnete er um 1550 in Annaberg auf.[8]

Einzelne Abschnitte aus Coß 1 wurden von Bruno Berlet im Jahre 1860 erstmals unter der Bezeichnung „Die Coß von Adam Riese“ veröffentlicht.[9] In der gesamten „Coß“, die jetzt im Erzgebirgsmuseum in Annaberg aufbewahrt wird, stehen viele Namen und Datierungen, vor allem bezieht sich Adam Ries oft auf einen von ihm hoch geschätzten Gelehrten namens Algebras oder auf das „Buch Algebre“, d.h. das Buch des Algebras. Dieser Algebras lebte nach seinen Ausführungen in unbestimmter Vorzeit. Es kann sich hierbei jedoch nur um den Muslimen Muḥammad ibn Mūsā al-Ḫwārizmī handeln, der im 9. Jahrhundert in Bagdad seine für die westliche Tradition

---

Ries[1] ist der Titel des dritten Rechenbuches, das 1611 eine zweite Auflage erfuhr.

[3]Hier unterrichtet etwa Peter Rochhaus: Adam Ries in Sachsen; in: *Adam Rieß vom Staffelstein*. Staffelsteiner Schriften, Band 1. Staffelstein 1992, S. 107-125.

[4]In Ries[2(2)], S. 28, heißt es: „Die wurtzel dem quadraten vnd cubic außzuzihen wil ich hie her beruen lassen, sonder zu seyner zeyt so ich das Visiern vnd etzlich regel der cossz ertzele gnugsam erclern.“ Dies war 1522, im nämlichen Jahr, als er sich vermutlich erstmals in Erfurt als Rechenmeister bezeichnete.

[5]Bruno Berlet: *Ueber Adam Riese*. Zwölfter Bericht über die Progymnasial- und Realschulanstalt zu Annaberg. Annaberg 1855, S. XXXI.

[6]Bruno Berlet: *Die Coß von Adam Riese*. Siebzehnter Bericht über die Progymnasial- und Realschulanstalt zu Annaberg. Annaberg 1860, S. 3.

[7]Dies folgt aus Eintragungen in der *Coß* auf den Seiten 1, 4 und 325, ferner aus Hinweisen auf Literatur, die ihm von Georg Stortz in Erfurt zur Verfügung gestellt worden war.

[8]In Coß 2 verweist Adam Ries etwa auf Michael Stifel und Girolamo Cardano.

[9]Berlet[6].

bedeutende Gleichungslehre aufgezeichnet hatte, oder um einen von deren späteren Bearbeitern.

Im Jahre 1955, also genau 100 Jahre nach dem Wiederentdecken der „Coß", erkannte Willi Saemann aus Halle in der Handschrift C 349 der Sächsischen Landesbibliothek Dresden dieses „Buch Algebre", also das Buch des Algebras, von dem in der „Coß" öfters gesprochen wird, und zwar deshalb, weil C 349 auch aus der Feder von Adam Ries stammt.[10]

Kodex Dresden C 349 umfaßt etwa 300 beschriebene Seiten, es handelt sich um eine deutsche algebraische Abhandlung, die in der Fachwelt die Bezeichnung: die „Algebra des Initius Algebras" erhielt.[11] Hierbei dreht es sich um ein laut Inhaltsverzeichnis ursprünglich aus acht oder neun Büchern aufgebautes mathematisches Werk,[12] von dem sich nur die Bücher I bis III erhalten haben. Diese „Algebra des Initius Algebras" ist nicht als eigene geistige Leistung von Adam Ries zu werten, sondern er hat sie von einer Vorlage, die heute vielleicht nicht mehr vorhanden ist, irgendwo abgeschrieben, und er beruft sich hierbei laufend auf den erwähnten Algebras. Als Verfasser des deutschen Traktates selbst kommt vermutlich ein Regensburger Mathematiker Andreas Alexander (um 1475 - nach 1504) in Betracht,[13] der von 1502 bis 1504 in Leipzig wirkte.

Kodex Dresden C 349 wurde von Adam Ries um das Jahr 1517 geschrieben, denn die vorhandenen Wasserzeichen wurden in hessischen und badischen Papiermühlen zwischen 1517 und 1519 verwendet.[14] Dies läßt den Rückschluß zu, daß Adam Ries damals in Hessen oder Baden lebte, vielleicht auch eine Universität in dieser Region besucht hatte; es könnte Köln oder Heidelberg gewesen sein.

Die „Algebra des Initius Algebras", wie sie in Kodex Dresden C 349 vorliegt, nahm Adam Ries als eine der Quellen für seine „Coß", und er bestätigt dies im Vorwort der „Coß" mit den Worten:[15]

> In Disem Nachgeschriebenn Buch werden vff das aller clerlichest auß getrucktt etzliche Algorithmi, Die vorlangest beschrieben vnd

[10] Willi Saemann: Neues zur Coß des Adam Rieß. *Kultur und Heimat, Kreis Annaberg*, Jg. 6, März 1959, S. 45. Willi Saemann: Adam Riese als Cossist. *Mathematik in der Schule* 2 (1964) 1, S. 64-68; 2 (1964) 3, S. 203-214; 2 (1964) 8, S. 632-639.

[11] Maximilian Curtze: Die Algebra des Initius Algebras ad Ylem geometram magistrum suum. *Abhandlungen zur Geschichte der mathematischen Wissenschaften mit Einschluß ihrer Anwendungen*, Heft 13. Leipzig 1902. Die Edition erfolgte nach der Handschrift Philos 30 der Universitätsbibliothek Göttingen von 1545, wobei die Handschriften C 8, C 349 und C 405 = KA 171 der Sächsischen Landesbibliothek Dresden mit herangezogen wurden.

[12] Dies erkennt man aus Eintragungen in den Handschriften Göttingen Philos 30, f. 1^v, und Dresden KA 171, f. 2^r.

[13] Dies folgt aus dem Text bei Fußn. 15 bzw. in Abb. 2.

[14] Es handelt sich um die Wasserzeichen „Buchstabe P" Nr. 153-156 gemäß Gerhard Piccard: *Wasserzeichen Buchstabe P*. Findbuch 4, Teil 1. Stuttgart 1977, Abt. V, S. 44; Teil 2. Stuttgart 1977, Abt. V, S. 142 und 143.

[15] Mit diesen Worten beginnt die *Coß* von Adam Ries, S. 1. Das genannte Werk liegt nun als Edition *Adam Ries. Coß*. Stuttgart/Leipzig 1992, vor.

nachgelasenn seint wurdenn Durch Algum, Bohecium, Archimedem etc. Vnd den berumbstenn In der Zall erfarnen Algebram, Den Arabischenn meister, Des gleichenn Inn Rechnung nihe gewesen, Auch schwerlich vber in eyner komen wirtt. Des Buch vom dem ding <Algebra> auß arabischer In krichisch gesacZt vonn Archimedo, Vnd alß dann auß der krichischn sprach in di lateinische Durch Apuleyum, vnnd Zum letzten Zu vnser Zeit <?> eins teyls vordeutscht Durch denn erfarnenn Mathematicum Magistrum Andream Alexandrum vnd dar nach vffs aller leichtest vnd gruntlichst wol Zu begreyfen gefertigett Durch Adam Riesenn Vom staffelstein Nach vnserß liebn Herrn geburtt Im 1524 Jhar (Abb. 2).

Aus heutiger Sicht handelt es sich um eine phantasievolle Abfolge zur Geschichte der Arithmetik – bei Adam Ries „etzliche Algorithmi" – und zur Geschichte der Algebra – hier bei Adam Ries das „Buch vom dem ding".

Die Geschichte des Cossisten Adam Ries muß also neu aufgezogen werden, zumindest anders, als dies bis vor einigen Jahren, nämlich bis zur Entdeckung von Kodex Dresden C 349 als Ries-Handschrift im Jahre 1955, der Fall gewesen war. Er hatte sich demnach, noch ehe er sich in Erfurt um 1518 als Rechenmeister und was sonst noch niederließ, schon irgendwo anders ausgiebig mit der neuen Algebra seiner Zeit, mit der Coß, auseinandergesetzt. Es könnte wie folgt gewesen sein:

Wie er in der „Coß" hervorhebt, rechnete er bereits im Jahre 1515 mit Hans Conrad Algebrabeispiele durch.[16] Er sagt nicht, wo dies war. Irgendwann um diese Zeit muß er also mit der algebraischen Lösung entsprechender Aufgaben vertraut gemacht worden sein. In Frage kämen autodidaktische Studien von irgendwelchen Handschriften, die er in Wien, Regensburg, Nürnberg, Leipzig, Erfurt, vielleicht auch in Bamberg betrieben haben könnte, denn gedruckte Algebrabücher gab es damals noch nicht. Es wäre auch denkbar, daß er bei Johannes Widmann aus Eger (um 1462 - nach 1500) direkt lernte, der um 1500, wie man neuerdings weiß, und vermutlich auch nachher, in Annaberg lebte.[17] Widmann hatte als erster überhaupt 1486 in Leipzig eine Algebravorlesung an einer Universität gehalten.[18] Denkbar wäre freilich auch, daß Adam Ries persönlich mit Andreas Alexander in Berührung kam, etwa in Köln, oder zumindest dessen algebraisches Werk kennenlernte. Denn er sagt ja ausdrücklich, daß Andreas Alexander einen Teil vom Werk des Algebras verdeutscht hatte,[19] und dies wurde nun von Adam Ries „vffs aller leichtest",[20] also auf einfachste Weise, deutsch wiedergegeben. Hiermit

[16] So berichtet Adam Ries in der *Coß*, S. 187 und 453 f.

[17] Dies folgt aus Eintragungen im 1. Häuserlehnbuch 1500/05, f. 20^{v} und 159^{r}, im Ratsarchiv Annaberg, auf die Herr Reinhart Unger, Annaberg, aufmerksam machte.

[18] Entsprechende Ankündigungen sind enthalten in Kodex C 80 der Sächsischen Landesbibliothek Dresden, f. 0^{v}, sowie in Kodex 1470 der Universitätsbibliothek Leipzig, f. 432^{r}; in Kodex 1470, f. 493^{v}, steht zu lesen, daß im Sommersemester 1486 von Johannes Widmann eine Algebravorlesung um die Gebühr von zwei Gulden abgehalten worden war.

[19] Im Vorwort der *Coß*, S. 1. Siehe Text bei Fußn. 15 bzw. Abb. 2.

[20] Im Vorwort der *Coß*, S. 1. Siehe Text bei Fußn. 15 bzw. Abb. 2.

meint er die Schrift, die wir heute als Kodex Dresden C 349 besitzen, und die zugehörige Vorlage. Adam Ries beschäftigte sich ausführlich mit dem fachlichen Inhalt der „Algebra des Initius Algebras", denn in seiner „Coß" findet sich vieles hiervon wieder.

Von der „Algebra des Initius Algebras" kennt man insgesamt vier Abschriften, drei in Dresden, eine in Göttingen.[21] Nach der Göttinger Handschrift Philos 30 wurde sie im Jahre 1902 von Maximilian Curtze herausgegeben,[22] ohne daß erkannt worden wäre, daß C 349 als ältestes der vier Manuskripte etwa 30 Jahre vor Philos 30 und von Adam Ries geschrieben wurde.[23]

Zum Inhalt der drei Bücher der „Algebra des Initius Algebras": In Buch I wird in 25 Kapiteln die Lehre von den linearen und quadratischen Gleichungen mit einer Unbekannten aufgebaut. Beachtung verdient, daß hierbei schließlich wohl die cossischen Symbole für die Potenzen der algebraischen Unbekannten gebraucht werden, nicht aber Plus- und Minuszeichen. Die sechs aus der muslimischen Tradition stammenden Gleichungsfälle $cx^2 = bx$; $cx^2 = a$; $bx = a$; $cx^2 + bx = a$; $cx^2 + a = bx$; $cx^2 = bx + a$ werden zu einzelnen Theoremen aus Euklid (um 300 v.d.Z.) Buch II in Beziehung gesetzt, anschließend werden die algebraischen Fachausdrücke und die damaligen Symbole für unser jetziges x^0, x, x^2 bis x^9 eingeführt (Abb. 3). Die Bezeichnungen *dragma*, *radix*, *census*, *cubus* bis *cubus de cubo* waren seinerzeit genormt; hierbei treten allerdings zwei unklare Begriffe *sursolidus* für x^5 und *bißsursolidus* für x^7 auf, die nach Adam Riesens Worten gesondert zu betrachten sind; es handelt sich nämlich nicht um „signa gegruntt vff die corpora vnd superficies",[24] denn sie können nicht aus Körpern und Flächen hergestellt werden, folglich nennt man sie *surdus*, also taub oder irrational, und darum heißen x^5 und x^7 richtig *surdussolidus* und *bißsurdussolidus*.[25]

In Kapitel I/17 wird für die Axiomatik der damaligen Gleichungslehre ein Kanon von acht Typen aufgestellt, und zwar jeweils aufsteigend bis zum Grad neun (Abb. 4).

Der achte Typ, der sonst nur noch in der „Coß", und zwar auch in der nämlichen Reihenfolge auftritt, beschäftigte die Mathematikhistoriker seit dem vergangenen Jahrhundert.[26] Er stammt also aus der „Algebra des Initius Algebras".

Zu bemerken ist, daß im Rahmen der „deutschen Coß" meist 24 Gleichungstypen zugrundegelegt wurden, die aus den sechs von al-Ḫwārizmī herrührenden abgeleitet sind; in der „Algebra des Initius Algebras" findet sich demgegenüber eine hiervon abweichende Anordnung.

---

[21]Siehe Fußn. 11.
[22]Curtze[11].
[23]Curtze[11], S. 440 und 447.
[24]Kodex Dresden C 349, f. XXVII[r]; *Adam Rieß vom Staffelstein*[3], S. 330.
[25]Kodex Dresden C 349, f. XXVII[v]; *Adam Rieß vom Staffelstein*[3], S. 331.
[26]Moritz Cantor: *Vorlesungen über Geschichte der Mathematik*. Band 2, 2. Auflage. Leipzig 1900, S. 423; Saemann[10(2)], S. 206-210.

Anschließend folgen Zahlbeispiele zu jedem Gleichungstyp. Hierbei ist interessant, daß der Gleichungstyp $cx^2 + a = bx$ wie auch sonst oft in der Periode der „deutschen Coß", so auch in der „Algebra des Initius Algebras" in der lateinischen Vorgabe zwar richtig angegeben ist, daß aber dann im Zahlbeispiel deutsch aus einem Übersetzungsfehler heraus $x_{1,2} \overset{!}{=} \sqrt{(\frac{b}{2c})^2 - \frac{a}{c}} \pm \frac{b}{2c}$ steht.[27] Genau dieser Tatsache in etlichen bekannten deutschen Texten des 15./16. Jahrhunderts nachzugehen, wäre eine langwierige, aber eine lohnende Aufgabe.

In Buch II der „Algebra des Initius Algebras" geht es vorwiegend um die Anwendung der vier Grundrechenarten auf Binomien der Form $ax^r \pm bx^s$. Hier werden Plus- und Minuszeichen eingeführt. Erstmals in einem deutschen algebraischen Text wird bis zur Potenz x^{18} aufgestiegen (Abb. 5). Beachtung verdienen die Namen der Primzahlpotenzen x^5, x^7, x^{11}, x^{13}, x^{17}. In einer weiteren Tafel stehen Beispiele zu Vorzeichenregeln (Abb. 6); diese stimmen zwar teilweise nicht, es ist aber einer der ersten deutschen Texte, der sie enthält.

Buch III befaßt sich großenteils mit Zahlentheorie und mit Wurzelrechnungen. Aufsteigend bis $(a + b)^9$, wird angedeutet, daß auf diese Weise die Wurzeln bis zur 9. gezogen werden könnten. Aufgrund der Ergebnisse kann man fragen, ob hier in der „Algebra des Initius Algebras" bereits die Gesetzmäßigkeiten des binomischen Dreiecks zu erkennen sind, und zwar mehr als 100 Jahre vor Blaise Pascal (1623-1662).

Plötzlich steht im Text die Bemerkung „So $x^0 + x^2 + x^3$ Werdenn Vorgleichtt x ... Ader so wir secZen x^3 vnd x werdenn vorgleichtt $x^2 + x^5$ vnnd Der ane Zal",[28] d.h., daß Gleichungen höheren Grades auch behandelt würden: „Wie ein itzliche vorgleichung In vnßer gebra vnnd almuchabola gebrachtt sol werden"[29] (Abb. 7). Falls also auf die Behandlung solcher Gleichungen höheren Grades eingegangen wurde, dann müßte die Erläuterung in den fehlenden Büchern IV bis VIII oder IX stehen.

In C 349 geht es weiter mit Zahlentheorie, etwa der Euklidische Algorithmus für den größten gemeinsamen Teiler zweier Zahlen, ferner folgt ein Restproblem der unbestimmten Analytik, dann werden die vollkommenen Zahlen unter 1000 – die Summe der Teiler ist gleich ihrem Produkt – besprochen. Bei der Behandlung des doppelten falschen Ansatzes – der *regula falsi* – spricht Adam Ries von einem „einfurendt leicht exempel", ferner bringt er „ein schlecht einfurend exempel", anschließend bringt er „ein kindisch exempel" und schließlich eines „Das dann Die kinder radende furschlagenn",[30] wie es also Kinder raten, und zwar: Man will 100 Stück Vieh um 100 Gulden kaufen, ein Ochse kostet drei Gulden, ein Esel einen Gulden, ein Schaf $\frac{1}{20}$ Gulden.

---

[27]Kodex Dresden C 349, f. XLII^v; *Adam Rieß vom Staffelstein*[3], S. 346. Hierauf hat schon Curtze[11], S. 495 Fußn. 1, mit Bezug auf Kodex Göttingen Philos 30 verwiesen.

[28]Kodex Dresden C 349, f. 94^r.

[29]Kodex Dresden C 349, f. 94^v.

[30]Kodex Dresden C 349, f. 123^v.

In der Folge findet man um die Behandlung von Wurzelausdrücken. Hier tritt der Wurzelhaken auf, den man Adam Ries bislang in der „Coß" zugeschrieben hatte, den er also schon vor seiner „Coß" verwendete. In einer sehr instruktiven Tabelle geht es um die Wurzelschreibweise (Abb. 8), diese Schreibart hat sich in der Periode der „deutschen Coß" eingebürgert. Die Darstellung in einer weiteren Tabelle soll folgendes besagen (Abb. 9): ein Quadrat oder eine 2. Potenz habe die Fläche 5, ein Würfel oder eine 3. Potenz habe den Rauminhalt 3, die vierte Potenz eines Ausdrucks sei 7, ..., die 9. Potenz eines Ausdrucks sei 13; diese Terminologie und die entsprechende Fragestellung läßt sich aus dem engen Zusammenhang mit der griechischen Irrationalitätenlehre heraus erklären, sie hatte für später keine Bedeutung.

Dann folgen Wurzelumformungen und Wurzelbeispiele, die zum Teil auf Euklid beruhen. Wie heißen $\sqrt[n]{a} : \sqrt[n]{b} = \sqrt[n]{a:b}$ und $\sqrt[m]{a} : \sqrt[n]{b} = \sqrt[mn]{a^n : b^m}$ in der „Algebra des Initius Algebras" bei Adam Ries (Abb. 10)?

Im folgenden werden die aus der Überlieferung stammenden Näherungsformeln $\sqrt{a^2+b} \approx a + \frac{b}{2a+1}$, $\sqrt[3]{a^3+b} \approx a + \frac{b}{3a^2+3a+1}$ angewandt, wobei sowohl die Theorie dasteht als auch Zahlbeispiele. So ergibt sich der Reihe nach für $\sqrt{5} = \ldots$ (Abb. 11). Die hiesige Näherung liefert 2,236067963, am Rechner folgt 2,236067978.

Dies umfaßt grob den Inhalt der noch vorhandenen drei ersten Bücher der „Algebra des Initius Algebras". Hätten Sie angenommen, daß der spätere Rechenmeister Adam Ries, der in Erfurt und in Annaberg eine Rechenschule inne hatte, der in Annaberg unserem Wissensstand nach nur in der Praxis tätig war, sich einst, Jahre vorher, mit derlei Problemen der theoretischen Mathematik beschäftigt hatte? Er hatte sich sogar intensiv damit beschäftigt, weil er hierauf große Teile seiner „Coß" aufbaute. Aber stimmt dies noch mit dem landläufigen Bild von Adam Ries überein, den man gerne bei $2 \times 2 = 4$ zitiert?

Es sei nochmals darauf verwiesen, daß die „Algebra des Initius Algebras" in Kodex C 349 aus der Feder von Adam Ries nicht sein eigenes geistiges Produkt ist, und er erwähnt dies auch laufend. Kodex Dresden C 349 ist aber der älteste derzeit bekannte Nachweis für diese umfassende deutsch geschriebene Abhandlung, die zu den großen Leistungen innerhalb der „deutschen Coß" zählt. Auf Schritt und Tritt wird dort auf einen *Algebras* verwiesen, und laufend wird die Bedeutung dieses Algebras herausgestellt. Ein erheblicher Teil des Wissens, das aus der „Algebra des Initius Algebras" stammt, wurde die Grundlage für Adam Riesens „Coß".

In seiner „Coß" zieht Adam Ries noch eine weitere Quelle heran, nämlich mehrere Abschnitte aus der jetzigen Handschrift Dresden C 80, ferner bringt er Beispiele, die nach seinen eigenen Worten von Hans Conrad, Münzprüfer in Annaberg und Eisleben, von Hans Bernecker, Rechenmeister in Leipzig, und von ihm selbst stammen.

In der neueren Adam-Ries-Forschung, die vor ungefähr 150 Jahren mit Bruno Berlet in Annaberg einsetzte, nahm seine „Coß" die herausragende Stelle ein, obwohl Berlet nur einige Abschnitte des etwa 530 Seiten umfassenden Werkes veröffentlichte. Noch offen ist bei der „Coß" die Frage, ob die jetzige Reihenfolge der darin enthaltenen Abhandlungen wirklich die zeitlich richtige ist.

Nach dem Vorwort folgt die Widmung von Coß 1 an Dr. Georg Stortz in Erfurt.

Vom fachlichen Inhalt her besteht die „Coß" von Adam Ries aus drei Teilen: Coß 1 wurde in Erfurt und Annaberg zwischen 1518 und 1524 zusammengestellt. Coß 2 wurde um 1550 aufgezeichnet. Als letzter Abschnitt ist enthalten ein Teil der „De numeris datis" des Jordanus Nemorarius (um 1220); hierbei dreht es sich um die erste nachgewiesene deutsche Bearbeitung dieses inhaltlich schwer zugänglichen Werkes aus dem 13. Jahrhundert. Alle in der „Coß" enthaltenen Abhandlungen sind entweder nicht vollständig erhalten, oder sie wurden von Adam Ries selbst nicht abgeschlossen. Gründe hierfür kennt man nicht. Zum Inhalt:

Coß 1 von S. 1-325 enthält eine lückenhafte und abgebrochene Einleitung in die Arithmetik. Wie gewöhnlich, legt Adam Ries großen Wert darauf, daß die Zahlenaufgaben geprobt werden; in der „Coß" stützt er sich hierbei u.a. auf die Neunerprobe oder die Siebenerprobe, die er in den Rechenbüchern in dieser Form nicht heranzieht, hier etwa im Beispiel (Abb. 12)

$$\begin{array}{r} 6789 \\ 3211 \\ \hline 10000 \end{array}$$

Die beim Multiplizieren und Dividieren von ihm in Worten aufgeführten Regeln stellen aus heutiger Sicht freilich eine wissenschaftliche Abhandlung dar, ebenso die Summenformeln für arithmetische und geometrische Reihen, bzw. die Formeln für Quadrat- oder Kubikwurzeln. Wir befinden uns in der Periode der „deutschen Coß", als eine grundlegende Symbolik erst geschaffen wurde. Bei Adam Ries ist wie bei seinen Vorgängern alles noch verbal ausgedrückt. Schon hier wird in der „Coß" einige Male auf Algebras verwiesen bzw. auf das „Buch Algebre", also das Buch des Algebras; auch dort, wo er den Euklidischen Algorithmus erläutert, um den größten gemeinsamen Teiler zweier Zahlen zu finden.[31]

Anschließend folgt in Coß 1 eine zweigliedrige algebraische Gleichungslehre, der die Erläuterung der Potenzen der Unbekannten bis zum Grad neun vorangeht (Abb. 13). Seine erste Anweisung verlangt, daß reduzierbare Gleichungsformen auf die niedrigst mögliche zurückzuführen sind. Wir finden der Reihe nach:

„Volgenn hernach die Acht equaciones Algebre", d.h. Adam Ries bringt die acht Gleichungstypen, wie er sie in der „Algebra des Initius Algebras" in Kodex Dresden C 349 aufgezeichnet hatte. Nur bei $cx^2 + a = bx$ gibt es die Doppellösung, die aber auch hier in Coß 1 aus einem Übersetzungsfehler lateinisch-deutsch heraus zu $x_{1,2} \stackrel{!}{=} \sqrt{(\frac{b}{2c})^2 - \frac{a}{c}} \pm \frac{b}{2c}$ führt. Wir haben also den nämlichen Fehler wie in der „Algebra des Initius Algebras". Nochmals:

[31] *Coß*, S. 50-52.

Bis auf die Zahlen und auf die Symbole für die Potenzen der algebraischen Unbekannten steht alles nur in Worten ausgedrückt da. Die geometrischen Beweise der Muslime wurden in den Texten vor 500 Jahren meist nicht mitgeliefert, somit ist es nicht verwunderlich, daß innerhalb der „deutschen Coß" der genannte Gleichungstyp meist falsch behandelt wird. Das Zahlbeispiel $3x^2 + 21 = 24x$ führt somit zu $x_{1,2} \stackrel{!}{=} \sqrt{16-7} \pm 4 = 3 \pm 4$, wobei die Probe für $x_1 = 7$ stimmt. Mehr brauchte er aber nicht.

Zum achten Gleichungstyp, der auf niedere Grade reduzierbar ist, bringt er das Beispiel $24 + 21x^3 = 3x^6$.

Aus diesen acht Gleichungstypen des Algebras (Abb. 4) – so sagt Adam Ries – werden die 24 hergeleitet, wie sie nun auch in der „Coß" folgen (Abb. 14) und wie sie zu einem Kennzeichen der „deutschen Coß" wurden. Die auf $cx^2 + a = bx$ rückführbaren werden hierbei von ihm durchgehend falsch behandelt.

Bei den Typen 19 und 20 tritt der Wurzelhaken auf; der erscheint aber schon von der Hand von Adam Ries, wie wir gesehen haben, in der „Algebra des Initius Algebras" in Kodex Dresden C 349, ferner in der noch älteren lateinischen Handschrift Leipzig 1696.

Anschließend folgt als Hauptteil von Coß 1 eine Sammlung von etwa 322 Textbeispielen, die alle vom ersten der acht Gleichungstypen $bx = a$ sind bzw. von Adam Ries hierauf zurückgeführt werden. Die ersten 137 hiervon hatte er nach seinen Worten „In eynem altenn lateinischn fur viel Jarnn geschribenn buch gefunden"[32] und aus dem Lateinischen ins Deutsche übersetzt; es handelt sich um Stellen aus dem heutigen Kodex Dresden C 80.

Ab Aufgabe 138 stammen die Vorgaben von Hans Conrad, Hans Bernecker, „vnd darZu etzliche von mir, Adam Riesenn".[33] Es sind die Fragestellungen, bei denen er etwa 25 Jahre später in Coß 2 bemerkt, daß er sie bereits im Jahre 1515 algebraisch löste.[34] Einfachste Beispiele wechseln mit äußerst komplizierten ab, auch mit solchen, die über- oder unterbestimmte Gleichungssysteme erfordern würden.

Wie lautet eine einfache Aufgabe (Abb. 15)? Nr. 141 auf S. 188 führt auf die Gleichung $\frac{5}{7}x = 29$.

Wie sieht demgegenüber eine komplizierte Aufgabe aus (Abb. 16)? Nr. 284 auf S. 289-291. Sieben Leute wollen ein Haus um 100 Gulden kaufen, keiner hat genügend Geld. Um kaufen zu können, bräuchte A $\frac{2}{3}$ des Geldes von B, B bräuchte $\frac{3}{4}$ des Geldes von C, C bräuchte $\frac{4}{5}$ von D, D bräuchte $\frac{5}{6}$ von E, E bräuchte $\frac{6}{7}$ von F, F bräuchte $\frac{7}{8}$ von G, G bräuchte $\frac{8}{9}$ von A. Adam Ries kommt mit einer einzigen Unbekannten, mit einer einzigen Coß, aus.

Als wahrer Meister zeigt er sich bei Mischungsaufgaben. So zieht sich etwa Beispiel 301 von S. 303-305 hin (Abb. 17).

---

[32] *Coß*, S. 122.
[33] *Coß*, S. 187.
[34] *Coß*, S. 453 f.

Wenn man nicht eingelesen ist, dann kann man diese Beispiele heute nicht berechnen, weil man mit den damaligen Maßeinheiten und den seinerzeitigen Begriffen nicht umzugehen gewohnt ist.

Coß 2 läuft von S. 329-499. Sie beginnt wieder mit einem Vorwort, in dem erneut auf die Bedeutung des sagenhaften Algebras und auf Andreas Alexander hingewiesen wird, aber auch auf Christoff Rudolff (Ende 15. Jh. bis 1. Hälfte 16. Jh.), Michael Stifel (ca. 1487 - 1567) und Girolamo Cardano (1501-1576); Coß 2 besteht eigentlich nur aus Fragmenten, sie ist jedoch – wieso, so fragt man sich – in zwei Abschriften von späterer Hand in den Kodizes Dresden C 461 und C 467 zugänglich, davon einmal in C 461 anscheinend im ursprünglichen Zustand.[35] Hierzu gibt es folglich noch viele Fragen.

Bei Coß 2 kann man von einem systematisch aufgebauten Lehrbuch sprechen. Der erste Teil besteht wiederum aus einführenden Algorithmen, also Rechenanleitungen, und aus Rechenvorschriften. Es handelt sich hierbei im wesentlichen nochmals um eine Bearbeitung der „Algebra des Initius Algebras“, wobei viele der aus der griechischen Zahlenlehre übernommenen Begriffe verwendet werden, die heutzutage auch den Fachleuten schon lange nicht mehr geläufig sind, die jedoch seinerzeit die volle Aufmerksamkeit der sich hiermit beschäftigenden Wissenschaftler beanspruchten. So werden etwa bei $a+\sqrt{b}= \ldots$ und $a-\sqrt{b}= \ldots$ – in Worten – die in Abb. 18 angegebenen Identitäten herangezogen.

Anschließend folgen in Coß 2 wiederum die acht Gleichungstypen des Algebras (Abb. 4). Diesmal jedoch ist $cx^2+a=bx$ richtig gelöst, so daß im Zahlbeispiel auch zwei Resultate erscheinen.

Nun geht er über zu Textaufgaben; zunächst folgt „Die Erst Equation“, d.h. $bx=a$. Erstmals läßt sich nun der bei Adam Ries in der „Coß“ einige Male zitierte „hocherfarne Mathematicus“ Andreas Alexander aus Regensburg näher fassen, denn hier stehen acht von ursprünglich elf Rechenbeispielen von ihm.[36] Das komplizierteste hieraus besagt:

Vier Leute wollen ein Pferd um 19 Gulden kaufen. Der erste bräuchte vom zweiten $\frac{1}{3}$, der zweite vom dritten $\frac{1}{5}$, der dritte vom vierten $\frac{1}{6}$, der vierte vom Geld der übrigen je $\frac{1}{8}$. Adam Ries bemerkt im Anschluß an die Probe, daß Hans Conrad dem Andreas Alexander für die Lösung dieser Aufgabe durch die Coß einen Goldgulden gab (Abb. 19).

Anschließend rechnete Adam Ries in Coß 2 etwa 170 der Textaufgaben aus Coß 1 nochmals durch, wovon die ersten 61 hier nicht mehr erhalten sind.

In Coß 2 beabsichtigte er ursprünglich – „Was ferner vnd weitter fur felh sich begebn werden, wil ich dir gnugsam in volgenden exempeln bericht thun“[37] – gemäß diesen Worten zu allen acht Gleichungstypen des Algebras Textbeispiele zu bringen. Er kam nicht dazu – aus irgendwelchen Gründen –,

[35] Saemann[10(2)], S. 632-637.
[36] *Coß*, S. 425-434.
[37] *Coß*, S. 424.

weil er schon beim ersten Typ $bx = a$ inmitten einer begonnenen Aufgabe abbricht.[38] Coß 2 wäre andernfalls das umfangreichste Algebralehrbuch mit Beispielen der damaligen Zeit geworden. Es handelt sich um ein ausgereifteres Werk als Coß 1, denn es liegen an die 25 Jahre dazwischen. In Coß 2 geht Adam Ries viel ungezwungener mit negativen Ausdrücken um als in Coß 1. Hier merkt man deutlich, wie in der Zwischenzeit der Bereich der natürlichen, rationalen und irrationalen um die negativen Zahlen erweitert wurde.

In einem Abschnitt des jetzigen Kodex Dresden C 80, f. 316^r-323^v, besaß Adam Ries auch die Vorlage für die „De numeris datis" oder die „Data" des Jordanus Nemorarius in seiner „Coß", denn in C 80 kommentierte er einzelne Fragestellungen am Rand. Ihm ging es hierbei darum, die schwer verständlichen „Data" des Jordanus ins Deutsche zu bringen und die vorhandenen Zahlbeispiele vermittels der symbolischen Algebra seiner Zeit, also mit der cossischen Methode, zu lösen. Dies führte meist auf quadratische Gleichungen, die er dann wie in der „Algebra des Initius Algebras" behandelte. Die Abfassung der „Data" in der „Coß" von Adam Ries, S. 507-531, dürfte zeitlich zwischen Coß 1 und Coß 2 liegen.

Weder die „Algebra des Initius Algebras", wie sie uns in Kodex C 349 von der Hand von Adam Ries vorliegt, noch seine „Coß" hatten breiten Einfluß auf die Entwicklung der Mathematik im 16. Jahrhundert, denn beide wurden seinerzeit nicht gedruckt. Abschriften dürften in Umlauf gewesen sein. Adam Ries zeigt in beiden Kodizes – der eine hat etwa 300 beschriebene Seiten, der andere etwa 530 –, daß er ein äußerst fleißiger und ein sensibler Mensch war, der das nötige Gespür für die Feinheiten der damaligen algebraischen Gleichungslehre mitbrachte. Es ist möglich, daß er ursprünglich ein anderes Berufsziel verfolgte, denn er befaßte sich mit dem Inhalt dieser beiden umfangreichen Werke bereits zu einer Zeit, ehe er sich um 1522 in Erfurt erstmals als Rechenmeister bezeichnete. In die Entstehungszeit von Coß 1, also 1518 bis 1524, fallen zusätzlich auch seine drei Rechenbücher, die bis zum Jahre 1525 zumindest im Konzept fertiggestellt waren.

Worin besteht nun das hier angesprochene Verdienst von Adam Ries? Was läßt sich nun zusammenfassend sagen bezüglich seiner Leistungen als Cossist, bezüglich seiner Leistungen, die nicht nur die seinerzeitige Mathematik beeinflußten, sondern die das Bild der Mathematik seither mitprägen? Wir kennen zwei deutsche Algebratexte aus dem 15. Jahrhundert. Einer hiervon wurde 1461 in Regensburg im Kloster St. Emmeram aufgezeichnet,[39] der an-

---

[38] Es handelt sich um die Aufgabe 171 in der Numerierung von Adam Ries in der *Coß*, S. 499.

[39] Jetzt Kodex Clm 14908 der Bayerischen Staatsbibliothek München, f. 133^v-134^v. Diese Textstelle ist abgedruckt bei Carl Immanuel Gerhardt: Zur Geschichte der Algebra in Deutschland. *Monatsberichte der königlich preußischen Akademie der Wissenschaften.* Berlin 1870, S. 142 f.; ferner abgedruckt bei Maximilian Curtze: Ein Beitrag zur Geschichte der Algebra in Deutschland im 15. Jahrhundert. *Abhandlungen zur Geschichte der Mathematik*, Heft 7. Leipzig 1895, S. 49 f.

dere 1481 in Leipzig.[40] Beim nächsten jetzt bekannten nachgewiesenen dürfte es sich bereits um die „Algebra des Initius Algebras" in Kodex Dresden C 349 handeln. Im Jahre 1521 erschien das erste gedruckte deutsche Algebrabuch[41] von Heinrich Schreyber (vor 1496 - Winter 1525/26), der vermutlich 1492 in Erfurt geboren wurde.

Es ist nicht anzunehmen, daß Adam Ries den jetzigen Kodex Dresden C 349 nur gedankenlos schrieb, um eben eine Kopie von einer ihm zugänglichen Vorlage zu erhalten, denn auf der dortigen „Algebra des Initius Algebras" ist ja ein beachtlicher Teil seiner vermutlich um 1518 in Erfurt begonnenen Coß 1 begründet, ebenso die noch vorhandene Einleitung von Coß 2.

Im Jubiläumsjahr 1992 wurde die „Coß" von Adam Ries bei Teubner in Leipzig vollständig zugänglich,[42] von Kodex Dresden C 349 Buch I in einer Edition in Staffelstein.[43] Hier in diesem Beitrag konnten nur vorläufige neue Erkenntnisse zum Inhalt dieser beiden Handschriften von Adam Ries vorgelegt werden. Es bleibt nun der künftigen Adam-Ries-Forschung vorbehalten, das Wirken dieses bedeutenden Renaissancemenschen ins rechte Licht zu rücken, denn Adam Ries hat sich zwar durch die Qualität seiner drei Rechenbücher und durch die Art seines Rechenunterrichts ins Gedächtnis der Menschen im ganzen deutschen Sprachraum eingegraben, um seine Bedeutung als Mathematiker, der in der vorderen Linie der seinerzeitigen Entwicklung stand, weiß man jedoch auch jetzt noch nicht vollständig Bescheid.

[40] Es ist die sogenannte „Deutsche Algebra" in Kodex Dresden C 80, f. 368^{r}-378^{v}. Hierzu Kurt Vogel: *Die erste deutsche Algebra aus dem Jahre 1481. Nach einer Handschrift aus C 80 Dresdensis herausgegeben und erläutert von Kurt Vogel*. Bayerische Akademie der Wissenschaften, Mathematisch-naturwissenschaftliche Klasse, Abhandlungen, Neue Folge, Heft 160. München 1981.

[41] Heinrich Schreyber: *Ayn new kunstlich Buech welches gar gewiß vnd behend lernet nach der gemainen regel Detre, welschen practic, regeln falsi vnd etlichen regeln Cosse mancherlay schöne vnd zuwissen notürfftig rechnung* ... Nürnberg 1521.

[42] Ries[15].

[43] *Adam Rieß vom Staffelstein*[3].

Rechenung nach der
lenge/ auff den Linihen
vnd Feder.

Darzu forteil vnd behendigkeit durch die Proportio-
nes/Practica genant/Mit grüntlichem
vnterricht des visierens.

Durch Adam Riesen.
im 1550. Jar.

Cum gratia & priuilegio
Cæsareo.

Abb. 1: Titelseite des dritten Rechenbuches von Adam Ries.

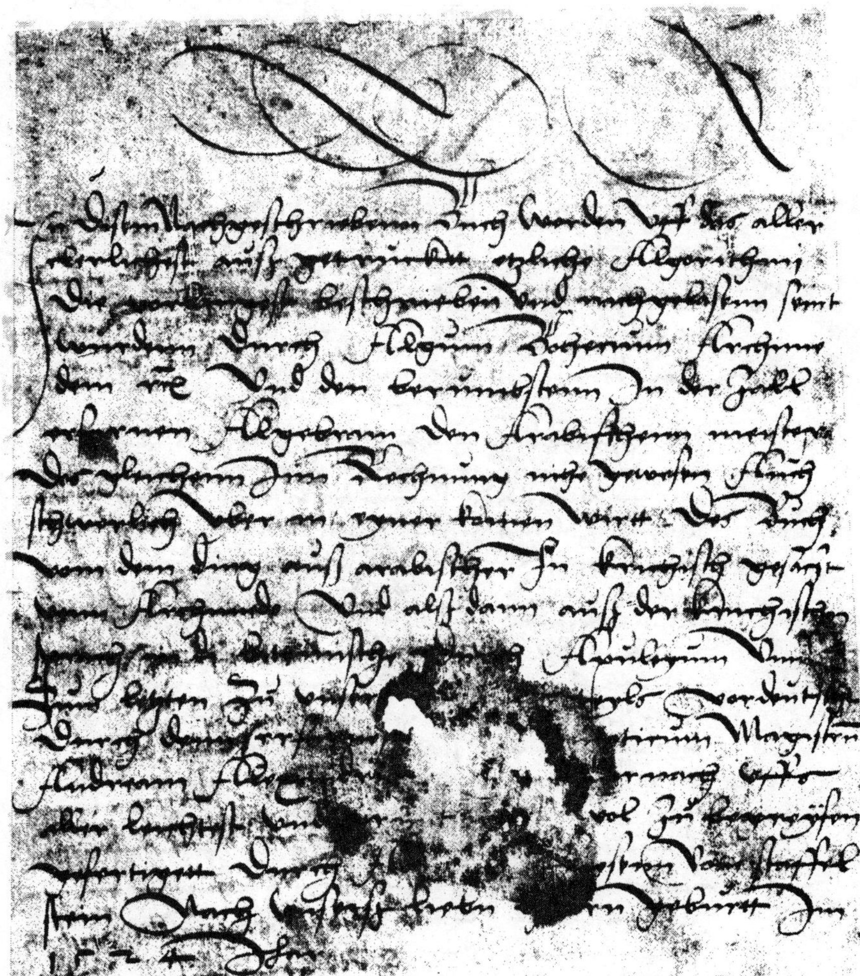

Abb. 2: Vorwort zur *Coß* von Adam Ries, S. 1. Signatur $0^{M}_{1}0$ im Erzgebirgsmuseum in Annaberg.

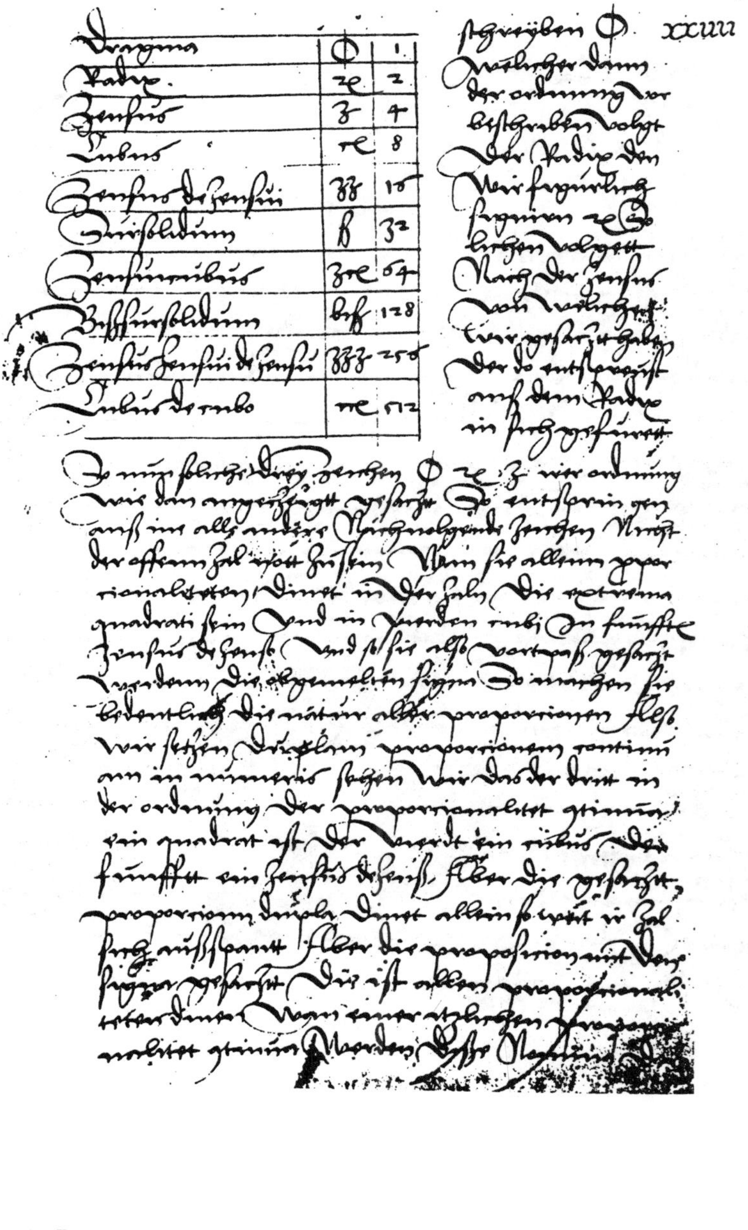

Dragma	φ	1
Radix	ꝛ	2
Zensus	ʒ	4
Cubus	ꝶ	8
Zensus de zensu	ʒʒ	16
Sursolidum	ß	32
Zensicubus	ʒꝶ	64
Bissursolidum	bß	128
Zensus zensui de zensu	ʒʒʒ	256
Cubus de cubo	ꝶꝶ	512

Abb. 3: Potenzen der algebraischen Unbekannten, Grad null bis neun, in Kodex Dresden C 349, f. XXIIII^r; *Adam Rieß vom Staffelstein*[3], S. 327.

Abb. 4:

1. $ax^n = bx^{n+1}$
2. $ax^n = cx^{n+2}$
3. $ax^n = dx^{n+3}$
4. $ax^n = ex^{n+4}$
5. $bx^{n+1} + cx^{n+2} = ax^n$
6. $ax^n + cx^{n+2} = bx^{n+1}$
7. $ax^n + bx^{n+1} = cx^{n+2}$
8. ax^n, bx^{n+k}, cx^{n+2k}
   mit Anordnungen wie in 5. - 7.

Diese acht Gleichungstypen bis einschließlich Grad neun sind ein Kennzeichen der „Algebra des Initius Algebras“, in Kodex Dresden C 349, f. XXIX^v bis XXXII^v; *Adam Rieß vom Staffelstein*[3], S. 333-336.

Der achte Typ sieht aus der Feder von Adam Ries wie folgt aus:

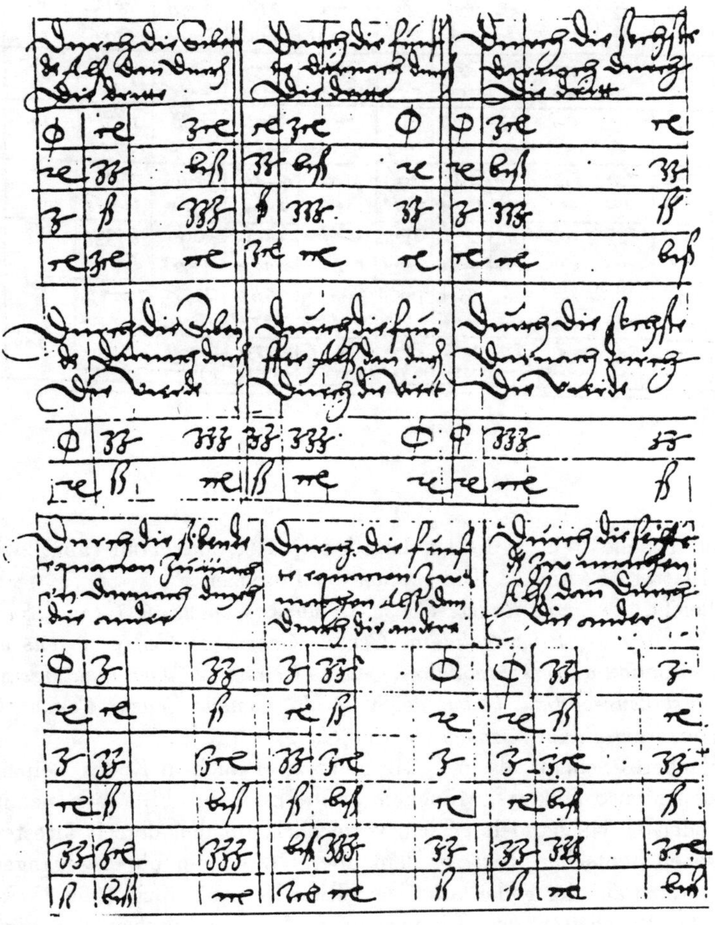

Grad
0, 3, 6
1, 4, 7
2, 5, 8
3, 6, 9

Grad
0, 4, 8
1, 5, 9

Grad
0, 2, 4
1, 3, 5
2, 4, 6
3, 5, 7
4, 6, 8
5, 7, 9

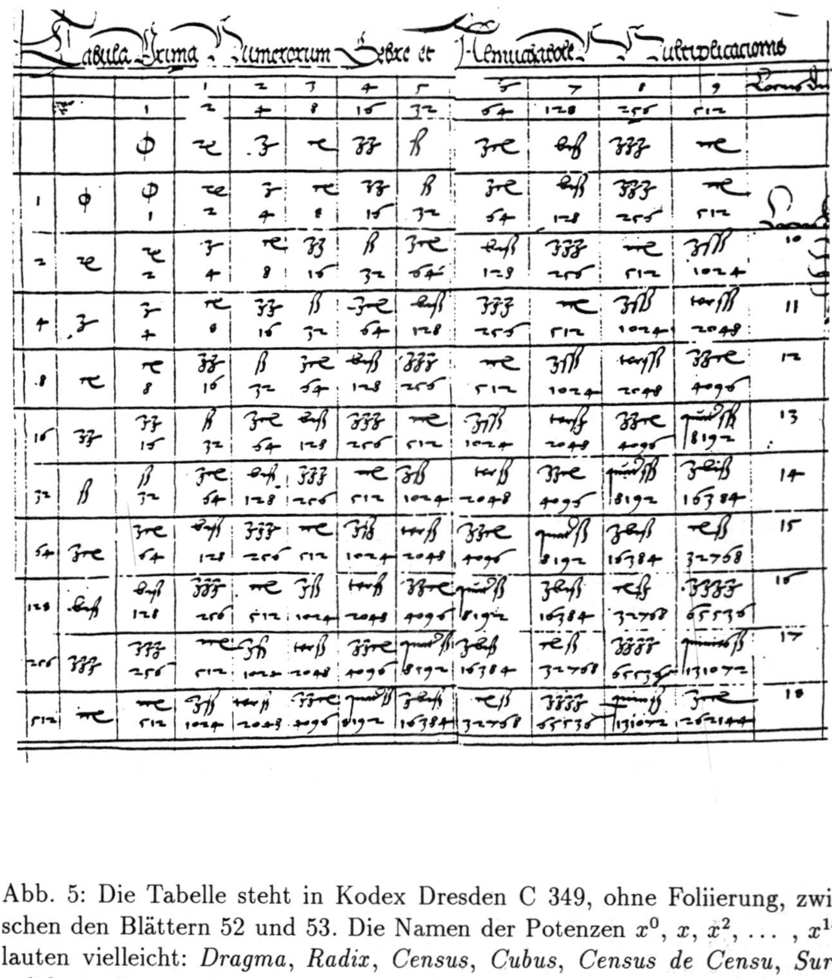

Abb. 5: Die Tabelle steht in Kodex Dresden C 349, ohne Foliierung, zwischen den Blättern 52 und 53. Die Namen der Potenzen x^0, x, x^2, ... , x^{18} lauten vielleicht: *Dragma*, *Radix*, *Census*, *Cubus*, *Census de Censu*, *Sursolidum*, *Censuicubus*, *Bißsursolidum*, *Census Censui de Censu*, *Cubus de Cubo*, *Censsursolidum*, *Terssursolidum*, *Census Censui de Cubo*, *Quaterssursolidum*, *Zensbißsursolidum*, *Cubus de Sursolido*, *Census Census Censui de Censu*, *Quiniessursolidum*, *Census Cubi de Cubo*.

Ein einfacheres Schema als in C 349 steht zwar auch in Kodex Leipzig 1696, auf den Menso Folkerts, München, aufmerksam machte, aber hier in Kodex Dresden C 349 handelt es sich vermutlich um den derzeit ältesten nachgewiesenen deutschen Text, in dem über Grad neun hinausgegangen wurde, noch dazu aus der Feder von Adam Ries. Kodex Leipzig 1696 dürfte ein lateinischer Paralleltext zur Originalschrift oder eine Abschrift der originalen „Algebra des Initius Algebras" sein.

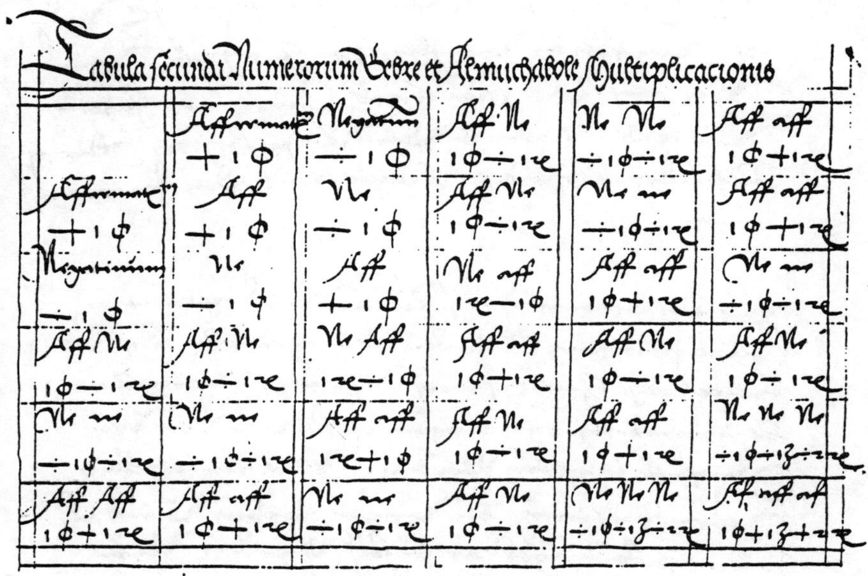

Tabula secundi Numerorum Cebre et Almuchabole Multiplicacionis

	Affirmat. +1φ	Negativum ÷1φ	Aff Ne 1φ÷1℞	Ne Ne ÷1φ÷1℞	Aff aff 1φ+1℞
Affirmat. +1φ	Aff +1φ	Ne ÷1φ	Aff Ne 1φ÷1℞	Ne ne ÷1φ÷1℞	Aff aff 1φ+1℞
Negativum ÷1φ	Ne ÷1φ	Aff +1φ	Ne aff 1℞—1φ	Aff aff 1φ+1℞	Ne ne ÷1φ÷1℞
Aff Ne 1φ÷1℞	Aff Ne 1φ÷1℞	Ne Aff 1℞÷1φ	Aff aff 1φ+1℞	Aff Ne 1φ÷1℞	Aff Ne 1φ÷1℞
Ne ne ÷1φ÷1℞	Ne ne ÷1φ÷1℞	Aff aff 1℞+1φ	Aff Ne 1φ÷1℞	Aff aff 1φ+1℞	Ne Ne Ne ÷1φ÷1ʒ÷2℞
Aff Aff 1φ+1℞	Aff aff 1φ+1℞	Ne ne ÷1φ÷1℞	Aff Ne 1φ÷1℞	Ne Ne Ne ÷1φ÷1ʒ÷2℞	Aff aff aff 1φ+1ʒ+2℞

Abb. 6: Das entsprechende Schema aus Kodex Leipzig 1696 ist bedeutend weniger instruktiv als das hier gezeigte aus Kodex Dresden C 349, ohne Foliierung, zwischen den Blättern 52 und 53, das allerdings auch nicht einwandfrei ist.

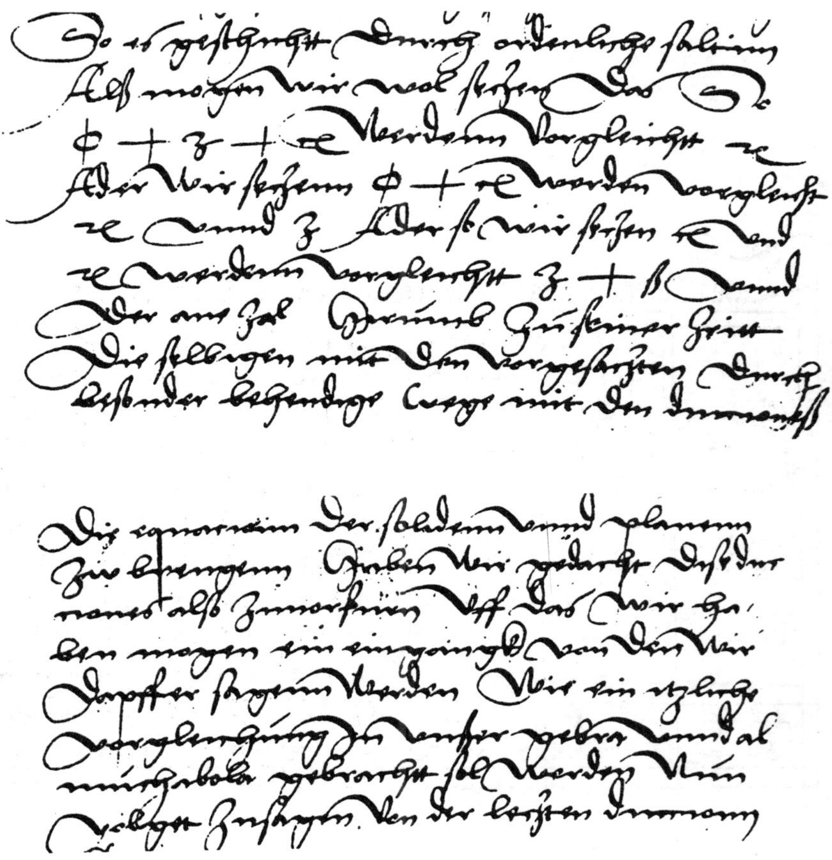

Abb. 7: Mit solchen Redewendungen wird in der „Algebra des Initius Algebras“ auf die Behandlung von Gleichungen höheren als zweiten Grades hingewiesen, wie hier in Kodex Dresden C 349, f. 94^{r}f.

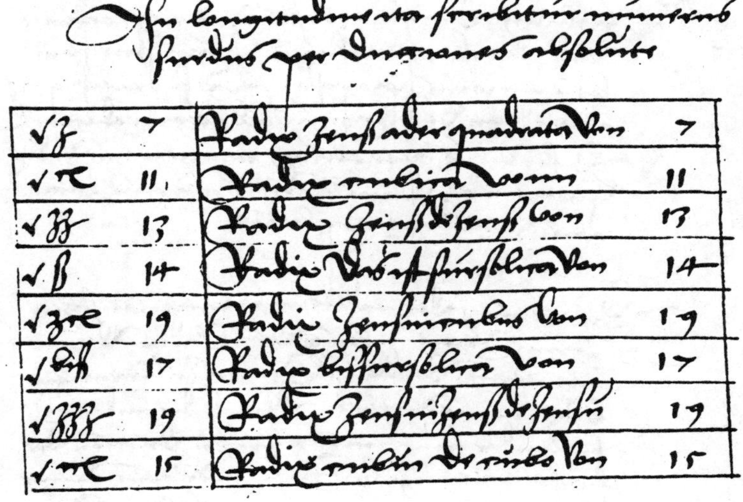

Abb. 8: In Kodex Dresden C 349, f. 126[v], stehen diese Bezeichnungen für $\sqrt{7}$, $\sqrt[3]{11}$, $\sqrt[4]{13}$, ..., $\sqrt[9]{15}$. Die nämliche Terminologie wurde z.B. von Michael Stifel angewandt.

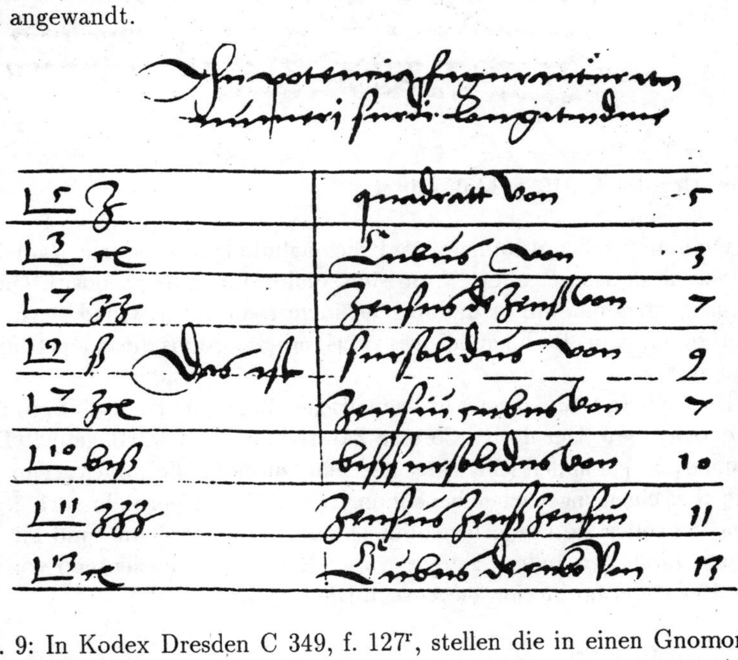

Abb. 9: In Kodex Dresden C 349, f. 127[r], stellen die in einen Gnomon gesetzten Zahlen die n-ten Potenzen ihrer n-ten Wurzeln dar, etwa $5 = (\sqrt{5})^2$, $3 = (\sqrt[3]{3})^3$, ..., $13 = (\sqrt[9]{13})^9$.

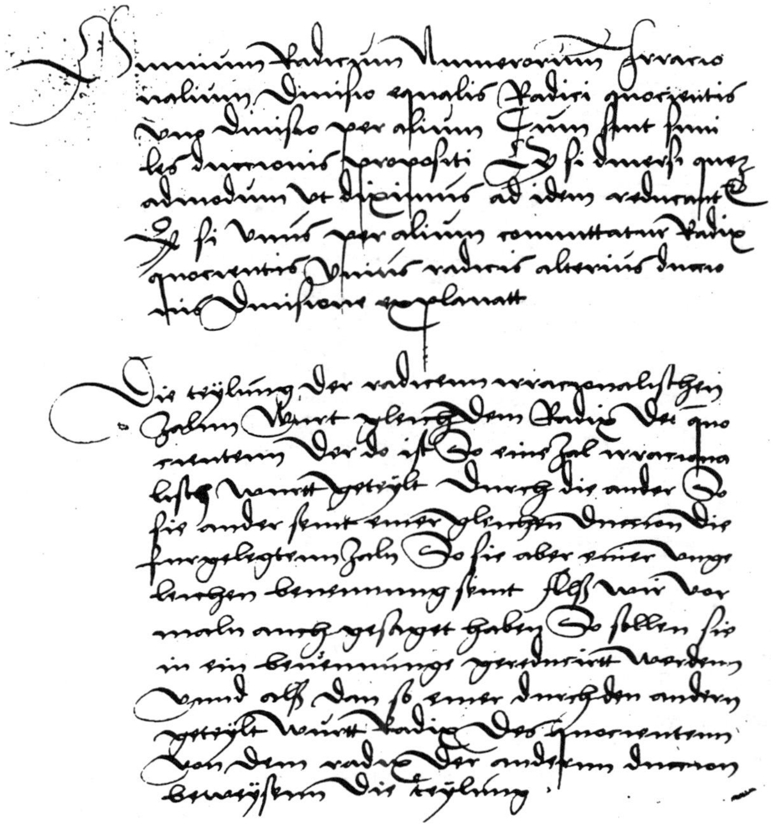

Abb. 10: Der Text für $\sqrt[n]{a} : \sqrt[n]{b} = \sqrt[n]{a : b}$ und $\sqrt[m]{a} : \sqrt[n]{b} = \sqrt[mn]{a^n : b^m}$ in Kodex Dresden C 349, f. 145^{v}, lautet:

„Omnium Radicum Numerorum Irracionalium Diuisio equalis Radici quocientis, vno diuiso per alium, Cum sint similes duccionis propositi. Quod si diuersi, quemadmodum vt diximus, ad idem reducantur. Quod si vnus per alium conmittatur, Radix quocientis vnius radicis alterius duccionis Diuisione explanatt".

„Die teylung Der radicenn irracionalischen Zalnn Wurt gleich Dem Radix Des quocientenn, Der do ist, So eine Zal irracionalisch wurtt geteylt Durch die ander, So sie ander seint einer gleichen Duccion, Die furgelegtenn Zaln. So sie aber einer vngeleichen benennung seint, Alß wir vormaln auch gesaget haben, So sollen sie in ein benennunge gereducirtt werdenn vnnd alß Dan, so einer durch den andern geteylt wurtt, Radix Des quocientenn von Dem radix Der andernn duccion beweysenn Die teylung".

Abb. 11: In Kodex Dresden C 349, f. 82[v], stünde in heutiger Schreibweise der Reihe nach:

$$\begin{aligned}\sqrt{5} &= \frac{\sqrt{4999696+304}}{1000} = \frac{\sqrt{2236^2+304}}{1000} \\ &\approx \frac{2236}{1000} + \frac{304}{(2\cdot 2236+1)\cdot 1000} = \frac{2236}{1000} + \frac{304}{4473\cdot 1000} \ .\end{aligned}$$

Abb. 12: In der *Coß*, S. 12, läuft die Siebenerprobe für $6789+3211=10000$ der Reihe nach wie folgt im Andreaskreuz ab:

$6789 \quad \bmod 7 = 6$
$3211 \quad \bmod 7 = 5$
$(6+5) \quad \bmod 7 = 4$
$10000 \quad \bmod 7 = 4$

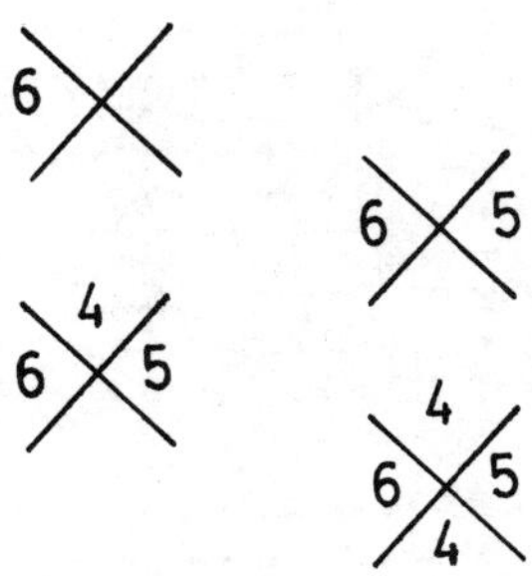

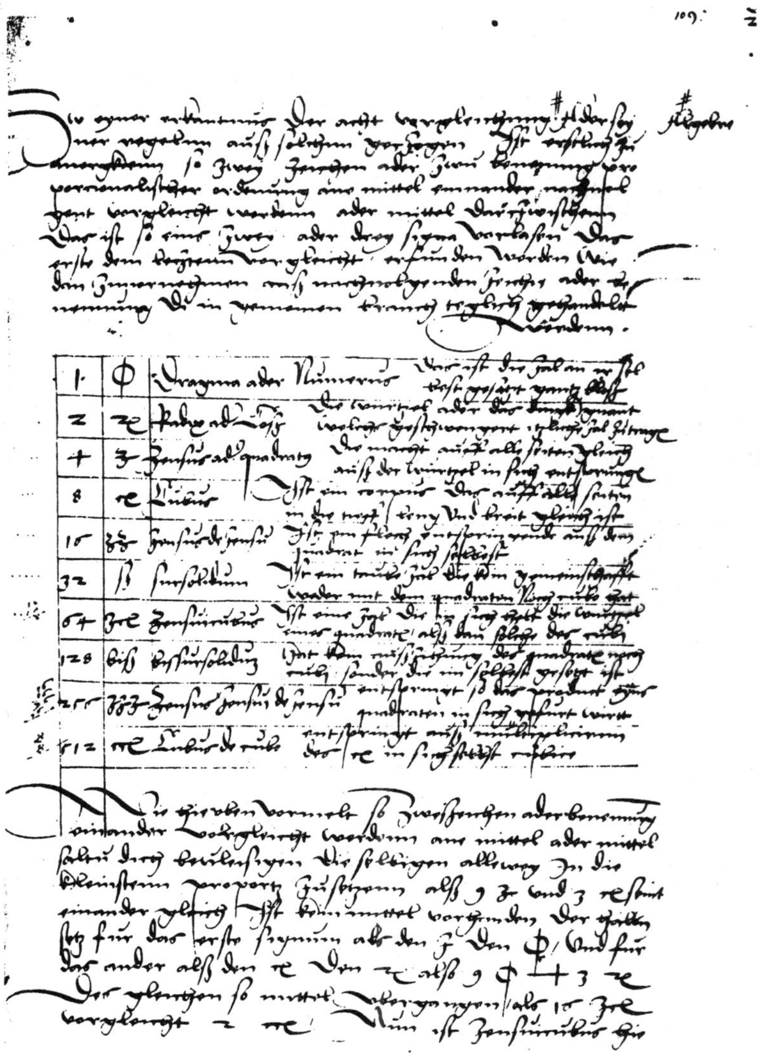

Abb. 13: Schreibart und Namensgebung der Potenzen der algebraischen Unbekannten, Grad null bis neun, sind in der *Coß*, S. 109, gegenüber Kodex Dresden C 349 stark erweitert – siehe Abb. 3.

Abb. 14: In der *Coß*, S. 115-121, stehen folgende 24 in der „deutschen Coß“ übliche Gleichungsformen, die Adam Ries vermutlich aus der jetzigen Handschrift Dresden C 80 entnahm:

1) $bx = a$
2) $cx^2 = a$
3) $cx^2 = bx$
4) $a = bx + cx^2$
5) $bx = a + cx^2$
6) $cx^2 = a + bx$
7) $dx^3 = cx^2$
8) $dx^3 = bx$
9) $dx^3 = a$
10) $bx = cx^2 + dx^3$
11) $cx^2 = bx + dx^3$
12) $dx^3 = bx + cx^2$
13) $ex^4 = dx^3$
14) $ex^4 = cx^2$
15) $ex^4 = bx$
16) $cx^2 = dx^3 + ex^4$
17) $dx^3 = cx^2 + ex^4$
18) $ex^4 = dx^3 + cx^2$
19) $cx^2 = \sqrt{bx}$
20) $cx^2 = \sqrt{c_1x^2}$
21) $ex^4 = a$
22) $a = cx^2 + ex^4$
23) $cx^2 = a + ex^4$
24) $ex^4 = a + cx^2$

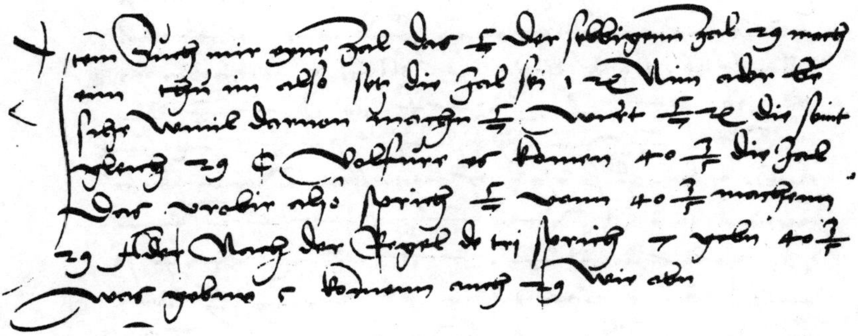

Abb. 15: Bei der Fragestellung Nr. 141 in der *Coß*, S. 188, handelt es sich um die Gleichung $\frac{5}{7}x = 29$.

Abb. 16: Adam Ries geht bei der Gesellschaftsrechnung Nr. 284 in der *Coß*, S. 289-291, wie folgt vor:

$A : x.$

$G : 100 - \frac{8}{9}x.$

$F : 100 - \frac{7}{8} \text{ von } G = 100 - \frac{7}{8}(100 - \frac{8}{9}x) \quad = \frac{100}{8} + \frac{7}{9}x.$

$E : 100 - \frac{6}{7} \text{ von } F = 100 - \frac{6}{7}(\frac{100}{8} + \frac{7}{9}x) \quad = \frac{625}{7} - \frac{2}{3}x.$

$D : 100 - \frac{5}{6} \text{ von } E = 100 - \frac{5}{6}(\frac{625}{7} - \frac{2}{3}x) \quad = \frac{1075}{42} + \frac{5}{9}x.$

$C : 100 - \frac{4}{5} \text{ von } D = 100 - \frac{4}{5}(\frac{1075}{42} + \frac{5}{9}x) \quad = \frac{1670}{21} - \frac{4}{9}x.$

$B : 100 - \frac{3}{4} \text{ von } C = 100 - \frac{3}{4}(\frac{1670}{21} - \frac{4}{9}x) \quad = \frac{565}{14} + \frac{1}{3}x.$

"Die weyl du nun itzlichem seynen geburenden teyl der Coß noch gefunden, Mustu ferner Zur bekanten Zal procedirnn": A begehrt $\frac{2}{3}$ von B :
$x + \frac{2}{3}(\frac{565}{14} + \frac{1}{3}x) = 100.$
„Ist recht gemachtt am sonabent Nach conuersionis pauli anno 1524".

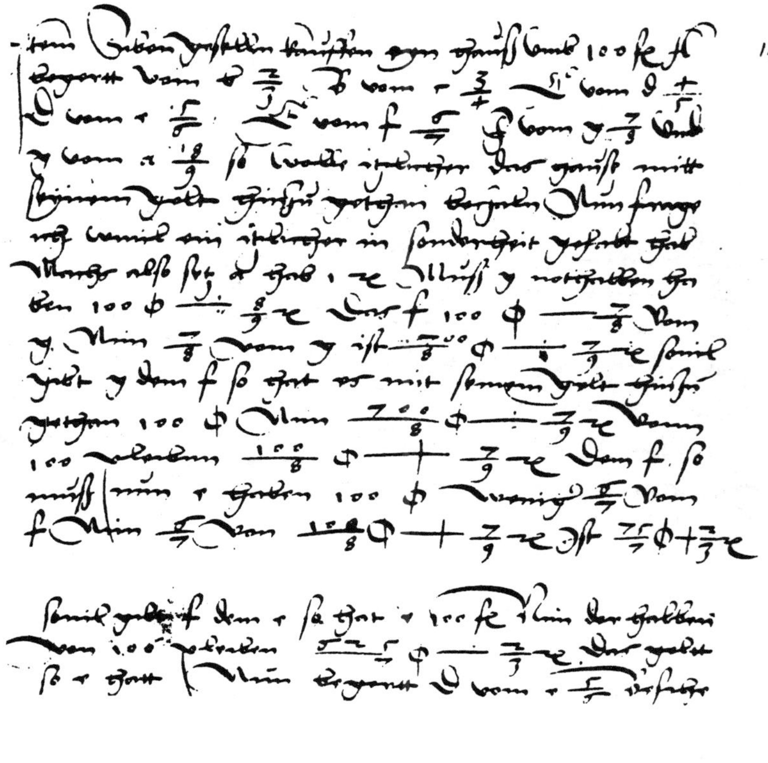

weniger 1/6 ist von 1 alß dann [illegible] φ − 2/3 ℞
macht [illegible] φ − 5/9 ℞ sovil gibt 1 dem
d ß hat d 100 Nun der halben [illegible] φ − 5/9 ℞
von 100 bleiben dem d [illegible] φ + 5/9 ℞ dar
von begert 1 [illegible] dieser weniger [illegible] ist von dem
gelt d macht [illegible] φ + 4/9 ℞ sovil gibt
d dem 1 Nun von 100 bleiben [illegible] φ −
4/9 ℞ sovil hat das 1 davon wil e haben 2/7
sprich 2/7 von 1 macht [illegible] φ − 1/3 ℞
Nun von 100 bleiben dem e [illegible] φ + 1/3 ℞
Die weil du nu etlichen seinen gebürenden teil
der ℞ noch gefunden Mustu ferner zur bekante
zal vorordern thu im also Nim 2/3 von di
alß von [illegible] φ + 1/3 ℞ ist [illegible] φ + 2/9 ℞
das adder zu dem gelt a alß 1 ℞ komen 1 2/9 ℞
+ [illegible] φ gleich 100 φ Nim beiderm
teylen [illegible] [illegible] φ bleibt einem teyl
1 2/9 ℞ und dem andern [illegible] φ teil
di φ in ℞ alß [illegible] in 1 2/9 ℞ komen 59 fl
und [illegible] sovil fl hat a ß habt e [illegible] φ
+ 1/3 ℞ macht 60 [illegible] fl ß hat 1 [illegible]
φ − 4/9 ℞ das ist 52 [illegible] fl ß hat d
[illegible] φ + 5/9 ℞ macht 58 [illegible] fl ß hat
1 [illegible] φ − 2/3 ℞ das ist 49 [illegible] fl ß
hat f 59 1/66 fl und das g hat 100 φ
− 8/9 ℞ macht 40 fl und [illegible] fl ist recht
gemacht am sonabent Nach conversionis pauli
anno 1524 das magstu proben also
Nim 2/3 von e das ist von 60 [illegible] macht 40
40 15/77 adder zu a gelt alß 59 62/77 komen 100

fl Also auch nim 2/7 von gelt 1 adder dem e ß
hastu 100 fl Also allenthalben komen dir 100 fl
welchem geselln ß der nechst nachvolget im stem
begeren [illegible]

Abb. 17: In der Mischungsrechnung Nr. 301 in der *Coß*, S. 303-305, verläuft der Rechengang wie folgt:

Ein Münzmeister hat ein erstes 5 Mark 12 Lot schweres Stück aus Gold und Kupfer; eine Mark hiervon kostet $75\frac{3}{5}$ fl, ein Karat Feingold gerechnet zu 3 fl 12 ß = $3\frac{3}{5}$ fl, das Kupfer kostet nichts. Hinzu kommt ein 6 Mark 4 Lot schweres zweites Stück aus Gold, Silber und Kupfer. Nach dem Schmelzen findet er in einer Mark 4 Karat Kupfer, ferner ist das nun in einer Mark Mischung vorhandene Gold und Silber zusammen 68 fl wert.
1 Mark = 24 Karat; 1 Mark = 16 Lot; 1 fl <Gulden> = 20 ß <Schilling>.
Eine Mark Feinsilber kostet $7\frac{1}{2}$ fl.
Für das 1. Stück folgt:
1 Mark ... $75\frac{3}{5}$ fl;
5 Mark 12 Lot ... $75\frac{3}{5} \cdot 5\frac{3}{4}$ fl = $434\frac{7}{10}$ fl;
$3\frac{3}{5}$ fl ... 1 Karat;
$75\frac{3}{5}$ fl ... 21 Karat.

24 Karat – 21 Karat = 3 Karat. Beim 1. Stück handelt es sich um 21karätiges Gold bzw. 3karätiges Kupfer. 1 Mark des 1. Stückes enthält 3 Karat Kupfer, das ganze 1. Stück zu 5 Mark 12 Lot enthält $17\frac{1}{4}$ Karat Kupfer.
Beim Schmelzen kommen zusammen 5 Mark 12 Lot und 6 Mark 4 Lot, insgesamt 12 Mark. Jede dieser 12 Mark enthält 4 Karat Kupfer, somit enthalten 12 Mark genau 48 Karat Kupfer. $48 - 17\frac{1}{4} = 30\frac{3}{4}$; $30\frac{3}{4}$ Karat Kupfer enthält das 2. Stück. Das 2. Stück zu 6 Mark 4 Lot enthält $30\frac{3}{4}$ Karat Kupfer, 1 Mark aus dem 2. Stück enthält folglich $4\frac{23}{25}$ Karat Kupfer. $24 - 4\frac{23}{25} = 19\frac{2}{25}$; das 2. Stück enthält $19\frac{2}{25}$ Karat Gold und Silber je Mark.
Nach dem Zusammenschmelzen:
1 Mark der Legierung kostet 68 fl, 12 Mark der Legierung kosten 816 fl. Hiervon geht der Preis des 1. Stückes ab: 816 fl – $434\frac{7}{10}$ fl = $381\frac{3}{10}$ fl.
Das zweite Stück wiegt 6 Mark 4 Lot und hat einen Wert von $381\frac{3}{10}$ fl; es handelt sich hierbei um eine $19\frac{2}{25}$ karätige Edelmetallmischung aus Gold und Silber. Somit kostet eine Mark hiervon ($381\frac{3}{10} : 6\frac{1}{4}$) fl = $61\frac{1}{125}$ fl.
Für das 2. Stück ergibt sich:

	Gold	Silber
Preis je Mark	$3\frac{3}{5} \cdot 24$ fl	$7\frac{1}{2}$ fl
Preis je Karat	$3\frac{3}{5}$ fl	$7\frac{1}{2} : 24$ fl = $\frac{15}{48}$ fl = $\frac{5}{16}$ fl
Karat	x	$19\frac{2}{25} - x$
Preis	$3\frac{3}{5}x$ fl	$(19\frac{2}{25} - x) \cdot \frac{5}{16}$ fl = $(5\frac{77}{80} - \frac{5}{16}x)$ fl

Nachdem eine Mark im 2. Stück $61\frac{1}{125}$ fl wert ist, folgt die Gleichung

$3\frac{3}{5}x + (5\frac{77}{80} - \frac{5}{16}x) = 61\frac{1}{125}$; $3x + \frac{48}{80}x - \frac{25}{80}x = 61\frac{16}{2000} - 5\frac{1925}{2000}$; $3\frac{23}{80}x = 55\frac{91}{2000}$;

$x = \frac{110091}{2000} \cdot \frac{80}{263} = \frac{110091 \cdot 2}{50 \cdot 263} = \frac{110091}{6575} = 16\frac{4891}{6575}$.

Im 2. Stück befinden sich $16\frac{4891}{6575}$ Karat Gold in jeder Mark.

$19\frac{2}{25} - 16\frac{4891}{6575} = 18\frac{7101}{6575} - 16\frac{4891}{6575} = 2\frac{2210}{6575}$; $2\frac{2210}{6575}$ Karat Silber finden sich in jeder Mark des 2. Stückes.

Adam Ries hat zum Schluß einen Rechenfehler, er erhält $2\frac{2226}{6575}$; als Facit schreibt er: „Ist recht gmachtt, hat keynen nehernn wegk".

Abb. 18: Die einzelnen Schritte von Adam Ries in der *Coß*, S. 380, bei Umformung von $\sqrt{4+\sqrt{7}}$:

$$a+\sqrt{b} = \left(\frac{\sqrt{a^2-b}}{2}+\frac{a}{2}\right)+\left(-\frac{\sqrt{a^2-b}}{2}+\frac{a}{2}\right)+2\sqrt{\frac{a^2}{4}-\frac{a^2-b}{4}} \quad ,$$

wobei gilt:

a^2-b	$4^2-7=9$
$\frac{a^2-b}{4}$	$\frac{9}{4}$
$\sqrt{\frac{a^2-b}{4}} = \frac{\sqrt{a^2-b}}{2}$	$\sqrt{\frac{9}{4}} = \frac{3}{2}$
$\frac{\sqrt{a^2-b}}{2}+\frac{a}{2} = \frac{a}{2}+\frac{\sqrt{a^2-b}}{2}$	$\frac{3}{2}+\frac{4}{2} = 3\frac{1}{2}$
$a-\left(\frac{\sqrt{a^2-b}}{2}+\frac{a}{2}\right) = \frac{a}{2}-\frac{\sqrt{a^2-b}}{2}$	$4-3\frac{1}{2} = \frac{1}{2}$
$\sqrt{\frac{a}{2}+\frac{\sqrt{a^2-b}}{2}}$	$\sqrt{3\frac{1}{2}}$
$\sqrt{\frac{a}{2}-\frac{\sqrt{a^2-b}}{2}}$	$\sqrt{\frac{1}{2}}$
$\sqrt{a+\sqrt{b}} = \sqrt{\frac{a}{2}+\frac{\sqrt{a^2-b}}{2}}+\sqrt{\frac{a}{2}-\frac{\sqrt{a^2-b}}{2}}$	$\sqrt{4+\sqrt{7}} = \sqrt{3\frac{1}{2}}+\sqrt{\frac{1}{2}}$
$a+\sqrt{b} = \left(\frac{a}{2}+\frac{\sqrt{a^2-b}}{2}\right)$	
$+2\sqrt{\left(\frac{a}{2}+\frac{\sqrt{a^2-b}}{2}\right)\left(\frac{a}{2}-\frac{\sqrt{a^2-b}}{2}\right)}$	$4+\sqrt{7} = (\sqrt{3\frac{1}{2}}+\sqrt{\frac{1}{2}})^2$
$+\left(\frac{a}{2}-\frac{\sqrt{a^2-b}}{2}\right) = a+2\,\frac{\sqrt{a^2-(a^2-b)}}{2}$.	

Entsprechendes gilt für $\sqrt{a-\sqrt{b}}$.

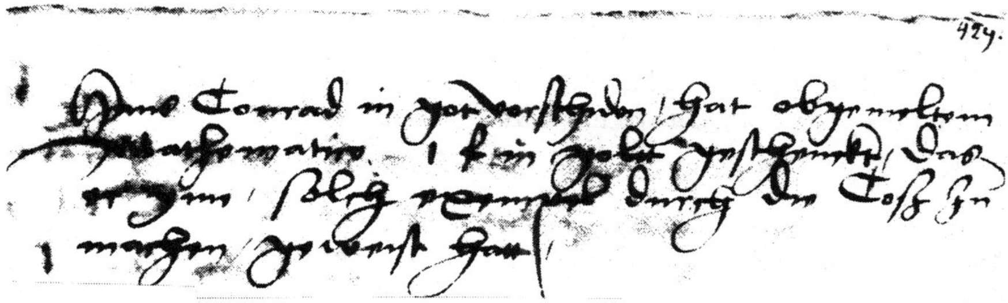

Abb. 19: Eine vielzitierte Bemerkung aus der *Coß*, S. 429: „Hans Conrad in got vorschiden, hat obgemeltem Mathematico 1 f <Gulden> in goltt geschenckt, Das er yme, soch exempel durch die Coß zu machen, geweist hatt."

"The Peacock's Tail": On the names of some theorems of Euclid's *Elements*

Paul Kunitzsch

In the course of the vast transmission of Euclid's *Elements* (Greek – Arabic – medieval Latin) some theorems were given individual names (Burnett 1985: 479: "nicknames"). In "Excursus II", Heath 1926: 415-418 discussed some of them without, however, arriving at the correct explanation in all instances. Since, through the almost complete collection of microfilms of involved manuscripts kept in the Institute of Prof. M. Folkerts, Munich, it was now possible to extend the investigation of these names to the entire Arabic-Latin stage of transmission, a new analysis could be tackled[1]. The results are here presented as a token of gratitude to H.L.L. Busard who has contributed so much to our knowledge of the medieval Western Euclidean tradition through his several valuable editions.

The names discussed in the following refer to five theorems: I,5; I,47; III,7; III,8 and IV,10.

While for one name ("the theorem of the Bride", I,47) Greek origin has been proved, the other four plus a second name for I,47 seem to be of Arabic origin. In the West, their introduction seems to be connected with Adelard of Bath. We find them in manuscripts of "Adelard I" (not mentioned in the apparatus of Busard's edition, 1983). Afterwards they occur mainly in "Adelard II" and the derived version of Campanus. Isolated quotations are also found in other versions (perhaps added in the manuscripts by later hands). In late medieval and Renaissance times speculations arose on the meanings of these names which, naturally, often failed to identify the true background. The names – mostly a choice of only one or some of them – are found in the following versions of the *Elements*[2]:

Arabic: Isḥāq–Thābit (F 6.2), 4 MSS.; an anonymous abbreviation of Book I in MS Paris BN ar. 2500[3]; the *Taḥrīr* of Naṣīr al-Dīn al-Ṭūsī[4]; the printed version ascribed to al-Ṭūsī, Rome 1594[4,5].

---

[1]I gratefully acknowledge the kind help of Prof. M. Folkerts and Dr. R. Lorch, Munich, in tracing some of the sources here used.

[2]Reference is made here to the exhaustive bibliographical list in Folkerts 1989. "F x" refers to the various sections in this list; the manuscripts are cited in abbreviated form. Below, in the discussion of the names, the manuscripts are cited with very short abbreviations.

[3]This MS has to be removed from F 6.2.

[4]Not listed in F; cf. Sezgin 1974: 113. The MSS inspected are: Istanbul, Saray, Ahmet III, 3453 (673 H = A.D. 1274-75); Munich, or. 848; Oxford, BL Marsh 621 (671 H = A.D. 1272-73).

[5]The version of Ibn Sīnā (980-1037) does not cite any of the names. See Ibn Sina, *Al-Shifa. Mathématiques, Géométrie (Usûl Al-Handasah)*. Ed.: I. Madkour. Texte établi par

Adelard I (F 7): Bruges 529 (s. XIII) [I,5]; Oxford, Trinity Coll. 47 (s. XII) [I,5; III,8; IV,10]; Paris BN lat. 16201 (s. XII) [I,5.47].

Gerard of Cremona (F 9): Madrid, Bibl. Univ. 102 [117-z-6] (s. XIV) [IV,10][6].

Adelard II (F 11): 14 MSS (see below, footnote 7).

Adelard III (F 12.1): Oxford, BL Digby 174 (s. XII) [III,7.8; IV,10]; Venice, BN Marciana f.a. 332 [1647] (s. XIII^2) [I,5; III,7.8; IV,10].

Anonymous commentary (F 12.3): Vat. Reg. lat. 1268 (s. XIV in.) [I,5].

Campanus (F 12.4): Cambridge, Univ. Lib. Add. 6866 (s. XIII/XIV) [I,5.47; III,7.8]; Erfurt, Amplon. fol. 37 (s. XIII/XIV) [I,5.(47?); III,7.8]; Florence, BN Magliabecchi XI 112 (A.D. 1259) [III,8]; Madrid, BN 8989 (s. XIII) [I,47]; Munich, Clm 28209 (s. XIII/XIV) [I,47; IV,10]; New York, Columbia University Plimpton 156 (ca. A.D. 1300) [I,47; III,7.8]; Vienna, Österr. Nationalbibl. 2465 (s. XIV in.) [I,5]. This is a choice of the oldest among the 130 MSS listed in F 12.4.

Albertus (Magnus?) (F 17): Vienna, Dominikanerkloster 80/45 (s. XIV) [I,5.47].

The manuscripts of the following versions were also examined, but did not contain any of these names: Greek-Latin translation (F 4); Hermann of Carinthia (F 8); "Mélanges" (F 10); Thirteenth century commentary (F 12.2); Reworkings (F 12.5 and 12.6; 12.7 does not contain the Books here involved); al-Nayrīzī (Latin transl., F 14; also not contained in the Arabic MS, Leiden or. 399, registered in F 6.1); Roger Bacon (F 18); Nicole Oresme (F 19).

In most of the texts, Arabic and Latin, the names are simply written in the margin, near the diagrams accompanying the theorems. Only in a few texts are they incorporated in the current text (Arabic: Paris 2500; al-Ṭūsī, MSS and edition of 1594; Latin: Albertus Magnus [F 17], Vienna, Domin. 80/45; Adelard III, Ven. Marc. f.a. 332 [only for I,5]).

I,5

Arabic: hādhā 'l-shakl yulaqqabu bi'l-ma'mūnī ("this theorem is named 'the Ma'mūnian'"): Rabat 1101; Paris 2500; al-Ṭūsī, MSS and 1594 / *al-shakl al-ma'mūnī* ("the Ma'mūnian theorem"): Oxf. Thurston 11 / *al-ma'mūnī* ("the Ma'mūnian"): Uppsala / *hādhā 'l-shakl yusammā 'l-nāfī* ("this theorem is called 'the One driving into exile'"): Cambr. Add. 1075.

Abd el-Hamid Sabra – Abd el-Hamid Lotfi, Cairo 1977 [on the Arabic title page: 1976]. Three manuscripts of the Isḥāq–Thābit version listed in F 6.2 were not available: Kabul, Kitābkhāne-i Wizārat-i Maᶜārif [297 is not the shelf-mark of the MS]; Kastamonu 607; and Rampur, Raza Library ᶜArshī 200.

[6]This manuscript was not used in Busard's edition of Gerard's version, 1983.

Latin: Adelard I: *eleufuga* (Bruges 529); *elnefea id est fuga* (Oxf. Trin. 47); *& lifuga* (Paris 16201) / Adelard II: *elnefea id est fuga* (Vo)[7]; *elnefeca*, corr. in *elnefea* (Ea); *elefuga* (M, Pc, Pe); *ellefuga* (Be); *eleufuga* (Ma) / Adelard III: *hec dicitur elnefea id est fuga miserorum, dicitur etiam eleufuga id est fuga miserorum quia eius difficultate multi absterriti aufugerunt* (Venice BN Marc. f.a. 332) / Anon. comm. [F 12.3]: *hanc vocant elefugam ab ele quod est miser et fuga* (?) *quasi fugam* ...(?) *miserum* (?) ...(?) *in antiquis commentis etiam elnefea*[8] (Vat. Reg. lat. 1268) / Campanus: *elefuga id est fuga captiui* (?)[8] (Cambr. Add. 6866); *Figura miserorum*, and – in another hand – *aliter* (?) *figura* ...(?) *fuga*[8] (Erfurt Amplon. fol. 37); *Ellefuga* (Vienna 2465) / Albertus (Magnus?): *Scias etiam quod quidam antiquorum hanc figuram vocaverunt elefugam, ab 'ele' quod est miser et 'fugare' quia fugat miserum desidiosum qui in disciplinalibus non intendit* (Vienna, Domin. 80/45, and ed. Tummers II 36, 16-18).

From the cited forms it is obvious that the theorem (or its diagram) was first called, in Arabic, *al-ma'mūnī*, "the Ma'mūnian", after the caliph al-Ma'mūn (reigned 813-833) under whose patronage many scientific Greek texts were translated into Arabic, among them Euclid's *Elements*. The reason behind this naming is not obvious. The name was corrupted, in the Arabic script (cf. Cambr. Add. 1075), into *al-nāfī* (which is easily possible with the omission of the syllable *mū* in the middle of the word), which would be the participle of the verb *nafā, yanfī*, "to drive into exile". In this form the name was read by the Latin transmitter, most probably Adelard of Bath, who tried somehow – though not fully correctly – to interpret the meaning of *al-nāfī*. The best Latin rendering is *elnefea* [= *al-nāfī*] *id est fuga* (perhaps originally the reading was *fugans* which would completely correspond to the Arabic). The other Western forms are corruptions thereof. A wrong derivation of the corrupt form *elefuga* (alluding to Greek ἔλεος, "pity, compassion")

---

[7]The names from this version are quoted according to the recent edition: ***Robert of Chester's (?) Redaction of Euclid's*** Elements, ***the so-called Adelard II Version***, ed. H.L.L. Busard and M. Folkerts, i-ii, Basel etc.: Birkhäuser Verlag, 1992 (Science Networks – Historical Studies Vols. 8-9); see the description of the manuscripts (vol. i, 33 ff.) and the apparatus (vol. ii). The symbols refer to the following 14 MSS: Be = Berlin lat. qu. 510, middle or 3rd quarter 13 c.; Ea = Erfurt Amplon. Q. 352, beg. 13 c.; F = Oxf. BL Auct. F.3.13, 2nd half 13 c.; J = Florence BN Conv. soppr. J.I.32 (San Marco 206), mid 13 c.; L = Leipzig, Univ. Lib. Rep. I,68c, 1st half or middle 13 c.; M = Munich Clm 14353, about 1300; Ma = *ib.* Clm 3523, middle 13 c.; N = Naples BN VIII.C.22, 2nd half 13 c.; O = Oxf. Corpus Christi Coll. 251, last third 13 c.; Oa = *ib.* BL Auct. F.5.28, 1st half 13 c.; Pc = Paris BN lat. 7374A, end 13 c.; Pe = *ib.* lat. 11245, 1st half 13 c.; T = Oxf. Trinity Coll. 47, 12 c., fol. 104^{v}-138^{r} [note: 139^{r}-180^{v} contain *Elem.*, version I, Books I-VIII,22]; Vo = Vat. Ottob. lat. 1862, ca. 1200. The names from the "mixed" MS Vf (= Venice BN Marciana f.a. 332, 2nd half 13 c.) are quoted under "Adelard III" [= F 12.1]. In another "mixed" MS, S (= London BL Sloane 285, middle 13 c.), version II starts on 14^{v} (VI,5 ff.); the earlier part is a still unidentified redaction of the *Elements*; in its proofs, III,7 and 8 are both directly called by their names: *Pes anseris quinque passiones habet* ...; *Cauda pauonis quinque passiones habet* ... (this section is quoted above as "S*").

[8]Not well readable on the film.

seems to have developed during the later middle ages, cf. above Albertus (Magnus?) and the Anonymous commentary [F 12.3]. Heath 1926: 416 cites the same etymology from Roger Bacon (*Opus Tertium,* ch. vi), but in his further considerations falls victim to the same erroneous associations as the medieval authors.

I,47 (a)

Greek: Heath 1926: 417 quotes the name θεώρημα τῆς νύμφης, "the theorem of the Bride", found by Tannery in a Paris manuscript of Georgius Pachymeres (1242-1310). To this should be added a scholion on I,47 in Euclid, *Elem.*, ed. Heiberg, V (Leipzig 1888), 217,6: ... τοῦτο τὸ τῆς νύμφης θεώρημα. As Heath demonstrates this name stems from an old Greek tradition.

Arabic: hādhā 'l-shakl yulaqqabu bi'l-ᶜarūs ("this theorem is named 'the Bride'"): Rabat 1101; Paris 2500; al-Ṭūsī, the MSS, but not 1594. This name was not taken over in the Latin versions.

I,47 (b)

Latin: The Latin texts mention for this theorem another name of doubtlessly Arabic origin which, however, has not been found so far in any of the Arabic sources. The name occurs in the following versions: Adelard I: *Ista figura vocatur dulcaron* (Paris 16021; both the diagram itself and the caption seem to stem from a later hand than the text) / Adelard II: *dulcarnon* (Be, F, J, Pe); *dulcarnen* (Ea); *dulcarnon vel nan* (O) / Campanus: *Dulcarnon id est bicornis* (Cambr. Add. 6866); *Dulcarnan* (Madrid BN 8989); *Dicta carmen* (Munich Clm 28209; the same hand as for the name of IV,10, but apparently different from the text itself); *nota quod ista figura nominata est Dulcarnun quod interpretatur immolatio bouis quia pitagoras cum peruenit ad demonstrationem huius figure immolauit bouem pro gaudio* (New York, Col. Univ. Plimpton 156); in Erfurt Amplon. fol. 37, there is a short note written beside the diagram, but it is illegible on the film / Albertus (Magnus?): beside the diagram: *Dulcarnon*, and in the current text: *Hec figura ab antiquis accepit nomen proprium quod arabice quidem dicitur dulcarnon, quod latine est cornuta. Quidam tamen ignari virtutis vocabuli dixerunt dulcarnon dictam quia inventa ea propter utilitatem sui gyometre dulciter canebant. Alii autem rudiores eos tunc dicebant dulces carnes comedisse et ideo sic vocatam. Sola autem prima ratio ab auctoritate habetur* (Vienna, Domin. 80/45, and ed. Tummers II 93, 1-6).

The Arabic name underlying the Latin *Dulcarnon* is *dhū 'l-qarnayn*, "the One with two horns"[9]. This was, basically, a well-known Arabic surname for

[9]Heath 1926: 418 erroneously thought of a combination of Persian *dū*, "two", and Arabic *qarn*, "horn".

Alexander the Great, also occurring in the Koran. In the *Elements*, however, it seems to aim at the two squares emanating from the upper left and right sides of the rectangular triangle, in the diagram accompanying the theorem. In the Arabic sources examined for this study we did not find this name, but it should have existed in some manuscripts; otherwise the word could not have been introduced into the Latin texts. The best Latin interpretation of the name is in Campanus, Cambr. Add. 6866: *bicornis*. The association with "horn(s)" is also echoed by the explanation given in Albertus (Magnus?), *cornuta*. A far echo of the "horns" (of an ox) may perhaps also be found in the explanation given in Plimpton 156 (one of the Campanus MSS), where the author (not necessarily Campanus himself, but some reader or so) remembers the tale that Pythagoras sacrificed an ox when he had found this theorem – but the association of the theorem with Pythagoras is in itself doubtful[10]. Albertus (Magnus?) cites two more Latin speculative etymologies and rejects them as not authorative: *dulcarnon* as being derived from *dulciter canebant* or from *dulces carnes* (*comedisse*)[11].

III,7

Latin: pes anseris. This name is found only in the Latin texts, in this form, without explanatory additions: Adelard II (Be, J, L, N, O, Oa, S*); Adelard III (Oxf. BL Digby 174; Venice BN Marc. f.a. 332); Campanus (Cambr. Add. 6866; Erfurt Amplon. fol. 37; New York, Col. Univ. Plimpton 156). The name was apparently inspired by the figure of the diagram accompanying the theorem. It strongly respires an Arabic aroma, though it could not be traced in the Arabic sources. In Arabic, it must have been something like *rijl al-baṭṭa*, or *rijl al-iwazza* ("the Goose's Foot")[12].

III,8

Latin: Adelard I: *θenep atoz. cauda pauonis* (Oxf. Trin. 47)[13] / Adelard II: *hec figura dicitur dhenep atoz id est cauda pauonis* (T); *scenepacoch cauda pauonis* (O); *Denebeltoz cauda pauonis* (L); *gezelde* (Ea); *cauda pauonis* (Be,

[10]See for details Heath 1921: 133f., 144f.

[11]In passing, it should be noted that in the wide-spread popular fables in the *Speculum stultorum* by Nigellus Wireker (12th cent.) there occurs a cow named *Bicornis* (cf. E. Sievers, in: *Zeitschrift für deutsches Altertum und deutsche Litteratur* **20**, 1876, 215f.– I owe the knowledge of this text to Dr. B.D. Haage, University of Mannheim). It may well be that this word, as a name for a cow or an ox, was also present in the minds of many scholarly readers of the *Elements*.

[12]Heath 1926: 99 and 418 mentions the names of III,7 and 8, but without detailed explanation.

[13]Burnett 1985: 479 quotes (from this MS) the names for III,8 and IV,10; he does not mention the name for I,5 in the same manuscript.

J, N, Oa, S*) / *cauda pauonis* further in Adelard III (Oxf. BL Digby 174; Venice BN Marc. f.a. 332); Campanus (Cambr. Add. 6866; Erfurt Amplon. fol. 37; Florence BN Magliab. XI 112; New York, Col. Univ. Plimpton 156).

Here, as in III,7, the name could not be traced in the Arabic sources. Again, it appears to be inspired by the figure of the diagram. However, in contrast to III,7, here the original Arabic form was preserved in the Latin texts, in transliteration; it can be identified as *dhanab al-ṭāwūs*, "the Peacock's Tail". The rendering of the Arabic *dh* as θ (in Adelard I) and *dh* (in Adelard II) points to Adelard as the translator of those Arabic names into Latin. In a list of Latin transliterations for certain Arabic consonants (in the margins of MS Bruges 529 [s. XIII], Adelard I), Δ is given for Arabic *dh* and θ is proposed for Arabic *th*[14]. The spellings θ and *dh* for our *dhanab* seem to be a loose reminiscence of that.

IV,10

Latin: Here, once more, the name was only found in Latin texts (including transliterated Arabic forms), but not in the Arabic sources. Adelard I: *hec est figura albeliz id est demonis* (Oxf. Trin. 47) / Gerard of Cremona: *Seclibiz* (Madrid Bibl. Univ. 102; uncertain whether in the same hand as the text or in a later hand) / Adelard II: *figura albeliz id est demonis* (T); *sesclalbliz* (Ea); *Segglebiz* (N); *equibisa vel elbeliz vel figura demonis* (J); *hec figura vocatur figura demonis id est scientis*[15] *propter eius difficultatem* (Pe); *Teckebyse* (S*, in a much younger hand) / Adelard III: *Equibise* (Oxf. BL Digby 174); *Seclabin* (Venice BN Marc. f.a. 332) / Campanus: *Figura demonis* (Munich Clm 28209; cf. above, *sub* I,47).

Some of the Latin forms contain the transliteration of the full Arabic name, *shakl Iblīs* ("the diagram/theorem of Iblīs [i.e. the Devil]")[16], while others only repeat the proper name of *Iblīs*. By simple distortion, or deliberately – in order to restore an assumed Arabic article *al* in front of the word –, they wrote *albeliz* etc. (instead of *abeliz* or the like; cf. the well-known *Albumasar* for Abū Maᶜshar; etc.). The reason for the naming is unknown, but apparently, this time, it was not inspired by the shape of the diagram. The argument brought forward by MS Pe (Adelard II) appears to be purely speculative.

*

[14] Cf. Kunitzsch 1991: 8, note 8.

[15] The question mark here added in ed. Busard–Folkerts is unnecessary. The medieval Western understanding of *demon* as *sciens* was discussed by Beaujouan 1981: 318.

[16] Burnett 1985: 479 writes *al-iblīs*; but *Iblīs* is used, in Arabic, without the article *al-* (and is inflected as a diptote). The word as such is assumed to derive from Greek διάβολος; see *Encyclopaedia of Islam*, new ed., iii (Leiden 1971), 668f., article "Iblīs". For *al-shakl*, MS Bruges 529 (Adelard I) presents, in the margin, the transliteration *elscekl*; cf. ed. Busard, 1983, p. 392, and Burnett 1985: 478.

Of these names one (I,47a) is of classical Greek origin and continued into the Arabic *Elements*, but it was not set forth in the Latin tradition. The others seem to have originated with the Arabs. As it seems, three of these (I,47b; III,7.8) were inspired by the shape of the diagrams accompanying the theorems, whereas for the other two (I,5; IV,10) the reason behind the naming remains unclear. In the Latin tradition they seem to have been rather popular, perhaps - as Roger Bacon argues (*apud* Heath 1926: 416) - as a pedagogical aid for memorizing at least some of the theorems. The introduction of the Arabic names into the Latin area is most probably connected with Adelard of Bath.

The transmission of the names in the Arabic tradition is less dense than in the Latin[17]. But also in the Latin it appears that practically no manuscript contains all the five names. Comparing all manuscripts, however, all the names are represented. It seems that the scribes proceeded at random, picking up one and omitting the other of the names. Altogether, from this state of the transmission we may conclude that what we now have in our hands is but a small portion from a much richer tradition, just a tip of an iceberg.

For the practical use of these names, outside the Euclidean tradition itself, a few specimens shall be mentioned.

MS Lüneburg, Ratsbibliothek Misc. D 4° 48 (presumably around 1200; cf. F 10: "Mélanges"), contains on fol. 1^r-7^v and 18^r-21^r fragments of an anonymous treatise on the construction of the astrolabe, *De artificio plane spere que dicitur walzacora,* Inc. *Omnia que de statu superioris machine* ...[18]. The author seems to have been a well-learned scholar, with a good library at his disposal. He cites Ptolemy's *Canones* and *Almagest* (the latter in the translation of Gerard of Cremona, Toledo, ca. 1150-1180); Albatenius; Alfraganus (in the Latin translation of John of Seville); and of Latin authors Vitruvius and Macrobius. Further, he knows and partly cites the astrolabe treatises of (Ps.-)Gerbertus and Herman the Lame. In addition, he makes ample use of Euclid's *Elements* for the mathematical demonstration of the construction of various circles etc. on the astrolabe. Here it is evident that he used the *Elements* in the version commonly called "Adelard II"[19]. In his mathematical argumentation he continuously refers to the *Elements* (much in the same way as, e.g., Jordanus de Nemore [1st half 13th c.?], "version 2"[20]), with those abbreviated terms common to thirteenth century mathematical texts: *per VI.*

[17]Of 15 MSS (including al-Nayrīzī, Leiden or. 399) of the Arabic *Elements* examined for our study only four contain some of the names: Cambr. add. 1075 (I,5); Oxf. BL Thurston 11 (I,5); Rabat 1101 (I,5.47a); Uppsala (I,5). Further, the abbreviation of Book I in Paris BN ar. 2500 and al-Ṭūsī's *Taḥrīr* (3 MSS seen) have I,5 and 47a.

[18]The text appears to be unknown until now; it is not mentioned in Thorndike-Kibre's *Incipits*.

[19]Cf. *inter alia* the words, or formulas, *elmuharifa* (2ra, -3); *theoreuma* (2rb, 5; cf. the *theoreumata* in the redaction of the *Elements*, 14ra ult.); *dicet aduersarius* ... (3vb, 7-8) and *Redarguatur itaque* ... (*ib.*, line 9).

[20]Cf. the ed. by R.B. Thomson: *Jordanus de Nemore and the Mathematics of Astrolabes: De plana spera,* Toronto 1978, pp. 87ff.

secundi (1vb, 4), etc. In some instances, he refers to Euclidean theorems by name. For I,5 we find: *per V. primi scilicet elnefeam* (5ra, -10), and for I,47: *per dulĉ* (1vb, 5) and *per dulcarnon* (2ra, -4). Since the three texts contained in the MS (astrolabe; *Elements*; planetary theory) show some relationship in the citation of sources and the mathematical style of argumentation, it would not seem improbable that they are all by the same (unknown) author.

Another author to be cited is Richard of Wallingford (d. 1336). In his *Albion* we find: ... *ex resolucione Dulkarnon* (North 1976, i: 250,13); and in the *Quadripartitum*: ... *per Dulcarnon, que est penultima primi* (*ib.*, i: 38, 39).

Third, we may mention Philippe Éléphant (middle 14th c.). From his *Mathematica*, Beaujouan cites *elefuga* (I,5), *cauda pavonis* (III,8) and *figura demonis sive intelligentis* (IV,10): Beaujouan 1957: 453, and *idem* 1981: 313[21].

Apart from these, there could be mentioned citations from Alexander Neckam (1157-1217), *De naturis rerum* II, 173: *dulcarnon* (in Latham 1989: 463), Roger Bacon: *elefuga* (in Heath 1926: 416) and Chaucer: *dulcarnon* (in Heath 1926: 416, 418; North 1976, i: 251 note 4).

A further development may be seen in Jean de Murs (first half 14th cent.) who calls certain figures formed in the diagrams of two propositions in his *De arte mensurandi*, *crista pauonis* ("the Peacock's Comb, or Crest")[22]. It seems probable that the choice of this name was provoked by the existing and well-known name *cauda pauonis* from Euclid's *Elements*.

Beyond the names of Greek and Arabic provenance discussed above, there existed several more names that seem to have originated in medieval Europe and which are outside the scope of this present study; see Heath 1926: 415f., 417f., 418, and Beaujouan 1975: 450, 453, 455f., and *idem* 1981: 312f., 318f.

[21]The citations from MS Oxf. Trinity Coll. 47 in Beaujouan 1981: 318 note 44 (based on excerpts made by Danielle Jacquart) are not fully exact. For I,5 read: 172^{v} (instead of 170^{v}); in III,8 read *dhenep atoz* (instead of *adhenepatoz*) – the wrong spelling also misled Beaujouan to explain the name from the Arabic plural *adhnāb* (instead of the singular *dhanab*) *al-ṭāwūs*; on fol. 167^{v} read *θeneb* (instead of *teneb*). For *elnefea*, Beaujouan's conjectured *al-nafy* (l'expulsion, le bannissement) comes quite near to the true *al-nāfī* (*"fugans"*) documented above.

[22]*De arte mensurandi*, cap. VI, prop. 23 and 24. I owe this information to H.L.L. Busard who is currently preparing an edition of that text.

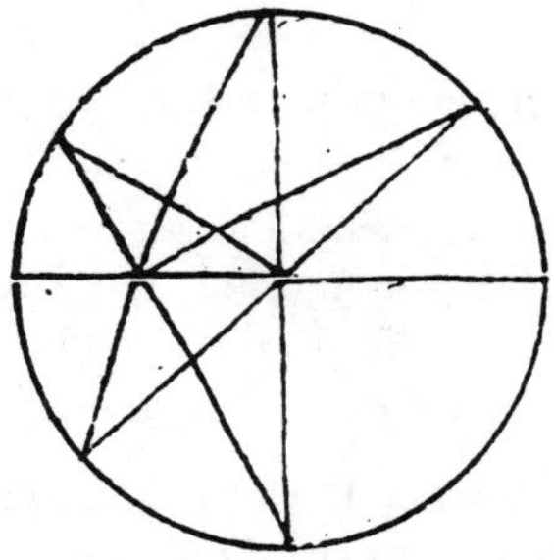

Diagram for III,7, adapted from MS Oxford, Trinity College 47, fol. 109v

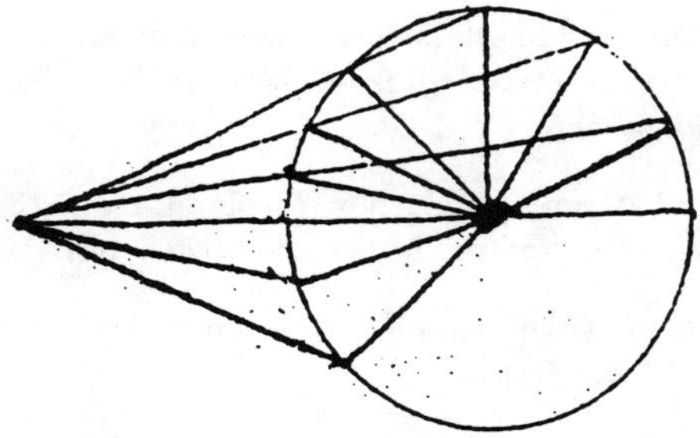

Diagram for III,8, adapted from MS Oxford, Trinity College 47, fol. 109v

Bibliography

(Editions of texts are cited in abbreviated form in the article; the full titles are given in Folkerts 1989)

Beaujouan 1975: G. Beaujouan, Réflexions sur les rapports entre théorie et pratique au Moyen Âge, in: *The Cultural Context of Medieval Learning,* ed. J.E. Murdoch and E.D. Sylla, Dordrecht/Boston, 437-484.

Beaujouan 1981: G. Beaujouan et P. Cattin, *Philippe Éléphant,* Paris (Extrait de l'Histoire littéraire de la France, t. xli, pp. 285-363).

Burnett 1985: C.S.F. Burnett, review of Busard's ed. of Adelard I, in: *Archives internationales d'histoire des sciences* **35**, 475-480.

Folkerts 1989: M. Folkerts, *Euclid in Medieval Europe*, Toronto (The Benjamin Catalogue for History of Science. Questio de rerum natura II).

Heath 1921: Th. Heath, *A History of Greek Mathematics*, I, Oxford.

Heath 1926: Th. Heath, *The Thirteen Books of Euclid's Elements*, transl., I (2nd. ed.), Cambridge.

Kunitzsch 1991: P. Kunitzsch, Letters in geometrical diagrams, Greek – Arabic – Latin, in: *Zeitschrift für Geschichte der Arabisch-Islamischen Wissenschaften* **7**, 1-20.

Latham 1989: J.D. Latham, Arabic into Medieval Latin (3): Letters D-F, *M.L.D.*, in: *Journal of Semitic Studies* **34**, 459-469.

North 1976: *Richard of Wallingford, An edition of his writings ...*, by J.D. North, i-iii, Oxford.

Sezgin 1974: F. Sezgin, *Geschichte des arabischen Schrifttums*, V: *Mathematik, bis ca. 430 H.*, Leiden.

Note added in proof: A late testimony (from 1747) for the name of I,5 is in Muḥammad ibn ᶜAlī al-Tahānuwī, *Kashshāf iṣṭilāḥāt al-funūn*, Istanbul 1318 [1900-01], I, p. 863. The author adds the – doubtlessly unhistorical – remark that the figure was called *al-Ma'mūnī,* because the caliph al-Ma'mūn was so much delighted with it that he had it embroidered on the sleeves of some garments. For this and three more names of geometrical theorems the author refers as a source to the work *Ashkāl al-Ta'sīs* of Shams al-Dīn al-Samarqandī (end of 13th century) and a commentary to it by Shams al-Dīn al-Būkhārī. (I am grateful to Dr. S. Brentjes, Leipzig, for bringing this source to my attention.)

Abū Kāmil on the Pentagon and Decagon

Richard Lorch

The Text

Abū Kāmil Shujā' b. Aslam lived after al-Khwārizmī, whom he quotes at the beginning of his *Algebra* and elsewhere, and before the *Fihrist* was compiled in the late tenth century[1]. His most famous work, the three-part *Algebra*, which is known in only one Arabic manuscript[2], Istanbul, Kara Mustafa Paşa 379 (dated 651/1253)[3], had considerable influence in the medieval Arabic world[4]. Of the Latin translation there is only one known manuscript, Paris, Bibliothèque Nationale, lat. 7377A, 71v–97r (14c.), which breaks off shortly after the beginning of the third part, on indeterminate equations. There was a further transmission into Latin in the form of the *Practica geometriae*, which was based on Arabic sources; it contains an elaborated version of the second part of the *Algebra*, on the pentagon and decagon[5]. A few citations of Abū Kāmil's procedures by the author of the *Liber Mahameleth* probably represent yet another independent transmission from the Arabic[6]. The whole treatise was translated into Hebrew in the fifteenth century by Mordecai Finzi; and of this translation there is a complete manuscript (Munich, cod. heb. 225) and a fragment in Paris (Bib. Nat., heb. 1029,7)[7].

In this paper an edition of the Latin translation of the second part[8] of the *Algebra* is presented with a note on the terminology of the translator. Since there are several modern translations of the work[9], only a translation of the

[1] For general references to Abū Kāmil see Sezgin 277-281, Levey (DSB), the introduction to Abū Kāmil (Hebrew). His dates are usually given as ca. 850 – ca. 930.

[2] Dr. Jacques Sesiano informs me (private communication) that the Meshed manuscript of the *Algebra* mentioned by Sezgin (p. 281) in fact contains another work. MS Teheran, Sanā 2672/6, ff. 189r–303r, *Misāḥat al-arāḍīn*, suggested *ibid* as being the treatise on the pentagon and decagon, has an *incipit* (see *Nashriye* II, 241) that does not occur in the Istanbul manuscript of the treatise; it has not been investigated.

[3] Published in facsimile by the Institut für Geschichte der Arabisch-Islamischen Wissenschaften in 1986.

[4] See e.g. Karpinski.

[5] For a detailed comparison with the Abū Kāmil text as transmitted through Hebrew, see Suter, pp. 38-42.

[6] See Sesiano, 73–74, 95–96. The first of these citations, at the beginning of the treatise, MS Paris BN lat. 7377A, f. 101r, quotes the titel of Abū Kāmil's book in transliteration, not in translation as in our text.

[7] The first part was translated from Hebrew into German by Josef Weinberg for his doctoral dissertation and was published by him in 1935; a Hebrew edition and English translation was published by Martin Levey in 1966.

[8] An edition of the Latin translation of the first and third part is to be found in the contribution of Dr. J. Sesiano in this volume.

[9] Sacerdote (from the Hebrew), Suter (from Sacerdote) and Yadegari and Levey (from

preface from the Arabic (for comparison with the Latin) and a summary of the content of the rest will be given here. The preface[10]:

> Abū Kāmil Shujā' said: since we have expounded what needs explanation of what was hard to understand in the calculation [called] ***al-jabr*** and *al-muqābala* for those advanced in the science and its application and the skilled geometers who are experts in the book of Euclid and other [books], and [since] we have explained and commented on that in this book of ours, we shall expound the amount of each of the sides of the equilateral and equiangular pentagon and decagon, each of which is contained in a known circle or contains a known circle, and the amount of the diameter of the circle that contains or is contained by a known equilateral and equiangular pentagon or decagon, and the amount of the chord of a fifteenth of the circumference of a known circle and the amount of each of the sides of the equilateral and equiangular pentagon and decagon of known area and the amount of the sides of triangles of known area when they are in an equilateral and equiangular pentagon or decagon, and other things that we have commented upon in this book of ours, the finding of much of which has been impossible for those of the calculators and geometers of our time whom we have met and for those advanced in the science of calculation and geometry of whom report has reached us. God (be He mighty and glorious) make easy for me to reach what remained inaccessible to them of it and to achieve what was difficult for them to understand of it. Praise, thanks, blessings and eternity [belong] to God, who has no companion.

In the following summary of the mathematical content of the treatise, the application of algebra to geometrical problems, the division into twenty propositions presented by Suter (and taken over from Sacerdote) will be retained; some propositions are subdivided. All polygons are regular and the figures referred to are those in the Latin text unless otherwise specified. x means the *res*, or thing sought; s_5 and S_5 are the sides of the inscribed and circumscribed regular pentagon, and similarly for the other regular figures; Ptol.($ABGD$) means Ptolemy's theorem applied to the cyclic quadrilateral $ABGD$; Pyth.(ABG) means Pythagoras' theorem applied to triangle ABG. Juxtaposed fractions are to be understood as added together. Numbers enclosed in square brackets after an enunciation indicate the previous propositions used in the problem. Only the essence of the argument is given, not necessarily in the order of the mathematical steps in the text. All references to book and proposition number of Euclid's *Elements* ("Eu.") and most of

the Arabic). Suter's translation, though from the Italian, translated from the Hebrew, which was translated probably direct from the Arabic, is the best, but is sometimes a paraphrase.

[10][Abū Kāmil], p. 134. It is a pleasure to thank Professor Paul Kunitzsch for considerable help with this translation and with many other parts of this paper.

the justifications for mathematical steps are supplied by the editor. The use of Ptolemy's theorem is justified only in a marginal note in the Latin translation: where it is first used (Prop. 1) "Ptolomeus in libro de arcu et corda" (presumably *Almagest* I 10) is quoted.

1. To find the side (x) of the pentagon inscribed in a circle of diameter 10[11]. (See p. 234)

   $EL = \frac{1}{10}x^2$ since $ED^2 = HE \cdot EL$

   Pyth.(ELD): $GL = \sqrt{x^2 - \frac{1}{10}\cdot\frac{1}{10}x^2}$

   Ptol.$(ABDG)$: $AB \cdot GD + AB^2 = GD^2$

   Now $GD^2 - AB^2 = 3x^2 - \frac{1}{25}x^2$

   $(3x^2 - \frac{1}{5}\cdot\frac{1}{5}x^2x^2) \div x = GD = \sqrt{4x^2 - \frac{1}{5}\cdot\frac{1}{5}x^2x^2}$

   Square, *Oppone*, $\div x^2$: $5 + \frac{1}{625}x^2x^2 = \frac{1}{5}x^2$

   Restaura to *census census*[12] ($\times 625$), solve by formula:

   $$x = \sqrt{62\tfrac{1}{2} - \sqrt{781\tfrac{1}{4}}}.$$

2. To find the side of a decagon (x) inscribed in a circle of diameter 10. [1] (See p. 236)

   Ptol.$(AGDH)$: $10x + s_5^2 = GH^2$; and $GH^2 + s_5^2 = 100$

   $125 + 10x - \sqrt{3125} = 100$

   Oppone, $\div 10$: $x = \sqrt{31\frac{1}{4}} - 2\frac{1}{2}$.

3. To find the side (x) of a pentagon circumscribed about a circle of diameter 10. [1] (See p. 236)

   Pyth.(ELT): $EL^2 = 9\frac{1}{4}\frac{1}{8} + \sqrt{48\frac{1}{2}\frac{1}{4}\frac{5}{8}\cdot\frac{1}{8}}$

   Similar $\triangle$s ETB, ELT: $x^2 : 25 = s_5^2 : EL^2$

   Cross-multiply, *reduc* to *census* ($\times\ \frac{1}{5}\frac{1}{25} - \sqrt{\frac{4}{125}}$):

   $x^2 = 375 + \sqrt{15625} - \sqrt{78125} - \sqrt{28125} = 500 - \sqrt{200000}$.

4. To find the side (x) of a decagon circumscribed about a circle of diameter 10. [2] (See p. 238)

   Similarly: $x^2 : s_{10}^2 = 25 : EL^2$. Hence $x^2 = 100 - \sqrt{8000}$.

---

[11] 10 was often used as a token value. It is clear from Prop. 16 that Abū Kāmil could work just as easily with an unknown instead of 10.

[12] These terms are explained below. Throughout the summary the Latin terms and the Latin diagrams are used.

5. To find the diameter of the circle in which a pentagon of side 10 is inscribed. (See p. 238)

   Put $x = EG$. Ptol.$(ABGE)$: $x \cdot 10 + 10^2 = x^2$

   Solve by formula: $x = 5 + \sqrt{125}$

   Pyth.(EDL): $DL^2 = 62\frac{1}{2} - \sqrt{781\frac{1}{4}}$

   Put $x = CD$.

   Similar $\triangle s\ CDE, EDL$: $x^2 : 10^2 = 10^2 : DL^2$

   Cross-multiply, *reduc* to *census* ($\times\ \frac{1}{50} + \sqrt{1\frac{1}{4} \cdot \frac{1}{15625}}$):

   $x^2 = 200 + \sqrt{8000}$.

6. To find the diameter of the circle circumscribed by a pentagon of side 10. [5] (See p. 239)

   $AM^2 = 50 + \sqrt{500}$

   Pyth.(AMC): $CM^2 = 25 + \sqrt{500}$

   (diam. inner $\odot)^2 = 100 + \sqrt{8000}$

   Or: (diam. outer $\odot)^2 - S_5^2 =$ (diam. inner $\odot)^2$ and similarly for other regular figures.

7. = Prop. 5 by another method. [1]

   Eu. within XII 2: chords of similar arcs $\propto$ diameters of the circles in which they stand; put $x =$ diam.:

   $100 : 62\frac{1}{2} - \sqrt{781\frac{1}{4}} = x^2 : 100$ etc., as in 5.

8. = Prop. 6 by another method. [3]

   $100 : 500 - \sqrt{200000} = x^2 : 100$

   Cross-multiply, *reduc* to *census* ($\times\ \frac{1}{10} \cdot \frac{1}{10} + \sqrt{\frac{1}{125} \cdot \frac{1}{100}}$):

   $x^2 = 200 + \sqrt{8000}$.

9. To find the diameter (x) of a circle when a decagon of side 10 is inscribed in it. [2]

   $x : 10 = 10 : \sqrt{31\frac{1}{4}} - 2\frac{1}{2}$

   Cross-multiply, *restauremus* with $2\frac{1}{2}x$: $100 + 2\frac{1}{2}x = \sqrt{31\frac{1}{4}x^2}$

   $(100 + 2\frac{1}{2})^2 = 10000 + 6\frac{1}{4}x^2 + 500x = 31\frac{1}{4}x^2$

   Opponamus, $\div 25$, and solve: $x = 10 + \sqrt{500}$.

10. *i.* To find the diameter (x) of a circle when a decagon of side 10 is circumscribed. [4]

$100 - \sqrt{8000} : 100 = 100 : x^2$

Cross-multiply, *reduc* to *census* $(\times\ \frac{1}{2} \cdot \frac{1}{10} + \sqrt{\frac{1}{100} \cdot \frac{1}{5}})$

$x^2 = 500 + \sqrt{200000}$.

10. *ii.* If AG is the side of a regular polygon and CY and AD are the diameters of the inscribed and circumscribed circles, $AG^2 + CY^2 = AD^2$. (See p. 242)

$\triangle AML \equiv \triangle DMZ : \quad LG = DZ;$

and since $A\hat{L}M = M\hat{Z}D, \quad ZD \parallel LG$

So $ZL \parallel$ and $= DG$

The result follows by Pyth.(AGD).

11. *i.* To find the side of the 15gon inscribed in a circle of diameter 10. [2] (See p. 243)

$BD = s_{10} = \sqrt{31\frac{1}{4}} - 2\frac{1}{2}, BG = s_6 = 5$, so that $GD = s_{15}$; AD and AG are found by Pyth.(ABD, ABG)

Ptol.($ABDG$): $AD \cdot BG - AG \cdot BD = AB \cdot GD$

Multiply out, $\div AB$ i.e. 10:

$GD = \sqrt{15\frac{5}{8} + \sqrt{48\frac{3}{4}\frac{5}{8}\frac{1}{8}}} + \sqrt{4\frac{1}{2}\frac{1}{8}\frac{1}{2}\frac{1}{8}} - \sqrt{23\frac{3}{8}\frac{1}{2}\frac{1}{8}}$.

11. *ii.* Approximately, $GD = 2;4,44,48$ in sexagesimals.

12. If the area of an equilateral triangle + its altitude = 10, to find its altitude (x). (See p. 244)

$DG = \sqrt{\frac{1}{3}x^2}$ and so area $= \sqrt{\frac{1}{3}x^2x^2}$

$x + \sqrt{\frac{1}{3}x^2x^2} = 10$

Restaura $\sqrt{\frac{1}{3}}$(*census census*) to $\sqrt{}$(*census census*) $(\times\sqrt{3})$:

$\sqrt{3x^2} + x^2 = \sqrt{300}$

Solve: $x = \sqrt{\frac{3}{4} + \sqrt{300}} - \sqrt{\frac{3}{4}}$.

13. If an equilateral triangle has side 10, to find the base of a rectangle of area 10 constructed within it[13]. (Propositions 13 and 14 are spoilt in Latin. The Arabic diagrams, with letters according to the system of

[13]There is a problem similar to Props. 13 and 14 in al-Khwārizmī: Rosen's edition, Arabic pp. 33–34, English pp. 84-85. They are not in the Latin translations by Robert of Chester and Gerard of Cremona.

transliteration used by the translator, are reproduced here. The Latin text jumps to "So area of $\triangle$" at the end of Prop. 14.)

$x = HE$. Then $HB = \sqrt{\frac{1}{3}x^2}$, etc., much as in Prop. 12.

$x = \sqrt{18\frac{1}{2}\frac{1}{4} - \sqrt{75}} + \sqrt{18\frac{1}{2}\frac{1}{4}}$.

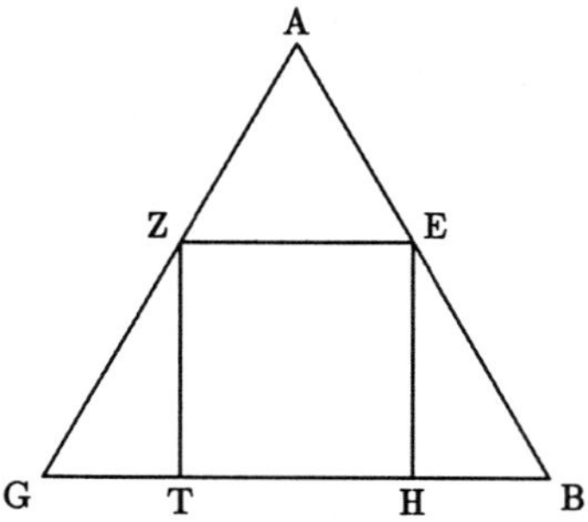

Diagram for Prop. 13 in Arabic

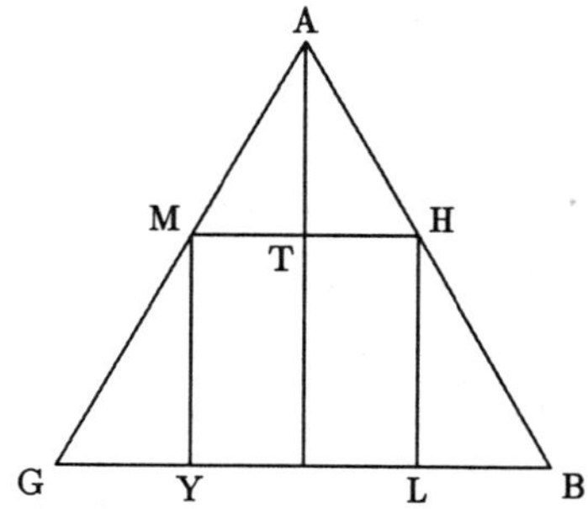

Diagram for Prop. 14 in Arabic

14. If the area of an equilateral triangle + the area of the square constructed within it is 10, to find the side (x) of the square.

$\triangle MYG + \triangle HLB = \sqrt{\frac{1}{3}x^2x^2}$; $\triangle HMA = \sqrt{\frac{1}{8}x^2x^2 + \frac{1}{2}\frac{1}{8}x^2x^2}$

$\sqrt{\frac{1}{3}x^2x^2} + \sqrt{\frac{1}{8}x^2x^2 + \frac{1}{2}\frac{1}{8}x^2x^2} = 10 - 2x^2$

$x^2 = \sqrt{45\frac{1395}{20449}} - \sqrt{11\frac{10261}{20449}}$

So area of $\triangle = 10 - x^2 = 3\frac{41}{143} + \sqrt{11\frac{10261}{20449}}$.

15. To find the side (x) of an equilateral pentagon constructed within a square of side 10. (See p. 245)

$BE = 10 - x$

From $\triangle HGZ$, $GH = \sqrt{\frac{1}{2}x^2}$; and so $HB = 10 - \sqrt{\frac{1}{2}x^2}$

$200 + 1\frac{1}{2}x^2 - 20x - \sqrt{200x^2} = BE^2 + HB^2 = x^2$

Oppone and solve: $x = 20 + \sqrt{200} - \sqrt{200 + \sqrt{320000}}$.

16. To find the side (x) of a pentagon of area 50. [5 or 7] (See p. 246)

From Prop. 5 or 7: for side x, diam.$^2 = 2x^2 + \sqrt{\frac{4}{5}x^2x^2}$

Pyth.(MEH): $MH^2 = \frac{1}{4}x^2 + \sqrt{\frac{1}{2}\frac{1}{10}x^2x^2}$

$\frac{1}{2}\frac{1}{8}x^2x^2 + \sqrt{\frac{1}{320}x^2x^2x^2x^2} = 100$

Restaura $\frac{1}{2}\cdot\frac{1}{8}$ *census census* to *census census* ($\times$ 16):

$x^2x^2 + \sqrt{\frac{4}{5}x^2x^2x^2x^2} = 1600$

Reduc to *census census* ($\times\ 5 - \sqrt{20}$) :

$x^2x^2 = 8000 + \sqrt{51200000}$.[14]

17. To find the side (x) of a decagon of area 100. [9] (See p. 247)

From Prop. 9: diam. $= x + \sqrt{5x^2}$

Eu. XIII 9 and 5[15]:

$ZM(s_6), ZU(s_{10})$ and their sum are in e.m.r.[16]

Eu. XIII 3: $(ZM - x + \frac{1}{2}x)^2 = 5(\frac{1}{2}x)^2$

$ZM^2 = 1\frac{1}{2}x + \sqrt{1\frac{1}{4}x^2x^2}$

Pyth.(MZS): $MS^2 = \sqrt{1\frac{1}{4}x^2} + \sqrt{1\frac{1}{4}x^2x^2}$

$MS^2 \cdot ZS^2 = \frac{1}{4}x^2x^2 + \frac{1}{2}\frac{1}{8}x^2x^2 + \sqrt{\frac{5}{64}x^2x^2x^2x^2} = 100$

Restaura $\frac{1}{4} + \frac{1}{2}\cdot\frac{1}{8}$ *census census* to *census census*: ($\times 3\frac{1}{5}$) :

$x^2x^2 + \sqrt{\frac{4}{5}x^2x^2x^2x^2} = 320$

Reduc to $x^2x^2(\times\ 25 - \sqrt{20})$: $x^2x^2 = 1600 - \sqrt{2048000}$.

18. If $\triangle BDE$ is 10, to find the side (x) of the pentagon. (See p. 249)

Eu. XIII 8: if M divides BE in e.m.r., $ME = DE$.

Eu. XIII 3: if L bisects ME, $LM^2 = \frac{1}{5}LB^2$

$BL = \sqrt{1\frac{1}{4}x^2}$ and so $BE = \frac{1}{2}x + \sqrt{1\frac{1}{4}x^2}$

Pyth.(BEH) : $BH^2 = 1\frac{1}{4}x^2 + \sqrt{1\frac{1}{4}x^2x^2}$

$BH^2 \cdot EH^2 = \frac{1}{4}x^2x^2 + \frac{1}{2}\frac{1}{8}x^2x^2 + \sqrt{\frac{5}{64}x^2x^2x^2x^2} = 100$

Restaura $\frac{1}{4} + \frac{1}{2}\cdot\frac{1}{8}$ *census census* to *census census* ($\times 3\frac{1}{5}$):

$x^2x^2 + \sqrt{\frac{4}{5}x^2x^2x^2x^2} = 320$

Reduc to *census census* ($\times\ 5 - \sqrt{20}$) :

$x^2x^2 = 1600 - \sqrt{2048000}$.

---

[14] This should be $8000 - \sqrt{51200000}$.

[15] XIII 9 and 5 are quoted *verbatim* and XIII 3 is later used. But the whole passage is not necessary to the argument.

[16] e.m.r. means "extreme and mean ratio".

19. If $\triangle ABE$ is 10, to find BE (x). (See p. 251)

Eu. XIII 1 and 8: $(ME + \frac{1}{2}BE)^2 = 5(\frac{1}{2}BE)^2$

Let $BE = x$. $AE = \sqrt{1\frac{1}{4}x^2} - \frac{1}{2}x$

Pyth.(AEY): $AY^2 = 1\frac{1}{4}x^2 - \sqrt{1\frac{1}{4}x^2x^2}$

$AE^2 \cdot YE^2 = \frac{1}{4}x^2x^2 + \frac{1}{2}\cdot\frac{1}{8}x^2x^2 - \sqrt{\frac{5}{64}x^2x^2x^2x^2} = 100$

Reintegra $\frac{1}{4} + \frac{1}{2} \cdot \frac{1}{8}$ *census census* to *census census* $(\times 3\frac{1}{5})$:

$x^2x^2 - \sqrt{\frac{4}{5}x^2x^2} = 320$

Reintegra so that *census census* becomes whole

$(\times\ 5 + \sqrt{20})$: $x^2x^2 = 1600 + \sqrt{2048000}$.

20. If $\triangle CTH$ is 10, to find CH (x). [5 or 7] (See p. 252)

As in Prop. 16, $(\text{diam.})^2 = 2x^2 + \sqrt{\frac{4}{5}x^2x^2}$

Eu. XIII 10: $s_5^2 = s_6^2 + s_{10}^2$ and so $CT^2 = \frac{1}{2}x^2 - \sqrt{\frac{1}{2}\cdot\frac{1}{10}x^2x^2}$

Pyth.(CTM): $TM^2 = \frac{1}{4}x^2 - \sqrt{\frac{1}{2}\frac{1}{10}x^2x^2}$

$CM^2 \cdot TM^2 = \frac{1}{2}\cdot\frac{1}{8}x^2x^2 - \sqrt{\frac{1}{320}x^2x^2x^2x^2} = 100$

Reintegra $\frac{1}{2} \cdot \frac{1}{8}$ *census census* to *census census* $(\times 16)$:

$x^2x^2 - \sqrt{\frac{4}{5}x^2x^2x^2x^2} = 1600$

Reintegra to *census census* $(\times\ 5 + \sqrt{20})$:

$x^2x^2 = 8000 + \sqrt{51200000}$.

As Suter has pointed out[17], the obtrusion of Euclid XIII 3 in Prop. 17 in the transmitted form of the text may well be due to a copyist. We may note that in the Latin translation of Prop. 18, which relies on XIII 8 and 3, XIII 5 is cited as well: this may well represent more interference of the same kind in the Arabic manuscript from which the translator was working. But we are concerned here only with the translation, not with the formation of the text itself.

The Latin Translation

The Latin terms used in the above summary should now be explained in modern terms. *Res*, *census*, *cubus* (Arabic: *shayʾ*, *māl*, *kaʿab*) are used for the unknown, here called x, and its square and cube. Higher powers of the unknown are expressed by combinations of them, e.g. *census census*. *Restaurare*, *opponere* (Arabic: *jbr*, *qbl* III) and their associated abstract nouns are

[17]Suter, p. 30, n. 6.

used in their tradition senses: *restaurare* in Prop. 9 means to take a negative term from one side of the equation to become a positive term on the other; *opponere* means to collect like terms and to express the result as one term, which is usually positive (there is an exception near the end of Prop. 2). But *restaurare* usually occurs as a translation of *kml* IV in the sense of making the coefficient of a specified term (here always *census census*) unity by multiplying all terms of the equation by a suitable factor, which may be a number or a root. By *reducere* (Arabic *rdd*) a similar reduction of a specified term (here always *census* or *census census*) is meant, but this time the factor is a number plus (or minus) a root. These reduction factors Abū Kāmil produces without calculation or explanation. In fact, to reduce a coefficient $a+\sqrt{b}$ to unity, we must multiply by $\frac{a-\sqrt{b}}{a^2-b}$ or, in the form given by Abū Kāmil, $\frac{a}{a^2-b}-\sqrt{\frac{b}{(a^2-b)^2}}$. A complication arises in the last two propositions: *reintegrare* appears in the text four times for *kml* IV in the sense of reducing coefficients to unity, twice when the factor is a number and twice when it is a number plus a root. Finally, *integra* is twice (Props. 17,18) written over *restaura* as an interlinear note, presumably to convert it to to *reintegra*.

The diagram letters are translated as follows:

ا A, ب B, ج G, د D, ه E, و U, ز Z, ح H, ط T, ك C, ل L, م M, ع Y, س S.

In Props. 4 and 20 the correspondence is not exact. The three letters E, F, U stand for the four letters ع, س, ص, ق in Prop. 4; and in Prop. 20 Z, I stand for the single letter و in the Arabic text and for و, ز in the diagram. But it is only a matter of letters mentioned once as the vertices of a decagon and not mentioned again. The representation of ع by Y is striking — Gerard of Cremona, for instance usually has Q and Adelard of Bath in at least part of his translation of Euclid has i or j —, but Hermann of Carinthia has Y in his translations of both the *Elements* and the *Planisphaerium*[18]. The representation of ك by C seems to be very unusual.

In the translation the diagrams themselves seem to have been reconstructed from the text. The fact that many are reversed, so that the vertices of a pentagon, say, will usually run anticlockwise in Arabic and clockwise in Latin, does not count for much. But the diagrams in the Latin manuscript certainly cannot have been copied from an Arabic manuscript such as Kara Mustafa Paşa 379, in which the diagrams in general correspond well with the text. In Prop. 3, for instance, we find that in the Arabic TH is parallel to AB; but this is not so in the Latin. Similar observations apply to Prop. 4. Further, the Latin manuscript has two diagrams for Props. 18 and 19, but in Arabic there is one composite diagram. Finally, both in the spoilt text of Props. 13–14 and in the explanatory note the figure is made to suit the new problem formulated in the note and does not exist in Arabic. Of course, these

[18]Kunitzsch, pp. 8, 15, 16.

differences and the implied reconstruction of the diagrams from the text may have occurred within the Arabic tradition.

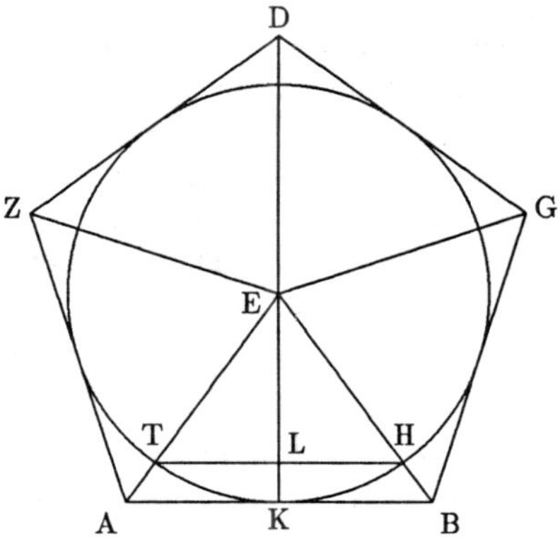

Diagram for Prop. 3 in Arabic

Numbers, which are written out in words in Arabic, are usually expressed in the Latin translation by Hindu-Arabic numerals, simple fractions usually appearing as in the modern notation. There are also some curious fractions, with two numerators and two denominators, which are also to be found in the works of Leonardo of Pisa. $\frac{e\ 1}{8\ 2}$ means $\frac{1}{2}\cdot\frac{1}{8}$. In such cases the notation usually follows the Arabic, which would in this case read "half of an eighth". There are exceptions; e.g. in Prop. 1 $\frac{e\ 1}{5\ 5}$ appears several times for the Arabic equivalent of $\frac{2}{5}\cdot\frac{1}{10}$. The "*e*" may be explained as representing the Latin word-ending (here of *octave*)[19]. $\frac{a\ b}{c\ d}$ at least once means $\frac{a}{c}+\frac{b}{c\cdot d}$, for $\frac{1\quad\ 1}{15625\ 4}$ appears in Prop. 5 as equivalent of the Arabic for $1\frac{1}{4}\cdot\frac{1}{15625}$. In a similar place in Prop. 7 the fraction has been corrected. Of the two occurrences (Props. 8,10) of fractions of form $\frac{1\ 0}{c\ d}$ meaning $\frac{1}{c\cdot d}$ at least one has been corrected. Accordingly, it is reasonable to suppose that the writer of MS BN 7277A was using the same convention as Leonardo, only reversed. For Leonardo has $\frac{5\ e}{8\ 8}$ for $\frac{5}{64}$ and $\frac{a\ b}{c\ d}$ for $\frac{b}{d}+\frac{a}{c\cdot d}$, and he commonly writes his numbers "backwards", i.e. fractions first[20]. Since these fractions also appear in Arabic writings[21], the question arises whether this notation was taken over from the Arabic manuscript used by the translator or whether it was introduced at translation stage or later. If it was not in the Arabic manuscript, the

[19] I am indebted to Dr. Jacques Sesiano for this explanation. It is supported by the appearance of $\frac{e}{10}$ for $\frac{1}{10}$ both in our manuscript (e.g. in Prop. 10*i*) and in Leonardo (e.g. in II 212).

[20] See Karpinski. p. 53, referring to *Liber abaci*, I p. 447, for $\frac{1\ e}{8\ 8}$ and to I pp. 24-25 for Leonardo's explanation of the system.

[21] See Vogel, pp. 3, 18; Suter (1901), pp. 19ff.

translator or some later copier must have found it in some other Arabic-based source. Unfortunately, no help is afforded by the Hebrew translation. For although $\frac{6}{8}\frac{5}{8}$ occurs for the value $\frac{53}{64}$ several times in Suter's translation of Props. 3 and 4, it is clear that he himself inserted most of them to patch up the text. There are nine places in these propositions where the value $48\frac{53}{64}$ occurs and in each of them both Arabic and Hebrew have the equivalent of $\frac{1}{2}+\frac{1}{4}+\frac{5}{8}\cdot\frac{1}{8}$ or $\frac{3}{4}+\frac{5}{8}\cdot\frac{1}{8}$ for the fraction[22].

Before leaving the matter of numerical notation we should notice that, although the "4" in most occurrences is the normal single loop usual in manuscripts of the fourteenth century, a form resembling the Arabic "4" occurs on f. 94r in the fraction $\frac{4}{125}$ and in the marginal note taken from Leonardo at the bottom of the same page[23].

The whole text is liberally supplied with interlinear notes, some giving extra information (e.g. "que est 10" not far from the beginning of Prop. 1), but most giving alternative renderings of the text. Of the latter, some bring the text nearer to the Arabic and some make the Latin easier to understand. Again, some are accompanied by crossings out or underscoring of the text and some are not. Further, there are several marginal notes. Almost all of these additions seem to be in the same hand as the text. A lengthy note in the bottom margin of f. 94r is an extract from Leonardo of Pisa and another — not in the same hand — on f. 94v appears to be a reworking of another such extract[24]. Some of the notes carries the identification"g."; in the first part of the algebra there are several notes with this sign and others with more of the name "Guilelmus" written out[25].

A note in the bottom margin of f. 95v is an attempt to repair a large omission in Props. 13–14. In the text the problem is stated for Prop. 13, but just as rectangle (*quadratum longum*) $EZHT$ is being described, the text jumps to the end of Prop. 14 at the words "the square $HMLY$", thus producing "quadratum longum $GMLY$", H (ح) being mistaken for G (جـ or ـجـ), no doubt within the Arabic transmission or at translation stage[26]. In the note an entirely different problem about the figure $GMLY$ (which becomes *rumbus*) constructed inside triangle ABG is posed and solved; and then a new problem bearing some similarity to the last few lines of the text and introduced by the same formula, "Et si volueris, dic ..." is stated, no doubt

[22]The places in the Hebrew manuscript, Munich, cod. heb. 225, are, with the corresponding passages in the Arabic manuscript in the pagination of the facsimile: 1. 156v12-13 [i.e. lines 12–13 of f. 156v] (137:4 [i.e. line 4 of p. 137]); 2. 156v15 (137:16); 3. 156v20 (137:20-21); 4. 156v22 (138:2); 5. 157r2 (138:7-8); 6. 157r22 (139:8); 7. 157r24 (139:10); 8. 157v2-3 (139:15); 9. 157v5-6 (139:10). There are omissions in the Hebrew, e.g. of $\frac{1}{2}$ in no. 1, and other mistakes, e.g. no. 4 reads "$\frac{6}{8}$ of $\frac{1}{8}$".

[23]For this form of "4" in early western manuscripts see Lemay, 455-456.

[24]The note on f. 94r is from Leonardo II 209, lines 31–39; the note on f. 94v is a reworking of II 211, lines 12–18.

[25]I am grateful to Dr. Jacques Sesiano for this information.

[26]In the facsimile of the Arabic text, everything between p. 147, line 19, word 2 and p. 149, line 16 (end) is omitted.

intended as an emendation[27].

Rather disappointing is the "g." note to Prop. 12 on the left margin of the same page. Abū Kāmil stated simply that if $AD = x$ (in modern terms), then $DG = \sqrt{\frac{1}{3}x^2}$. In fact this must follow by applying Pythagoras' theorem to $\triangle AGD$: $x^2 + DG^2 = (2DG)^2$ — a procedure that involves two unknowns. But the note assumes that $AD \cdot \frac{1}{3}AD = DG^2$ and appears to explain only the appearance of *census*.

No conclusion has been reached on the identity of the translator. In the main the translation is literal; and the few passages of great deviation are probably to be explained by the vagaries of transmission. This might suggest Gerard of Cremona[28] as translator — and indeed the vocabulary seems in general more like that of Gerard of Cremona than that of Robert of Chester in their translations of al-Khwārizmī's *Algebra*. But a similar vocabulary is also to be found in the works of Leonardo of Pisa; and neither the inconsistencies of terminology nor the translation of diagram letters favours Gerard as translator. Identification of the translator by style must await detailed analysis; and only a brief note on terminology is added here.

Terminology

Before we come to the technical vocabulary, a few generalities about the translation and about non-technical words may be made. First, we may observe that although the translated Latin usually mimics the original Arabic — e.g. by translating قد regularly as *iam* , ثم as *deinde* or (more ludicrously) باب by *porta* — and although fixed phrases in Arabic are rendered by (more or less) fixed phrases in Latin — so فإن قيل لك becomes *et si dicemus tibi* —, the translator has evidently felt at liberty not only to translate several different Arabic words by the same Latin word — e.g. أراد and شاء by *velle*; قال, ذكر (also translated as *rememorare*) and وصف (which is also translated by *exponere*, *rememorari* and *memorare*) by *dicere*, or ضلع and جانب by *latus* — but even to translate one Arabic word by two or more Latin words. Examples of this are إذا , which becomes *cum*, *quando* or even *si*, and جعل in the phrase جعل ... شيئاً, which is usually translated by *facere*, but occasionally by *ponere*. The pronouns هو, ذلك etc. and the suffixes ـه, ـها are all rendered by *is*, *ille*, etc., apparently at will. The conjunction أن is mostly usually rendered by *quia*, though typically by *ut* in such phrases as وهو أن, but *quod* occasionally appears for it. The particle فـ is sometimes represented by *et*,

[27]Note that $6\frac{82}{123} - \sqrt{11\frac{1681}{15129}}$ is simply $6\frac{2\cdot 41}{3\cdot 41} - \sqrt{11\frac{41^2}{41^2\cdot 9}}$.

[28]In Steinschneider's discussion (pp. 585, 588) of the text he mentions the suppositions that Gerard of Cremona or Abraham bar Ḥiyya translated it.

igitur or similar, but is often suppressed, particularly when coupled with a form of the imperfect translated as an imperative — so فأسقط *prohice*, فاقسم *divide*, etc. It should be noticed that a Latin imperative sometimes corresponds to an Arabic imperative with ف, so فرد *reduc*, فأكمل *restaura*, فاضرب *duc*.

فيكون[29] introducing the result of a calculation or operation is usually translated as *et provenī*[30] or *et erit/erunt*. تبيّن (verb) or بيّن (adjective) is rendered by *manifestum est* (typically فقد تبين أن *iam manifestum est quia*). The second form is usually translated *exponere* (so وقد كنا بيّنّا أن *et iam exposuimus quia* and similar, occasionally with *quod*, often without كنا). In the "Q.E.D." formula at the end of most of the propositions we find وذلك ما أردنا أن نبين *et hoc est quod voluimus exponere*.

As for the commonalty of non-technical terms we might mention بدأ *incipere*, علة *causa* in the sense of "general principle", علم and عرف *scire* (and معرفة *scientia*), فهم *intelligere*, تقدّم *premittere*, كل *unusquisque* or *quisque* meaning each of a specified and limited number and *omnis* in a more general sense, كله after a noun *totus*, يجب *necesse est*, موضع *locus*.

The *termini technici* are here listed alphabetically according to the root of the word (or of the principal word in a phrase). Not all variants are given, but it is intended that the principal Latin equivalents are sufficiently indicated. References to specific examples are only given when they are unique or are needed to point up some difference of usage.

أخذ *accipere*; مأخوذ جذره *accepta radice eius* (prop. 11*i*), but *accipiamus radicem eius* (Prop. 11*i*).

إلا in contexts where English would use "minus" is expressed by means of *diminutus*. Although this word is seldom written out, but abbreviated to *dim̄* or *dimin̄*, it is clear that it is constructed as an ablative absolute, e.g. جذر إلا *diminuta radice*, "minus the root of [what follows]".

وبرهان ذلك أن *cuius demonstratio est quod* (Prop. 10(ii)).

بقى *remanere*, typically فيبقى *et remane(n)t*, *et remanebit*, etc.; فما بقى *residuum*.

تامّ *integer* (Prop. 20).

[29]The prefix letter in an imperfect form of the verb is sometimes deliberately left without dots, so that the verb can be either masculine or feminine. In the manuscript the masculine form is usual in the present case. A similar convention is also used when it is not desired to specify second person singular or first person plural.

[30]The word appears like this in the manuscript. From similar passages at the beginning of the whole work it appears that the future is probably meant (I am indebted to Jacques Sesiano for this information).

مثلث triangulus.

جبر: فنجبره بـ *restauremus cum* (Prop. 9). حساب الجبر ومقابلة *computatio restaurationis et oppositionis* (Intro.).

جذر *radix*.

جزء "part" used to express fractions translated by symbols, *pars*, *portio* (Prop. 7, in quotation from Euclid).

جمع *agregare*; ما اجتمع *agregatum*; من جميع ذلك *ex toto hoc*.

جانب *latus*.

مجهول *qui non est scitus* (Prop. 7), *ignotus* (Prop. 8).

حساب see جبر.

حفظ *servare*.

أحاط بـ *circumdare in* + ablative. The Latin verb always appears as a participle. A typical example is الذى يحيط به *circumdans in eo* (note the pedantic, and misleading, rendering of the preposition بـ). Sometimes this is rendered as *circumdans* + accusative (e.g. in Prop. 3), more often in the suprascript emendations than in the text, though on two occasions (Prop. 3,4) the text with the accusative is "corrected" into the *in* construction. Other examples are تحيط به الدائرة *circumdans circulum* (Prop. 1), *circumdata a circulo* (Prop. 11*i*); التى يحيط بها هذا المعشر *circumdans hunc decagonum* (Prop. 10(i)); المخمس يحيط به *contentus a pentagono* (Prop. 8). Another translation of أحاط بـ is *continere* + accusative or (Prop. 4) with *in* + ablative. محيط (n.) *circumferentia*, *circumdans*.

خرج *exire* (فيخرج *exibit*) for the result of an operation. خارج *extra* (of figures) (Prop. 10(i)). أخرج *producere*, *ducere* (lines). استخراج *exitus* corrected into *inventio* (Intro.), *inventio* (Prop. 1), *exiens* (Prop. 11(ii)).

خط *linea*.

مخمس *pentagonus*. Sometimes *pentagonus* appears in a phrase translating an Arabic phrase containing خمس.

داخل *infra* (of figures).

درهم *dragma* (almost always abbreviated to *drag*).

دور: يدير (of drawing a circle), *facere*, *addere* (Prop. 2,3) and *adiungere* (Prop. 4), presumably by confusion with يزيد. دائرة *circulus*.

مربع *quadratum* (in algebraic sense or four-sided figure); مربع مستطيل *quadratum longum* (Prop. 13-14).

ردّ: فردّ *reduc*.

رسم *facere* in geometrical constructions; رسمنا *diximus* (Prop. 7).

مركز *centrum.*

زاوية *angulus*; على زاوية قائمة *secundum angulum rectum.*

زاد ... على *addere ... ad* (+ dat. in Prop. 11*i*).

مسدس *exagonus.*

أسقط ... من *prohicere ... ex.*

مساوٍ لـ *equalis* + dative; متساوى الأضلاع *equilaterus* (Prop. 15), متساوى الأضلاع والزوايا *equilaterus et equiangulus* (up to Prop. 12) and *equalium laterum et angulorum* (frequent from Prop. 13-14 on).

شكل *figura*; شكل له خمس عشرة زاوية *figura quindecagona.*

شيء *res*, i.e. the unknown quantity to be found, x.

ضرب ... بـ (vb.) *ducere ... in* (multiply), *multiplicare ... in*; ضرب ... بـ (n.) *multiplicatio ... in*; مضروب *ductus* (usual), *productum* (Prop. 6).

ضعّف *duplare.*

طرف *extremitas*; and see نسبة.

مطلوب *quesitus.*

طول *longitudo.*

من العدد *ex numero* usually, but also *numero* (Prop. 2,3).

عدل (vb.) *est equalis*, *equatur*, in sentences of structure "there results A *equal to* B" usually *equalis.*

معشر *decagonus.* Sometimes *decagonus* appears in a phrase translating an Arabic phrase containing عشر.

معلوم *scitus, notus*; معلوم أن *scitum est quia.*

عمل ... على *facere ... super* of circumscribing one figure about another and similarly with فى for inscribing.

قبل: فنقابل بـ *oppone in* (Prop. 1,15), *opponamus* (Prop. 2,9); مقابلة *oppositio* (Prop. 2 and see جبر).

قدر *quantitas* (Prop. 1,4), *mensura* (Prop. 2); كمية قدر *quantus* (Prop. 3) and there are other such variants. قدر ... من *proportio ... ad* (usual) or *proportio ... ex* (Prop. 4); قدر ... عند *proportio ... ad* (Prop. 7,8). *v.* كم.

بالتقريب *in propinquitate* (Prop. 11(ii)), *secundum propinquitatem* (Prop. 11(ii)), *fere* (Prop. 11(ii)).

قسم ... على *dividere ... per.* انقسم dividere (various constructions) (Prop. 17).

القسم الأعظم *maior pars* (of a divided line).

قطر *diameter.*

قاعدة *basis*.

قائمة *rectus* (of angles). قائم الزوايا *rectorum angulorum* (of figures). استقامة على *secundum rectitudinem*.

قياسه أن *exemplum*, *verbi gratia*.

كعب *cubus*, x^3.

كل شيء معك and كل ما معك *totum quod habes*.

كم *quantus*; كم مقدار *quantus* (Prop. 1).

كمل: فأكمل ... حتى يكون *restaura ... ut sit* (Prop. 1,12,16,17,18), *reintegra ... ut sit* (Props. 19,20); وإكمالك إياه ... ان *et restaurabis illud si* ... (Prop. 12).

كمية see قدر.

ألقى ... من *prohicere ... ex*.

مثل in the sense of "equal" *similis*; في مثله in multiplication *in se*. ذلك مثال *exemplum* (Prop. 10(ii)), *verbi gratia* (Prop. 17).

ماسّ *contingere*.

مساحة *mensura*.

مال *census*, the square of the unknown, x^2.

على نسبة ذات وسط وطرفين *per proportionem per medium et 2 partes* corrected into *secundum proportionem medii et duo extrema*, *secundum proportionem medii et extrema*, *secundum proportionem habentem medium et duo extrema*, ditto with *duoque* (all Prop. 17)., *secundum proportionem medii et extremorum* (Prop. 17,18,19).

نصف *medietas*, *medium* (of a line), $\frac{1}{2}$. نصّف *dimidare*.

في نفسه in multiplication *in se*.

نقص ... من *subtrahere ... ex*; منقوص *subtractus* (Prop. 15), *sublatus* (Prop. 9,11*i*,11*ii*); نقصته *minus eam* (Prop. 17).

وتر *corda*. وتّر *est corda* etc.

وقع *cadere* (of one figure inside or outside another).

موازٍ ... لـ *equidistans* + dative.

وصل ... بـ *coniungere*; وصل فيما بين *coniungere quod est inter*.

The Manuscript

The contents of MS Paris, BN lat. 7377A may be itemized as follows:

1. ff. 1r–33v: Muḥammad ibn 'Abd al-Bāqī's commentary on Euclid's *Elements*, book X, in the translation by Gerard of Cremona; ed. Curtze, 1899, in his edition of Anaritius, pp. 252–384.
2. ff. 34r–43v: al-Khwārizmī's *De algebra et almuchabala* in the translation by Gerard of Cremona; ed. Hughes, 1986.
3. ff. 43r–56v: Abū Bakr's *Liber mensurationum* in the translation of Gerard of Cremona; ed. Busard, 1968.
4. ff. 56v–57v: *Liber Saydi Abuothmi* on mensuration; ed. Busard, 1969.
5. ff. 57v–58v: *Liber Aderameti* on mensuration; ed. Busard, 1969.
6. ff. 58v–68r: *Liber augmenti et diminucionis* (anon.); ed. Libri, 1838, pp. 304–376[31].
7. ff. 68r–70v: Pappus' commentary on Euclid's *Elements*, book X, probably translated by Gerard of Cremona; ed. Junge 1936.
8. 71v–97r: Abū Kāmil, *Algebra*.
9. 99r–203r: *Liber Mahameleth*, a Latin compilation of problems in arithmetic and algebra, based on Arabic sources[32].

Most, if not all, of the first seven items are translations by Gerard of Cremona. Items 2–6 appear in the same order in MSS Paris, BN lat. 9335 and other manuscripts[33]. The writing appears to be of the fourteenth century.

Editing conventions

For the purposes of the edition it has been assumed that the literal style of the translator was smoothed in the process of copying and that both literal and smooth readings often appear in text and interlinear note (or in note and text), as will often happen when a scribe has two manuscripts before him[34]. It is intended that the edited text be as near to the original translation as the evidence allows. Where appropriate, the Arabic has been used to decide on readings. Occasionally an Arabic reading in transliteration or translation is given in the apparatus; diagram letters are quoted according to the translator's system (see above). Emendations, notice of omission, trivial corrections made in the manuscript and the usual small editorial corrections

[31] See also Tannery.

[32] See Sesiano.

[33] Busard (1969), 163–164.

[34] For a completely different view of the matter, see the article by Dr. Jacques Sesiano in this volume.

appear in the first apparatus. The second apparatus is to accommodate the numerous interlinear and marginal notes.

In either apparatus, if a word or phrase in the accepted text is commented upon, it is written in the apparatus followed by a square bracket,]. A word followed by a shallow angle-bracket, ⟩, is used simply to indicate the position for the note that follows; additions in the text come after the word so indicated. In the case of a phrase the line-numbers given in the apparatus refer to its last word. Text added by the editor appears in angle-brackets, < >. Text between square brackets is not legible because of physical damage to the manuscript[35]; difficult readings in the note to Prop. 3 on f. 94r, have been resolved with the help of Boncompagni's edition of Leonardo's *Practica geometriae*.

Sometimes the text confuses the diagram letters G and H. Since this mistake can reasonably only be made in Arabic or by the translator, the "incorrect" reading has been retained, as being what the translator wrote. But to help the reader a dash is put on the letter. Thus G', H' respectively mean G, H in the Latin manuscript, which stand for the equivalent of H, G in Arabic.

Several abbreviations have been left unexpanded: *dim̄* or *dimin̄*, *drag'*, *proven̄*, *rad'* for a form of *diminutus*, *dragma*, *provenire*, *radix*, respectively. With the exception of *drag'*, which usually has no sign of abbreviation, these resemble the writing of the manuscript.

The diagrams have been redrawn much as they are in the manuscript, albeit a little idealised. The two Es in each of the diagrams for Props. 3 and 4, as well as the corresponding mistakes in the text, have been left (in each case the E referred to in the mathematical argument is the centre of the circle).

Bibliography

[Abū Kāmil] Abū Kāmil Shujā' ibn Aslam, *The Book of Algebra. Kitāb al-Jabr wa 'l-muqābala*, Frankfurt am Main 1986.

[Abū Kāmil (Hebrew)] *The* Algebra *of Abū Kāmil*, Kitāb fī al-jabr wa'l-muqābala *in a Commentary by, Mordecai Finzi: Hebrew Text, Translation, and Commentary with Special Reference to the Arabic Text*, edited with introduction, by Martin Levey, Madison etc. 1966.

[Anaritius] M. Curtze, *Anaritii in decem libros priores Elementorum Euclidis*, in *Euclidis Opera Omnia*, ed. I. L. Heiberg et H. Menge, *Supplementum*, Leipzig 1899.

H. L. L. Busard, "L'algèbre au moyen âge: le 'Liber mensurationum d'Abū Bekr'", *Journal des Savants*, (1968), 65–214.

H. L. L. Busard, "Die Vermessungstraktate *Liber Saydi Abuothmi* und *Liber Aderameti*", *Janus*, 16 (1969), 161-174.

[35] Dr. Jacques Sesiano very kindly eliminated a large class of uncertain readings – words that were caught in the binding and could not be read on the film available to me – by direct inspection of the manuscript. It is a pleasure to acknowledge my debt to him.

Barnabus Hughes, "Gerard of Cremona's Translation of al-Khwārizmī's *al-Jabr*: a critical edition", *Medieval Studies*, 48 (1986), 211-263.

Barnabus Hughes, *Robert of Chester's Latin Translation of al-Khwārizmī's al-Jabr*, Stuttgart 1989.

G. Junge, "Das Fragment der lateinischen Übersetzung des Pappus-Kommentars zum 10. Buch Euklids", *Quellen und Studien zur Geschichte der Mathematik, Astronomie und Physik*, Abt. B, 3, Studien (1936), 1–17.

L. C. Karpinski, "The algebra of Abu Kamil Shoja' ben Aslam", *Bibliotheca mathematica*, 3. F., 12 (1911-12), 40–55.

Paul Kunitzsch, "Letters in Geometrical Diagrams, Greek – Arabic – Latin", *Zeitschrift für Geschichte der Arabisch-Islamischen Wissenschaften*, 7 (1991/92), 1–20.

[al-Khwārizmī] *The Algebra of Mohanned ben Musa*, ed. and trans. Frederic Rosen, London 1831 (repr. Hildesheim-Zurich-New York 1986).

Richard Lemay, "The Hispanic Origin of Our Present Numeral Forms", *Viator*, 8 (1987), 435–462 + 15 pp. of ilustrations.

[Leonardo of Pisa] *Scritti di Leonardo Pisano matematico del secolo decimoterzo*, ed. Baldassarre Boncompagni, 2 vol., Rome 1862.

Martin Levey, "Abū Kāmil Shujā' ibn Aslam ibn Muḥammad ibn Shujā'", *Dictionary of Scientific Biography*, I, 30-32, New York 1970.

Libri, *Histoire des sciences mathématiques en Italie depuis la renaissance des lettres jusqu'à la fin du dix-septième siècle*, 4 vol., Paris 1838-1841.

Nashriya II (1340-1341).

G. Sacerdote, "Il trattato del pentagono e del decagono di Abu Kámil Shogia' ben Aslam ben Muhammed", *Festschrift zum achtzigsten Geburtstag Moritz Steinschneiders*, Leipzig 1896, 169–194.

J. Sesiano, "Le Liber Mahameleth, un traité mathématique latin composé au XIIe siècle en Espagne", Premier Colloque Internationale sur l'Histoire des Mathématiques Arabes, *Actes*, Algiers 1988.

Fuat Sezgin, *Geschichte des arabischen Schrifttums*, V, Leiden 1974.

M. Steinschneider, *Dir hebraeischen Übersetzungen des Mittelalters und die Juden als Dolmetscher*, 1893 (repr. Graz, 1956).

[Suter 1901] Heinrich Suter, "Das Rechenbuch des Abū Zakarījā", *Bibliotheca mathematica*, 3.F 2 (1901), 12-40.

[Suter] Heinrich Suter, "Die Abhandlung des Abū Kāmil Šoǵā ben Aslam 'Über das Fünfeck und Zehneck', *Bibliotheca mathematica*, 3. F, 10 (1909-1910), 15–42.

P. Tannery, "Sur le 'Liber augmenti et diminutionis' compilé par Abraham", *Bibiotheca mathematica*, 3.F, 2 (1901), 45–47.

Josef Weinberg, *Die Algebra des Abū Kāmil Šoǵā ben Aslam*, Munich 1935.

Kurt Vogel, "Zur Geschichte der Stammbrüche und der aufsteigenden Kettenbrüche", *Sudhoffs Archiv*, 66 (1982), 1–19.

Mohammad Yadegari and Martin Levey, "Abū Kāmil's 'On the Pentagon and Decagon'", *Japanese Studies in the History of Science*, Supplement 2 (1971), 54 pp.

Dixit Abucamel ssagia filius abrahim agregator istius libri: postquam exposuimus quod indiget expositione in eo quod est occultum ex computatione restaurationis et oppositionis ex modis antiquorum in scientia et in aggravatione in eo et intelligentium ex geometris et sapientibus in libro euclidis et aliis et glosavimus illud et discooperuimus in libro nostro isto, an nobis est nunc ad exponendum quantitatem cuiusque lateris de lateribus cuiusque figure pentagoni et decagoni equilateri et equianguli descripti intra circulum notum vel extra circulum notum, et quantitatem omnis diametri circuli circumdantis pentagonum vel decagonum scitos equilateros et equiangulos vel circumdati ab eis, et mensuram corde partis ex 4 partibus circumferentie circuli sciti, et mensuram cuiusque lateris de lateribus decagoni equilateri et equianguli cum scitur mensura eius et quantitatem laterum triangulorum mensura scitorum cum fuerint in pentagonis vel decagonis equilateris et equiangulis, et alia ex eis que exponemus in libro nostro isto ex illo quod est exitus multus ex eo. Et ego iudico in bonum quos vidi compotistas et geometras nostri temporis huius et antiquos de quibus audivi sapientes in numero et mensura. Creator tamen erexit me ad posse applicare ad id quod eis fuit impossibile contingere.

< 1 > Incipiamus in inventione quantitatis corde $\frac{1}{5}$ circuli ex diametro eius. Et illud est ut faciamus circulum scitum $ABDEG$ cuius diameter sit 10 ex

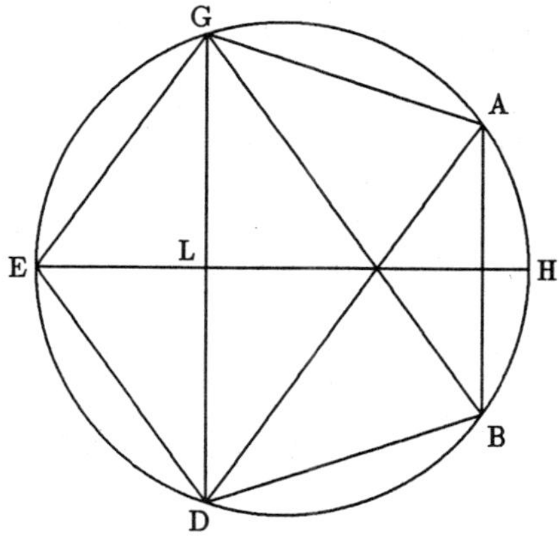

Prop. 1

---

5 glosavimus] *corr. ex* eglosavimus *9* equilateros] equilatos *9* vel⟩ c'ade *add. et del.* *12* eius⟩ s *add. et del.* et mensuram corde ... mensura eius *om. Ar.* *15* Et ego ... bonum] *Ar. yata'adhdharu*

---

3 ex modis] *corr. in* respicientium *6* an nobis est] *corr. in* pervenimus *6* quantitatem⟩ vel mensuram *supra* *7* pentagoni] *supra* *11* equilateri] *corr. ex* equilateris *14* illo⟩ per *supra* *15* exitus multus ex eo] *corr. in* inventio multa *18* contingere⟩ et ad id quod occultum fuit eis, et Ipsi sit gloria et honor Qui est unum et secundum non habet *marg.*

numero, sitque linea HE, et in eo sit pentagonus equilaterus et equiangulus circumdans circulum sitque pentagonus $ABGD$. Cum autem voluerimus scire quantum est quodlibet latus de lateribus huius pentagoni, producemus lineam DLG, que est corda $\frac{2}{5}$ circuli, et faciamus lineam ED rem. Scitum est quia linea EL est $\frac{1}{10}$ census, quia multiplicatio ED in se est sicut multiplicatio HE in EL; ergo linea GL est radix census diminꝰ $\frac{1}{10}$ decime census census; et linea GL est sicut linea LD: linea GD est radix 4 censuum diminꝰ $\frac{e\,1}{5\,5}$ census census. Et scitum est quia multiplicatio AB in GD et AB in se est sicut multiplicatio GD in se, quia multiplicatio AB in GD et AG in BD est sicut multiplicatio AD in BG, et AG in BD est sicut multiplicatio AB in se et AD in BG sicut GD in se. Et GD in se est 4 census diminꝰ $\frac{e\,1}{5\,5}$ census census. Prohice ex eo multiplicationem AB in se, que est census, et remanebunt 3 census diminꝰ $\frac{e\,1}{5\,5}$ census census equalis AB in GD. Divide 3 census diminꝰ $\frac{e\,1}{5\,5}$ census census per lineam AB, que est res, exibit linea GD 3 res diminꝰ $\frac{e\,1}{5\,5}$ cubi. Et iam exposuimus quia linea GD est radix 4 censuum dimꝰ $\frac{e\,1}{5\,5}$ census census. Ducamus 3 res diminꝰ $\frac{1}{25}$ cubi in se, et provenꝰ 9 census et $\frac{1}{625}$ cubi cubi diminꝰ $\frac{6}{25}$ census census equalis 4 censibus diminꝰ $\frac{e\,1}{5\,5}$ census census. Oppone in ipsis et erit $\frac{1}{5}$ census census equalis 5 censibus et $\frac{1}{625}$ cubi cubi. Divide totum quod habes per censum, et exibunt 5 drag′ et $\frac{1}{625}$ census census equalis $\frac{1}{5}$ census. Restaura ut sit census census, quod est ut ducas eum in 625, et ducas totum quod habes in 625, et provenꝰ census census et 3125 drag′ equalis 125 censibus. Dimidia ergo census, et exibit 62 et $\frac{1}{2}$; duc ea in se et provenꝰ 3906 et $\frac{1}{4}$; ex quibus prohice 3125, et remanebit 781 et $\frac{1}{4}$; radix eius subtracta ex 62 et $\frac{1}{2}$; accepta rad′ residui est linea ED, que est unum ex lateribus pentagoni. Et hoc est quod voluimus exponere.

< 2 > Et quando voluerimus scire mensuram quantitatis corde $\frac{1}{10}$ circuli sciti et iam scivimus $\frac{1}{5}$ eius cordam, quod addemus circulum super quem sint $ABGDEG'$ et faciemus diametrum eius 10 numero, que sit linea AG', et faciamus in medietate eius 5 cordas equales, que sint AB et BH' et linea $H'D$ et linea DE et linea EG'. Manifestum est quia linea DG' est corda $\frac{1}{5}$ huius circuli; et iam exposuimus quia ducta in se est 62 et $\frac{1}{2}$ diminuta radice 781 et $\frac{1}{4}$; et linea AH' sicut ipsa. Et multiplicatio AG in DG' et GD in AH est sicut AD in GH. Set AD sicut GH et AG sicut DH, multiplicatio $G'D$

21 equilaterus] equilatus 26 GL] *Ar.* DL 31 in se[1]] *repet. et del.* 32 in] *repet.* 33 diminꝰ[1]⟩ *seq. ras.* 33 GD] BG 33 diminꝰ[2]⟩ $\frac{e\,1}{10\,5}$ census census *add.* 46 scire⟩ quantit *add. et del.* 48 et⟩ 10 *add. et del.* 49 faciamus⟩ i[a] *add.* 50 $\frac{1}{5}$⟩ seq. ras.

22 circumdans⟩ intra *supra* 22 $ABGD$] GD *del., corr. in* $ABDEG$ 24 DLG] *supra* 25 se⟩ que producit censum *supra* 26 HE⟩ que est 10 *supra* 26 EL⟩ quam $\frac{1}{10}$ census esse ab hoc necesse est *supra* 26 est] *supra* 26 decime] *supra* 27 linea[3]⟩ igitur *supra* 30 BG⟩ ex eo quod dicit ptolomeus in libro de arcu et corda cum fuerit quadrangulum in circulo *marg.* 31 $\frac{e\,1}{5\,5}$ census] *corr. ex* 25. censibus 34 census census] *corr. ex* censibus 35 cubi] *corr. ex* cubis 36 cubi] *corr. ex* cubis 37 census[3]] *corr. ex* censibus 41 ducas eum in 625] *is underlined in the MS, but is present in Arabic* 46 quantitatis] *del. in textu* 46 corde] *supra* 47 cordam] *supra* 48 $ABGDEG'$] *corr. in* $ABHDEG$, *Ar.* $ABGDEH$ 48 faciemus] *supra* 48 10] *supra* 52 DG'] *corr. in* DH 53 sicut[2]] *corr. in* in 53 et AG sicut DH] *del. in textu*, H in $D \ldots G$ (?) *add. et del. supra*, est sicut GH in se Et *supra*

| in AH' est sicut AH' in se: ergo multiplicatio DH in AG et AH' in se est sicut multiplicatio GH in se. Et linea $H'D$ est corda $\frac{e}{10}$ huius circuli; faciamus eum rem. Et ducamus eam in lineam AG', que est diameter circuli et est 10, et proveñ 10 res; et ducamus lineam DH in se et proveñ 62 et $\frac{1}{2}$ dimiñ radice 781 et $\frac{1}{4}$. Ergo multiplicatio linee GH in se 62 et $\frac{1}{2}$ et 10 res dimiñ radice 781 et $\frac{1}{4}$. Addamus ei multiplicationem AH' in se, que est 62 et $\frac{1}{2}$ dimiñ radice 781 et $\frac{1}{4}$: erit 125 drag′ et 10 res dimiñ radice 3125 drag′ equalis multiplicationi AG' in se, que est 100. Opponamus ergo in ipsis, et erit post oppositionem radix 3125 drag′ dim̃ 25 drag′ equalis 10 rebus ergo res equatur radici 31 et $\frac{1}{4}$ dim̃ 2 et $\frac{1}{2}$, et hoc est linea GD que est corda $\frac{1}{10}$ circuli. Et hoc est quod voluimus exponere. 94r

< 3 > Cum voluerimus scire quanta sit corda unius lateris pentagoni equilateri et equianguli circumdantis in circulo scito, addamus circulum scitum, et faciamus diametrum eius 10; et faciamus super ipsum pentagonum equilaterum et equiangulum, qui sit pentagonus $ABGDE$. Manifestum est quia corda TH corda pentagoni circumdantis circulum, et iam exposuimus quia ducta in se est 62 et $\frac{1}{2}$ dim̃ rad′ 781 et $\frac{1}{4}$ et quadratum linee TL 15 et $\frac{5}{8}$ dim̃ rad′ 48 et $\frac{1}{2}$ et $\frac{1}{4}$ et $\frac{e}{8}\frac{5}{8}$. Prohice ergo ipsum ex quadrato linee TE, que est $\frac{1}{2}$ diametri, quod est 25, et remanebit quadratum linee EL 9 et $\frac{1}{4}$ et $\frac{1}{8}$ et radix 48 et $\frac{1}{2}$ et $\frac{1}{4}$ et $\frac{e}{8}\frac{5}{8}$. Deinde faciamus lineam AB, que est una ex

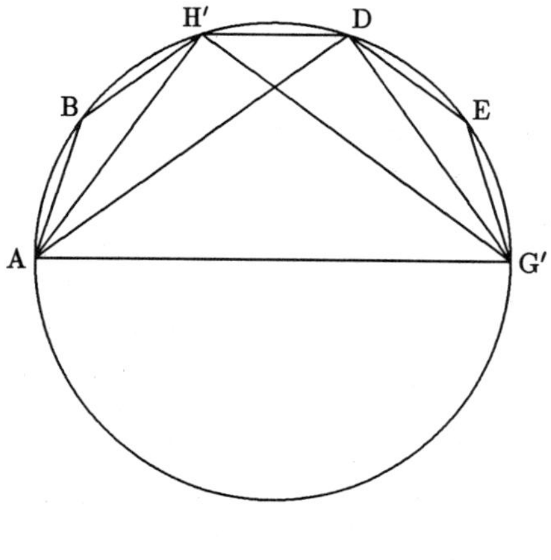

Prop. 2

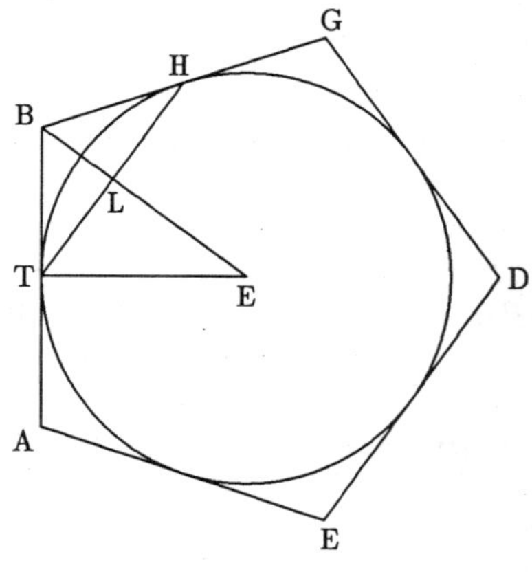

Prop. 3

54 multiplicatio GD in ... se] *om. Ar.* 56 in⟩ se *add. et del.* 65 Cum] Et non 71 rad′⟩ 8 *add. et del.*

55 et AH' in se est] *supra* 57 DH] *corr. in* GD 59 781] *corr. ex* 581 60 781] *corr. ex* 981 63 radici] *supra* 66 addamus] *corr. in* faciamus 69 circumdantis⟩ vel continentis *supra* 69 circulum⟩ intra *supra* 72 et^2] *supra* 73 $\frac{1}{4}$ et $\frac{1}{8}$] $\frac{e}{8}$ (*corr.*) $\frac{1}{4}$

lateribus pentagoni circumdantis in circulo, rem et quadratum eius censum, et proportio quadrati AB, quod est census, ad quadratum TE, quod est 25, sicut proportio quadrati TH, quod est corda $\frac{1}{5}$ circumdantis circulum, quod est 62 et $\frac{1}{2}$ dim̄ radice 781 et $\frac{1}{4}$, ad quadratum EL, que est 9 et $\frac{1}{4}$ et $\frac{1}{8}$ et radix 48 et $\frac{1}{2}$ et $\frac{1}{4}$ et $\frac{e\ 5}{8\ 8}$. Duc censum in 9 et $\frac{3}{8}$ et radix 48 et $\frac{1}{2}$ et $\frac{1}{4}$ et $\frac{e\ 5}{8\ 8}$, et proveñ 9 census et $\frac{3}{8}$ census et radix 48 census census et $\frac{3}{4}$ census census et $\frac{e\ 5}{8\ 8}$ census census: hoc est equalis quadrato TE in quadratum TH, quod est 1562 et $\frac{1}{2}$ dim̄ radice 488281 et $\frac{1}{4}$. Reduc totum quod habes ad censum, quod est ut ducas totum quod habes in $\frac{1}{5}$ et $\frac{e\ 1}{5\ 5}$ dim̄ radice $\frac{4}{125}$ drag′. Duc 9 census et $\frac{3}{8}$ census et radicem 48 census census et $\frac{3}{4}$ census census et $\frac{e\ 5}{8\ 8}$ census census <in $\frac{1}{5}$> et $\frac{e\ 1}{5\ 5}$ dimiñ rad′ $\frac{4}{125}$ unius, et proveñ census. Deinde duc 1562 et $\frac{1}{2}$ dim̄ radice 488281 et $\frac{1}{4}$ in $\frac{1}{5}$ et $\frac{e\ 1}{5\ 5}$ dim̄ radice $\frac{4}{125}$ unius, et proveñ hoc 375 drag′ et rad′ 15625 drag′ et dim̄ rad′ 78125 drag′ et dim̄ rad′ 28125 drag′, et est 500 dim̄ rad′ 200000. Ergo census equatur 500 dim̄ rad′ 200000; accipe radicem eius, AB, que est unum de lateribus pentagoni equilateri et equianguli circumdantis in circulo scito cuius diameter est 10 numero. Iam manifestum est quia quadratum linee AB, que est unum de lateribus pentagoni, est 500 dim̄ rad′ 200000. Et hoc est quod voluimus exponere.

< 4 > Et si dicemus tibi, nos volumus scire quanta sit mensura corde decagoni equilateri et equianguli circumdantis in circulo scito, adiungamus circulum et faciamus diametrum eius 10 ex numero, et faciamus super eo circumdantem in ipso decagonum $ABGDEFUZM$; et volumus scire quanta est AB, que est unum ex lateribus decagoni. Manifestum est quia linea TH facta est corda decagoni continentis circulum. Et iam exposuimus quia ipsa est radix 31 et $\frac{1}{4}$ dim̄ 2 et $\frac{1}{2}$, et linea TL est $\frac{1}{2}$ eius et est radix 7 et $\frac{1}{2}$ et $\frac{1}{4}$ et $\frac{e\ 1}{8\ 2}$ dimiñ 1 et $\frac{1}{4}$; quadratum eius 9 et $\frac{1}{4}$ et $\frac{1}{8}$ dim̄ radice 48 et $\frac{1}{2}$ et $\frac{1}{4}$ et $\frac{e\ 5}{8\ 8}$. Prohice illud ex quadrato TE, quod est 25: remanebit quadratum EL 15 et $\frac{5}{8}$ et rad′ 48 et $\frac{1}{2}$ et $\frac{1}{4}$ et $\frac{e\ 5}{8\ 8}$. Et quadratum TH 37 et $\frac{1}{2}$ dim̄ radice 781 et $\frac{1}{4}$. Deinde fac lineam AB rem, que est unum de lateribus decagoni circumdantis in circulo, et quadratum eius censum. Et erit proportio census ex quadrato TH, quod est 37 et $\frac{1}{2}$ dim̄ radice 781 et $\frac{1}{4}$, sicut proportio quadrati CE, et est

---

75 TE] *Ar.* KE *80* TE] *Ar.* KE *89* circumdantis] circumdatis *97* facta] ? *101* 37] 39 *104* est[2]⟩ *seq. ras.*

---

74 in circulo] *corr. ex* circulum *75* quadratum] *corr. ex* quadrati *77* ad] *supra* *77* quadratum] *corr. ex* quadrato *79* census[4]] *supra* *80* TE⟩ ducto *supra* *82* radice] *supra* *88* eius⟩ que est *supra* *91* exponere⟩ [Aliter] promtius quia in orthogonio trigono ETB ab angulo recto producta est cathetus [TL erit prop]ortio quadrati BE ad quadratum ET sicut quadratum ET ad quadratum [EL Quare multip]licatio quadrati BE in quadratum EL est sicut multiplicatio quadrati ET [in se quod est 6]25 Quare ponam BE rem et ducam eam in se exibit census quem du[cam in qua]dratum EL quod est $\frac{3}{8}$9 radix $\frac{5\ 6}{8\ 8}$48 venient (census *add. et del.*) $\frac{3}{8}$9 et ra[dix $\frac{5}{8}$]$\frac{6}{8}$48 census census que equatur 625 dragmis. Reduc hec ad censum [hoc est multip]lica 625 per $\frac{6}{25}$ dragme diminuta radice $\frac{4}{125}$ ue[nient] 150 minus radice 12520 que equatur censui hoc est quadrato BE [de quo si au]feratur ... [quad]rato [T]B ... *marg.* *95* decagonum] *corr. ex* pentagonum *95* $ABGDEFUZM$⟩ N *add. supra* *97* decagoni] *corr. ex* pentagoni *100* EL] ? *corr. in* TL; T *supra* *101* et[4]] *supra* *103* in circulo] *corr. ex* circulum *103* Et erit proportio] *supra*; rad′ *add. et del.* *103* ex quadrato] *corr. supra in* ad quadratum ; eius *add. et del.*

25, ex quadrato EL, et est 15 et $\frac{5}{8}$ et rad′ 48 et $\frac{1}{2}$ et $\frac{1}{4}$ et $\frac{e\ 5}{8\ 8}$. Ducamus eum in quadratum AB, quod est census, et proveñ 15 census et $\frac{5}{8}$ census et rad′ 48 census census et $\frac{3}{4}$ census census et $\frac{e\ 5}{8\ 8}$ census census equalis quadrato CE in quadatum TH, et est 937 et $\frac{1}{2}$ dim̄ radice 488281 et $\frac{1}{4}$. Reduc totum quod habes ad censum, quod est ut ducas totum quod habes in $\frac{e\ 2}{5\ 5}$ unius dim̄ rad′ $\frac{e\ \ 4}{625\ 5}$ unius, et erit census equalis 75 drag′ et rad′ 625 drag′ dim̄ radice 3125 drag′ et dim̄ rad′ 1125 drag′. Et illud est 100 drag′ dim̄ rad′ 8000, et illud est quadratum linee AB, quod est unum de lateribus decagoni continentis in circulo scito cuius diameter est 10 ex numero. Et hoc est quod voluimus exponere.

| < 5 > Et cum voluerimus scire quanta est diameter circuli circumdan- 94
tis in pentagono scito equilatero et equiangulo, faciamus pentagonum notum $ABGDE$ et faciamus unumquodque laterum eius 10 ex numero et circulum super eo circumdantem in eo cuius diameter linea CL, et scientia diametri eius est ut coniungamus EG et ponamus eum rem. Manifestum est quia multiplicatio eius in AB, et est 10, et AB in se est sicut multiplicatio EG in se. Sed multiplicatio EB, que est res, in GD, que est 10, proveñ 10 res; et DG in se producit 100; et EG in se producit censum. Erit census equalis 10 rebus et 100 drag′. Ergo res equatur 5 et rad′ 125, que est linea EG; medietas eius 2 drag′ et $\frac{1}{2}$ et rad′ 31 et $\frac{1}{4}$, et hoc est linea EL. Duc eam in

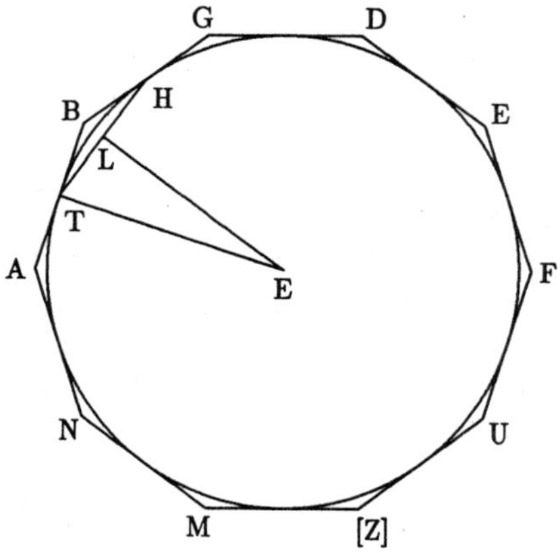

Prop. 4

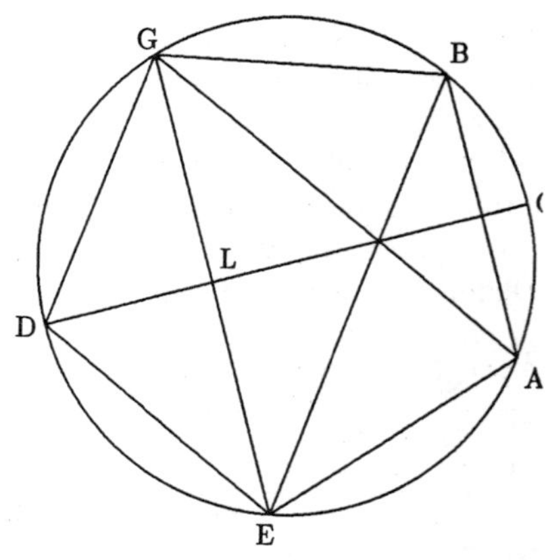

Prop. 5

106 et²⟩ ig ? *add. et del.* *109* rad′⟩ $\frac{4}{5}$ partis *add.* *110* et¹] *supra* *121* se⟩ erit 10 res *add. et del.*

105 ex quadrato] *corr. supra in* ad quadratum *112* continentis⟩ circumdantis *supra* *115* cum] ?, *corr. in* si *116* in pentagono ... equiangulo] *corr. in* pentagonum scitum equilaterum et equiangulum *117* 10] *corr. ex* 16 *118* circumdantem] *corr. ex* circumdantis *118* in eo] *corr. in* eum *119* *EG*] *BH corr. in BE*, *Ar. EG corr. ex* ? *120* *AB*¹] *corr. in DG* *120* 10⟩ res *supra* *120* *AB*²] *corr. in BG* *121* Sed] Sed *supra* *121* proveñ] *corr. in* producit *122* censum⟩ cum sit equalis *EB marg.*

se et provenit 37 et $\frac{1}{2}$ et radix 781 et $\frac{1}{4}$. Prohice eum ex ED in se, quod est 100, et remanebit 62 et $\frac{1}{2}$ dim̄ rad′ 781 et $\frac{1}{4}$, et hoc est quadratum linee DL. Deinde faciamus diametrum circuli, que est linea CLD, rem et quadratum eius censum. Proportio census ad quadratum DE, quod est 100, est sicut proportio quadrati DE, quod est 100, ad quadratum DL, quod est 62 et $\frac{1}{2}$ dim̄ rad′ 781 et $\frac{1}{4}$. Duc eum in censum, et proveñ 62 census et $\frac{1}{2}$ census dim̄ rad′ 781 et $\frac{1}{4}$ census census equalis multiplicationi quadrati ED in se, quod est 10000. Reduc totum quod habes ad censum, quod est ut ducas totum quod habes in $\frac{e}{10}\frac{1}{5}$ unius et rad′ $\frac{1}{15625}\frac{1}{4}$ unius, et proveñ census equalis 200 et rad′ 8000. Iam manifestum est quia quadratum diametri circuli, que est CLD, 200 et radix 8000. Et hoc quod voluerimus exponere.

< 6 > Et si voluerimus scire diametrum circuli contenti a pentagono $ABGDE$ noto equilatero et equiangulo cuius quodlibet latus est 10 ex numero, et centrum circuli M < ... > et medietas diametri circuli contenti a pentagono linea AM. Et iam exposuimus quod quadratum diametri eius 200 et radix 8000. Ergo quadratum linee AM, quod est $\frac{1}{2}$ diametri eius,

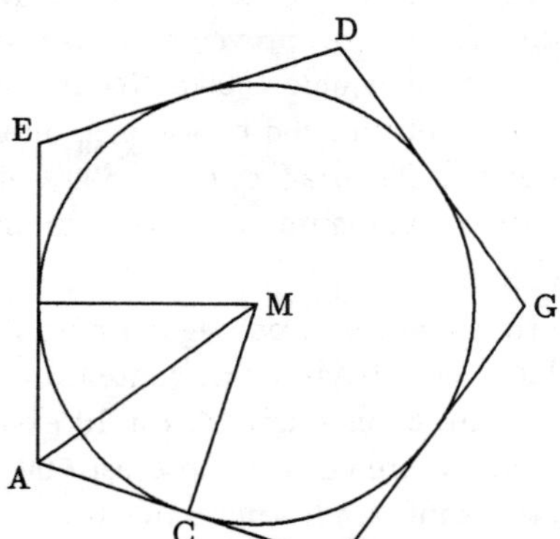

Prop. 6

129 DE] *corr. ex D* *133* $\frac{e}{10}\frac{1}{5}$⟩ uni *add. et del.* *135* exponere⟩ Aliter promtius multiplica *EB* que est 5 et radix de 125 in se quia est corda anguli pentagonici erunt 150 diminutis [ ] et 10 radices de 125 quibus adde quadratum lateris pentagonici quod est 100 erunt 250 et 10 radices de[ ] quorum quinta pars que est 50 et 2 (*ras.* radicis *ras.* de 125 est quadratum lateris exago[ ] silicetmedietatis diametri ut in quartodecimo libro euclidis habetur quadruplum vero eius silicet200 Et 8 radices de 125 que sunt radix 8000 est quadratum diametri *DC* ut superius in ... *marg.* *138* < ... >] *Ar. wa-niṣf quṭrihā khaṭṭ KM* *140* radix⟩ 8 quadra *add. et del.*

139 contenti in pentagono] *corr. in* tan<g>entis pentagonum *139* eius⟩ est *supra* *140* Ergo] *supra*

50 et radix 500. Prohice ex eo quadratum AC, quod est 25, et remanebit quadratum CM, quod est $\frac{1}{2}$ diametri circuli cadentis in hoc pentagono, 25 et radix 500, quorum quadruplum scilicet 100 et radix 8000 est quadratum totius diametri . Et hoc est quod voluimus exponere.

Et si volueris, duc unum de lateribus pentagoni in se et prohice illud ex producto diametri circuli circumdantis cadentis < ... > infra pentagonum. Et hec porta ducitur in omni figura equilatera et equiangula. Et intellige hoc.

< 7 > Et quando voluerimus scire quanta sit diameter circuli continentis in pentagono noto equilatero et equiangulo cuius quodlibet latus sit 10 ex numero, secundum modum operationis quam exposuimus in illo. Et est manifestum ex eo quod dixit euclides quia omnis corde portionis circuli proportio ad cordam similis portionis alterius circuli est sicut proportio diametri illius ad diametrum illius. Et iam exposuimus quia circuli cuius diameter est 10 numero pentagoni lateris quadratum est 62 et $\frac{1}{2}$ dimñ rad′ 781 et $\frac{1}{4}$. Manifestum est ex eo quod diximus quod proportio centum ad 62 et $\frac{1}{2}$ dim̄ rad′ 781 et $\frac{1}{4}$ est sicut proportio census, qui est quadratum diametri circuli qui non est scitus circumdantis in pentagono cuius quodlibet latus est 10, quem fecimus rem, ex 100 et illud est ut quadratum diametri circuli noti circumdantis in pentagono noto. Ducamus 100 in 100, et provenñ 10000. Deinde ducamus censum in 62 et $\frac{1}{2}$ dim̄ radice 781 et $\frac{1}{4}$ et provenñ 62 census et $\frac{1}{2}$ census dim̄ rad′ 781 et $\frac{1}{4}$ census census: hoc est equale 10000. Reduc totum quod habes ad censum, et illud est ut ducas totum quod habes in $\frac{1}{50}$ unius et radicem $\frac{1}{15625}\frac{1}{4}$ unius, et provenñ census equalis 200 drag′ et radici 8000, et hoc est quadratum diametri circuli continentis pentagonum quando erit unumquodque latus eius 10.

< 8 > Et sic si voluerimus facere in pentagono noto equilatero et equiangulo cuius quodlibet latus est 10 circulum contentum a pentagono — et iam exposuimus quod circuli, cuius diameter est 10 ex numero, quadratum cuiusque lateris pentagoni circumdantis in ipso est 500 dim̄ rad′ 200000 — dicemus proportio 100, que sunt quadratum unius laterum pentagoni, ad 500 dim̄ rad′ 200000 est sicut proportio census, qui est quadratum circuli ignoti

---

141 AC] *corr. ex ABC* *145* lateribus⟩ eius *add. et del.* *146* diametri] diametro *146* circuli cadentis < ... > infra pentagonum] *Ar. al-dā'ira allatī taqa'u 'alā hādhā al-mukhammas fī mithlihi wa-khudh jidhr mā baqiya fa-mā kāna fa-huwa quṭr al-dā'ira allatī taqa'u dākhil al-mukhammas* *147* ducitur] duc̄ *150* secundum modum operationis] *Ar. bi-ghayr al-'amal* *154* quadratum] *corr. ex* quadraztum (?) *157* circumdantis] circumdatis *157* in⟩ circulum *add. et del.* *158* 100] 10 *166* pentagono] petagono *corr. ex* pedtagono

---

141 50] *corr. ex* $\frac{1}{5}$ *144* scilicet 100 ... diametri] *supra For* 8000 *MS has* 800; quorum quadruplum ... totius diametri] *Ar. fa-murabba' quṭr al-dā'ira allatī taqa'u fī hādhā 'l-mukhammas mi'a wa-jidhr thamāniyat ālāf* (*MS 'lf*) *146* cadentis] productum ex diametro circuli ca *marg.*, -dentis *in textu* *146* infra⟩ ipsum *supra* *148* continentis] *corr. in* circumdantis *152* cordam] *supra* *152* circuli] *corr. ex* circulus *153* quia⟩ diameter *add. et del.* *153* circuli] *corr. ex* circulus *154* pentagoni⟩ ei contenti ab eo *supra* *158* ex] *corr. in* ad *158* et⟩ et ut ? *supra* *158* illud] *supra* *158* ut] *supra* *162* $\frac{1}{50}$] *corr. ex* ? *162* $\frac{1}{15625}\frac{1}{4}$] ? *corr. in* $\frac{1}{3125}\frac{1}{4}$, *Ar.* $1\frac{1}{4}\frac{1}{15625}$ *163* 200] *supra*; *ras. in textu* *171* circuli⟩ diametri *supra*

quesiti, ad quadratum diametri circuli sciti, quod est 100. Duc 100 in 100 et proven̄ 10000. Deinde duc censum in 500 dim̄ rad′ 200000, et proven̄ 500 census dim̄ rad′ 200000 census census equalis 10000 drag′. Reduc totum quod habes ad censum, quod est ut ducas eum in $\frac{e}{10}\frac{1}{10}$ unius et rad′ $\frac{1}{100}\frac{0}{125}$ unius, et proven̄ census equalis 100 et rad′ 8000, | et illud est quadratum diametri 95r circuli cadentis in pentagono cuius quodlibet latus est 10.

< 9 > Et similiter si voluerimus facere super decagono equilatero et equiangulo cuius quodlibet latus sit 10 circulum circumdantem in eo. Et iam exposuimus quod circuli cuius diameter est 10 corda decagoni cadentis in eo est radix 31 $\frac{1}{4}$ dim̄ 2 et $\frac{1}{2}$ faciamus diametrum circuli rem. Deinde dicemus proportionem rei ad 10 sicut est proportio 10 ad radicem 31 et $\frac{1}{4}$ dim̄ 2 et $\frac{1}{2}$. Ducamus igitur 10 in 10, et proven̄ 100. Deinde ducamus rem in radicem 31 et $\frac{1}{4}$ dim̄ 2 et $\frac{1}{2}$, et proven̄ rad′ 31 census et $\frac{1}{4}$ census sublatis ex eo 2 rebus et $\frac{1}{2}$, quod equatur 100. Restauremus cum 2 rebus et $\frac{1}{2}$ et addamus illas ad 100, erunt 100 drag′ et 2 res et $\frac{1}{2}$ rei equalis radici 31 et $\frac{1}{4}$ census. Duc 100 drag′ et 2 res et $\frac{1}{2}$ rei in se, et proven̄ 10000 drag′ et 6 census et $\frac{1}{4}$ census et 500 res equales 31 censibus et $\frac{1}{4}$ census. Opponamus ergo in ipsis, et erit res 10 et radix 500, quod est diameter circuli circumdantis in pentagono cuius quodlibet latus est 10.

< 10*i* > Et similiter si volueris facere in decagono noto equilatero et equiangulo cuius quodlibet latus est 10 circulum circumdantem decagonum. Et iam exposuimus quod circuli cuius diameter est 10 corda $\frac{e}{10}$ cadentis super eo est 100 dim̄ radice 8000. Dicamus quia proportio 100 dim̄ rad′ 8000 ad 100 est sicut proportio 100 ad censum. Duc 100 in 100, et proven̄ 10000. Deinde duc 100 dim̄ rad′ 8000 in censum, et proven̄ 100 census dim̄ rad′ 8000 censuum census equalis 10000. Reduc totum quod habes ad censum, et illud est ut ducas eum in $\frac{1}{2}$ $\frac{e}{10}$ unius et rad′ $\frac{1}{5}\frac{0}{100}$ unius, et erit census equalis 500 et radici 200000, et hoc est quadratum diametri circuli cadentis infra decagonum cuius quodlibet latus est 10.

Et si volueris, duc diametrum circuli cadentis extra hunc decagonum in se, et iam exposuimus quod <est> 10 et rad′ 500, erit 600 et rad′ 200000. Prohice itaque ex eo 100, quod est multiplicatio lateris decagoni in se, et remanebit 500 et rad′ 200000. Radix eius est diameter circuli circumdantis hunc decagonum.

< 10*ii* > Et ego ponam causam huius porte cum adiutorio dei. Omnis figure equilatere circumdantis circulum vel circumdate in ipso circulo multiplicatio

175 $\frac{1}{100}\frac{0}{125}$] *corr. ex* ?, *Ar.* $\frac{4}{5}\frac{1}{10000}$ *175* unius] unus *177* in⟩ circulo (?) *add.* *181* radix⟩ unius *add.* *186* radici⟩ unius *add. et del.* *189* circuli⟩ conten̄ *add. et del.* *193* corda] *lege* quadratum corde *193* cadentis] cadetis *193* super⟩ s *add. et del.* *198* $\frac{e}{10}$] $\frac{e}{10}\frac{1}{12}$ *198* rad′⟩ *seq. ras.* *198* $\frac{1}{5}\frac{0}{100}$] *manu altera?Ar.* $\frac{4}{5}\cdot\frac{1}{400}$

174 equalis] *corr. ex* equale *179* circulum circumdantem] *corr. ex* circulus circumdans *180* corda] *corr. ex* cordan *181* $\frac{1}{4}$] *supra* *182* proportio] *corr. in* proportionem *186* 100¹⟩ et *supra* *186* et $\frac{1}{4}$ census] *om.*, $\frac{1}{4}$ censuum *supra*; Restauremus etiam censum et $\frac{1}{4}$ census *add. et del.* *192* circumdantem⟩ ia (?) *supra* *194* 8000¹] *corr. ex* 8 *207* in] *supra*; ab *add.*

lateris in se et diametri circuli circumdantis in ipso in se agregate sunt sicut multiplicatio diametri circuli circumdantis figuram in se. Exemplum: faciamus triangulum equilaterum et equiangulum sitque triangulus *ABG*, et in ipso circulum circumdantem triangulum sitque *HZL*, extra triangulum vero circulum circumdantem in ipso sitque *ABG*. Et diameter eius sit *AD*, diameter vero circuli circumdantis triangulum sit linea *CY*. Dico quia multiplicatio *AG* in se et *CY* in se est sicut multiplicatio *AD* in se. Cuius demonstratio est quod producamus super lineam *AG* ex puncto *L*, qui est locus in quo contingit circulus lineam *AG*, lineam secundum angulum rectum. Manifestum est quod in ipsa est centrum duorum circulorum, et applicent secundum extremitates ipsius ad circumferentiam circuli *HZL* sitque linea *LZ*: linea *LZ* est diameter circuli *HZL* que est equalis linee *CY*. Et coniungemus punctum *Z* cum puncto *D* cum linea *ZD*, et punctum *D* cum puncto *G* cum linea *DG*. Linea *AM* est sicut linea *MD* et linea *ML* est sicut linea *MZ*, et due linee *AM* et *ML* sunt sicut 2 linee *DM* et *MZ*; et angulus *AML* est sicut angulus *DMZ*: et basis *DZ* est sicut basis *AL*. Et *AL* est sicut *LG*: ergo *LG* est sicut *DZ* et triangulus *AML* est sicut triangulus *DMZ* et anguli eius sicut anguli illius. Angulus *ALM* sicut angulus *MZD*; et linea *ZD* equidistans linee *LG* et iam exposuimus quia ipsa equalis illi; et iam coniunctum fuit quod est inter duas quasque extremitates ipsarum lineis *ZL* et *DG*: linea *ZL*

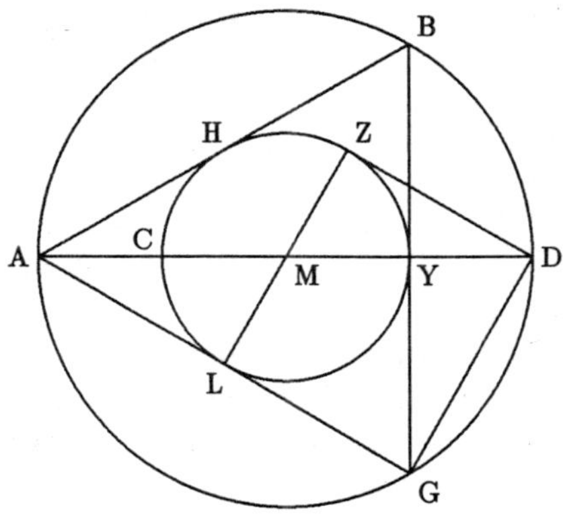

Prop. 10*ii*

208 circuli⟩ cont *add. et del.* *212* in ipso] *corr. in* ipsum *212* sitque] t *supra* *215* quod⟩ sciamus *add. et del.* *221* Linea⟩ *DG add. et del.*

209 circumdantis] circumdatis, continentis *supra* *211* triangulum[1]] *underlined* *212* circumdantem] *corr. ex* circumdantes *213* triangulum] *corr. in* in triangulo *216* contingit⟩ eum *add. et del.*; applicat ad *supra* *221 AM*] *corr. ex AG* *225* angulus] *supra* *227* quod est inter duas extremitates ipsarum] *corr. ex* cum duabus extremitatibus ipsas, *Ar. fīmā bayna aṭrāfihimā*; in *add. et del.* *227 DG*⟩ ergo *supra* equidistans et equalis linee DG. Sed multiplicatio AG in se et GD in se est sicut multiplicatio AD in se, quia angulus AGD est rectus; et DG est sicut LZ et LZ est sicut CY. Ergo multiplicatio AG in se et CY in se est sicut multiplicatio AD in se. Et hoc est quod voluimus exponere.

< 11*i* > Si fuerit figura quindecagona circumdata a circulo et diameter eius sit 10 sitque equilatera et equiangula et voluerimus scire unumquodque latus eius, faciemus circulum notum AGB, cuius diameter sit AB, et faciemus in eo latus decagoni, quod erit corda partis et $\frac{1}{2}$ partis, sitque linea BD. Et iam exposuimus quia ipsa est radix 31 et $\frac{1}{4}$ dim̄ 2 et $\frac{1}{2}$ corda AD est corda $\frac{2}{5}$ circuli et est 62 et $\frac{1}{2}$ et radix 781 et $\frac{1}{4}$ accepta radice eius. Et faciamus in ea cordam $\frac{1}{6}$, sitque BG, que corda 2 partium et $\frac{1}{2}$ et <est> 5, et linea AG est corda $\frac{1}{3}$ et est radix 75. Ducamus lineam AG in lineam BD et proveñ rad′ 2343 et $\frac{1}{2}$ et $\frac{1}{4}$ dim̄ rad′ 468 et $\frac{1}{2}$ et $\frac{1}{4}$. Deinde prohice eum ex multiplicatione AD in lineam BG, | que est 1<562 et $\frac{1}{2}$ et rad′ 488>281 et $\frac{1}{4}$ (accipiamus 95v radicem eius), remanebit 1562 et $\frac{1}{2}$ et radix 488281 et $\frac{1}{4}$ accepta rad′ eius. Addemus ei rad′ 468 et $\frac{1}{2}$ et $\frac{1}{4}$, sublata ex toto hoc rad′ 2343 et $\frac{1}{2}$ et $\frac{1}{4}$: et hoc est sicut multiplicatio linee AB in lineam GD. Dividamus per lineam AB, que est 10, et exibit linea GD 15 et $\frac{5}{8}$ <et> rad′ 48 et $\frac{3}{4}$ et $\frac{e\ 5}{8\ 8}$ accepta rad′ eius et rad′ 4 drag′ et $\frac{1}{2}$ et $\frac{1}{8}$ et $\frac{e\ 1}{8\ 2}$ dim̄ rad′ 23 et $\frac{3}{8}$ et $\frac{e\ 1}{8\ 2}$.

< 11*ii* > Scietur exiens corda GD ad intelligendum in acceptione radicis 48 et $\frac{3}{4}$ et $\frac{e\ 5}{8\ 8}$ in propinquitate, exitus eius 6 et 59 minuta et 15 secunda et

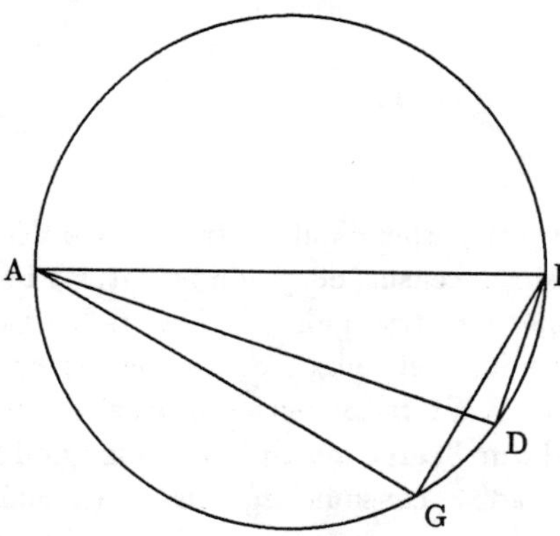

Prop. 11*i*

236 2] 62 *236* corda] corde *236* *AD*⟩ seq.ras. *238* *AG*⟩ 8 que (?) *add. et del.* *239* Ducamus … *BD*] *om. Ar.* *242* accepta] acceptam *242* eius²] *corr. ex* oeius *243* 468] 462 *245* 15] 25 *247* Scietur … acceptione] *Ar. fa-nqrb istikhrāj watar GD li-tafham kayfa ya'khudhu* *248* 59] *Ar.* 56

228 equalis⟩ est *supra* *234* *AB*] *corr. ex AG* *238* sitque] *corr. in* et est *240* 468] *corr. ex* 488 *244* Dividamus] *corr. ex* divisa

46 tertia. Addimus ea ad 15 et $\frac{5}{8}$, et erit 22 et 36 minuta et 45 secunda et 46 tertia; quorum accipe radicem et invenies eam fere 4 et 45 partes et 19 secunda et $\frac{1}{3}$ unius; cui adde radicem 4 et $\frac{5}{8}$ et $\frac{e}{8}\frac{1}{2}$ fere, et erit 2 et 9 minuta et 54 secunda et 12 tertia, et erit 56 et $\frac{2}{5}$ partes et 13 secunda et 13 tertia. Sublata ex eo rad′ 23 et $\frac{3}{8}$ et $\frac{e}{8}\frac{1}{2}$ secundum propinquitatem, quod est 4 drag′ et 50 minuta et 28 secunda et 25 tertia, et remanebunt 2 et 4 minuta et 44 secunda et 48 tertia, et hoc corda GD secundum propinquitatem. Et hoc est quod voluimus exponere.

< 12 > Et si dicemus tibi trianguli equilateri et equianguli mensura cum perpendiculari ipsius est 10 ex numero, quanta est perpendicularis. Faciamus triangulum ABG cuius perpendicularis AD. Cum autem voluerimus scire

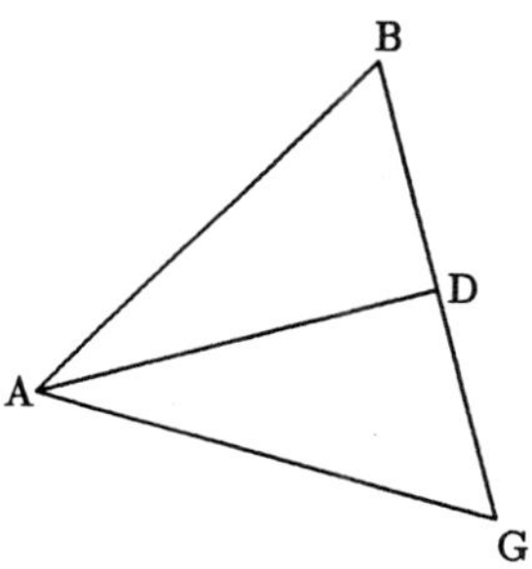

Prop. 12

quanta sit linea AD, que est perpendicularis trianguli ABG, faciamus ipsum rem. Igitur linea DG radix $\frac{1}{3}$ census, et quodlibet latus trianguli ABG radix census et $\frac{1}{3}$ census, et mensura trianguli ABG est radix $\frac{1}{3}$ census census, et perpendicularis eius res: erit res et radix $\frac{1}{3}$ census census equalis 10. Restaura radicem $\frac{1}{3}$ census census ut sit radix census census, et restaurabis illud si multiplicabis illud in radicem 3: ergo multiplica totum quod habes in radicem 3, et proveñ census et rad′ 3 censuum equalis radici 300. Dimidia igitur radicem 3 censuum, et exibit radix $\frac{3}{4}$; quam duc in se. Et proveñ $\frac{3}{4}$; quas adde ad radicem 300, et erunt $\frac{3}{4}$ et radix 300; quorum accipe radicem et prohice ex

250 invenies] *corr. ex insuenies* *250* et²⟩ $\frac{1}{5}$ et *add. et del.* *251* et $\frac{1}{3}$ unius] *lege* et tertium unum, *Ar wa-thālitha wāḥida* *252* et¹⟩ 4 *add. et del.* *252* 56 et $\frac{2}{5}$ partes] *lege* 6 et 55 minuta, *Ar. sitta wa-khams wa-khamsīn daqīqa* *262* *ABG*⟩ radix *ABG add. et del.* *267* proveñ⟩ *seq. ras.*

252 et³] *supra* *257* mensura⟩ est *supra* *258* perpendiculari⟩ cum (?) *supra* *260* *AD*] *supra*, *DG in textu* *261* *DG*⟩ *DG supra* *261* census⟩ g. quia tanta erit multplicatio linee *AD*, qui est res, in tertiam sui, que multiplicatio est tertia census, quanta (est *supra*) multplicatio linee *DG* in se. Ergo *DG* est radix tertie census. Intellige. *263* erit] *corr. in* ergo *263* equalis⟩ sunt *supra* *267* censuum] *supra* *267* quas] *corr. ex* que

eo rad′ $\frac{3}{4}$, et residuum est linea AD, que est perpendicularis trianguli ABG. Et hoc est quod voluimus exponere.

< 13 & 14 > Si dicemus tibi trianguli equalium laterum et angulorum unumquodque latus est 10 et in ipso est quadratum longum rectorum angulorum cuius mensura est 10, quanta est longitudo quadrati. Exemplum: faciamus triangulum triangulum ABG et quadratum longum < ... > $GMLY$, quod est in triangulo ABG. Prohice ergo eum ex 10 et quod remanet est mensura trianguli ABG. Et si volueris, dic: mensura quadrati $6\frac{102}{143}$ unius dim̄ rad′ 11 drag′ et $\frac{10261}{20449}$ par<tis> drag′, remanebit mensura trianguli 3 et $\frac{41}{143}$ unius et rad′ 11 et $\frac{10261}{20449}$ unius.

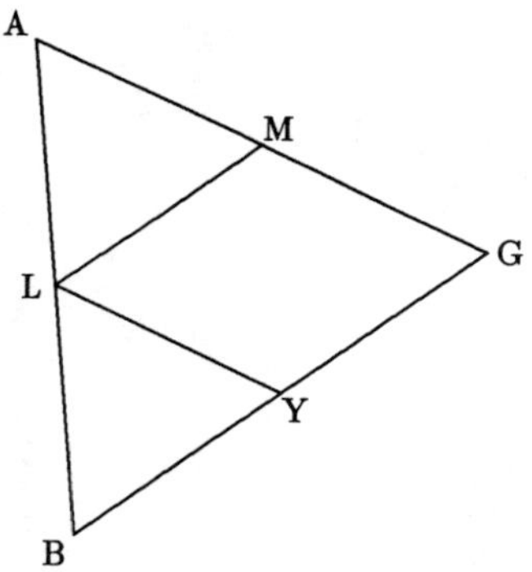

Props. 13-14

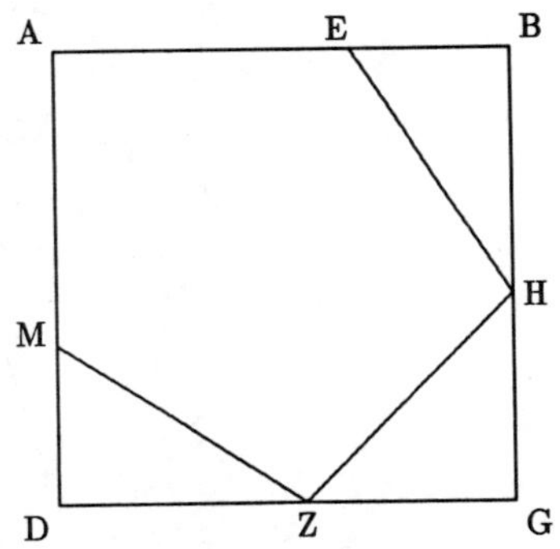

Prop. 15

< 15 > Et si dicemus tibi quadrati equilateri et rectorum angulorum quodlibet latus est 10, sitque quadratum $ABGD$, in quo fecimus pentagonum secundum hanc figuram, et illud pentagonum est $AEHZM$, scientia cuiusque laterum eius pentagoni est ut faciamus unum de lateribus pentagoni, et est linea AE, rem. Remanebit linea EB 10 dim̄ re. Et linea GH est rad′ $\frac{1}{2}$ census; remanebit linea $G'B$ 10 diminuta radice $\frac{1}{2}$ census. Et linea BE 10 diminuta re. Duc unumquodque ipsorum in se et agrega ea, et erunt 200 drag′

276 $\frac{102}{143}$] $\frac{101}{143}$ *277* et[1]⟩ 20461 ***add. et del.*** *277* $\frac{10261}{20449}$] $\frac{20261}{20449}$ *277* mensura⟩ quadrati ***add. et del.*** *278* $\frac{10261}{20449}$] $\frac{20261}{20449}$; 20449 ***add. et del. supra*** *282* eius] ***del. in textu***

271 < 13 & 14 >⟩ G (?) mihi videtur quod sic debeat esse lecta. Si dicemus tibi trianguli equalium laterum et angulorum et quad[rati] longi id est rumbi et in triangulo contenti mensura est 10 ex numeris tunc quanta est mensura rumbi predicti et trianguli. verbi gratia sit triangulus equilaterus ABG et rumbus intra ipsum rumbus $GMLY$ fac et invenies rumbum esse medietatem trianguli. quare ipse erit $\frac{1}{3}$ de 10 id est $3\frac{1}{3}$ trianguli vero 6 et $\frac{2}{3}$ vel si volueris dic quod si (?) rumbum est 6 (de rad′ ... que est $3\frac{1}{3}$ ***add. et del.***) et $\frac{82}{123}$ diminuta rad′ 11 et $\frac{1681}{15129}$ que est [ ] et radix [ ] et $\frac{1681}{15129}$ ***marg.*** *273* quadrati⟩ guli ***supra*** *275* eum] ***supra*** *275* quod[2]] ***supra*** *276* dic] ***supra*** *276* quadrati⟩ est ***supra*** *278* 11] ***corr. ex*** unius *280* in quo] ***supra*** *283* est[2]] ***supra*** *284* census[1]⟩ .g. quia GH et GZ sunt equales ***supra*** *284* $\frac{1}{2}$⟩ tis ***supra*** et census et $\frac{1}{2}$ census dim̄ 20 partibus et dim̄ radice 200 censuum equalis censui. Oppone in ipso sicut exposui tibi: et exibit linea AE, que est unum de lateribus pentagoni, 200 et rad′ 320000 accepta radice illorum subtracta ex 20 et radice 200. Et hoc est quod voluimus exponere.

< 16 > Et si dicemus tibi pentagoni equilateri et equianguli mensura est 50 drag′, quantum est quodlibet latus eius. Verbi gratia faciamus pentagonum pentagonum $ABGDE$ et centrum circuli circumdantis in eo pentagonio M; et exeant linee AM MB GM DM ME. Manifestum est quia iam dividitur triangulis equalibus, qui sunt triangulus AMB et BMG et GDM et DME et EMA: triangulus EMD 10 ex numero. Et iam exposuimus in eo quod

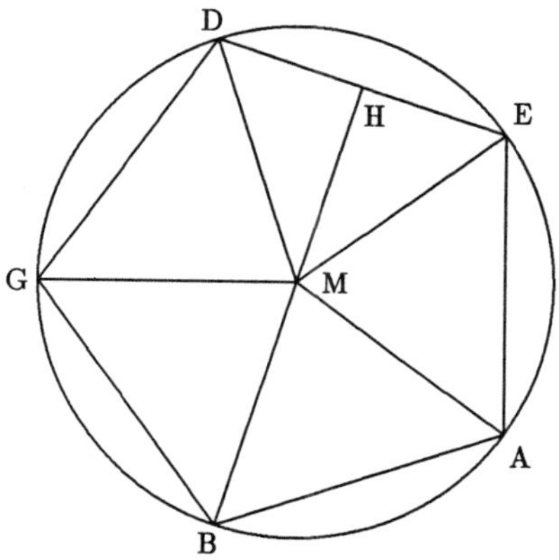

Prop. 16

premisimus ex libro nostro hoc quia, quando erit pentagonus equilaterus et equiangulus quodlibet latus eius est 10 ex numero, quadratum diametri circuli circumdantis ipsum est 200 et radix 8000. Manifestum est ex eo quod diximus quia, quando volueris scire diametrum circumdanti a pentagono noto equalium laterum et angulorum, duces unum de lateribus pentagoni in se, deinde duplabis eum et servas eum; deinde ducis etiam similiter unum de lateribus pentagoni in se, deinde agregatum in se, deinde accipias $\frac{4}{5}$ eius et id quod iam accepisti radicem eius. Quod exit addes | ad id quod servasti, et quod habebis accipe radicem, et quod exibit est diameter circuli. Et postquam est hoc ita, faciamus lineam ED rem. Et illud est unum de lateribus pen- 9

292 circumdantis] circumdatis *293 MB*⟩ et *B add. et del.*

288 accepta radice ipsorum] *corr. ex* accipiamus radicem eius *292* M] *supra* *293 DM*] *corr. ex DC* (?) *295 EMD*⟩ est *supra* *297* equiangulus⟩ et *supra* *301* servas] *corr. ex* serva *303* et id quod iam accepisti] *Ar. fa-mā kāna akhadhta*; ad accipionem *supra* *303* et⟩ illius *supra* *304* postquam⟩ ita *add. et del.*; fuit *supra*

tagoni: quadratum diametri circuli circumdantis in eo pentagono, sicut iam exposuimus, est 2 census et rad′ $\frac{4}{5}$ census census, et quadratum $\frac{1}{2}$ diametri eius, que est linea EM, est $\frac{1}{2}$ census et radix $\frac{e}{10}\frac{1}{2}$ census census. Deinde produc perpendicularem trianguli EMD — manifestum quia ipsa cadet super medium linee ED — et ipsa est linea MH. Linea EH est $\frac{1}{2}$ rei: duc eam in se, et proveniet $\frac{1}{4}$ census; prohice ergo eam ex quadrato linee EM, que est $\frac{1}{2}$ et radix $\frac{1}{2}$ decime census census, et remanebit quadratum linee MG' $\frac{1}{4}$ census census et radix $\frac{e}{10}\frac{1}{2}$ census census. Duc eum in lineam EG', que est $\frac{1}{2}$ rei, et proveñ $\frac{e}{8}\frac{1}{2}$ census census et radix $\frac{1}{320}$ census census census census equalis 10 ductis in se, que sunt 100. Restaura $\frac{e}{8}\frac{1}{2}$ census census ut sit census census, quod est ut ducas eum in 16. Duc ergo totum quod habes in 16, et proveñ census census et radix $\frac{4}{5}$ census census census census equalis 1600. Reduc totum quod habes ad censum census, et illud est ut ducas totum quod habes in 5 dim̃ rad′ 20, et proveñ census census equalis 8000 et radici 51200000. Rad′ eius est quodlibet latus pentagoni.

< 17 > Et si dicemus tibi decagoni equalium laterum et angulorum mensura est 100 ex numero, quantum est unumquodque latus eius. Verbi gratia faciamus decagonum $ABGDEUZHTL$ et centrum circuli circumdantis in ipso decagono M. Et manifestum est quia ipse dividitur 10 triangulis equalibus; triangulus ZMU 10. Iam exposuimus in eo quod premisimus ex verbis

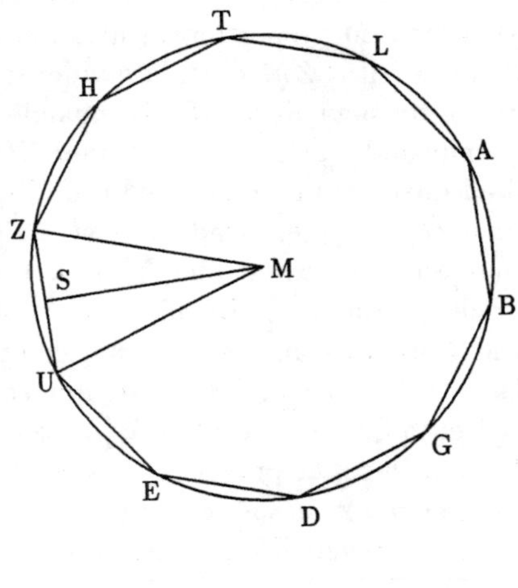

Prop. 17

308 Deinde⟩ fa *add. et del.* *309* trianguli⟩ EM *add. et del.* *310* est¹⟩ linea END et est *add. et del.* *314* radix⟩ 3 *add. et del.* *314* 10⟩ censibus *add. et del.* *317* et radix $\frac{4}{5}$ census census] *om. Ar.* *321* tibi⟩ pen *add. et del.* *323* circumdantis] circumdantes *325* equalibus⟩ trian e *add. et del.*; similibus *marg.*

311 $\frac{1}{2}$⟩ census *supra* (*māl supra in Ar.*) *312* MG'] *corr. in* MH *317* census⁶] *supra* *323* decagonum] *corr. ex* pentagonum *325* exposuimus⟩ autem *supra* nostris quia, quando erit decagonus equilaterus et equiangulus unumquodque latus cuius sit 10, diameter circuli circumdantis eum 10 et radix 500. Manifestum est ex eo quod exposuimus quia, quando voluerimus scire diametrum circuli circumdantis in decagono noto, ducas unum de lateribus decagoni in se, deinde in 5 deinde; accipias radicem agregati et adde illud ad unum de lateribus decagoni, et quod exierit erit diameter circuli. Et postquam manifestum est ex eo quod diximus quod faciamus lineam *ZU* rem, et illud unum de lateribus decagoni. Necesse est quod sit diameter circuli circumdantis in eodem decagono, sicut exposuimus, res et radix 5 censuum; et $\frac{1}{2}$ diametri eius, que est linea *ZM*, $\frac{1}{2}$ rei et radix census et $\frac{1}{4}$ census et manifestum est etiam similiter ex hoc quod diximus ex eo quod rememoravit euclides, et est quia, quando dividetur linea per proportionem per medium et 2 partes et additum in longitudine linee sicut maior pars eius, tota linea addita divisa erit secundum proportionem medii et duo extrema maiorque pars eius est linea prima. Manifestum est ex eo quod diximus quia quando dividetur linea secundum proportionem habentem medium et duo extrema et dividetur maior pars secundum proportionem habentem medium duoque extrema, maior pars linee sic divise erit minor pars linee totius scilicet prime. Et dixit in alio loco: quando coniungetur latus exagoni cum latere decagoni unius circuli secundum rectitudinem, tota linea divisa est secundum proportionem medii et extremorum maiorque pars eius est linea exagoni. Set linea *ZU* est latus decagoni et linea *ZM* est latus exagoni. Et quando fuerint hee 2 linee linea una, iam divisa est secundum proportionem medii et extremorum et maior pars est linea *ZM*; linea igitur *ZM* quando dividetur secundum proportionem medii et extremorum sicut linea *ZU*. Et quando facies *ZU* rem et minues eam ex *ZM*, deinde addes ad residuum $\frac{1}{2}$ linee *ZU* et duces agregatum in se: erit sicut 5 ad quadratum medietatis linee *ZU*, sicut exposuit euclides. Duc medietatem linee *ZU* in se, deinde in 5, et accipe radicem eius et adde ad $\frac{1}{2}$ linee *ZU*, que est $\frac{1}{2}$ rei: exibit linea *ZM*. Iam manifestum est quia linea *ZM* $\frac{1}{2}$ rei et radix census et $\frac{1}{4}$. Deinde produc perpendicularem trianguli *ZMU*, que est *MS*; manifestum est quia ipsa cadit super medium linee *ZU*, quod est *S*. Deinde duc lineam *ZM* in se, que est $\frac{1}{2}$ rei et radix census et $\frac{1}{4}$: erit census et $\frac{1}{2}$ et radix census census et $\frac{1}{4}$ census census. Deinde duc lineam *ZS*, que est $\frac{1}{2}$ rei, in se, et proveñ $\frac{1}{4}$ census. Prohice ipsam ex eo, et remanebit census et $\frac{1}{4}$ et radix census census et $\frac{1}{4}$ census census. Rad' huius est linea *MS*: duc eam in lineam *ZS* que est $\frac{1}{2}$ rei, et proveñ $\frac{1}{4}$ census

---

330 deinde⟩ dic *add. et del.* *336* quod[2]⟩ dixit e *add. et del.* *343* sic divise] *om. Ar.* *343* erit⟩ linea *add. et del.* *346* linea⟩ *ZU add. et del.* *347 ZM*⟩ est latus *add. et del.* *349* est linea *ZM*; linea igitur *ZM*] *lege* linee *ZM*, *Ar. min khaṭṭ ZM* *356 ZMU*] *corr. ex ZM* *359* ipsam] ipsa

---

337 per[1]⟩ secundum *supra* *337* partes⟩ medii (*corr. ex* mediu) et duo extrema *supra* *338* et⟩ erit *supra* *338* in⟩ se *marg.* *338* eius⟩ quia *supra* *338* tota] *corr. ex* totam *339* secundum⟩ item *supra* *343* linee[1]⟩ illius *supra* *345* rectitudinem⟩ quia *supra* *350* extremorum⟩ est maior pars *supra* *352* ad quadratum] *corr. ex* cum quadrato *357 ZU*] *corr. ex HT* *357* quod est *S*] *corr. ex* quod est *ZH*, *Ar. wa-huwa* ZS; in *supra* *359* census⟩ et *supra*; extrap (?) *add. et del.* *360* eo et] eo (?) et *supra*, et *in textu*

census et $\frac{e}{8}\frac{1}{2}$ census census et radix $\frac{5}{64}$ census census census census equalis 100 drag′. Restaura $\frac{1}{4}$ census census et $\frac{1}{2}$ octave census census ut sit census census, quod est ut ducas eum in 3 et $\frac{1}{5}$: duc ergo totum quod habes in 3 et $\frac{1}{5}$, et provenī census census et radix $\frac{4}{5}$ census census census census equalis 320 drag′. Reduc totum quod habes ad censum census, et illud est ut ducas totum quod habes in 5 dim̄ rad′ 20, et erit census census equalis 1600 dim̄ rad′ 2048000, quorum accipe radicem, et hec est linea ZU, et illud est unum de lateribus decagoni. Et hoc est quod voluimus exponere.

| < 18 > Si dicemus tibi pentagoni $ABGDE$ equalium laterum et an- 96v
gulorum mensura trianguli BDE est 10, quanta est linea DE, que est basis trianguli, unum de lateribus pentagoni. Exemplum: faciamus lineam DE rem. Et iam exposuit euclides quod linea BE quando dividetur secundum proportionem medii et extremorum, maior eius pars erit equalis linee DE; et exposuit etiam quia, quando dividetur linea secundum proportionem medii et

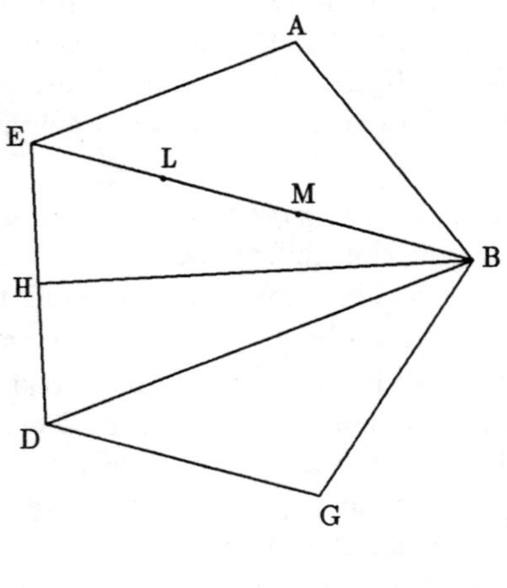

Prop. 18

extremorum et addetur toti linee sicut maior pars eius, tota addita divisa erit secundum proportionem eandem et maior pars erit linea prima; et similiter exposuit quod, quando dividetur linea secundum proportionem medii et extremorum et addetur minori parti medietas maioris et multiplicabitur addita in se, erit quadratum productum 5plum quaddrati medietatis linee longioris. Ergo dividamus lineam BE secundum proportionem medii et extremorum in

377 et exposuit etiam ... prima] *om. Ar.* *381* extremorum⟩ sm *add. et del.*

363 Restaura] staura *underlined*, integra *supra* *365* census[5] census[6]] *supra* *367* erit⟩ provenit *supra* *371* mensura] *supra* *371* *BDE*] *E supra* *372* trianguli⟩ et est *supra*

punctum M, et faciamus maiorem partem lineam ME: linea ME equalis est linee DE. Et dividamus lineam ME in duas partes in puncto L: erit multiplicatio linee LM in se $\frac{1}{5}$ multiplicationis LB in se. Et multiplicatio LE in se est $\frac{1}{4}$ multiplicationis DE in se, ergo est $\frac{1}{4}$ census, quia fecimus lineam DE rem et lineam ME rem et LE $\frac{1}{2}$ rei. Ergo multiplicatio BL in se est census et $\frac{1}{4}$; ergo linea BL est radix census et $\frac{1}{4}$. Et linea LE est $\frac{1}{2}$ rei: linea BE $\frac{1}{2}$ rei et radix census et $\frac{1}{4}$. Exibit perpendicularis trianguli BED, que est BH: <linea EH est> $\frac{1}{2}$ rei. Duc igitur eam in se, et provenī $\frac{1}{4}$ census; et prohice eum ex multiplicatione BE in se, quod est census et $\frac{1}{2}$ et radix census <census> et $\frac{1}{4}$ census census, et remanebit census et $\frac{1}{4}$ et radix census census et $\frac{1}{4}$ census census; radix huius est linea BH. Duc eam in lineam EH, que est $\frac{1}{2}$ rei, et provenī $\frac{1}{4}$ census census et $\frac{1}{2}$ octave census census et radix $\frac{5}{64}$ census census census <census> equalis 100 dragmis. Restaura $\frac{1}{4}$ census ex censu tuo et $\frac{e\ 1}{8\ 2}$ census census tui ut sit census census, quod est ut ducas eum in 3 et $\frac{1}{5}$: duc ergo totum quod habes in 3 et $\frac{1}{5}$, et provenī census census et radix 4 censuum census census <census> equalis 3 censibus et 20 drag′. Reduc totum quod habes ad censum census, quod est ut ducas eum in 5 dim̄ rad′ 20, et provenī census census equalis 1600 dim̄ rad′ 2048000; cuius radix est linea ED. Et manifestum est quod triangulus BED ex isto pentagono cuius rememorati sumus similis trianguli ZM ad decagonum quem memoravimus in figura que est superius ab hac.

< 19 > Si dicemus tibi mensura trianguli ABE est 10 ex numero, quanta est linea BE. Verbi gratia faciamus lineam BE rem. Et iam exposuit euclides quia, quando dividetur linea secundum proportionem medii et extremorum, pars maior quando addetur in longitudinem ipsius sicut medietas linee totius et ducetur in se, erit quadratum illius 5^{plum} quadrati quod erit ex $\frac{1}{2}$ linee in se. Faciamus igitur lineam BE rem, et cum ducetur $\frac{1}{2}$ eius in se, deinde in 5, provenī census et $\frac{1}{4}$. Manifestum est ex eo quod diximus quia linea AE est radix census et $\frac{1}{4}$ dim̄ $\frac{1}{2}$ rei. Exibit perpendicularis super lineam BE, et est AY: ducamus lineam AE in se, et provenī census et $\frac{1}{2}$ dim̄ rad′ census census et $\frac{1}{4}$ census census. Deinde duc lineam EY in se et est $\frac{1}{2}$ rei et provenī $\frac{1}{4}$ census; prohice ergo eum ex eo, et remanebit census et $\frac{1}{4}$ dim̄ rad′ census census et $\frac{1}{4}$ census census: erit census census. Radix huius est linea AY, que est trianguli ABE perpendicularis. Duc ergo eam in lineam EY, que est $\frac{1}{2}$ rei, et provenī $\frac{1}{4}$ census census et $\frac{e\ 1}{8\ 2}$ census census dim̄ rad′ $\frac{5}{64}$ census census census: radix huius equatur triangulo ABE, que est 10. Duc ergo 10 in

384 erit multiplicatio ... in se] *Ar. fa-ḍarb BL fī mithlihi amthāl ḍarb LE fī nafsihi* *385* est $\frac{1}{4}$ ... ergo] *om. Ar.* *387* et⟩ c̄ *add. et del.* *396* 4] *Ar.* $\frac{4}{5}$ *401* similis] similitudo *409* linea] repet. *414* erit census census] this appears misplaced, but is present here in the Arabic

383 dividamus] *corr. ex* divisa est *385* $\frac{1}{4}^{1}$] *corr. ex* quadruplum *386* *LE*] *corr. ex* *LNE* *387* *BE*] *corr. ex* *ZE* *389* rei⟩ exibit ... rei *underlined*; medietas autem linee *ED* est $\frac{1}{2}$ rei et est comunis *supra* *394* Restaura⟩ integra *supra* *394* censu] *corr. ex* censui *403* dicemus] *corr. in* dicit *404* exposuit] *corr. ex* exposuimus *406* pars maior quando] *del.*; et *supra* *406* ipsius⟩ parti maiori *supra* *410* Exibit⟩ exit *supra* *410* super⟩ per *supra* *415* perpendicularis] *marg.*

se, et proveñ 100 equales $\frac{1}{4}$ census et $\frac{e\ 1}{8\ 2}$ census census dim̄ rad′ $\frac{5}{64}$ census census census <census>. Reintegra $\frac{1}{4}$ census census tui et $\frac{e\ 1}{8\ 2}$ census census tui ut sit census census, quod est ut ducas eum in 3 et $\frac{1}{5}$: duc ergo totum quod

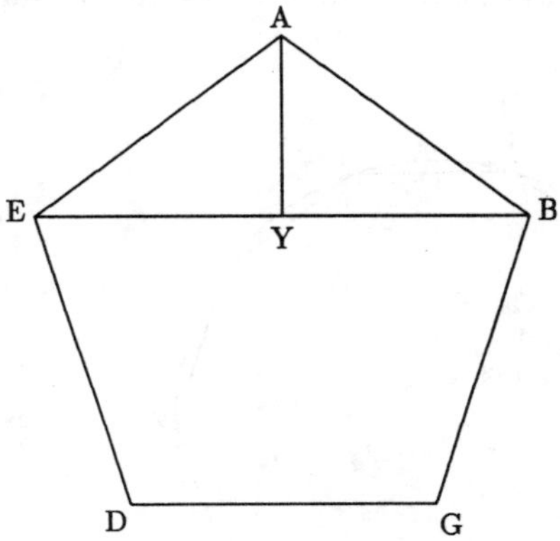

Prop. 19

habes in 3 et $\frac{1}{5}$, et proveñ census census dim̄ rad′ $\frac{4}{5}$ census census equalis 320. Reintegra ergo illud ut sit census census integer, quod est ut ducas eum in 5 et radicem 20: duc ergo totum quod habes in 5 et rad′ 20, et proveñ census census equalis 1600 drag′ et rad′ 2048000; cuius radix est radix linee BE. Et hoc est quod voluimus exponere.

< 20 > Et si dicemus tibi decagoni $ABGDEZIHTC$ equilateri et equianguli mensura trianguli CTH est 10 ex numero, quanta est linea CH, que est latus decagoni $\frac{1}{5}$ circuli circumdantis hunc decagonum. Exemplum: faciamus lineam CH rem et iam exposuimus quod quando erit corda $\frac{1}{5}$ circuli res, erit quadratum diametri eius 2 census et radix $\frac{4}{5}$ census census: ergo est quadratum $\frac{1}{2}$ diametri eius $\frac{1}{2}$ census et radix $\frac{e\ 1}{10\ 2}$ census census. Prohice ergo illud ex quadrato corde $\frac{1}{5}$, quod est census, et remanebit quadratum CT $\frac{1}{2}$ census dim̄ radice $\frac{e\ 1}{10\ 2}$ census census. Et quod cogit nos facere hoc est quia euclides iam exposuit quia corda $\frac{1}{5}$ potest super cordam $\frac{1}{6}$ et $\frac{1}{10}$, quando erit circulus unus. Prohiciamus ex eo quadratum CM in se quod est $\frac{1}{4}$ census et remanebit quadratum TM, que est perpendicularis trianguli

421 equalis] *corr. ex* equales *424* radix^2] *Ar. om.* *424* cuius radix ... *BE*] *lege* cuius radix radicis est linea *BE* ? *432* quadratum] *lege* quarta

431 ergo est quadratum] *corr. ex* ex *431* diametri⟩ ergo est semidiametrum ***supra*** *435* unus⟩ continens eos ***supra*** *436* $\frac{1}{4}$] ***supra***, ***lege*** quadratum *436* que] *corr. ex* quod

CTE, quadratum census dim̄ rad′ census et $\frac{1}{2}$ $\frac{e}{10}$ census census. Duc ergo eum in quadratum CM, quod est $\frac{1}{4}$ quadrati census: | erit $\frac{e}{8}\frac{1}{2}$ census census 97r dim̄ rad′ $\frac{1}{320}$ census census <census census>, cuius radix equatur mensure MTH, que est 10 ex numero. Duc ergo 10 in se, et proveñ 100 equalis $\frac{e}{8}\frac{1}{2}$ census census dim̄ rad′ $\frac{1}{320}$ census census census census. Reintegra $\frac{1}{2}$ octave census census ut sit census census, et illud est ut ducas eum in 16. Duc ergo

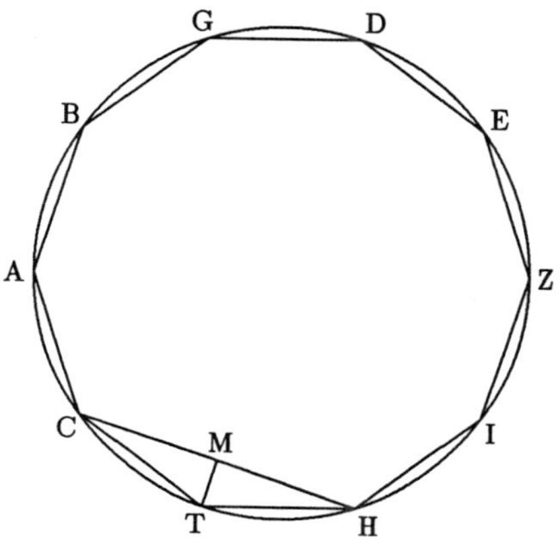

Prop. 20

totum quod habes in 16, et proveñ census census dim̄ radice $\frac{4}{5}$ census census census <census> equalis 1600. Reintegra etiam illud ut sit census census integer, quod est ut ducas eum in 5 et radicem 20: et duc totum quod habes in 5 et radicem 20, et proveñ census census equalis 8000 et radici 51200000 cuius radix est linea CH, que est corda $\frac{1}{5}$ circuli circumdantis decagonum. Et hoc est quod voluimus exponere.

437 *CTE*] *lege CTH* 438 in⟩ se *add. et del.*

The 'double-meaning' method for dating mathematical texts

Jan van Maanen

1 Introduction

When questions arise about the date and the authorship of a text sometimes only a tentative answer can be given. In this article I shall show how passages in mathematical texts which have a double meaning can be used for dating and attributing these texts. This 'double-meaning' method (for short: DM-method) will be applied to several texts from the sixteenth up to the eighteenth century. The DM-method is unlikely to be useful for dating *medieval* texts, but I am pleased to have the opportunity here to propose the checking of this idea to those who are specialized in medieval mathematics.

In dating texts the historian of mathematics can use the same methods as any other historian. If the date of composition is stated in the document explicitly, e.g. in the *explicit* of a manuscript or on the title page of a book, the text can be dated directly. If the date of composition is not stated, the document may contain references to another text, or it may be referred to in another text, whence a *terminus post quem* or a *terminus ante quem* can be given. In the same manner limits for the date can be deduced when the name of a person or a specific technical term occurs in the text. Studies by Busard provide numerous examples of such indirect dating. A recent example is the so-called Adelard II version of Euclid's *Elements*, which Busard and Folkerts now suppose was composed by Robert of Chester. An important argument for this tentative attribution is the date of the text (shortly before 1140), which can be accurately determined since the text uses Hermann's Euclid-translation, which was written shortly before 1140, and the oldest extant copy of the text was written by Thierry of Chartres 'probably not later than 1141'. Busard and Folkerts argue that Robert of Chester, who was Hermann's friend and associate in Spain, was one of the very few people, who could have had access to Hermann's translation at such an early stage, and in addition they point out that Hermann was a pupil of Thierry.[1]

For dating a document the writing material may present evidence as well. Bones with carfs may be dated roughly by the C14-method, and watermarks give an indication about the period in which the paper was produced. This survey of dating methods does not aim to be complete. It simply serves as an introduction to the 'double-meaning' method, which is based on the observation that there are mathematicians who, in composing a text, tend

[1] H.L.L. Busard; M. Folkerts (ed.), *Robert of Chester's (?) Redaction of Euclid's* Elements, *the so-called Adelard II Version*, Basel etc. (Birkhäuser) 1992.

to use information which refers to themselves (e.g. numbers: the year in which they are writing, the year of their birth, or – but not in the Middle Ages – their telephone number or house number). Sometimes they do so deliberately. The problems they pose may for instance be about a regular polygon of 1991 sides (a 'Mathematical Olympiad' problem in 1991), or they may ask in 1992: "The year 1995 can be written as $5 \cdot 7 \cdot 57$. What is the last year before 1995 which can be written as $a \cdot b \cdot (10a + b)$, in which a and b are integers taken from $\{1, 2, 3, \ldots, 9\}$?"[2] A good example too, although it may not be strictly mathematical, is the famous *Melancholia* engraving, which Dürer signed with his monogram and dated 1514 in the lower right-hand corner; the engraving features a magical square of dimensions 4×4. The two numbers at the centre of the bottom row, 15 and 14 from left to right, are generally considered to be an allusion to the year in which Dürer produced the *Melancholia*.

A preliminary conclusion is that a mathematical text may contain numerical data which conceal information *about* the text or its author. Sometimes this information is easily recognizable, but in other cases it has to be decoded before it can be interpreted. Examples of these two cases are given in the following section. The mathematician may have introduced such information into the text deliberately (just like the play on numbers in the Olympiad problems), but sometimes authors make an unconscious use of numbers which refer to the text, although this seems to be less frequent. In the latter case applying the DM-method is rather like giving a psychoanalytical interpretation of the text. In the first case, where the numbers were put into the text deliberately, the DM-method responds to the challenge that the mathematician has attached to his text, and tries to solve the puzzle. Clearly the method has risks (just like psychoanalysis), since it depends heavily on the interpreter. If the interpreter sees hidden information where there is none, or uses a wrong key to decode the information, a fallacy is easily created. I shall return to this problem at the end of this article. For the time being, let us give the method the benefit of the doubt and see where it leads us.

---

[2] I had just hit upon this particular factorization of 1992 and 1995 when, at the end of 1991, the New Year card of the Utrecht Freudenthal Institute arrived; it depicted a child discovering that $1992 = 8 \times 3 \times 83$.

2 Some examples in which the DM-method can be applied

2.1 Arithmetic texts

In sixteenth and seventeenth century arithmetic texts numerical examples often use year-numbers. In the *Arithmetica* of Peter van Halle (Brussels, Koninklijke Biblotheek, Hs. 3552;[3] we are lucky to find the explicit year of composition: "This arithmetic full of beautiful questions, Peter van Halle has set, written and completed to his liking, now aged about eighteen years, in Mechelen in the city (hear my explanation!), when one wrote sixty and eight." The Flemish original is in verse:

"Desen arithmetica vol schoonder vraghen
heeft Peter van Halle naer sijnen behaghen
nu out wesende ontren den iaren achtien
te mechelen inder steden hoort myn bedien
alsmen screef tsestich ende acht
ghestelt ghescreven ende volbracht." (f. 1^v)

On f. 15^v Van Halle sets the problem: "How many days, hours and minutes is it from the moment that our Lord became flesh, until the Year 1567." The solution consists of the multiplication $1567\times365\times24\times60 = 823,615,200$. On f. 16^r he computes that the hour-clock has struck 89,285,976 times since the "moment that our Lord was born". (The reader who wants to check these results should know that in the latter computation Van Halle includes 391 intercalary days according to the Gregorian calendar, whereas in the first computation all years have the same length, namely 365 days, according to the Julian calendar).

There is one year of difference between the explicit date "68", which because of the style of the manuscript can only be interpreted as 1568, and the DM-date 1567. An explanation is directly at hand: Van Halle was already working on his arithmetic in 1567, and he completed the manuscript in 1568.

A second manuscript which can be dated directly and also with the DM-method is the *Arithmetica* written by Cornelis Pijck (Brussels Stadsarchief, *Oud Archief* 2179). The manuscript can be dated directly, for 1^r says: "This cipherbook belongs to Cornelis Pijck 1.5.8.4. written 13 September 1584". Pijck, however, asks "how many months, weeks, days, hours and minutes and so on it is since our dear Lord came to save mankind, namely within the time of 1585 years; do as follows, which is the common way of doing" (f. 9^r). Pijck even computed the number of thirds (sixtieth parts of seconds) in 1585 Julian years, and found 2,990,856,960,000 thirds. The manuscript gives 13 september 1584 as the explicit date when Pijck started his work,

[3]This manuscript and several others that I shall discuss below were brought to my attention by Marjolein Kool, to whom I wish to express my gratitude.

and DM-dating leads to the conclusion (or should I say hypothesis?) that he was still working on it in 1585.

A third example of this type can be found in *Eerste Deel Vande Nieuwe Telkonst, ...* (the Dutch translation, Gouda 1626, of Napier's *Rabdologiae, Seu Numerationis Per Virgulas Libri Duo ...*, Edinburgh 1617). As an example of division, which probably also featured in the original edition, which was not consulted, Napier gives $589,475 : 365 = 1615$. The numbers are not explicitly described as the number of days, the number of days in one year and the number of years, but the occurrence of 365 strongly suggests that they refer to days and years. Here DM-dating at least makes it probable that the original text was composed in 1615. The year of publication of the original edition (1617) fits in well with this date.

The last example of dates in arithmetical texts is in an arithmetic written by Frans van Schooten Jr. (Groningen, University Library, Hs. 437). Here the number 1646 occurs on f. 1^r, where Van Schooten explains how numbers are pronounced "1646 Facit one thousand six hundred and forty six", and also on f. 2^r. There Van Schooten discusses "how many Years have passed from the Creation of the world up till now." Fortunately Van Schooten knew that "the world had existed for 3949 Years up to the birth of Christ" and (more important for our purpose) "for 1646 Years from the birth of Christ". In the year 1646 Frans Jr. succeeded his father Frans Sr. at Leiden University, so he *may very well have* produced new educational texts in this year and in following years, and arithmetic was one of the introductory courses. Or are we to conclude that Van Schooten *did indeed* compose this text in 1646?

In order to indicate how common the usage was to work with the year number, I list some other arithmetic texts (manuscripts and books) in which this feature appears:[4]

1584 Arithmetica manuscript by Josse Verniers (Gent University Library, Ms. 694).
The manuscript has a direct date on the title page ("This book belongs to me, Josse Verniers, Anno 1.5.8.4", both in French and in Flemish), and on f. 7^r Verniers asks how many hours there are in 1582 years.

1569, 1594 Adriaen vander Gucht, *Cijferbouck ...*, Brugge 1569 (copy: Gent University Library, Acc 1463), reprinted Amsterdam 1594 (copy: Gent University Library, 175 P 2).
On ff. $30^v - 31^r$ both 1565 and 1570 appear as year numbers. In the 1594 edition these numbers were left unchanged.

1591, 1600 Martin van den Dijcke, *Chijfer-Boeck ...*, Antwerpen 1591 (copy: Antwerp Museum Plantin Moretus, R 50.24), revised edition 1600 (copy: Amsterdam University Library).

[4]Again, I wish to express my gratitude to Marjolein Kool, who supplied the examples presented in the list that follows.

On p. 33 the author asks how many days, hours and minutes there are in 1591 years. In the 1600 edition the number in the text of this problem (p. 24) is changed to 1600!

1598 Marthin Wentsel, *T'Fondament van arithmetica*, 1599 (copy: Amsterdam University Library, 967 C 16).
On f. 12^{v} Wentsel asks how many days, hours and minutes there are between the birth of our Lord and Saviour Jesus Christ and the year 1598.

undated, but probably 1635 Anonymous and undated manuscript "Volumen lingua Belgica, continens institutiones Arithmeticae, Geometriae et Architecturae Militaris", Copenhagen Kongelige Bibliotheket, *gl. Kgl. S.* 2°279.
The manuscript includes an undated arithmetic, which on f. 11^{r} discusses the problem: "1635 Years, how many months, weeks, days, hours and minutes do they make?"

It was still common to use the year number in arithmetic texts in 19th century England and United States, as the following examples show:[5]

?, 1811 Francis Walkingame, *The tutors assistant*, new edition by T. Crosby, York 1811.
In the 1811 edition the reader is asked "How many days is it since the birth of our Saviour to Christmas, 1794?" The answer, $655258\frac{1}{2}$, probably takes into account that 1794 is half way between two leap years. The reader is invited to find and check a DM-date for the first edition of this book.

1806 William Butler, *Arithmetical questions, on a new plan*, 4th edition, London 1806.
Apparently new editions of this book were revised, since on p. 46 we read: "Ireland was formerly divided among a number of petty sovereigns; a circumstance which facilitated Henry the second's conquest of that valuable island in 1172, when it became an appendage to the English crown. (...) How long has Ireland been annexed to the British dominions this present year 1805?"

1835 Daniel Adams, *Arithmetic ... designed for the use of schools and academies in the United States*, Keene, New Hampshire 1835.
Again (on p. 18) a question about invasion, but with a less political dimension: "The invasion of Greece by Xerxes took place 481 years before Christ; how long ago is that this current year 1827?" Did publication of this book take eight years?

[5]With thanks to John Fauvel for supplying them.

?, 1876 Bishop Colenso, *A shilling arithmetic*, New edition, London 1876. According to the Bishop (p. 7) the Flood took place in 2348 B.C., and he wants to know how many years separated the Flood from "the end of A.D. 1868". Was the first edition published in or shortly after 1868?

2.2 An anonymous manuscript about cryptography

The library of Utrecht University contains an anonymous manuscript about cryptography. The manuscript is written partly in Dutch and partly in Latin and is entitled "Short instruction to write in covert manner, in French called cifres, and to unravel the same" ("Corte onderrichtinge om bedectelijck te schrijven, int fransch cifres genaemt, ende om 't selve te ontdoen"; Utrecht University Library, Hs. 1372). If there is one place where one expects an author to conceal information about himself or herself, it is in a text about cryptography. And indeed, the text does contain a key which reveals the name of a person. For at f. 39 the author describes the following coding technique. Take a page from a book, which contains a considerable amount of text. Say you want to code the message 'long live Busard' (the manuscript of course uses a different example). Put a dot or small line under the first 'l' on the page. Then put a dot under the first 'o' that follows, a dot under the first 'n' that comes after this 'o' and a dot under the first 'g' after this 'n. In this manner the word 'long' has been coded. If the page has a sufficiently rich variation of letters the message is readily coded. Once you know the key, decoding is quickly done: you track the dots one by one, and write down the associated letters.

Then a copy of a page from the book follows on f. 40^{r} (see Figure 1), which is described in the caption as "Opera Virgilii printed at Amsterdam by Guili. Jans. Caesium [= Willem Jansz. Bleau] A°M · D · C · XXIV." According to a marginal note on the same page (see Figure 1) the characters underlined reveal the name, familyname and city of birth of the author of the text. The underlined text reads "Joannes Baptista Houwaert Bruxellensis". The name Houwaert also appears in an earlier passage, which discusses a coat-of-arms depicted on f. 3^{r}. This coat-of-arms, says the text (f. 5^{r}), "has been chosen by the Houwaert family, as it can be seen clearly, together with its explanation, in several books by my grandfather under the motto KEEP TO THE MIDDLE, which I translated into Latin as INTER UTRUMQUE TENE." This grandfather must have been the famous poet Joannes Baptista Houwaert[6] (1533-1599), and the author of the anonymous cryptography text must have been his grandson, who was named after his grandfather.

[6] *Biogr. Nat. de Belgique* 9, 554–560.

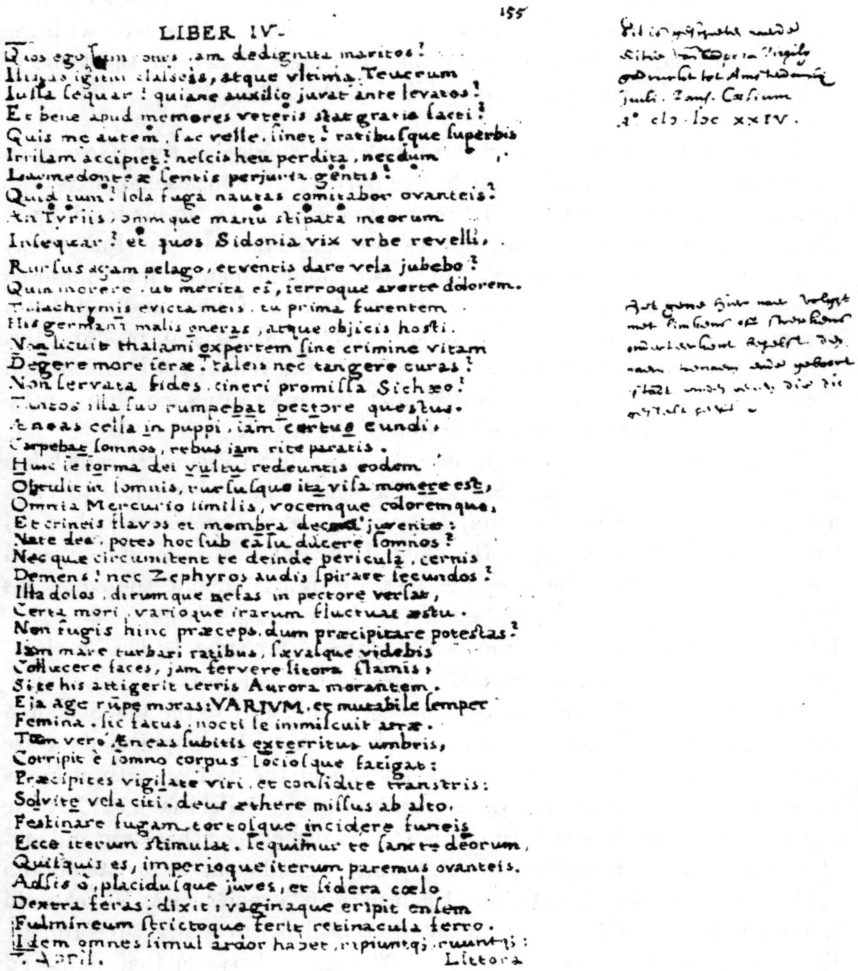

Figure 1: The author reveals his identity. Cryptography manuscript "Corte onderrichtinge om bedectelijck te schrijven, int fransch cifres genaemt, ende om 't selve te ontdoen"; Utrecht University Library, Hs. 1372, 40r. Courtesy Utrecht University Library.

2.3 A system of equations and the mother of Frans Jr.

> "A man, wife and child are together 96 years, that is to say the man and child together 2 years more than the wife and the wife with the child together 14 years more than the man. I ask for the years of each." [Groningen University Library, Hs. 443 f. 54^{r}]

Frans van Schooten Sr. posed this problem in an algebra text (Groningen University Library, Hs. 443), which, like most of his educational texts, has remained in manuscript form.[7] He probably used the text in his teaching at the Leiden Engineering School, and his two sons Frans Jr. and Pieter may have used it as well, since the problem is also extant in manuscripts composed by Frans Jr. (Groningen University Library, Ms. 437 f. 65^{v}) and in a Latin version by Pieter (Groningen University Library, Ms. 112 p. 6: "Maritus, Uxor & Infans simul habent 96 annos, ..."). In the versions of Frans Jr. and Pieter the problem appears with one small but interesting modification, to which we shall return in the final section of this article.

In his solution of the problem (see the facsimile in Figure 2), which has a rather modern form, Frans Sr. puts the age of the father equal to $1\mathfrak{r}$ (a notation taken from the Cossists) and the age of the mother equal to $1A$ (following Viète). The age of the child is then expressed as $96 - 1\mathfrak{r} - A$ and the given relations between the ages are translated into the equations $1\mathfrak{r}$ (man) $+ 96 - 1\mathfrak{r} - 1A$ (child) equal to $1A + 2$ and $1A$ (wife) $+ 96 - 1\mathfrak{r} - 1A$ (child) equal to $1\mathfrak{r} + 14$, from which $\mathfrak{r}$ and A are solved. The result is: the wife is 47 years old, the man 41 years and the child 8 years.

I wondered whether this family situation could apply to the Van Schooten family, which also began with one son (Frans Jr.). Biographical data are scarce. The only information that we have about the date of birth of Frans Sr. is that he matriculated at Leiden University on 15 February 1612 at the age of 30, so Frans Sr. was born on 16-02-1581 or between that date and and 15-02-1582. Frans Jr. was "old about 9 years" when his father registered him in the Leiden Orphan Register on 14 February 1625[8] and he was 16 years old when, on 15 May 1631, he matriculated at Leiden University. Accordingly Frans Jr. must have been born on 15-02-1615 or between that date and 15-05-1615. No dates have been found concerning Jannetgen Haermansdr. van Hogenvorst, the first wife of Frans Sr. and mother of Frans Jr. She was born at Delft, but the Delft register of baptisms starts only in 1616 whereas

[7] In his catalogue of the Groningen manuscripts (1898, p. 244) Brugmans describes this manuscript as "Petrus van Schooten, De Cosinus-rekening, benevens een Verhandeling over de irrationeele grootheden." In the first place the manuscript is written in two different hands; ff. $2^{r} - 37^{v}$ and $41^{r} - 63^{v}$ are in the hand of Frans van Schooten Sr. and ff. $64^{r} - 81^{r}$ and $38^{r} - 40^{v}$, to be read in this order according to a note on f. 81^{r}, are in the hand of his second son Pieter. And in the second place the text discusses the "Regel Coss", and is therefore about algebra and not about trigonometry.

[8] *Nieuw Nederlands Biografisch Woordenboek* Vol. 7, 1110.

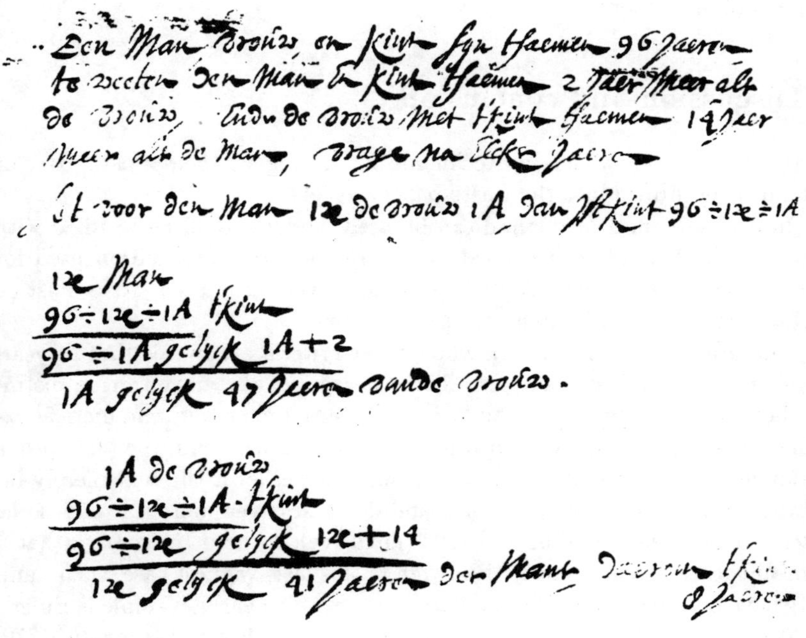
Een Man, vrouw, en kint syn t'saemen 96 jaeren
te weeten den Man en kint t'saemen 2 jaer meer als
de vrouw, ende de vrouw met t'kint t'saemen 14 jaer
meer als de Man, vrage na elcks jaeren

Stelt voor den Man 1x de vrouw 1A dan sal t'kint 96 ÷ 1x ÷ 1A

1x Man
96 ÷ 1x ÷ 1A t'kint
96 ÷ 1A gelyck 1A + 2
1A gelyck 47 jaeren van de vrouw.

1A de vrouw
96 ÷ 1x ÷ 1A t'kint
96 ÷ 1x gelyck 1x + 14
1x gelyck 41 jaeren den Man, jaeren t'kint 8 jaeren

Figure 2: Problem written by Frans van Schooten Sr. in: Groningen University Library, Hs. 443 f. 54^{r}. Courtesy Groningen university library.

she must have been born much earlier since she gave birth to a son in 1615.

Since no information about Jannetgen is available, the question "Could the family situation in the system of equations apply to the Van Schooten family" reduces to the question "was there a time when Frans Sr. was 41 and Frans Jr. was 8 years old?" The answer is positive: Frans Sr. may have been 41 on 16-02-1622 or between that date and 14-02-1624. Frans Jr. may have been 8 on 15-02-1623 or between that date and 14-05-1624, and these periods overlap from 15 February 1623 to 14 February 1624 (inclusive).

A possible conclusion is that at least part of Hs. 443 has a DM-date between 15 February 1623 and 14 February 1624, and that Frans Jr. had a relatively old mother. She was 47 when her son was 8. This was just before she died, since at the age of 9 Frans Jr. was registered as an orphan, at the request of his father.

3 Discussion and conclusion

Is the DM-method a serious technique for dating mathematical texts, or should it be thrown into the waste-paper basket?

In my view the DM-method can be used, but it should be handled with care. In the first place one must be certain that the information used for dating does indeed have a double meaning. In the second place one must be certain that one has the right key for decoding.

The least disputable case in which these criteria are met is that of year-numbers in numerical examples. Three degrees of certainty can be distinguished. There is greater certainty that a year-number n in a numerical example is identical to the year in which the text was written if the text speaks about n as "this present year n". Certainty decreases if the text merely indicates that n is some year-number, and decreases even more if it uses some integer n between, say, 1200 and 1993 (inclusive), without the remark that n is meant to be a year-number. But even in this last, vaguest case, such numbers have been used to estabish a date for a text. An early example is Euler's *Vollständige Anleitung zur Algebra*, which was published in German in 1770, after it had already appeared in Russian translation in two volumes in 1767 and 1768. Weber, who edited the *Algebra* for the *Opera Omnia*, in which it is Volume 1 of Series 1 (published 1911), remarked that the numbers 1765 and 1766 appear remarkably often and prominently in numerical examples, so he stated: one can "conclude with certainty that the work had already been started by 1765, so [when Euler was] still in Berlin".[9] Weber cites two examples, the first (p. 87) where Euler in the chapter on the logarithm explains that 1765 means "Ein Tausend, Sieben Hundert, Sechzig und Fünf", and the other (p. 158) in which $\sum_1^{1766} n$ is calculated to be $\frac{1}{2} \cdot 1766 \cdot 1767$. So, although Euler does not explicitly present these numbers as year-numbers,

[9] Euler, *Opera Omnia* Series 1, Vol. 1, p. 3.

the origin of the *Algebra* in the years 1765–1766 is accepted by Weber. And who would disagree?[10]

If DM-dates are accepted in cases which have this low degree of certainty, then we should not expect much resistance to DM-dating in cases which have a higher degree of certainty, and definitely not if there is a specific reference to "this present year *n*". And indeed, this is confirmed by the example about the [**1635**] arithmetic in the list above, which Marjolein Kool uses to reject the manuscript from her forthcoming inventory of 15th and 16th century arithmetics. John Fauvel[11] in the same manner dates a copy of Bonnycastle's *Arithmetic*, which lacked its title page, at 1776, on the basis of the problem (p. 8): "The Arabian method of notation was first known in England about the year 1150, how long is it since, to this present year 1776." On the other hand the examples of second and later editions of books show that you must be careful, since without further information it is impossible to say whether "this present year" refers to the year of composition of the original text, or to the year of preparation of the later edition.

That the two criteria are met in the cryptography manuscript (§2.2) is immediately clear from the marginal note which states that the page contains coded information and which gives the key.

The most disputable case is the age of Van Schooten's mother (§2.3). However, in this case the described family situation (relatively old parents with one son, the mother six years older than the father) is so peculiar, that a mere coincidence of the situation described in the problem with the structure of the Van Schooten family is rather improbable. Apparently the situation described in the problem was also thought peculiar in the seventeenth century, for Frans van Schooten Jr. and his half-brother Peter who also included the problem in their mathematics texts, first reversed the ages of the father and the mother: with them the father is six years older than the mother. This observation yields such convincing extra evidence for the DM-dating of the problem that I conclude that at least a part of the manuscript Groningen, University Library, Hs. 443 was written in the period 1623–1624, and that Jannetgen Haermansdr. van Hogenvorst, the first wife of Frans van Schooten Sr. and mother of Frans Jr. was born between 16-02-1575 and 14-02-1577. She was about 39 when Frans Jr. was born. So Frans Jr. had a relatively old mother, and in fact, this is almost the only new fact that arises from this long pondering about double meanings and dates.

The DM-method is open for discussion!

---

[10]One of the reasons for paying attention to this example is that I want to show that the DM-method has been in use for a long time. However, it never seems to have been given any special attention.

[11]Private communication.

Roger Bacon's *Geometria Speculativa**

George Molland

Introduction

In 1940 Robert Steele published his edition of Roger Bacon's *Communia Mathematica* from the two manuscripts, London, British Library, Sloane 2156, ff. 72r-95v (= **S**), and Oxford, Bodleian, Digby 76, ff. 48r-69r (= **D**). **S**, which he regarded as containing a later version of the work, lacks the end of the *Communia*, and **D** the beginning, but they overlap in the middle. Steele remarked, rather cryptically in the context, that, "The remainder of [the] earlier form is held over for the final fascicule [of the *Opera hactenus inedita Rogeri Baconi*]." This in fact never appeared, but the reference is to ff. 69r sqq. of **D**, of which ff. 69r-76r (paragraphs 1-40 below) are clearly a continuation of the *Communia mathematica*, in which Bacon moves on to the special subject of theoretical geometry.[1] F. 76v is blank, and ff. 77r-78bv are fragments of a Euclid commentary: again the Baconian style is unmistakable, and they were probably intended to form parts of the same geometrical treatise as the earlier folios. All of them are edited here, with English translation, under the title *Geometria Speculativa.*

In *Archives Internationales d'Histoire des Sciences* for 1974, H.L.L. Busard edited some geometrical fragments relating to Euclid from MS Florence, Bib. Naz. Conv. Soppr. J.IX.26, and argued that they should be ascribed to Roger Bacon.[2] Despite Busard's persuasive arguments, I am not fully convinced of the Baconian authorship of all the fragments, but the present document adds some support to his case, for there are close verbal correspondences between the discussions of the postulates in the two manuscripts.[3]

The readings of the single manuscript known to me (**D**) of the *Geometria Speculativa* are often corrupt. I have adopted a policy of minimal emendation, and this has sometimes resulted in translations that are either conjectural or verge on the senseless. I have enclosed material added by myself in triangular brackets < >. Unless otherwise stated, footnotes to the text may be taken as

* A welcome opportunity for much of the work on this text was provided by my tenure of a Rockefeller Foundation Fellowship in the University of Oklahoma.

[1] A.G. Little, "Roger Bacon's Works", *Roger Bacon: Essays*, ed. A.G. Little (Oxford, 1914), 375-426, at 403-4, regarded these as representing "Pars II" of the *Communia mathematica*, but the text printed below makes clear that he is moving on from common things to proper ones; also that the complete mathematical work would comprise nine parts (one common and eight proper) rather than the six mentioned by Little.

[2] H.L.L. Busard, "Ein mittelalterlicher Euklid-Kommentar, der Roger Bacon zugeschrieben werden kann", *Archives Internationales d'Histoire des Sciences*, **24** (1974), 199-218.

[3] Paragraphs (49) - (51) below; Busard, 211-212.

referring to manuscript **D**. Footnotes to the translation are mainly devoted to citations of Bacon's sources. For ease of reference I have numbered serially the paragraphs into which I divide the work.

For various reasons, not least those of space, I do not here attempt a full analysis of the *Geometria Speculativa*, but it may be useful to provide a brief conspectus of its sources.

I. Euclid. Unsurprisingly the *Elements* is Bacon's major source. He seems usually to use the Adelard II version[4], which in their recent edition[5] H.L.L. Busard and M. Folkerts ascribe to Robert of Chester. Bacon also refers (16, 17) to *alia translatio*. This seems clearly to be the text accompanying the commentary of Anaritius (al-Nayrīzī) in Gerard of Cremona's translation,[6] and Bacon often appeals to the commentary itself. Bacon also makes use of the Adelard III version of the *Elements*, which he cites (16, 42, 44) as Adelard's *editio specialis*,[7] and on one occasion (35) he refers to Euclid's *Data*.[8]

II. Ptolemy. Bacon (22, 28) appeals to Ptolemy's *Optics* under the titles *Perspectiva* and *De aspectibus*.[9]

III. Alhazen (Ibn al-Haitham). Bacon uses his *De speculis comburentibus*,[10] and in (19) this is joined by *Liber Piramidum Apollonii*, a fragment on conic sections that Gerard of Cremona prefaced to his translation of the former

[4] I retain the notation established in M. Clagett's classic article, "The Medieval Latin Translations from the Arabic of the *Elements* of Euclid, with Special Emphasis on the Versions of Adelard of Bath", *Isis*, **44** (1953), 16-42.

[5] *Robert of Chester's (?) Redaction of Euclid's Elements: The So-Called Adelard II Version* (Basel, 1992) [= Busard & Folkerts].

[6] I use the edition of Book I in P.M.J.E. Tummers, *Albertus (Magnus)' Commentaar op Euclides' Elementen der Geometrie* (Nijmegen, 1984), II, 103-90 [= Tummers].

[7] The prologue of this is edited in M. Clagett, "King Alfred and the Elements of Euclid", *Isis*, **45** (1954), 269-277 [= Clagett – Alfred], and the whole of Book I in M.St M. Van Ryzin, *The Arabic-Latin Tradition of Euclid's Elements in the Twelfth Century* (Ph.D. thesis, Univ. of Wisconsin, 1960) [= Van Ryzin]. Bacon's readings seem to be nearer to Clagett's manuscript **V** (Venice, Biblioteca Nazionale Marciana, fondo antico 332) than to his **O** (Oxford, Bodleian, Digby 174). Both Clagett, "Medieval Translations", 23-24 n.18, and T.J. Cunningham, *Book V of Euclid's Elements in the Twelfth Century: The Arabic-Latin Traditions* (Ph.D. thesis, Univ. of Wisconsin, 1972), 29, 199, have suggested, on the basis of the manuscript's colophon, that Bacon may have been using MS Paris, Bibliothèque Nationale, Lat. 16648. Clagett writes of this that "although it now has only the last five books" of Adelard III, it "was no doubt originally intact". However, A. Birkenmajer, *Études d'Histoire des Sciences et de la Philosophie du Moyen Âge* (= Studia Copernicana I) (Wroclaw 1970), 162-3, has identified the manuscript with Item 40 of Richard of Fournival's library, and that codex did not include the earlier books. Bacon must therefore have had some other access to them.

[8] Edited in Shuntaro Ito, *The Medieval Latin Translation of the Data of Euclid* (Tokyo & Basel, 1980) [=Ito].

[9] *L'Optique de Claude Ptolémée dans la Version Latine d'après l'Arabe de l'Emir Eugène de Sicile*, ed. Albert Lejeune (Louvain, 1956) [= Lejeune].

[10] Latin translation edited in J.L. Heiberg & E. Wiedemann, "Ibn al-Haitams Schrift über parabolische Hohlspiegel", *Bibliotheca Mathematica*, 3. Folge, **10** (1910), 201-237 [= Heiberg & Wiedemann].

work.[11]

IV. Gernardus. In (5) Bacon cites a *Liber Demonstrationis Algoristice.* This may well be the *Algorismus Demonstratus* of a certain Magister Gernardus.[12]

V. Al-Fārābī. His *Liber de scientiis* is mentioned in (41) and is used elsewhere.[13]

VI. Boethius. Bacon refers (12) to his commentary on Aristotle's *Categories* and (44) to a *Geometria Practica* ascribed to him.[14]

VII. Josephus. Passages from his *Antiquities* were important for Bacon for establishing the antiquity and genealogy of scientific and philosophical knowledge, and the work is appealed to in (44).[15]

VIII. Aristotle. The influence of Aristotle and of scholasticism is pervasive. Specific references are made to *Posterior Analytics* (2, 48), *Metaphysics* (11), *De coelo* (37), *Physics* (54). Elsewhere in his corpus and here at least once (1) Bacon makes use of a florilegium[16] rather than the Aristotelian text itself.

IX. Roger Bacon. Bacon makes frequent reference upwards to his *Communia Mathematica,*[17] and this in turn is related to what were probably earlier lectures by Bacon.[18] Bacon's citations here (11, 30), and in the *Communia Mathematica,* of a treatise on metaphysics remain rather opaque.[19] I make here no systematic comparison with other items in the Baconian corpus, but do refer to the *Opus Tertium*[20] and the *De multiplicatione specierum.*[21]

---

[11]Cf. M. Clagett, *Archimedes in the Middle Ages* (Vol.1, Madison, 1964; Vols 2-5, Philadelphia, 1976-84), IV, 3-13.

[12]Edited in G. Eneström, "Der 'Algorismus de integris' des Meisters Gernardus", *Bibliotheca Mathematica*, 3. Folge, **13** (1912-13), 289-332, and "Der 'Algorismus de minutiis' des Meisters Gernardus", *Bibliotheca Mathematica*, 3. Folge, **14** (1913-14), 99-149.

[13]Al-Fārābī, Catálogo de las Ciencias, ed. & tr. Ángel Gonzalez Palencia (2nd edn, Madrid, 1953) [= Gonzalez Palencia] includes the text of Gerard of Cremona's translation.

[14]They may be found in J.P. Migne, *Patrologiae Cursus Completus ... Series prima* (Paris, 1844-93) [= Migne], LXIV, 159-294 and LXIII, 1352-64 respectively.

[15]Latin translation partially edited in F. Blatt, *The Latin Josephus. V.I: Introduction and Text: The Antiquities I-V* (Copenhagen, 1958).

[16]J. Hamesse, *Les Auctoritates Aristotelis: Un Florilège Médiévale: Étude Historique et Édition Critique* (Louvain & Paris, 1974) [= Hamesse].

[17]*Communia Mathematica Fratris Rogeri*, ed. R. Steele, Opera Hactenus Inedita Rogeri Baconi, XVI (Oxford: Clarendon, 1940) [= CM].

[18]Edited in R. Steele, "Roger Bacon as Professor: A Student's Notes", *Isis*, **20** (1933), 53-71 [= Steele-Professor].

[19]I give a brief discussion in my "Roger Bacon's Appropriation of Past Mathematics", forthcoming in a volume edited by S.J. Livesey and F.J. Ragep.

[20]The relevant part is edited in Roger Bacon, *Opera Quaedam Hactenus Inedita. Vol.I. Containing I. - Opus tertium. II. - Opus minus. III. - Compendium philosophiae*, ed. J.S. Brewer (London, 1859) [= Brewer]. I am preparing a new edition, with English translation, of the *Opus Tertium*.

[21]Edited in D.C. Lindberg, *Roger Bacon's Philosophy of Nature: A Critical Edition, with English Translation, Introduction, and Notes, of De multiplicatione specierum and De speculis comburentibus*, (Oxford, 1983) [= Lindberg].

Text

(1) Determinato de parte mathematice prima in qua premisi que necessaria sunt omnibus posterioribus partibus mathematice et specialibus, et hec communia sunt quia nata est nobis via a communibus ad propria, nunc sequitur pars secunda principalis, in qua de specialibus temptabo ut possum tractare.

(2) Hec omnia que sequuntur vadunt in infinitum in post asumendo et in latus, ut Aristoteles utatur euloquio, nam possunt multiplicari in infinitum conclusiones et demonstrationes circa species quantitatis, prehabitas accipiendo unam semper post aliam ordine naturali, et nichilominus sub principalibus conclusionibus sic ordinatis contingit accipere ex latere conclusiones in infinitum, et huiusmodi infinitatis causa est divisio quantitatis continue in infinitum, et appositio discreta infinita per divisionem continui. Non oportet autem iri in infinitum, nec expedit ad hoc laborari, sed standum est ad veritates meliores et utiliores rebus et scientiis, et ille sunt eligende que etiam possunt vocari radices et elementa respectu ramorum et foliorum, quorum productio vadit in infinitum. Et huiusmodi maxime sunt ille conclusiones in libro Euclidis Pictagorici, propter quod vocatur Liber Elementorum et Radicum. Quia tamen ibi non continentur tantum ea que communia partibus mathematice specialibus sed ad specialia geometrie[1] et arismetice[2] descenditur, non meretur dici ita proprie Liber Elementorum sicut tractatus qui solum habet radices que omnibus partibus mathematice sunt communes, et non descendit ad conclusiones alicuius partis specialis, propter quod illa pars communis merito debet vocari Liber Radicum et Elementorum respectu omnium partium mathematicarum.

(3) Et iterum scitum est a principio quot sunt partes mathematice, quoniam preter communem que iam finita est, oportet poni quatuor de quantitate continua et quatuor de discreta, nam geometria speculativa et[3] practica, et astrologia[4] speculativa et practica sunt de quantitate continua, et arsmetrica speculativa et practica, atque musica speculativa et practica sunt de quantitate discreta, ut declaratum est. Sunt ergo 8 scientie speciales preter nonam, que communis est, nam proculdubio quelibet practica est omnino alia scientia et separata a sua speculativa, propter modum diversum suum et scientie pro se diversitatem, nam illud est demonstrare veritates absolutas, et istud

---

[1]geometris
[2]armetice
[3]*om.*
[4]astrolabia: *deleted in text, marginal correction cut off*

Translation

(1) Having determined concerning the first part of mathematics, in which I prefaced the things that are necessary for all the later and special parts of mathematics (and these are common because a way from common to proper things is inborn in us,[1]) there now follows the second principal part, in which I shall address the treatment of special things so far as I can.

(2) All the things that follow go off to infinity by being assumed afterwards, and also in breadth, as Aristotle puts it,[2] for conclusions and demonstrations about the species of quantity can be multiplied to infinity by taking up those previously held one after another in the natural order, and nevertheless under the principal conclusions so ordered there may be taken up conclusions on the side to infinity, and the cause of this kind of infinity is the division of continuous quantity to infinity, and the infinite discrete reckoning through the division of the continuous. But it is not necessary to go to infinity, nor is it worth working for this, but one should stop at the better truths and those more useful in affairs and sciences, and those should be chosen that can also be called roots and elements with respect to the branches and leaves, whose production goes off to infinity. And most of all like this are the conclusions in the book of Euclid the Pythagorean, on which account it is called the *Book of Elements and Roots*.[3] But because there are not contained there only those things that are common to the special parts of mathematics, but there is descent to the special things of geometry and arithmetic, it does not deserve so properly to be called *Book of Elements* as a treatise that only has roots that are common to all parts of mathematics, and does not descend to the conclusions of any special part, on which account that common part must deservedly be called *Book of Roots and Elements* with respect to all the parts of mathematics.

(3) And again, it was understood from the beginning how many are the parts of mathematics, because, beside the common part that has already been finished, it is necessary to posit four concerning continuous quantity and four concerning discrete, because theoretical and practical geometry, and theoretical and practical astronomy concern continuous quantity, and theoretical and practical arithmetic, and theoretical and practical music concern discrete quantity, as has been explained.[4] Therefore there are eight special sciences, beside the ninth, which is common, for without doubt each practical science is altogether other and separate from its theoretical, on account of its different procedure and the difference of the science in itself, for the latter is to demonstrate independent truths, and the former to demonstrate

[1]Hamesse, 140. Cf. Aristotle, *Physics* I.1, 184a16-18; Molland, "Roger Bacon's Appropriation".

[2]*Anal. Post.* I.12, 78a15-22; cf. Hamesse, 316.

[3]Cf. CM, 38; Gonzalez Palencia, 148.

[4]CM, 38; cf. Molland, "Roger Bacon's Appropriation".

est demonstrare utilitates in opere, et alius modus est demonstrandi, et omnino alterius generis veritates sunt hinc inde.

(4) Quod facile potest cuilibet patere in geometria speculativa et practica, nam speculativa probat conclusiones circa figuras superficiales et corporales et lineas absolutas ab opere et abstractas a materia determinata, non curans in quo corpore sint nec in qua materia; sed practica considerat huiusmodi quantitates in materia determinata, et demonstrat in opere, ut de compositione speculorum comburentium et aliorum 5 generum speculorum, et instrumentis quibus utitur astronomia practica, cuiusmodi sunt astrolabia et armille et huiusmodi, que geometria practica componit[5] et demonstrat opera ipsa, et multa alia que habent tractatus speciales. Propter quod magna est scientia geometria practica, et sufficit per se vocari scientia.

(5) Et similiter arsmetrica practica est scientia magna et subtilis, ut patuit prius, cuius una pars est tractatus algoristicus, qui docet componere et distinguere numeros et fractiones secundum omnes species algorismi, qui tractatus datur rudibus per modum narrationis et per exempla, sed provectiores considerant demonstrationem horum operum, unde in Libro Demonstrationis Algoristice demonstrantur omnes species algorismi tam in fractionibus quam in integris, et sic est de aliis speciebus arismetice practice.

(6) Et similiter est astrologia practica, que utitur tabulis et canonibus et instrumentis ad inveniendum ea que in celo sunt, et que in aere apparent ex secundis stellarum, que sunt comete, et cetere impressiones calide, et iris, et multa. Et habet considerationes sufficientes scientie separate, propter multitudinem veritatum operabilium et principiorum earum sub quibus. Constituti sunt similiter per modum narrationis et secundum modum non probationis ac demonstrationis, ut patet inquirenti per se, et habet varias partes, et plures quam aliqua practicarum et nobiliores et utiliores, sicut patuit ex[6] musice.

(7) Et ideo oportet quod scientia < una > constitutatur de speculativa cuiuslibet per se et alia de eius practica. Et probatum est prius quod geometria precedit omnes, quapropter prius determinandum est de geometria, et primo de speculativa quam de practica, nam prior est speculatio quam operatio, et facilius est speculari veritates absolutas quam opera circa quantitates, et in utrisque cadit demonstratio, sed fortiores sunt et difficiliores demonstrationes operum quam veritatum nudarum, sicut apparet in Libro de Speculis Comburentibus, que est una de practicis (?) geometrie, que habet

[5]componunt

[6]dimensione *deleted in text, correction probably cut off*

utilities in practice, and it is another mode of demonstrating, and the truths are altogether of another genus in one and the other.

(4) This can easily be evident to anyone in theoretical and practical geometry, for theoretical proves conclusions about superficial and corporeal figures and about lines that are independent of practice and abstracted from determinate matter, not caring in which body they are nor in which matter; but practical considers such quantities in determinate matter, and demonstrates in practice, as in the composition of burning mirrors and the other five genera of mirrors,[5] and in the instruments which practical astronomy uses, such as astrolabes and armillaries and the like, which practical geometry composes and demonstrates in their practice, and many others which have special treatises. On account of this practical geometry is a great science and suffices by itself to be called a science.

(5) And similarly practical arithmetic is a great and subtle science, as was evident earlier.[6] One part of it is algoristic discourse, which teaches how to compose and distinguish numbers and fractions according to all species of algorism. This discourse is given to tyros by way of narration and by examples, but the more advanced consider the demonstration of these works, and accordingly in the *Book of Algoristic Demonstration*[7] all species of algorism, in both fractions and integers, are demonstrated, and so it is with other species of practical arithmetic.

(6) And practical astronomy is similar, which uses tables and canons and instruments for finding what there are in the heavens, and those that appear in the air with the succour of the stars. These are comets, and other impressions of heat, and the rainbow, and many others.[8] And it has sufficient considerations for a separate science, on account of the multitude of useful truths and of their underlying principles. They are constituted similarly by way of narration and not by way of proof and demonstration, as is evident in itself to the enquirer, and it has various parts and more numerous, noble and useful than any of the other practical sciences, as was evident from music.

(7) And so it is necessary that one science be constituted about the theory of something in itself, and another about its practice. And it was proved earlier that geometry precedes all of them,[9] on which account geometry is to be determined about before the others, and first theory rather than practice, for theory is prior to practice, and it is easier to theorise about absolute truths than works concerning quantities. Demonstration occurs in both, but demonstrations of works are stronger and more difficult than of bare truths, as appears in the *Book of Burning Mirrors*,[10] which is one of the practical parts of geometry, which has the subtlest and most beautiful demonstra-

---

[5]Cf. Lindberg, 326.
[6]CM, 47-49.
[7]See Introduction, IV.
[8]Cf. Brewer, 106-7.
[9]CM, 40-41.
[10]See Introduction, III.

subtilissimas operum demonstrationes et pulcherrimas, et sic de aliis practicis geometrie, de quibus prius exemplificavi in capitulo de hoc facto.

(8) Quoniam igitur dicendum est de geometria speculativa, et principia sunt priora principiatis, oportet dari principia specialia in principio; et, cum iam in precedentibus tacta sunt in universali aliqua de lineis, figuris et angulis, nunc in speciali sunt cetera exponenda.

(9) Inter igitur lineam rectam et curvam est linea media que vocatur tortuosa vel flexuosa, que necessario habet angulum, et est composita ex pluribus rectis sic , vel ex pluribus curvis sic , vel sic  , vel ex recta et curva sic . Et si ex pluribus curvis, tunc potest esse vel ex pluribus concavis, vel ex pluribus convexis, vel ex concavo et ex convexo, sicut in eadem lineatione, nam A angulus est ex pluribus convexis, et B ex convexo et concavo, et C ex pluribus concavis.

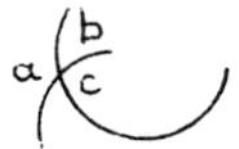

(10) Et si ex pluribus rectis, tunc non habet magnam diversitatem geometricam absolute consideratam, sed ut in perspectivam descendit,[7] nam dividitur in lineam fractam et reflexam. Linea reflexa vocatur cuius angulus comprehenditur a duabus lineis in eandem partem a puncto angulari protensis, sic ; linea fracta dicitur cuius angulus comprehenditur a duabus lineis a puncto angulari in diversas partes protractis,[8] sic . Linea etiam incidentie anexa est istis, nam linea que protenditur antequam frangatur vel reflectatur dicitur linea incidentie, quando enim facta est reflexio vel fractio dicitur reflexa vel fracta. Unde prima pars linee reflexe vel fracte dicitur linea incidentie, secunda autem proprie dicitur reflexa vel fracta, et tota linea composita denominatur reflexa vel fracta propter secundam eius partem. Sufficit autem hoc modo loqui ad presens prout geometrie absolute competunt, nam aliter oportet has lineas exponi secundum proprietatem reflexionis et fractionis, sed hec requiri debent a perspectiva.

(11) Nulla vero istarum linearum est una vere nec vere plures, sed medio modo inter unam et plures, propter hoc quod habet angulum, sicut Aristoteles vult Quinto Methaphisice, sicut ex Tractatu Methaphisice patebit.

---

[7]decendit

[8]proroctis (?) *deleted in text, marginal correction cut off*

tions. And similarly for the other practical parts of geometry, which I have exemplified earlier in the chapter about this.

(8) Therefore because theoretical geometry is to be spoken of, and principles are prior to derivatives from them, it is necessary to give the special principles at the beginning; and since already above some things have been mentioned in general about lines, figures and angles, others must now be expounded specially.

(9) Between straight line and curved, there is a middle line which is called tortuous or flexuous, which necessarily has an angle, and it is composed from multiple straight lines thus , or from multiple curved, thus , or thus , or from a straight and a curved, thus . And if from multiple curved lines, then it can be from concave lines, or convex lines, or from a concave and a convex, as in the same figure, for angle *A* is from convex lines, and *B* from a convex and a concave, and *C* from concave lines

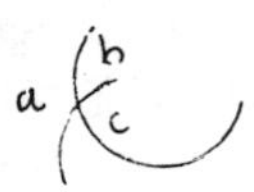

(10) And if it is from multiple straight lines, then it does not have much geometric diversity absolutely considered, but only as it descends into perspective, for it is divided into refracted and reflected line. A line is called reflected whose angle is comprehended by two lines protended in the same direction from the angular point, thus ; a line is called refracted whose angle is comprehended by two lines produced in different directions from the angular point, thus . Also line of incidence is related to these, for a protended line is called line of incidence before it is refracted or reflected, and is called reflected or refracted after reflection or refraction has been made. Thus the first part of a reflected or refracted line is called line of incidence, while the second is properly called reflected or refracted, and the whole composite line is denominated reflected or refracted on account of its second part. It suffices to speak in this way for the present as concerns geometry absolutely, for otherwise it is necessary to expound these lines according to the property of reflection and refraction, but these must be demanded of perspective.

(11) But none of these lines is truly single or truly multiple, but in a middle state between single and multiple, on account of its having an angle, as Aristotle holds in *Metaphysics* V,[11] as will be evident from the *Treatise of Metaphysics*.[12] Similarly there is a certain middle surface between plane and

[11] *Metaph.* Δ. 6, 1016a12-17.
[12] See Introduction, IX.

Similiter superficies quedam media inter planam et curvam, que scilicet componitur ex pluribus planis, vel pluribus curvis, vel pluribus convexis, vel ex plana et convexa, vel ex plana et concava, vel ex concava et convexa. Similiter corpus medium est inter planum et curvum, et habet tot diversitates quot nunc dicte sunt de superficie media, nam planities corporis, et curvitas, et medium inter hec non est nisi a superficiebus, et possunt exempla in imaginatione cuiuslibet et per corpora vulgata que videmus. Et sicut linea media inter rectam et curvam non est vere una propter angulum, sic nec superficies media vel corpus habet veram unitatem propter naturam anguli, sed secundum veritatem huiusmodi superficies est composita ex pluribus superficiebus et corpus ex pluribus corporibus, continuatis tamen per angulum.

(12) Ex necessitate quadam prius diffinita est figura, que sub termino vel terminis continetur. Est enim figura uno modo spatium inclusum linea vel lineis, vel superficie vel superficiebus, et sic est quantitas; alio modo est ipsa clausio facta per lineas vel superficies, et sic est passio quantitatis et proprietas eius, et est in predicamento qualitatis, scilicet in quarta specie. Sic distinguit Boetius in commento super Predicamenta.

(13) Dictum est autem quod quedam est figura superficialis, quedam corporalis[9] et inter superficiales. Illa que una linea, que omnino est una sine puncto et angulo, contineatur est circulus, qui est figura plana una linea contenta in cuius medio punctus est a quo omnes linee ducte ad circumferentiam sunt equales. Et punctus ille vocatur centrum circuli, et linea circumscribens vocatur circumferentia circuli. Nam alia figura potest una linea contineri que non est vere una: in medio quodammodo habet punctum et angulum; nullo[10] omnes linee ducte a puncto aliquo in puncta (?) ipsam sunt equales, ut in hac  , que accidit in sectionibus piramidis, de quibus postea explicatur. Diameter vero circuli est linea recta transiens per centrum, applicata ad circumferentiam in eius extremitatibus, dividens circulum in duas partes equales. Et utraque illarum partium vocatur semicirculus. Corda circuli vocatur linea recta que non transit per centrum, nec dividit circulum in duas partes equales, sed inequales. Et vocantur portiones circuli maior et minor: maior est que semicirculum excedit, minor que a semicirculo exceditur. Arcus est aliqua pars circumferentie, et hic cum corda vel diameter facit portionem vel semicirculum. Equivoce tamen portio dicitur aliquando pars circumferentie, et est idem cum arcu, secundum quod Euclides dicit in Tertio Libro: Portio vero circumferentie arcus nuncupatur. Unde dicimus pluries porcionem circumferentie et cuiuslibet rei partem determinatam. Hiis etiam comminandum est sector circuli, et est sector circuli figura que sub

[9]curvalis *deleted in text, marginal correction cut off*

[10]n°

curved, which namely is composed from multiple plane or multiple curved or multiple convex surfaces, or from plane and convex, or from plane and concave, or from concave and convex. Similarly there is a middle body between plane and curved, and it has as many differences as are now said about middle surface, for the planeness of body, or the curvitude, or the middle between these comes only from surfaces, and examples can be in anyone's imagination or from the common bodies that we see. And just as a middle line between straight and curved is not truly one on account of the angle, neither does a middle surface or body have true unity on account of the nature of angle, but according to truth a surface like this is composed from multiple surfaces, and a body from multiple bodies, but continued by angle.

(12) From a certain necessity figure, which is contained by a bound or bounds, was defined earlier. For in one way a figure is a space included by a line or lines, or by a surface or surfaces, and thus is quantity; in another way it is the enclosure itself made by the lines or surfaces, and is thus a passion and property of quantity, and is in the category of quality, namely in the fourth species. So Boethius distinguishes it in a comment on the *Categories*.[13]

(13) It was said that some figure is superficial, some corporeal and between superficial figures. That which is contained by one line which is altogether one without point or angle is a circle, which is a plane figure contained by one line in whose middle is a point from which all lines drawn to the circumference are equal. And that point is called the centre of the circle, and the circumscribing line is called the circumference of the circle. For another figure can be contained by one line which is not truly one: in a certain way it has in the middle a point and angle; in no way are all lines drawn from a certain point to its points equal, as in this  , which happens in sections of a pyramid, which are expounded later. A diameter of a circle is a straight line passing through the centre, joined to the circumference at its extremities, and dividing the circle into two equal parts. And each of the parts is called a semicircle. A straight line which does not pass through the centre, nor divide the circle into two equal parts, but into unequal ones, is called a chord of a circle. And the parts are called greater and lesser segments of the circle: a greater is that which exceeds a semicircle, a lesser that which is exceeded by it. An arc is a part of the circumference, and this with a chord or diameter makes a segment or semicircle. But segment is sometimes said equivocally for a part of the circumference, and it is the same as arc, according as Euclid says in Book III: "A segment of the circumference is named an arc."[14] Therefore we often call a segment a determinate part of the circumference and of anything. With these should be included sector of a circle, and a sector of a circle is a figure contained by two lines drawn

[13]Migne, LXIV, 251.

[14]Busard & Folkerts, I, 137.

duabus a centro ductis lineis et sub arcu qui ab eis comprehenditur continetur  . Et licet sic dicatur sector circuli, et sectio dicitur. Loquendo communius de sectione, dicitur omne illud circulum vel aliquam aliam figuram secare quod partem illius figure abscindit. Omnia hec sunt eiusdem generis, et ideo simul poni deberent: si igitur fiat figura ex una linea, necesse est quod sit curva, et vocatur circumferentia, et figura hec est circulus; si vero preter curvam est recta una, tunc sunt semicirculus et portiones circuli; si vero due recte cum curva, tunc est sectio circuli.

(14) Possunt autem figure fieri ex pluribus curvis lineis et ex pluribus rectis. Si ex curvis, tunc vocantur curvilinee figure, ut si coniungantur arcus circulorum adinvicem due vel tres vel plures, sic  , et hiis indigemus in corporibus rotundis. Sed alie sunt figure ex pluribus rectis lineis, et vocantur figure rectilinee, quarum quedam trilatere, qui tribus rectis lineis, quedam quadrilatere ex quatuor rectis lineis continentur.

(15) Figurarum trilaterum, alia est triangulus tria habens equalia latera, alia est triangulus duo habens equalia latera, alia triangulus trium inequalium laterum. Earum iterum, alia est ortogonium, unum scilicet habens angulum rectum, alia ambligonium, unum obtusum angulum habens, tertia est oxigonium, in quo tres anguli acuti.

(16) Figurarum quadrilaterum, alia est quadratum equilaterum atque rectangulum, alia est tetragonus longus, estque figura rectangula sed equilatera non est, < alia elmuahim, estque equilaterum sed rectangulum non est, alia simile >[11] elmuahim, quod opposita[12] latera habet equalia et angulos oppositos equales, idem[13] tamen nec rectis angulis nec equis lateribus continetur. Sed alia translatio elmuahim vocat rumbum, et simile elmuahim vocat simile rumbo. Et dicit Anaritius quod rumbum est quadratum a duabus partibus comprehensum et coangustatum in angulos acutos, et ideo habet duos angulos acutos et duos obtusos, quantum acuuntur isti expanduntur illi, et similis rumbo est tetragonus longus a duabus partibus comprehensus, et unumquodque laterum eius est inequale dubous lateribus ei proinquioribus, et unusquisque angulus illius est inequale duobus sibi vicinioribus. Preter has figure omnes quadrangule helmuarife vocantur.

(17) Dicit A. Bathoniensis in editione speciali quod helmuhariphe sunt irrationales.[14] Sed tamen rationale distinguit scilicet in illud quod rationi artis subiectum est, et sic figure et linee superius dicte irrationales, propter hoc quod non denominantur ab aliqua mensura vel numero, sunt rationales hoc modo, id est rationi huius artis subiecte, nam alio modo est rationale quod ab aliqua mensura geometrica secundum aliquem numerum

[11] Bracketed passage supplied on the basis of Busard & Folkerts, I, 114.
[12] composita
[13] inde
[14] irregulares

from the centre and by the arc that is comprehended by them  .

And although a sector of a circle is thus called, it is also called a section. In speaking commonly of section, everything is said to cut (*sc.* section) a circle or some other figure that separates off a part of that figure. All these are of the same genus, and so should be placed together: if therefore a figure be made from one line, it must be curved, and it is called circumference, and the figure is a circle; if beside the curve there is a straight line, then there are semicircle and segments of a circle; if there are two straight lines with a curve, then it is a section of a circle.

(14) But figures can be made from multiple curved lines and from multiple straight ones. If from curves, then they are called curvilinear figures, as when two or three or more arcs of circles are mutually joined, thus  , and we need these for round bodies. But there are other figures from multiple straight lines, and they are called rectilinear figures, and some are trilateral, which are contained by three straight lines, and some are quadrilateral, which are contained by four straight lines.

(15) Of trilateral figures, one is a triangle having three equal sides, another is a triangle having two equal sides, and another is a triangle of three unequal sides. Again, one of them is orthogonal, having one right angle, another ambligonal, having an obtuse angle, and a third oxygonal, in which there are three acute angles.

(16) Of quadrilateral figures, one is an equilateral and rectangular square, another is a long tetragon, and it is a rectangular figure, but is not equilateral, another is elmuhaim, which is equilateral but not rectangular, another is similar to elmuhaim, which has opposite sides equal and opposite angles equal, but is contained neither by right angles nor by equal sides. But another translation[15] calls elmuhaim rhombus, and it calls similar to elmuhaim, similar to rhombus. And Anaritius says that a rhombus is a square comprehended on two sides and narrowed into acute angles, and so it has two acute angles and two obtuse, for as much as the former are sharpened the latter are expanded, and "similar to a rhombus" is a long tetragon comprehended on two sides, and each one of its sides is unequal to the two sides nearer to it, and each one of its angles is unequal to the two neighbouring it.[16] Beside these, all other quadrangular figures are called helmuarife.

(17) Adelard of Bath says in his special edition that helmuarife are irrational.[17] But rational divides into that which is subjected to the rationale of an art, and so the figures and lines called irrational above, on account of their not been denominated by some measure or number, are rational in this way, that is in being subject to the rationale of the art, for in another way rational is what is denominated by some geometrical measure accord-

[15]Tummers, 142.

[16]Tummers 142.

[17]Clagett – Alfred, 274.

denominatur, ut bipedalis vel cubitalis et huiusmodi, et sic quantitates continue sunt irrationales, ut superius expositum[15] est. Helmuariphe igitur dicuntur figure irrationales non ut in precedentibus, sed quia non sunt rationi huius artis subiecte. Secundum vero aliam translationem vocantur trapezeie, et Salmabelicus dicit quod sunt trapezeie inordinate, et hoc est quia sub ordine artis non cadunt, et ideo geometre vocant eas irregulares. Sed hic tunc sumitur irregularitas quia non sunt regulis artis subiecte, quoniam aliter dicuntur figure regulares, ut tangetur post, et similiter de trapezeis fiet alias inventum.

(18) Que vero habent plura latera quam 4 multiplicantur in infinitum secundum numerum laterum, ut quintangula, que et vocatur grece pentagonus, et sexangula que vocatur exagonus, et septangula que vocatur eptagonus, et sic in infinitum.

(19) De figuris corporalibus dicendum est quod quedam est spera, quedam alia, sed hee figure sunt incomplete omnes preter speram. Potest igitur figura compleri ex duabus superficiebus, et vocatur piramis rotunda, quia habet basim circularem et eius grossicies incipiens a basi semper diminuitur usque in punctum conalem coartetur. Habet autem hec piramis rotunda suum sectorem proprium tripliciter, sicut ex Libro Piramidum Apollonii et ex Libro Speculorum Comburentium patet, qui ut dicit hic Anaritius non continetur una linea simplici, ut circulus, sed composita et tortuosa in parte, quia non contingit signare punctum in sectore a quo omnes linee ducte a centro usque in circumferentiam sunt equales. Sed de hiis tortuis dicetur inferius. Si ex tribus, tunc est columpna rotunda, quia duas habet extremitates circulares inter quas extenditur corpus columpne, sed de hiis figuris curvis fiet expositio suo loco, et ideo solum transeo nunc in universali et exemplariter. Sunt tamen preter has corpora que non habent ita figuram determinatam corporalem, nec quas ars geometrica speculativa considerat, sicut est ovalis figura anexa piramidi rotunde, et lenticularis que est annexa spere, ut figura lentis, quod est genus leguminis.

(20) Si vero sit ex pluribus planis superficiebus, tunc est corpus angulare et multilaterum, et oportet quod sit ex 4 superficiebus ad minus, nam omnis angulus solidus ut prius dictum est constat ex tribus superficiebus ad minus, et ideo oportet quod tres ad minus sint superficies laterales, quarum anguli tres constituant unum angulum solidum, et tunc, si perficiatur corpus per basim, oportet quod quarta sit ibi superficies, aut non erit clausum undique sed apertum. Si igitur compleatur corpus, erit ex 4 superficiebus ad minus,

[15]compositum

ing to some number, as two-foot, or cubital, and the like, and in this way continuous quantities are irrational, as was explained above.[18] Helmuarife therefore are called irrational figures not as in the above, but because they are not subjected to the rationale of the art. But according to another translation they are called trapezia,[19] and Simplicius says that they are unordered trapezia, and this is because they do not fall under the order of the art, and so geometers call them irregular. But here then irregularity is taken for their not being subject to the rules of the art, because otherwise they are called regular figures, as will be mentioned later, and as will be found similarly in trapezia elsewhere.

(18) Those that have more than four sides are multiplied to infinity according to the number of sides, like quintangle, which is also called in the Greek manner pentagon, and sexangle which is called hexagon, and septangle which is called heptagon, and so to infinity.

(19) Of corporeal figures it should be said that one is a sphere, and another other, but except for the sphere all these figures are incomplete. A figure can be completed by two surfaces, and it is called a round pyramid, because it has a circular base and its thickness beginning from the base always diminishes until it is narrowed to a vertical point. This round pyramid has its own proper sector of three kinds, as is evident from Apollonius's *Book of Pyramids* and from the *Book of Burning Mirrors*,[20] which sector, as Anaritius says here,[21] is not contained by one simple line, like a circle, but one composite and tortuous in part, because a point in the sector cannot be designated from which all lines drawn from the centre to the circumference are equal. But these tortuous lines will be spoken about below. If it is completed by three surfaces, then it is a round cylinder, because it has two circular extremities between which a cylindrical body is extended, but an exposition of these curved figures will be made in its own place, and so now I only pass over them in general and by way of example. But there are, beside these, bodies that do not thus have a determinate corporeal figure, which the theoretical geometrical art does not consider, such as the oval figure associated with the round pyramid and the lenticular associated with the sphere, like the figure of a lentil, which is a genus of pulse.

(20) But if it be from multiple plane surfaces, then it is an angular and multilateral body, and it must be from at least four surfaces, for every solid angle is established from at least three surfaces, as was said before,[22] and so there must be at least three lateral surfaces, whose three angles constitute one solid angle, and then, if the body be perfected by a base, there must be a fourth surface there, or it will not be everywhere closed but open. Therefore, if the body be completed, it will be from at least four surfaces, and it can be

---

[18]CM, 79-80.

[19]Tummers, 142.

[20]See Introduction, III.

[21]Tummers, 136.

[22]CM, 28.

et potest esse ex 5 vel ex 6, et sic de infinitis, nec est modo tempus explicandi ista sed in sequentibus.

(21) Et iam patet quod linea recta magis est una quam curva, sed econverso est de corpore, nam corpus curvum magis est unum quam planum. Et Anaritius in commento Euclidis reddit causam huius, dicens quod una linea curva ut circumferentia comprehendit figuram, ut < non > potest fieri nisi figura fiat cum ea. Sed recta non sic, propter quod habet puram et absolutam naturam linee et nullo modo rationem superficies concernit nec ponit. Quapropter maior diversitas est in linea curva quam recta. Preterea habet concavitatem et convexitatem, quas non habet linea recta, et propter easdem rationes superficies plana est magis una quam curva. Sed corpus curvum ut spericum est maxime unum quia unam solam habet superficiem; omnia alia habent plures, et quanto magis recedunt a natura spere tanto maiorem habent diversitatem.

(22) Cum speciebus quantitatis continue et figuris oportet sciri angulum et modos eius. Sed quoniam contingentia in aliquibus circa angulos circa sequentia est necessaria, ideo oportet quod sciatur quod hic expedit. Et linea dicitur contingere circulum que cum circulum tangat in utramque partem erecta circulum non secat, sic  . Et circuli sese invicem contingere dicuntur qui tangentes sese non se secant, sic  . Et lineam tangere speram dicit Tholomeus in pluribus locis, ut in 3 Perspective et alibi, quapropter linea contingere speram dicitur que in utramque partem erecta speram non secat. Et spere eius contingere se dicuntur que sese tangentes non se secant, sicut de circulis. Et huiusmodi linea que sic tangit speras vel circulos vocatur linea contingentie. Superficies etiam plana contingere speram vel circulum dicitur que in utram partem erecta non speram neque circulum secat. Et universaliter potest dici de omni superficie plana contingente quodcumque corpus concavum seu convexum illum quod autor Speculorum Comburentium intendit, scilicet quod illa est inter quam et corpus tale est punctum unum tantum. Hec sufficunt ad presens.

(23) De angulis aliqua prius dicta sunt, et hic adduntur que superius sunt omissa. Angulus rectilineus est qui ex rectis lineis continetur. Curvilineus est qui curvis lineis includitur. Angulus medius inter curvilineum et rectilineum est qui ex cursu linee recte et curve producitur, de quibus aliqui sunt plani et superficiales, alii corporales, et aliqui recti, aliqui acuti,[16] aliqui obtusi,

[16] curvi

from five or from six, and so infinitely, but now is not the time for explaining these things, but in what follows.

(21) And it is now evident that a straight line is more one than is a curve, but contrarily for a body, for a curved body is more one than a plane. And Anaritius in a comment on Euclid gives the cause of this,[23] saying that one curved line, such as a circumference, comprehends a figure, as cannot be done unless a figure is made with it. But a straight line is not thus, on account of its having the pure and independent nature of a line and in no way having reference to or assuming the rationale of surface. Therefore there is greater diversity in curved line than in straight. Moreover, it has concavity and convexity, which straight line does not have, and for the same reasons plane surface is more one than curved. But curved body, like spherical, is most of all one because it only has one surface; all others have more, and the more they recede from the nature of sphere the greater diversity they have.

(22) Together with the species of continuous quantity and figures, angle and its modes must be understood. But because tangency is necessary for what follows in some things concerning angles, it must be understood what is expedient here. A line is said to be tangent to a circle, which, when it touches the circle and is set up in either direction, does not cut the circle, thus  . And circles are said to be mutually tangent to each other which touch each other, but do not cut each other, thus  . And in several places, such as *Perspective* III and elsewhere, Ptolemy says that a line touches a sphere,[24] on which account a line is said to be tangent to a sphere which when set up in either direction does not cut the sphere. And spheres are said to be tangent to each other which when touching do not cut, as with circles. And a line of this kind which thus touches spheres or circle is called a line of tangency. Also a plane surface is said to be tangent to a sphere or circle which when set up in either direction does not cut the sphere or circle. And in general it can be said of every plane surface tangent to any concave or convex body what the author of *On Burning Mirrors* holds, namely that it is that between which and the body there is only one point.[25] And these things are sufficient for the present.

(23) Some things were said about angles earlier,[26] and here are added what were omitted above. A rectilinear angle is that which is contained by straight lines. A curvilinear is that which is included by curved lines. A middle angle between curvilinear and rectilinear is that which is produced by the meeting of a straight line and a curved, of which some are plane and superficial, others corporeal, and some are right, others acute, others obtuse,

[23]Tummers, 137-8.
[24]Lejeune, 88.
[25]Heiberg & Wiedemann, 220.
[26]CM, 27.

ut superius explicatur. Angulus superficialis curvus concavus est[17] expansio duarum linearum concavarum quarum applicatio in superficie concava est non recta. Angulus superficialis curvus convexus est duarum linearum convexarum quarum applicatio in superficie convexa est non directa. Et intelligendum est quod angulus planus potest esse curvilineus sicut rectilineus, nam in plano possunt discerni linee concave et convexe sicut recte, et ideo angulus quicumque fiat in plano est planus, quamvis fit ex lineis curvis concavis et convexis. Unde angulus curvus non dicitur talis propter lineas curvas sed propter spatium curvum, in quo spatio nullus etiam angulus rectilineus potest fieri, quia in superficie concava vel convexa impossibile est lineam rectam protrahi, quamvis in plano possint concave et convexe produci. Cuius causa est quod curvitas linee habetur ex se et ex curvitate superficei. Et quia ex se ipsa potest habere propter hoc quod curvitas primo naturaliter est in linea quam in superficie, ideo in plana superficie possunt linee curve deduci. Sed linee recte in curva superficie non nascuntur, quia rectum et curvum naturaliter opponuntur, et ideo anguli rectilinei sunt in sola superficie plana. Et possunt anguli curvilinei concavi et convexi esse in plana superficie, et tunc plani dicuntur propter superficiem planam, quia in veritate angulus superficialis est superficies et figura superficialis incompleta, quia non est undique conclusa nec terminata. Et superficies et figura et quantitas referuntur ad spatium, et ideo angulus quicumque sit spacio plano erit planus, quamvis includatur lineis curvis vel concavis.

(24) Anguli vero contra se positi dicuntur qui ex intersectione duarum linearum nascentes ex oppositis partibus se mutuo respiciunt. Anguli coalterni dicuntur quilibet duo quos linea cadens super duas constituunt instrinsecas cum eis ex oppositis partibus illius linee incidentis.

(25) Angulus semicirculi dicitur qui continetur ex diametro et circumferentia. Angulus portionis dicitur qui comprehenditur ex corda et arcu, sive sit angulus portionis maioris sive minoris. Angulus contingentie dicitur qui per circumferentiam circuli et per lineam contingentie continetur.

(26) Angulus corporalis planus est qui constituitur ex planis superficiebus. Angulus corporalis curvus est qui constituitur ex angulis superficialibus curvis. Angulus corporalis convexus qui fit ex angulis superficialibus convexis. Angulus corporalis concavus qui fit ex angulis superficialibus concavis.

(27) Cognitis modis anguli possumus diffinire perpendicularem, sed non est facile invenire diffinitiones perpendicularium precisas, quia geometre parum tangunt de eis. Potest quod linea comparari ad lineam vel ad superficiem vel corpus secundum rationem perpendicularitatis. Linea autem recta perpendicularis super aliam rectam dicitur quando steterit super eam ad angulos

[17]est est

as is explained above.[27] A concave curved superficial angle is the expansion of two concave lines whose application in a concave surface is not direct. A convex curved superficial angle is of two convex lines whose application in a convex surface is not direct. And it should be understood that a plane angle can be curvilinear as well as rectilinear, for in a plane there can be discerned concave and convex lines as well as straight ones, and so whatever angle is made in a plane is planar, even if it is made from concave and convex curves. Therefore a curved angle is not so called on account of curved lines but on account of curved space, in which space also a rectilinear angle cannot be made, because in a concave or convex surface it is impossible for a straight line to be protracted, although concave and convex lines can be produced in a plane. The cause of this is that a line's curvitude comes from itself or from a surface's curvitude. And, because it can have it from itself on account of curvitude being naturally first in a line rather than in a surface, curved lines can accordingly be drawn in a plane surface. But straight lines do not arise in a curved surface, because straight and curved are naturally opposed, and so rectilinear angles are only in a plane surface. And concave and convex curvilinear angles can be in a plane surface, and then they are called plane on account of the plane surface, because in truth a superficial angle is a surface and an incomplete superficial figure, because it is not everywhere included and terminated. And surface and figure and quantity are referred to space, and so whatever angle is in a plane space will be plane, even if it be included by curved or concave lines.

(24) Angles are called opposite each other which, arising from the intersection of two lines, face each other mutually from opposite sides. Any two angles are called alternate which a line falling on two lines constitutes internally with those lines on opposite sides of the incident line.

(25) The angle of a semicircle is said to be that which is contained by the diameter and the circumference. The angle of a segment is said to be that which is comprehended by a chord and an arc, whether it be the angle of a greater segment or of a lesser. The angle of tangency is said to be that which is contained by the circumference of the circle and the line of tangency.

(26) A plane corporeal angle is that which is constituted from plane surfaces. A curved corporeal angle is that which is constituted from curved superficial angles. A convex corporeal angle is that which is made from convex superficial angles. A concave corporeal angle that which is made from concave superficial angles.

(27) With the modes of angle known we can define perpendicular, but it is not easy to find precise definitions of perpendiculars, because geometers do not touch on them enough.[28] It can be that a line be compared to a line or to a surface or a body according to the rationale of perpendicularity. A straight line is said to be perpendicular on another straight line when it stands on it

---

[27]CM, 28.
[28]Cf. Steele-Professor, 58; CM, 28; Molland, "Roger Bacon's Appropriation".

equales, et quia isti anguli sunt recti ideo linea hec dicitur ortogonaliter super aliam stare, nam orton est rectum et gonia angulus. Linea vero perpendicularis respectu circuli duobus modis potest intelligi, ut primo comparetur ad circumferentiam, deinde ad superficiem circuli. Quando igitur linea recta cadit super circumferentiam ad angulos equales, perpendicularis est ad eam, et hec transit in centrum circuli. Sed hec non est ortogonaliter super circumferentiam nec super circulum, nam anguli sunt obliqui et obtusi, sicut patet per angulos contingentie, quia hec linea cadens a circumferentia in centrum facit angulos rectos cum linea contingentie, quapropter angulorum uterque quos linea cadens in centrum constituit cum circumferentia addit angulum contingentie super rectum. Ergo uterque eorum est maior recto sicut patet in figura, quare uterque est obtusus necessario, licet pauci mathematicorum hoc considerunt, cum tamen possint de facili, et ideo abiciuntur vocabulis et dicunt quod sunt anguli recti speciales vel circulares, licet non sint recti ad lineam rectam. Si vero linea recta comparetur ad superficiem circuli, tunc est perpendicularis quando cum singulis sibi conterminabilibus lineis in eadem superficie expansis angulos rectos facit. Et ideo dicitur ortogonaliter insistere super eam, non quia cum superficie faciat angulos rectos, sed cum lineis ductis. Unde vere est perpendicularis ad superficiem, et erecta dicitur super eam. Et primo et per se hec verificantur, sed secundario verificatur quod dicatur ortogonaliter stare, quia hec est ratione linearum describendarum. Et sicut ad superficiem circuli linea dicitur perpendicularis, sic ad omnem superficiem, nec est diversitas, licet priori modo non omnino est simile, quia, quando superficies terminatur lineis rectis, et linea cadat super aliquam earum perpendiculariter, illi anguli sunt equales, et sunt recti per diffinitionem recti anguli, et non sunt obtusi. In equalitate igitur conveniunt, sed in rectitudine et obtusitate differunt.

(28) Potest autem linea dici perpendicularis super corpora, et si respectu spere volumus dicere, tunc accipienda est diffinitio quam Ptolomeus dicit in libro tertio de Aspectibus, nam ibi Ptolomeus hec verba dicit: Vocatur autem perpendicularis in superficie spere communi nomine linea que circumfertur cum omnibus lineis que tangunt speram exeuntes a puncto communi qui est in superficie spere ad angulos rectos. Et ex hoc concludit dicens: Unde necesse ut omnes perpendiculares cadentes super sperarum superficies cum penetraverint transeant per centrum spere. Et secundum hanc viam similis est ratio perpendicularis a parte spere et a parte circumferentie circuli, quia, descripto circulo in spera, et ducatur linea contingens, et transeat linea ad circumferentiam circuli et ad speram per punctum contactus: erit perpendicularis ad utramque, et cadet ad angulos obtusos et in centrum utriusque.

at equal angles, and because these angles are right the line is on that account said to stand on them orthogonally, for *orthon* is right and *gonia* angle. But a line perpendicular with respect to a circle can be understood in two ways, as it is first compared to the circumference and then to the surface of the circle. When therefore a straight line falls on the circumference at equal angles, it is perpendicular to it, and it passes into the centre of the circle. But it is not orthogonally on the circumference nor on the circle, for the angles are oblique and obtuse, as is evident from angles of tangency, because this line falling from the circumference into the centre makes right angles with the line of tangency, on which account either of the angles which the line falling into the centre constitutes with the circumference adds an angle of tangency over a right angle. Therefore each of them is greater than a right angle, as is evident in the figure, wherefore each of them is necessarily obtuse, although few mathematicians consider this, although they could easily, and so they abuse words, and say that they are special or circular right angles, although they are not of a straight (*sc.* right) line to a straight line. But if a straight line be compared to the surface of a circle, then it is perpendicular when it makes right angles with individual lines conterminous with it and extended in the same surface. And on that account it is said to impinge on it orthogonally, not because it makes right angles with the surface, but with the drawn lines. Therefore it is truly perpendicular to the surface, and is said to be erected on it. And first these things are verified in themselves, but secondarily it is verified that it be said to cut orthogonally, because this is in the rationale of the lines to be described. And just as a line is said to be perpendicular to the surface of a circle, so to any surface, and there is no difference, although in the earlier way the cases are not altogether similar, because, when a surface is bounded by straight lines, and a line falls perpendicularly on one of them, the angles are equal, and they are right by the definition of a right angle, and they are not obtuse. Therefore the cases agree in equality, but differ in rightness and obtuseness.

(28) But a line can be said to be perpendicular on bodies, and if we wish to say this with respect to a sphere, then the definition that Ptolemy gives in *On Optics* III is to be taken, for there Ptolemy speaks these words:[29] "A line is called by the common name perpendicular on the surface of a sphere which is surrounded by all the lines that touch the sphere going out from a common point in the surface of the sphere at right angles." And from this he concludes, saying, "Wherefore, it is necessary that all perpendiculars falling on surfaces of spheres should when they have penetrated pass through the sphere's centre." And according to this way there is a similar rationale of perpendicular on the part of a sphere and on the part of a circle's circumference, because, with a circle described in a sphere, let a tangent line be drawn, and let a line pass to the circle's circumference and to the sphere through the point of contact: it will be perpendicular to each, and will fall at

[29]Lejeune, 88.

Considerandum tamen quod linea recta transiens cadens perpendiculariter in superficie spere, si non intelligatur ibi circulus, quod non facit aliquos obliquos, quia anguli non sunt nisi ex concursu linearum. Sed tamen eadem est ratio perpendicularitatis, que scilicet consistit in casu linee istius ad lineam rectam contingentem speram vel circulum circumductum in spera. Nam sic est linea perpendiculariter super circulum sicut super speram, et similiter quantum ad hoc conveniunt quod transit per centra eorum. Sed non est ad angulos rectos, nisi velimus multum loqui per accidens, propter hoc quod alia linea facit angulos rectos cum linea contingente circulum et speram, et sic ortogonaliter ducatur ab eos. Sed hoc est omnino per accidens, vel etiam non per se nec per accidens, sed per extensionem vocabuli et transsumptionem, nam non est simile de hoc et de linea perpendiculari ad superficiem, quoniam linee ille ad quas est perpendicularis sunt in superficie ipsa, et ideo bene secundario attribuitur rectitudo angulorum superficiei, quia in ea[18] sunt illi anguli descripti. Sed hec linea continue nec est in spera nec in circulo, et ideo nec anguli recti de quibus loquimur, quia non potest attribui ortogonalitas a parte ista sicut a parte superficiei plane.

(29) Si vero linea comparetur ad corpus planum, tunc linea est perpendicularis ei sicut superficiei plane. Rursus de perpendicularitate linee possunt intelligi. Sed similiter considerandum est de perpendicularitate superficiei, que cito expeditur, nam superficies ad aliam dicitur perpendiculariter erecta quando lineis terminalibus earum in quibus se tangunt due perpendiculares superstant conterminales que spatium recto angulo equale inter se continentes in eisdem superficiebus site sunt. Et si nos ymaginaverimus unam lineam pro communi termino utriusque superficiei, et linee perpendiculariter ad eam contineant angulum rectum inter se, erit superficies una erecta super aliam perpendiculariter. Hunc modum secundum quantum ad substantiam diffinitionis habent mathematici in usu, sed tamen aliter mutant verba, in quibus est dubitatio, et videtur esse falsitas si ex potestate methaphisice fuerint examinanda, nam methaphisici est verificare principia geometrie et omnium scientiarum. Et quoniam volo movere questionem vel dubitationem in sequentibus grandem de diffinitione perpendicularis superficiei secundum vocabula consueta, et in quibusdam aliis consimilibus, ideo posui hanc diffinitionem sub aliis modis Ioandi (?) planioribus, in quibus non est calumpnia. Superficies erigitur super corpus planum sicut ad superficiem planam, sed super speram potest superficies plana dici perpendicularis quando dividit eam in partes equales, et cum ymaginata fuerit transire in speram transibit per centrum spere.

(30) De equedistantia vero linearum et aliorum dicendum post hec. Ad hoc vero quod linee equedistent oportet quod sic spatium inter eas sit et distantia sicut nomen equedistantie sonat. Et ideo due linee exeuntes a punctis diversis, coniuncte[19] simul sine medio per coniunctionem superficierum

---

[18] eo

[19] coniunctis

obtuse angles and into the centre of each. But it should be considered that a transient straight line falling perpendicularly on the surface of a sphere does not make oblique angles unless a circle be understood there, because angles only arise from the concourse of lines. But yet there is the same rationale of perpendicularity, namely what consists in the falling of a line on a straight line tangent to the sphere or on a circle drawn around in the sphere. For it is the same for a line being perpendicularly on a circle as on a sphere, and the cases agree in this, that it passes thorugh their centres. But this is not at right angles, unless we wish to speak very much *per accidens* on account of another line making right angles with the line tangent to the circle and the sphere, and so being drawn orthogonally to them. But this is altogether *per accidens*, or else neither *per se* nor *per accidens*, but by the extension or transumption of a word, for this is not similar to the case of a line perpendicular to a surface, because the lines to which it is perpendicular are in the surface itself, and so rightness of angles is well attributed secondarily to the surface, because the angles are described in it. But this line is continually neither in the sphere not in the circle, and so neither the right angles of which we speak, because orthogonality cannot be attributed in this case as in the case of a plane surface.

(29) But if a line be compared to a plane body, then a line is perpendicular to it just as to a plane surface. Again these things can be understood in terms of perpendicularity to a line. But the perpendicularity of a surface should be considered similarly, which is swiftly dealt with, for a surface is said to be erected perpendicularly to another when, on their terminal lines in which they touch each other, there stand two conterminous perpendiculars which, containing between them a space equal to a right angle, are situated in the surfaces themselves. And if we imagine one line as the common bound of either surface, and lines perpendicular to it contain a right angle between themselves, then one surface will be erected on the other perpendicularly. Mathematicians have this second way in use as regards the substance of the definition, but they change the words in another fashion, and there is doubt in them, and there seems to be falsity if they are examined with the power of metaphysics, for it is for the metaphysician to verify the principles of geometry and of all sciences. And because I wish to move an important question or doubt in what follows about the definition of a perpendicular surface according to the customary words, and in some other similar cases, I have therefore posited this definition in the other plainer modes of Joandus (?), in which there is no chicanery. A surface is erected over a plane body just as over a surface, but a surface can be said to be perpendicular over a sphere when it divides it into equal parts, and when it is imagined to pass into the sphere, it will pass through the centre of the sphere.

(30) After these things the equidistance of lines and other objects should be spoken about. For two lines to be equidistant it is necessary that the space and the distance between them be such as the name equidistance means. And so two lines that go out from different points and are joined together without a

et corporum in quibus sunt illa puncta et ille linee, non possunt trahi per equedistantiam quia non distant. Cum tamen hoc petitur multototiens et supponitur, et patet adhuc, quod non sunt equedistantes quia in veritate concurrunt ad punctum idem in tertia superficie supposita, sicut in Methaphisicis erit manifestum, quia, licet in verificatione istorum principiorum oportet aliqua dicere, plena declaratio exigitur a methaphisico, ut sepe declaratum est. Debetur igitur spatium exitus. Hoc erit brevissimum quod potest esse inter lineas que equedistare debent, et iterum debet esse inter eas. Spatium autem brevissimum designatur hic per lineam brevissimam que potest inter eas trahi. Et cum debet esse utrimque hoc spatium, non concurrent, et ideo equedistantes necesse sunt que non concurrunt, etiam si in infinitum protrahantur. Sic Anaricius et Sambelichus et Aganius investigant equedistantiam, et addunt quod hoc spatium est perpendiculare super utramque, nam si sunt equedistantes habent spatium perpendiculare, et econverso. Sed hec duo sunt demonstrabilia, sicut ipsi docent, et ideo, ut vult Anaritius, dicuntur hec duo hic per se propter necessitatem expositionis ipsius diffinitionis, sed ad maiorem evidentiam et docet post 26^{am} propositionem illa duo demonstrari.

(31) Secundum vero Euclidem non sufficit quod non concurrent quod sint in eadem superficie, quia due linee quarum altera est in uno et alia in altero intersequantes se nunquam concurrunt, et non sunt equedistantes cum non habeant idem spatium utrimque inter se, ut volunt dicti expositores. Sed si essent in eadem superficie haberent idem spatium. Propter quod estimat Euclides quod equedistantes linee sunt ille que in eadem superficie collocate atque in eandem partem protracte non convenient, etiam si in infinitum protraherentur. Quomodolibet in corporibus et superficiebus diversis inter que est spatium uniforme brevissimum vel unum sit in uno secundum in altero considerentur linee per equedistantiam, secundum aliam considerationem sunt in eadem superficie que ymaginabitur erigi perpendiculariter super utramque superficiem nunc dictam (?), cuius extremitates erunt linee que posite sunt equedistare.

(32) Hec igitur gratia expositionis hic dicta sunt secundum opiniones commentatorum, quamvis oportet aliter in Methaphisicis tractari et ostendi quid secundum puram veritatem tenendum est, supponentes quod licet hec vera sint tamen oportet certius loqui de hac materia.

(33) Superficies vero equedistantes dicuntur que in utramque partem protracte non conveniunt, et si in infinitum protrahantur. Ita dicit Euclides, sed que necessitas est ut linee equedistantes sint in una superficie, eadem necessitas est ut superficies equedistantes sint in eadem corpore, et ideo hoc

mean by the joining of the surfaces and bodies in which those points and lines are situated cannot be drawn with equidistance because they are not distant. But with this it is often demanded and supposed, and it is also evident, that they are not equidistant because in truth they come together at the same point in a third supposed surface, as will be manifest in *Metaphysics*, because, although it is necessary to say some things in verification of the principles, a full declaration is required of the metaphysician, as has often been declared. There must therefore be a space of exit. This will be the shortest that there can be between the lines that have to be equidistant, and again it must be between them. The shortest space is designated here by the shortest line that can be drawn between them. And since there must be this space on each side, they will not meet, and so equidistant lines are necessarily those that do not meet, even if they are produced to infinity. In this way Anaritius and Simplicius and Agapius[30] investigate equidistance, and they add that this space is perpendicular on each, for if they are equidistant they have a perpendicular space, and conversely. But these two are demonstrable, as they themselves teach, and so, as Anaritius holds, they are said here in themselves on account of the need to expound the definition, but for greater evidence he also teaches their demonstration after Proposition 26.[31]

(31) But according to Euclid, it is not sufficient that they do not meet but that they be in the same surface, because two lines of which one in is in one and the other in the other intersecting each other do not meet, and they are not equidistant because they do not have the same space on each side between them, as the said expositors wish. But if they were in the same surface they would have the same space. On this account Euclid reckons that equidistant lines are those which, being located in the same surface and produced in the same direction, will not meet, even if they were produced to infinity. In whatever way in different bodies and surfaces between which there is a uniform shortest space (or one is in one and the second in the other) lines are considered for equidistance, according to another consideration they are in the same surface which is imagined to be erected perpendicularly over each surface now mentioned, and of which the extremities are lines which are posited to be equidistant.

(32) These things are said here for the sake of exposition according to the opinions of the commentators, although they must be treated otherwise in *Metaphysics* and it must be shown what should be held according to pure truth, supposing that, although these things be true, it is still necessary to speak more certainly of the matter.

(33) Surfaces are said to be equidistant which when produced in either direction do not meet, even if they are produced to infinity. Thus says Euclid, but what necessity there is for equidistant lines to be in the same surface, the same necessity there is for equidistant surfaces to be in the same

[30]Cf. A.I. Sabra in *Dictionary of Scientific Biography*, X, 6, *s.v.* al-Nayrīzī.

[31]Tummers, 144.

forsitan intellexit. Unde sua sententia est quod idemptitas superficiei requiritur ad equedistantiam linearum, et idemptitas corporis ad equedistantiam superficierum, nam si una superficies aliam intersecans, et una lateat in uno et alia in altero, non sunt equedistantes quia non sunt in eodem corpore, et due equedistare possunt que sunt in diversis corporibus habentibus inter se idem spatium quorum una[20] est in uno et alia in altero, quia secundum aliam viam sunt in eodem corpore, ut dictum est de lineis. Et debent superficies equedistantes habere spatium brevissimum et idem utrimque, sicut de lineis, et est perpendiculare super utramque, nam si sunt equedistantes oportet quod spatium sit perpendiculare super eas, et si est perpendiculare super eas sunt equedistantes, et hec sunt demonstrabilia non solum de lineis sed de superficiebus.

(34) Corpora autem equedistare necesse est si spatium brevissimum est utrimque inter ea. Et quia intellectus huius rei patet per equedistantiam superficierum et linearum, ideo parum loquuntur geometre de equedistantia corporum, sicut parum loquuntur de equedistantia superficierum, dico quantum ad diffinitionem, quia consimilis intellectus est in eis et in lineis. Corpora igitur equedistant que cum fuerint in eodem corpore etiam si in infinitum protrahantur non concurrent, quia superficies eorum equedistantes, quamvis sint in illis diversis corporibus, tamen per aliam considerationem sunt in eodem corpore, sicut linee illarum superficierum equedistantium et illorum corporum equedistantium equedistantes sunt in eadem superficie. Sed hee superficies in aliquo corpore interiacente et superficies equedistantes sunt, que sunt corporibus equedistantibus sicut in eodem corpore, et ideo corpora equedistantia sunt in quibus sunt eedem superficies equedistantes, et linee similiter sunt in eodem corpore ymaginabili inter corpora illa, cuius intelliguntur partes extreme.

(35) Preter autem diffinitiones has occurrunt quedam quas supponit Euclides. Et alii in libris geometrie vulgate faciunt mentionem de diffinitis, sed aut non ponunt diffinitiones aut non suo loco. Et est una date linee, de qua in prima propositione et alias tangit. Dicit vero A.Batoniensis in editione speciali quod, Data hic dicuntur quibus equalia habundanter poteris invenire, media scilicet inter immensa et brevissima, que scilicet media humana ratio indagatrix sue ratiocinationi supposuit. Et istud sumitur ex prima diffinitione Libri Euclidis de Datis, que (?) est hec, Data magnitudine dicuntur et spatia et linee et anguli quibus possumus equalia asignare.

(36) Designare superficiem super lineam aliquam est lineam ipsam facere

[20]unum

body, and so he perhaps understood this. Accordingly his meaning is that identity of surface is required for equidistance of lines, and identity of body for equidistance of surfaces, for if one surface intersects another, and one is concealed in one body and the other in another, they are not equidistant because they are not in the same body, but two can be equidistant that are in different bodies that have between them the same space, of which bodies one surface is in one and the other in the other, because according to another way they are in the same body, as was said about lines. And equidistant surfaces must have the shortest space and the same on each side, as for lines, and it is perpendicular on each, for if they are equidistant the space must be perpendicular on them, and if it is perpendicular on them they are equidistant, and these things are demonstrable not only of lines but of surfaces.

(34) Bodies are necessarily equidistant if there is the shortest space on each side between them. And because the understanding of this is evident from the equidistance of surfaces and lines, on that account geometers say too little about the equidistance of bodies, just as they say too little about the equidistance of surfaces (this I say with regard to the definition), because there is a similar understanding in them and in lines. Bodies are therefore equidistant which when they are in the same body do not meet even if they are produced to infinity, because, although their equidistant surfaces are in the different bodies, they are yet by another consideration in the same body, just as the equidistant lines of the equidistant surfaces and of the equidistant bodies are in the same surface. But these surfaces in some intervening body are also equidistant surfaces, which are in equidistant bodies just as in the same body, and so equidistant bodies are those in which there are the same equidistant surfaces, and lines similarly are in the same imaginable body between those bodies, of which body they are understood as the extreme parts.

(35) Beside these definitions there occur some which Euclid supposes. And others in books of common geometry make mention of the things defined, but they either do not posit the definitions or not in their own place. And one is that of a given line, on which he touches in Proposition 1 and elsewhere. But in the special edition Adelard of Bath says that, "Given are here said for those to which you can find equals in abundance, means namely between the immense and the shortest, and these means human reason the investigator has assumed for her reasoning."[32] And it is taken from Definition 1 of Euclid's *Book of Data*, which is this: "Spaces and lines and angles are said to be given in magnitude to which we can assign equals."[33]

(36) To describe a surface on a line is to make that line the side of

[32]MS Oxford, Bodleian, Digby 174, f.99v. "Data dicuntur quibus equalia habundanter invenire poterimus media scilicet inter immensa et brevissima que scilicet media humana ratio veritatis indagatrix sue rationi supposuit." Cf. Van Ryzin, 202, 255; Clagett, "Medieval Translations", 36.

[33]Ito, 54.

latus illius superficiei quod proprie basis vocatur, sicut supponit Euclides in 44 propositione Primi Libri.

(37) Diameter superficiei quadrangule dicitur linea[21] que ab angulo in angulum oppositum protrahitur, et hec dicitur linea dividens superficiem per medium. Et sic sumitur in usu communi, nam strictius et non ita vulgariter dicitur diameter in figura equedistantium laterum que, dividens ipsam figuram in duas partes equales, equedistat utrique laterum oppositorum. Et secundum hoc sunt duo diametri in tali figura. Et in tetragono longo una longior et alia brevior, super quas possumus ymaginare tetragonum revolui, sicut possumus intelligere in ovali figura et in lenticulari, in quibus non equantur omnes diametri sicut[22] in spera, similiter nec in pyramidali nec columpnari. Et hec valde consideranda sunt in figura celi et totius universi, propter quod Aristoteles vult quod non potest esse ovalis figure nec lenticularis. Sed de hiis exigitur[23] sermo perfectior.

(38) Paralelle sunt linee equedistantes. Paralellogramum quasi paralellis inscriptum, scilicet superficies equedistantium laterum, sicut est quadratum et tetragonus longus. Sed rumbus habet equedistantia latera opposita et simile rumbo similiter, et ideo quia non omnia latera habent equedistantia non sunt paralellograma. Et declinatur hoc paralellogramum huiusmodi mi, secundum quod est substantivum. Potest quod aiective declinari, ut paralellogramus ma mum, sed raro accipitur nisi in neutro genere, aliquando tamen accipitur in feminino, ut superficies paralellograma, sed in masculino non utimur; potest tamen dici quod tetragonus est paralellogramus. Et quia genitive pluralis est paralellogramorum et singularis paralellogrami secundum actores, si aliquis scribat vel dicat paralellogramata in plurali nichil est, debet enim dicere paralellograma secundum quod in omnibus antiquis invenitur. Omnis paralellogrami spatii ea quidem que diametros secat per medium paralellograma circa diametrum consistere dicuntur. Eorum vero paralellogramorum que circa diametrum consistunt quodlibet unum cum supplementis duobus gnomo vocatur. Et hec due diffinitiones necessarie sunt in 43 primi libri, quapropter nescio quare non preposuit primo libro. Et oportet diffinitionem tertiam addere, et est hec: Supplementa dicuntur superficies paralellograme quas diameter paralellogrami divisi in quatuor superficies non secat. Et hec antecedit naturaliter diffinitionem gnomonis. Et tres diffinitiones ultime oportet quod ponantur ante primum librum Geometrie, quoniam omnes exiguntur in 43 eiusdem.

(39) Cathetus vocatur linea erecta super basim *AD* ad angulos rectos, et ypothenusa vocatur illa que opponitur angulo recto, et includit superficiem trianguli. Hec tres linee in commentis inveniuntur, et ideo volui hic eas recitare.

---

[21] latus

[22] *om.*

[23] exitur

the surface which is properly called the base, as Euclid assumes in Book I, Proposition 44.

(37) A diameter of a quadrangular surface is said to be a line which is protracted from one angle to the opposite angle, and it is called a line dividing the surface through the middle. And thus it is taken in common use, but more strictly and not so commonly a diameter in a figure of equidistant sides is said to be a line which, while dividing it into two equal parts, is equidistant to each of the opposite sides. And according to this there are two diameters in such a figure. And in a long tetragon one is longer and the other shorter, on which we can imagine the tetragon to be revolved, just as we can understand in the case of an oval and a lenticular figure, in which not all the diameters are equal as in a sphere, and similarly not in a pyramidal or cylindrical figure. And this is much to be considered in the figure of the heavens and of the whole universe, on which account Aristotle holds that it cannot be of an oval or lenticular figure.[34] But a more complete discussion of these matters is required.

(38) Parallels are equidistant lines. A parallelogram is as if inscribed in parallels, namely a surface of equidistant lines, like a square or long tetragon. But a rhombus has equidistant opposite sides, and similarly a "similar to rhombus", and so, because they do not have all sides equidistant, they are not parallelograms. And it is declined, *paralellogramum mi*, as a substantive. It can be adjectivally declined as *paralellogramus ma mum*, but it is rarely found except in the neuter gender, although it is sometimes found in the feminine, like *superficies paralellograma*, but we do not use it in the masculine, although it can be said that a *tetragonus* is *paralellogramus.* And because the genitive plural is *paralellogramorum* and the singular *paralellogrami* according to the authors, if anyone should write or say *paralellogramata* in the plural, it is nothing, for one must say *paralellograma* according to what is found in all the ancients. In any parallelogramic space those parallelograms that a diameter cuts through the middle are said to stand about the diameter. Either one of the parallelograms that stand about the diameter together with the two complements is called a gnomon. And these two definitions are necessary in Proposition I.43, on which account I do not know why he did not preface them to Book I. And it is necessary to add a third definition, and it is this: The paralellogramic surface that the diameter of the parallelogram which is divided into four surfaces does not cut are called complements. And this naturally precedes the definition of the gnomon. And the three last definitions ought to be placed before Book I of *Geometry*, because they are all needed in its Proposition 43.

(39) The line erected at right angles on base *AD* is called the cathetus, and that which is opposite the right angle is called the hypotenuse, and it includes the surface of a triangle. These three lines are found in the comments, and so I wished to speak of them here.

---

[34] *De coelo* II.4, 287a19-24.

(40) Post omnes has sciendum quod figura regularis dicitur cuius latera omnia sunt equalia et anguli equales, et irregularis dicitur que non huiusmodi. Sic enim regulare et irregulare sumuntur communiter in geometria, licet superius dictum est quod irregulares dicuntur que sub regulis artis geometrie non cadunt, scilicet de quibus non determinatur in ea quare; et irregulares dicuntur, quia rationacioni artis non subiacent. ...[24] et trapezei, ut dictum est, secundum actores multos, ut Anaritius et Sambelichum commentatores[25] Euclidis. Et tamen, ut inveni apud Anaritius, Euclidis dicit in Libro Divisionum, Non[26] nominamus figuras quadrilateras trapezeias[27] nisi illas quarum duo latera que sibi oponuntur equantur. Alia sunt similia trapezeis.

(41) Titulus autem istius libri secundum auctores est huiusmodi: Incipit Liber Elementorum seu Radicum Mathematice Euclidis ab Euclide Picthagorico compositus, continens 464 propositiones et proposita que theoreumata vocantur et 12 porismata, preter anxiomata et petitiones et conceptiones que singulis libris premittuntur. Alpharabius enim in Libro de Scientiis dicit hunc librum esse Euclidis Pythagorici et esse librum elementorum et radicum, nam hec finita sunt, et ab eis alia ramificantur que vadunt in infinitum, ut ipse vult, finitas enim eorum patet ex numero signato.

(42) Propositiones vero, ut dicit A. Bathoniensis in editione speciali, per indicativum, proposita per infinitivum explicantur, et addit quod propositiones proponunt ut aliquid est vel non est, proposita ut aliquid est faciendum vel non faciendum. Anaritius tamen, volens hoc opus reducere ad speculationem et distinguere a geometria practica, dicit quod demonstrationes geometrie speculative vel sunt ad operandum, ut prima primi, cum dicit, Triangulum equilaterum super datam lineam collocare, in cuius fine et consimilium dicimus, Et hoc est quod operari voluimus; vel sunt ad inveniendum, ut prima tertii, Circuli propositi centrum invenire, in cuius fine et consimilium dicimus, Et hoc est quod invenire voluimus; vel sunt ad ostendendum et sciendum, sine inventione et opere, ut omnes alie adiectis (?), in quarum fine dicimus, Et hoc est quod scire voluimus vel demonstrare vel ostendere. Et tamen, ut ipse determinat, inventio et operatio non faciunt nisi scientiam quam geometria speculativa intendit, quia non est ibi operatio manualis ut in practica, quoniam multi sciunt operari per demonstrationem triangulum equilaterum super datam lineam, qui nescirent huiusmodi triangulum facere de lingno vel lapide vel alia materia per artem geometrie practice.

[24]The scribe leaves a space here.

[25]commentatoris

[26]*om*

[27]*om*

(40) After all these it should be known that a figure is called regular of which all the sides are equal and the angles are equal, and irregular which is not like this. For thus are regular and irregular commonly received in geometry, although it was said above that those are called irregular which do not fall under the rules of the art of geometry, namely those for which the "wherefore" is not determined in geometry, and they are called irregular because they are not subject to the rationale of the art. ...and trapezia, as has been said, according to many authors, like Anaritius and Simplicius, commentators of Euclid. And yet, as I found in Anaritius,[35] Euclid says in the *Book of Divisions*, "We do not call quadrilateral figures trapezia except those of which two sides opposite to each other are equal. The others are 'similar to trapezia'."

(41) The title of the book according to the authors is like this: Here begins Euclid's Book of the Elements or Roots of Mathematics composed by Euclid the Pythagorean, containing 464 propositions and proposita, which are called theorems,[36] and 12 porisms, beside the axioms and postulates and conceptions that are prefaced to the individual books. For in the Book of Sciences Alfarabi says that this book is by Euclid the Pythagorean and is the Book of Elements and Roots, since these are finite, and others branch out from them and go off to infinity, as he holds,[37] for their finitude is evident from the designated number.

(42) Propositions, as Adelard of Bath says in his special edition, are explicated in the indicative mood, proposita in the infinitive, and he adds that propositions propose that something is or is not, proposita that something is to be done or not to be done.[38] But Anaritius, wishing to reduce this doing to theory and distinguish it from practical geometry, says that the demonstrations of theoretical geometry are either towards doing, like I.1, when he says, "To construct an equilateral triangle on a given straight line", at the end of which and similar ones we say, "And this is what we wished to do"; or they are towards finding, like III.1, "To find the centre of a given circle", at the end of which and similar ones we say, "And this is what we wished to find"; or they are for showing and knowing, without finding and doing, like all the others, at the end of which we say, "And this is what we wished to know, or to demonstrate, or to show". And yet, as he determines, finding and doing only make the knowledge that is the intention of theoretical geometry, because there is not here manual operation as in practical geometry, because many know how to produce by demonstration an equilateral triangle on a given line, who would not know how to make such a triangle from wood or stone or other material by the art of practical

[35]Tummers, 143.
[36]Cf. Clagett – Alfred, 275.
[37]Gonzalez-Palencia, 148.
[38]Clagett – Alfred, 275.

Porismata vero sunt corelaria, nam in quartodecimo libro invenitur rubrica talis, Epilogat ut prisma irroget id est corelarium inferat, sicut manifestum est ibi.

(43) Anxioma vero ut dicit A. Baconiensis in editione dignitas interpretatur, explicat enim diffinitiones rerum. Et hoc est verum specialiter sumendo anxioma et stricte, quamvis large sumendo omnia principia vocantur anxiomata, secundum quod epilogus A. Bathoniensis in fine libri supponit. Et sic utitur Aristoteles dignitate in Posterioribus. Petitio vero est, ut dicit A. Bathoniensis, quo concesso ex ypothesi nullum sequitur inconveniens. Conceptio est, ut idem dicit, que ultro occurrit humane intelligentie, in qua non est exigendum propter quid.

(44) Quod autem dicitur ab Euclide Picthagorico hanc scientiam esse compositam, cum tamen multi ante eum elaboraverunt in ea. Pulcre A. Bathoniensis in editione speciali determinat hanc dubitationem secundum sententias vulgatas auctorum. Dicit igitur: Geometria sicut relique discipline precessit prius in usu quam in arte. Cuius usus apud Egiptos inolevit, quorum limites agrorum cum Nilus supereffluens singulis annis obduceret, instituerunt quasdam certas rationes mensurandi, quibus prioris mensure agros recuperarent. Has igitur rationes philosophi retractantes et aprobantes, inde ocasione assumpta, mensurationi operam adhibuerunt. Sed propter rei difficultatem singuli citra perfectionem operis subsistentes, quidam inicialia, ut Pithagoras et Architas, alii progressiva, ut Plato et Aristoteles et Alexander, instituunt, quousque tandem Euclides confirmationem et ordines superaddiciens, artem propositam contexuit, et in memoriam posteris monumento signavit. Et licet hiis multi et fere omnes auctores vulgati concedent, vocantes opera eis nota, tamen si quis revolvat[28] historias usque ad tempora primitiva, inveniet in primo Antiquitatum libro quod Iosephus determinat quod a principio mundi fuit hec scientia inventa a filiis Ade ante diluvium. Et huic concedat Varro peritissimus Latinorum. Atque Boetius in principio Geometrie sue Practice et recitat rationem inventionis eius vulgatam propter Nilum, ut tactum est, sed approbat sententiam[29] Varronis, qui eam fuisse habitam ante hoc per secula longa testatur.

(45) Et licet geometria dicatur a gi Greco vocabulo quod est terra sed Latini mutant[30] i in e, sicut in multis faciunt, nam Pentecosti dicit Grecus sed

[28]renvolvat
[29]sententia
[30]imitant

geometry.[39] Porisms are corollaries, for in Book XIV this rubric is found, "It ends so that a porism may be proposed, that is a corollary inferred",[40] as is manifest there.

(43) An axiom, as Adelard of Bath says in his edition, is interpreted as a dignity, for it explicates the definitions of things.[41] And this is specially true when axiom is taken strictly, although in a wide sense all principles are called axioms, as Adelard of Bath's epilogue at the end of the book supposes. And Aristotle employs dignity thus in the *Posterior Analytics.*[42] A postulate is, as Adelard of Bath says, that which being conceded nothing inconvenient follows from the hypothesis. A conception is, as the same says, that which also occurs to the human intelligence, in which the "wherefore" is not to be demanded.[43]

(44) < A doubt: > That this science is said to have been composed by Euclid the Pythagorean, when, however, many worked at it before him. In his special edition, Adelard of Bath determines this doubt admirably according to the common opinions of the authors. He says thus: "Geometry like the other disciplines went on in use before it did in art. Its use grew up with the Egyptians. Since the flooding Nile covered over the boundaries of their fields each year, they instituted certain procedures of measurement, by which they recovered fields of the previous measure. Then philosophers, reviewing and approving these procedures, and taking thence the occasion, applied their efforts to mensuration. But because of the difficulty of the thing, individuals stopped short of perfecting the work; some, like Pythagoras and Archytas, instituted the rudiments, and others, like Plato, Aristotle and Alexander, more advanced parts, until at length Euclid by superadding confirmation and order wove together the proposed art, and marked it as a monument for the memory of posterity."[44] And although many, and almost all the common authors, grant this, citing works known to them, yet if one turns over the histories back to the earliest times, one will find that in Book I of the *Antiquities* Josephus determines that this science was discovered at the beginning of the world by the sons of Adam, before the Flood.[45] And Varro the most skilled of the Latins grants this. And Boethius, at the beginning of his *Practical Geometry*, relates the common rationale of its discovery, as given above, on account of the Nile, but he approves Varro's opinion, who testifies that it had been possessed for long ages before this.[46]

(45) And although "geometry" is so called from the Greek word *gi*, which is earth (but the Latins change *i* into *e* , as they do in many cases, for the

[39]Tummers, 154-155.

[40]I have not yet been able to identify this reference; cf. CM, 69.

[41]Clagett - Alfred, 275.

[42]E.g. I.7, 75a39-b1.

[43]Clagett - Alfred, 275.

[44]Clagett - Alfred, 273-274.

[45]Blatt, 132.

[46]Migne, LXIII, 1352-53.

Latinus abutitur dicens Pentecoste, et a metron quod est mensura, quasi de mensuratione terre dicta, tamen denominationes non sunt semper a digniori nec a priori secundum naturam, sed a notiori nobis et a magis vulgato, et ideo apud vulgum dicitur geometria, quia actus mensurandi est magis in usu vulgi quam aliquod aliud quod pertinet ad geometriam.

(46) Solebant vero auctores in processu geometrie demonstrantes quedam asignare, sed alii plura alii pauciora, quorum sententie debent reduci ad 4, que sunt propositio, exemplum, dispositio, ratio. Propositio ut dicit Adelardus est theoreumatis expositio, quantum ad vocabula que indigent declaratione. Exemplum, ut dicit, est alicuius figure suppositio propter theoreumatis plenam evidentiam. Et aliquotiens ponuntur exempla de numeris propter maiorem evidentiam, ut in secundo Euclidis et aliis locis, et hec referuntur ad theoreumata secundum se ante demonstrationem, et sicut melius est multotiens dari intellectum theoreumatis quam per difficilem et longam et inutilem demonstrationem, propter causas superius anotatas. Deinde, quando oportet hanelari ad demonstrationem, prima est dispositio, que est linearum et figurarum ordinatio, prout competit demonstratori. Qua facta, adesse debet ratio sive argumentum demonstrativum quod iuxta formam dispositionis convenienter per principia elicit conclusionem. Et sic perficitur intentio demonstrantis. Si vero aliquando plura inveniantur poni inter hec, ut falsigrafie vel instantie dissolutio, patet quod sub ratione comprehenditur, nam ratio potest esse ostensiva vel ad impossibile, et illa que est ad impossibile est dissolutio instantie.

(47) De figuris vero superficialibus rectilineis prius determinavit Euclides quam de circulis, nam ut dicit Anaritius et certum est quod rectilinee notiores et certiores sunt, eo quod cum aliquis circulos metiri voluerit, non potest eos mensurare nisi cum figuris rectilineis, eo quod proportio unius circuli ad alium est sicut proportio quadrati unius diametrorum ad quadratum diametri alterius, sicut ex 12 Euclidis patet.

(48) Datis diffinitionibus subiunguntur petitiones. Et prima, propter irregularitatem seu tortuositatem vel difformitatem materie corporalis in hiis inferioribus, non potest ad opus reduci per hominem, aut vix et cum magna difficultate. Possibile tamen est nature operanti, ut in multiplicatione virtutis et speciei in rebus mundi, ut in diffusione lucis et radiorum, que multiplicius fit per lineas rectas in corpore uno, et etiam perpendiculares super prima corpora fiunt recte. Geometer igitur considerat vias possibiles nature, quia

Greek says *Pentecosti*, but the Latin abuses this, saying *Pentecoste*), and from *metron*, which is measure, as if called from mensuration of the earth, yet denominations are not always from the more worthy or from what is prior according to nature, but from what is more known to us and from the more common, and so with the crowd it is called geometry, because the act of measuring is more in the use of the crowd than anything else that pertains to geometry.

(46) Authors demonstrating in the course of geometry used to assign certain things, but some gave more and others less, and their meanings ought to be reduced to four, which are proposition, example, disposition, reasoning. Proposition, as Adelard says, is the exposition of a theorem, as regards the words that need explanation.[47] An example, as he says, is the supposition of a figure for the full clarity of the theorem. And sometimes examples are given about numbers for greater clarity, as in Book II of Euclid and in other places, and these are referred to theorems in themselves and before the demonstration, and just as it is often better for giving the understanding of the theorem than by a difficult and long and useless demonstration, on account of the causes noted above. Then, when it is necessary to pant after a demonstration, disposition is first, which is the ordering of lines and figures, so far as concerns the demonstrator. This being done, reasoning, or demonstrative argument, must attest that according to the form of the disposition it elicits the conclusion conformably from the principles. And thus the intention of the demonstrator is accomplished. But if sometimes more be found to be posited among these, such as the dissolution of a false position or of an instance, it is evident that it is comprehended under reasoning, for reasoning can be either ostensive or to the impossible, and that which is to the impossible is the dissolution of an instance.

(47) Euclid determined about superficial rectilinear figures before circles, for, as Anaritius says,[48] it is also certain that rectilinear figures are better known and more certain, since, if one wishes to measure circles, one can only measure them with rectilinear figures, since the ratio of one circle to another is as the ratio of the square of one of the diameters to the square of the other diameter, as is evident from Book XII of Euclid.

(48) The definitions being given, the postulates are subjoined. And on account of the irregularity or tortuosity or difformity of corporeal matter in these inferior things, the first cannot be reduced by man to operation, or scarcely and with great difficulty. But it is possible for operative nature, as in the multiplication of virtue and species in the things of the world, as in the diffusion of light and rays, which is made multiplicatively by straight lines in a single body, and also the perpendiculars to the first bodies are made in a straight fashion. The geometer therefore considers the possible paths of

[47]Clagett – Alfred, 274.

[48]Tummers, 138.

propter opera nature certificanda constituta est[31] geometria primo et per se, deinde propter opera humana. Auctores enim perspective nobis ostendunt quod linee et figure declarant nobis totam operationem nature, et principia et effectus. Et similiter patet per celestia, de quibus naturalis et astronomia communicant. Geometer igitur non curat de materia sensibili tortuosa, sed intelligit naturam regularem sicut est in celestibus et sicut in hiis inferioribus natura scit invenire in suis operationibus, et vias nature imitatur. Et sic non erat in imaginatione linearum rectarum, ut Aristoteles dicit in Posterioribus, quod geometer non loquitur de linea sensibili sed de ea que per eam intelligitur.

(49) Petitiones sunt V. A quolibet puncto in quemlibet punctum rectam lineam ducere, atque lineam diffinitam in continuum rectumque quantumlibet protrahere. Item: Super centrum quodlibet quantumlibet occupando spatium circulum designare. Item: Omnes rectos angulos sibi invicem esse equales.

(50) Anaritius et omnes antiqui commentatores Arabici et Greci volunt quod hec tertia petitio est demonstrabilis, nam de angulis rectis coniunctis que sunt A, B, patet per diffinitionem perpendicularis. Item alio modo: Si A est maior, separetur differentia eius ad B per lineam CD. Ergo residuum ipsius A equatur B, quare acutum equatur recto contra diffinitionem acuti anguli. Item CBD angulus est obtusus, sed equatur toti A Ergo obtusus equatur recto contra diffinitionem obtusi. Sic leviter demonstrari potest, sed efficacius demonstrabitur inferius post 32^am^, quia ex ea et 5 potest manifesta demonstratio huius petitionis. De angulis vero divisis Anaritius dat demonstrationem, ut sit unus EXH, alius cadat[32] super EX. Si igitur angulus ABC sit minor, et ponatur ille super alium ut linea BC, tunc linea BA non cadit super HX, quare declinabit. Sit ergo eius declinatio sicut est LXH (?). Protrahatur igitur HX usque C in continuum et directum. Tunc angulus CXE est equalis angulo HXE, sicut prius demonstratum est, quare erit maior angulo LXE sive ABC. Protrahatur igitur LX in continuum et directum usque K. Tunc cum angulus EXL sit rectus, et ei coniungitur angulus KXE, erit rectus, per diffinitionem perpendicularis, et tunc erunt equales per demonstrationem dictam et per diffinitionem perpendicularis. Sed CXE[33] est maior EXL. Ergo est maior quam EXK, pars scilicet suo toto. Considerandum vero quod eadem ratio est de divisis sicut de coniunctis, nam coniungi possunt, et tunc prime demonstrationes et illa que inferius dicetur communes sunt angulis[34] divisis et coniunctis.

(51) 4. Si linea recta super duas rectas ceciderit, duoque anguli ex una parte duobus angulis rectis minores fuerunt, illas duas rectas in eam partem

---

[31] ex

[32] cadas

[33] XCE

[34] angulus dimissis et contentis

nature, because geometry was first and essentially constituted for the the sake of certifying the works of nature, and thereafter for human works. For the authors of perspective show us that lines and figures declare to us the whole operation of nature, its principles and effects. And this is similarly evident by celestial things, which both natural philosophy and astronomy consider. The geometer therefore does not attend to tortuous sensible matter, but he understands regular nature as it is in celestials and as nature knows how to find in its operations in these inferiors, and he imitates the ways of nature. And thus it was not in the imagination of straight lines, as Aristotle says in the *Posterior Analytics*,[49] that the geometer does not speak of the sensible line but of what is understood through it.

(49) The postulates are five. To draw a straight line from any point to any point, and to produce a finite line continuously and in a straight line so far as is wished. Again: To describe a circle on any centre and occupying any amount of space. Again: All right angles to be mutually equal.

(50) Anaritius and all ancient commentators both Arabic and Greek hold that this third postulate is demonstrable,[50] for with the conjoined right angles A, B, it is evident by the definition of perpendicular. Again in another way: If A is greater, let its difference to B be separated by the line CD. Then the residue of A is equal to B, wherefore an acute angle is equal to a right angle, against the definition of acute angle. Again, angle CBD is obtuse, but it is equal to the whole of A. Therefore an obtuse angle is equal to a right angle, against the definition of obtuse. It can thus be demonstrated easily, but it will be demonstrated more effectively below after Proposition 32, because from it and Proposition 5 the demonstration of this postulate can be manifest. For divided angles Anaritius gives a demonstration, so that let one be EXH, and another falls on EX. If therefore angle ABC is less, and it is placed on the other as line BC, then line BA does not fall on HX, wherefore it will decline. Therefore let its declination be as LXH. Therefore let HX be produced comntinuously and in a straight line to C. Then angle CXE is equal to angle HXE, as was demonstrated earlier, wherefore it will be greater than angle LXE or ABC. Therefore let LX be produced continuously and in a straight line to K. Then since angle EXL is right, and there is joined to it angle KXE, it too will be right, by the definition of perpendicular, and then they will be equal, by the said demonstration and the definition of perpendicular. But CXE is greater than EXL. Therefore it is greater than EXK, the part to its whole. But it should be considered that there is the same rationale for divided as for conjoined, for they can be conjoined, and then the first demonstrations and that which will be given below are common to divided and to conjoined angles.

(51) 4. If a straight line fall on two straight lines, and the two angles on one side be less than two right angles, those two straight lines produced on

[49]I.10, 76b40-77a3.
[50]Tummers, 150-1.

protractas procul dubio coniuncti iri. Hec quarta est demonstrabilis, sicut docent Ptolomeus, Aganius, Anaritius, et Sambelichius, et ideo non voluerunt uti ea in demonstratione 29[e], ad quam Euclides adduxit eam, sed aliter probaverunt eam. Et hec patebunt suo loco, nam hic non, quia oportet uti 15[a] et 16[a] et quibusdam aliis ad probationem istius petitionis.

(52) Quinta. Inter duas lineas rectas superficiem nullam concludere. Hec est contra quod prius dictum est rectangulum contineri duabus lineis, quia aliud est contineri, aliud concludi. Quid vero sic contineri prius dictum est, et quid sic concludi figuram manifestum est. Sed hanc Anaritius demonstrat sic: Si falsigraphicus[35] dicat duas lineas rectas superficiem includere, sit igitur *AD* una et *AGD*[36] alia. Describatur igitur circulus super *AD*, et perficiatur *AD* diameter usque *B*; similiter *AGD*[37] trahatur ad circumferentiam usque *E*. Quare *AGE* erit diameter sicut *ADB*. Per diffinitionem diametri separant semicirculos et partes arcus equales. Ergo *AE* arcus equabitur *AB* arcui, pars toti, quod est impossibile.

(53) Quapropter hec tres ultime petitiones sunt conclusiones secundum veritatem, sed quia propinqua sunt veris principiis et possunt in multas conclusiones, ideo Euclides reposuit ea inter principia.

(54) Conceptiones proprie in geometria pauce sunt, quoniam fere omnes sunt communes quantitati continue et discrete, et ideo prius posite sunt in tractatu de communibus totius mathematice. Due tamen hic poni possunt: Si aliqua quantitas continua alii supponatur appliceturque ei, nec excedat altera altera, ille sibi[38] invicem erunt equales. Hec est Euclidis, et propter suppositionem et applicationem unius ad alterum, patet quod pertinet tantum ad continua non discreta, quoniam conceptio de equalitate conveniens discretis et continuis prius habita est in communibus, et est prima diffinitio ibi assignata. Et A. Bathonensis in commento addit aliam, que est hec: Quanta est aliqua quantitas ad aliam tantam esse quamlibet tertiam ad aliquam quartam. In quantitatibus quidem continuis hoc observandum est sive antecedentes sint maiores seu minores, nam magnitudo decrescit in infinitum per divisionem, ut allegat, et potest secundum voluntatem geometre imaginari crescere in infinitum tum quantum vult, sicut Aristoteles docet tertio Phisicorum.

---

[35]falsi convenit phicus

[36]*AGA*

[37]*AGA*

[38]*Notes on f.78br:* "unus est quantitas discreta permanens nam numerus est quantitas discreta vero permanens" "universis presentes litteras inscriptitur Guido"

that side without doubt to meet. This fourth is demonstrable, as Ptolemy, Agapius, Anaritius and Simplicius teach, and so they did not wish to use it in the demonstration of Proposition 29, for which Euclid adduced it, but they proved it in another way.[51] And these things will be evident in their place, but not here, because it is necessary to use Propositions 15 and 16 and some others for the proof of this postulate.

(52) Fifth. Between two straight lines to enclose no surface. This is against what was said earlier for a rectangle to be contained by two lines, because it is one thing to be contained, another to be enclosed. What it is to be thus contained was said earlier, and what it is for a figure to be thus enclosed is manifest. But Anaritius demonstrates it thus:[52] If a pseudographer should say that two straight lines enclose a surface, then let AD be one and AGD the other. Then let a circle be described on AD, and let AD be completed as a diameter so far as B; similarly let AGD be drawn to the circumference so far as E. Then AGE will be a diameter just like ADB. By the definition of diameter they separate semicircles and equal parts of arc. Therefore arc AE will be equal to arc AB, the part to the whole, which is impossible.

(53) Thus these three last postulates are conclusions according to truth, but, because they are near to true principles and are potent in many conclusions, on that account Euclid placed them among the principles.

(54) Conceptions proper to geometry are few, because almost all are common to continuous and to discrete quantity, and so were placed earlier, in the treatise on the common things of the whole of mathematics. But two can be placed here: If one continuous quantity be superposed to another and applied to it, and neither exceed the other, they will be mutually equal. This is Euclid's, and on account of the superposition and application of one to the other, it is evident that it belongs only to continua and not to discreta, because the conception of equality appropriate to discreta and continua was had earlier in the *Communia*, and was the first definition assigned there. And Adelard of Bath in a comment adds another, which is this:[53] "As much as one quantity is to another, so much any third to be to some fourth. This is to be observed in continuous quantities, whether the antecedents be greater or less, for magnitude decreases infinitely" by division, as he alleges, and by the geometer's will it can also be imagined to increase infinitely so far as he wishes, as Aristotle teaches in *Physics* III.[54]

---

[51] Cf. Tummers, 151, 173-180.

[52] Tummers, 151.

[53] Van Ryzin, 201, from Adelard III.

[54] See the long discussion in *Physics* III, 4-8.

"Geometric trigonometry" in treatises of al-Khwārizmī, al-Māhānī and Ibn al-Haytham

B.A. Rosenfeld

Many medieval Arabic mathematicians used rules equivalent to theorems of spherical trigonometry in their astronomical and geographical treatises. For instance, following Indian astronomers, Muḥammad ibn Mūsā al-Khwārizmī (ca. 780-ca. 850) formulated the rules

$$\sin\theta = \frac{\sin\delta}{\cos\phi} \tag{1}$$

and

$$\cos\alpha = \frac{\sin\theta - \tan\phi\sin h}{\cos h} \tag{2}$$

in his treatises *Determination of the ortive amplitude in each town* (Maʿrifa siʿa al-mashriq fī kull balad) and *Determination of the azimuth according to the altitude* (Maʿrifa samt min qibal al-irtifāʿ). The rule (1) determines the *ortive amplitude* θ (the arc of the horizon between its Eastern point E – the point of its intersection with the celestial equator – and the point G of the sunrise on the given day) in terms of the declination δ of the sun and the latitude ϕ of the town. The rule (2) determines the azimuth α of the sun in terms of the ortive amplitude θ, the altitude h of the sun and the latitude ϕ of the locality [1, p. 15]. The rule (1) is equivalent to the *spherical sine theorem*

$$\frac{\sin A}{\sin a} = \frac{\sin B}{\sin b} = \frac{\sin C}{\sin c} \tag{3}$$

for the rectangular triangle EFG, where F is the base of the perpendicular dropped from G onto the celestial equator; if θ in rule (2) is expressed according to rule (1) in terms of δ and ϕ, (2) is equivalent to the *spherical cosine theorem*

$$\cos a = \cos b\cos c + \sin b\sin c\cos A \tag{4}$$

for the spherical triangle NSZ where N is the celestial North pole, S is the sun and Z is the zenith [1, pp. 11-13].[1]

In the treatise *Geometric construction of the ortive amplitude for every (zodiacal) sign and for every latitude* (ʿAmal siʿa ayy mashriq shiʾta min al-burūj fī ayy arḍ shiʾta biʾl-handasa) al-Khwārizmī gives the geometric construction for finding the ortive amplitude according to the rule (1), and thus he founds a new direction in spherical trigonometry which by analogy with Euclid's *geometric algebra* can be called *geometric trigonometry*.

---

[1] $A = \alpha,\ b = 90^0 - \phi,\ c = 90^0 - h,\ a = 90^0 - \delta.$

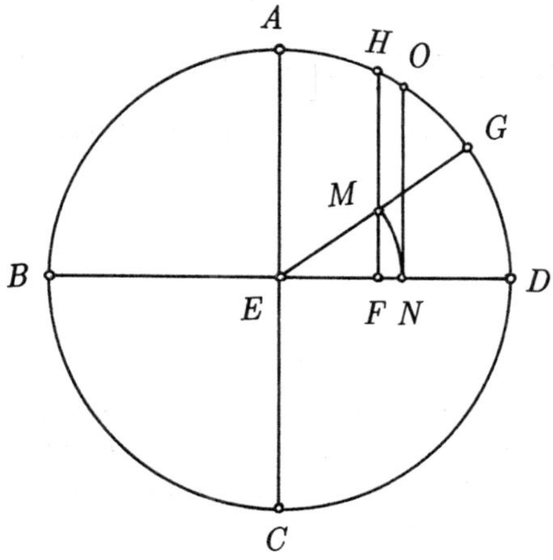

Figure 1

In this treatise al-Khwārizmī writes:

> Draw a circle $ABCD$ (Figure 1) and divide its arc AD into 90 parts. From the point D on the circle lay off the arc DG having the magnitude of the latitude of (your) town and draw a line from this (point) G to the centre (of the circle), that is the line EG. Then from the point A on the circle lay off the arc AH having the magnitude of the declination of that zodiacal sign for which you want (to make) the construction. From (the end of) this (arc) draw a line parallel to the line AE. It is the line HF, which reaches the line BD which is a diameter. Then look where it cuts the line EG; let it be the point M. Then take the pair of compasses with opening ME, put its leg in the centre point, that is in the point E, and find the place where the other leg cuts the line DE. Let this be the point N. From the point N raise the perpendicular reaching the arc AG, it is the perpendicular ON. From (the point O) where (the perpendicular) reaches count out (the arc) to the point A. The thing you get is the ortive amplitude for the sign you wanted. [2, p. 180; 3, pp. 190-191]

In his *Book on the determination of the azimuth at any time and in any place* (Maqāla fī maʿrifa al-samt li-ayy sāʿatin aradta wa-fī ayy mawḍiʿ aradta) Muḥammad al-Māhānī (ca. 825-885) gives an analogous construction for finding the azimuth α of the sun according to the rule (2) [4, p. 501].

In the lost *Book on the azimuth of the qibla by means of calculation* (Maqāla fī samt al-qibla bi'l-ḥisāb), which is quoted in the *Short book on the azimuth of the qibla* (Maqāla mukhtaṣara fī samt al-qibla), Abū ʿAlī ibn al-Haytham (965 – ca. 1040) found the rule for the determination of the azimuth of the qibla – that is the angle α between the direction from a given town to Mecca and the meridian of the given town – from the longitudes λ_1 and λ_2 and the latitudes ϕ_1 and ϕ_2 of the given town and Mecca. In the second treatise he gives the geometrical construction of this rule. The rule of Ibn al-Haytham is

$$\cot \alpha = \left| \frac{\sin \phi_1 \cos \lambda - \cos \phi_1 \cot \phi_2}{\sin \lambda} \right| \tag{5}$$

where $\lambda = |\lambda_2 - \lambda_1|$. This rule can be obtained by means of theorems (3) and (4) for the spherical triangle MNT where M is Mecca, N is the terrestrial North pole and T is the given town; by means of these rules arc MT is expressed in ϕ_1, ϕ_2, λ and α and this arc is eliminated by means of the relation $\cos^2 MT + \sin^2 MT = 1$.

Ibn al-Haytham formulates his geometrical construction as follows:

> If you want to determine the azimuth of the qibla according to the meridian line in any town with Northern latitude, take a copper plate or a wooden board, hammer out on it a circle and divide the circumference of this circle into 360 equal parts. Let it be the circle $ABCD$ with the centre E (Figure 2). Through this point draw two diameters AC and BD, intersecting at right angles. From the end of one of these diameters lay off on the circumference an arc equal to the latitude of the town, let it be the arc CG. Draw the radius EG. From the other end of the same diameter lay off on the circumference the latitude of Mecca, let it be the arc AH (and draw the radius EH). Draw from the point H a line parallel to the radius BE, it is perpendicular to the diameter AC, let it be HF. Lay off on the radius EG from the point E a line equal to HF, let it be EK. In the point K raise a perpendicular to EK, it will intersect EB in L. Lay off on the circumference of the circle from the point A an arc equal to the difference of the longitudes of Mecca and of the given town in which there is the azimuth (of the qibla), if these longitudes are different. The arc of the question could be equal to AN. The point N can be between the points A and B, in the point B itself or between the points B and C. In every case draw the radius NE and around the centre E the circle with distance[2] EF until this arc will intersect the radius NE in the point O. Draw through the point O a line parallel to BE, let it be the line OU. If the point N is between A and B, take the lines UE and KL and if KL is smaller than UE extend KL on the side of L in its direction and cut off a line equal to UE, and if KL is bigger than UE lay off on it (a line equal to) UE and let the obtained line be KM. Then draw through M a line parallel to AC, which will meet BE in Q. Then cut off from UC a line equal to MQ, it will be UX. Draw also XO, then the angle AXO is equal to the angle of the azimuth (of the qibla). [5, pp. 244-245]

[2]i.e. radius

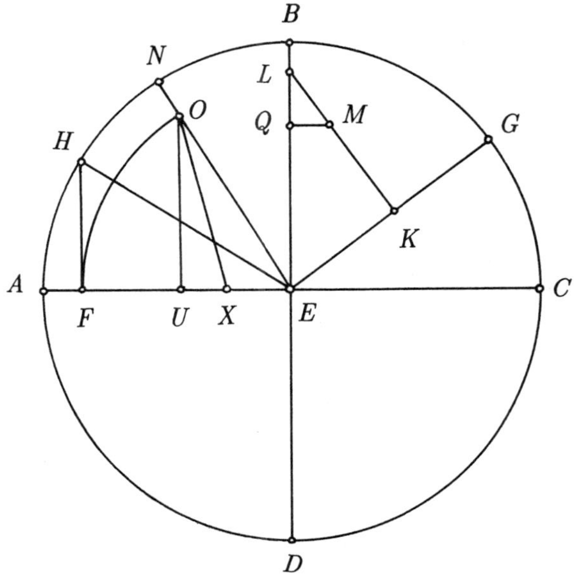

Figure 2

Then Ibn al-Haytham considers the case where $KL = UE$ and the azimuth of the qibla is 90°, and the case where N lies between the points B and C and U lies between E and C.

The angles $\angle CEG$ and $\angle AEH$ are equal to the latitudes ϕ_1 and ϕ_2 of the given town and Mecca respectively, and $\angle AEN$ is equal to the difference $\lambda = |\lambda_2 - \lambda_1|$ between the longitudes of these towns. Therefore, if the radius of the circle $ABCD$ is 1, the segments EF and $FH = EK$ are equal to $\cos\phi_2$ and $\sin\phi_2$ respectively, the segments $EU = KM$ and OU are equal to $\cos\phi_2\cos\lambda$ and $\cos\phi_2\sin\lambda$ respectively, $KL = EK\cot\phi_1 = \sin\phi_2\cot\phi_1$. Hence $ML = |KM - KL| = |\cos\phi_2\cos\lambda - \sin\phi_2\cot\phi_1|$. It follows by the similarity of triangles LMQ and LEK that

$$UX = MQ = ML\sin\phi_1 = |\cos\phi_2\sin\phi_1\cos\lambda - \sin\phi_2\cos\phi_1|,$$

and

$$\cot\alpha = \frac{UX}{UO} = \left|\frac{\cos\phi_2\sin\phi_1\cos\lambda - \sin\phi_2\cos\phi_1}{\cos\phi_2\sin\lambda}\right|.$$

References

1. Rosenfeld, B.A., *A history of Non-Euclidean geometry.* New York etc. (Springer) 1988.

2. Ahmedov, A.A., ad-Dabbagh, J., Rosenfeld, B.A., Istanbul manuscripts of al-Khwārizmī's treatises. *Erdem* **7** (1987) pp. 153-186.

3. Ahmedov, A., Traktaty al-Khorezmi ob opredelenii azimutov i vremenii. *Materialy po istorii i istorii nauki i kultury narodov Srednei Azii*, Tashkent, "Fan", 1991, pp. 155-200.

4. Luckey, P., Beiträge zur Erforschung der islamischen Mathematik. *Orientalia* N.S. **17** (1948), pp. 490-510, 22 (1953), 166-189.

5. Schoy, C., Abhandlung des al-Ḥasan ibn al-Ḥasan ibn al-Haiṯam (Alhazen) über die Bestimmung der Richtung der Qibla. *Zeitschrift der Deutschen Morgenländischen Gesellschaft* **75** (1921), pp. 242-253.

Zur Aufgabe 86 des Byzantinischen Rechenbuchs Cod. Vindob. Phil. Gr. 65

Christoph J. Scriba

1 Einleitung

Unter den hundert von Herbert Hunger und Kurt Vogel 1963 veröffentlichten Aufgaben aus einem byzantinischen Rechenbuch des 15. Jahrhunderts[1] befindet sich als Aufgabe 86 auch ein Resteproblem, eingekleidet in die Erzählung der Eierverkäuferin auf dem Markt, deren Eier durch ein Mißgeschick zerbrochen werden. Die Anzahl der Eier soll aus den Bedingungen

$$\left.\begin{array}{lll} N & \equiv & 1 \quad (\mathrm{mod}\, 2,\, 3,\, 4,\, 5 \text{ und } 6) \\ N & \equiv & 0 \quad (\mathrm{mod}\, 7) \end{array}\right\} \qquad (1)$$

ermittelt werden.

Das Problem selbst ist wesentlich älter. Band I der 4. Auflage des Tropfke[2] enthält eine Zusammenstellung über sein Vorkommen. Danach ist es erstmals bei Bhāskara I (um 600 n. Chr.) nachgewiesen, außerdem bei Ibn al-Haiṯam (965?-1039), im Abendland in vielen Werken seit Leonardo von Pisa (1180?-1250?) und eben auch in Byzanz. Die bemerkenswerte Tatsache, daß genau die gleichen obengenannten Bedingungen bei Bhāskara I, Ibn al-Haiṯam, Leonardo sowie im *Algorismus Ratisbonensis* (um 1450), bei Tartaglia (1500?-1557) und bei Bachet de Méziriac (1581-1638) auftreten (während in anderen Schriften andere Restebedingungen genannt sind), läßt auf eine Wanderung des Problems durch die Kulturen schließen.

Das bedeutet jedoch nicht unbedingt, daß auch die Lösungsverfahren das Problem auf seiner Reise begleitet haben. Darauf deutet zum Beispiel die Tatsache hin, daß die aufgeführten Autoren häufig nur eine oder einige wenige spezielle Lösungen angeben und nicht die allgemeine, ja nicht einmal immer die kleinste (siehe die Übersicht in [Tropfke 1980]).

Hier soll das eigenwillige Lösungsverfahren analysiert werden, das der Verfasser des *Byzantinischen Rechenbuches* aus dem 15. Jahrhundert in so unbeholfener Weise beschrieb, daß die Herausgeber desselben es für unsinnig hielten.

[1] Hunger, Herbert und Vogel, Kurt: *Ein byzantinisches Rechenbuch des 15. Jahrhunderts. 100 Aufgaben aus dem Codex Vindobonensis Phil. Gr. 65. Text, Übersetzung und Kommentar.* Österreichische Akademie der Wissenschaften, Philosophisch-Historische Klasse, Denkschriften, 78. Band, 2. Abhandlung. Wien 1963. — Zitiert als [Hunger/Vogel 1963].

[2] [Tropfke 1980], S. 640-641.

2 Die Eierfrau im Byzantinischen Rechenbuch

In der Übersetzung der Editoren lautet die Formulierung der Aufgabe 86:[3]

> Eine alte Frau verkaufte Eier auf dem Markt, und da kamen zufällig zwei Männer daher und stießen die Frau an und zerbrachen ihre Eier. Sie führte sie vor den Richter, und das Urteil bestimmte, daß sie der Frau ihre Eier ersetzen sollten. Und sie fragen die Alte, wieviel Eier es waren, um sie zu bezahlen, und die Frau sagt: Ich weiß es nicht; ich weiß nur das, daß ich zu zwei ⟨und⟩ zwei abgezählt habe, und es blieb eines übrig, dann ⟨zu⟩ drei ⟨und⟩ drei, und es blieb 1, ⟨zu⟩ 4 ⟨und⟩ 4, und es blieb 1, ⟨zu⟩ 5 ⟨und⟩ 5, und es blieb 1, ⟨zu⟩ 6 ⟨und⟩ 6, und es blieb 1, ⟨zu⟩ 7 ⟨und⟩ 7, und es blieb nichts, nicht ein einziges. Und so viele Eier der alten Frau waren es. Ich frage dich, wie viele es waren.

Zum anschließend an diese Aufgabenstellung beschriebenen Verfahren bemerken die Herausgeber in ihrer Fußnote 1:

> Für diese Abzählungsaufgabe hat sich der Rechner eine Formel ausgedacht, die durch nichts begründet ist, nämlich $(7-1)\cdot(7\cdot 7)+1$, die zufällig den richtigen Wert ergibt.[4]

Nun ist auffällig, daß in dieser Aufgabe zweimal (sowohl vor wie nach der vorgenommenen Berechnung) betont wird, derartige Rechnungen würden zielbewußt und nicht mit viel Theorie bzw. mit einer Methode durchgeführt. Daher stellt sich die Frage, ob das beschriebene Vorgehen wirklich nur zufällig zum richtigen Wert führt oder ob sich nicht doch sinnvolle Überlegungen hinter den angegebenen Rechenschritten entdecken lassen. Im folgenden wird ein Erklärungsvorschlag gemacht, der der deutschen Übersetzung der Anweisungen des Schreibers folgt. Zunächst der Text der Rechenanweisung:[5]

> Merke, die derartigen Rechnungen werden zielbewußt und nicht mit viel Theorie gemacht. Und ich sage, daß die Zahl der Eier der Frau 301 war. Und wenn du eine solche Rechnung machen willst, sprich so: Da man ⟨zu⟩ 7 ⟨und⟩ 7 abzählte und nicht ein einziges übrigblieb, multipliziere 7 mit den 7, und es wird 49, und du behalte 50. Und da, sooft du eine ein⟨stellige Zahl⟩ mit der anderen multiplizierst, keine Zehnerzahl herauskommt, so erreichst du sie ⟨künstlich⟩, und es wurden 50. Nun multipliziere mit der anderen Zahl, die hinter der 7 kommt, und diese ist 6, und sprich: 6 mal 50 wird 300; nimm noch die 1 dazu, die erste, die du vernachlässigst*, und es gibt 301. Und soviel Eier der

[3] [Hunger/Vogel 1963], S. 73.
[4] Die Formel sollte lauten $(7-1)\cdot[(7\cdot 7)+1]+1$. [CJS]
[5] [Hunger/Vogel 1963], S. 73.
* Vielleicht: die beim Abzählen nicht vorkam. [Anmerkung 2 der Herausgeber]

alten Frau waren es.

```
      2
,3    0   1  ]  43
     ,7   7
```

Zum Vergleich sei hier der Text der entsprechenden Aufgabe (Nr. 349) aus dem *Algorismus Ratisbonensis*[6] zitiert:

> Item mulier vadens ad forum portans oua, obuians sibi rusticus trudens eam et casu fracta sunt oua. Mulier querit rusticum coram iudice. Qui dicit, se velle soluere oua. Iudex querit, quot fuerunt oua. Mulier simplex respondit: primo. quando conputaui oua per 2, tunc omnibus sic conputatis remansit vnum. Secundo quando conputavi per 3, item remansit 1, tercio per 4 iterum remansit 1 in superfluo et sic de alijs. Sed quando conputaui per 7, tunc fuerunt equalia et nichil remansit in superfluo. Queritur ergo quot fuerunt oua. Respondetur: Si vis scire, multiplica illos numeros, in quibus fuit imparitas (!) ut 2, 3, 4, 5, 6 et habebis 720, quibus adde 1 et habebis numerium ouorum. Et multiplica 2 ter et sunt 6, multiplica 4 [per 6] et sunt 24, multiplica quinquies et 6 etc.

Man erkennt die Ähnlichkeit der Aufgabenstellung, in der die gleichen Zahlenwerte wie im *Byzantinischen Rechenbuch* verwendet werden, sieht aber auch, daß die Lösungsvorschrift mit der dortigen nichts gemein hat.

Wie könnte der Byzantiner überlegt haben? Da die gesuchte Zahl N durch 7 teilbar sein soll, muß $N = 7n$ gelten. Nun könnte man (im Stil eines „falschen Ansatzes") probeweise einmal mit $n_1 = 7$ beginnen:

$$N_1 = 7 \cdot n_1 = 7 \cdot 7 \equiv 0 \,(\mathrm{mod}\, 7)\,.$$

Dann wird

$$\begin{aligned} N_1 + 1 &= 50 \equiv 0 \,(\mathrm{mod}\, 2) \quad \text{und} \quad \equiv 0 \,(\mathrm{mod}\, 5)\,, \\ \text{folglich} \quad 6(N_1 + 1) &= 300 \equiv 0 \,(\mathrm{mod}\, 60)\,. \end{aligned}$$

Durch Addition von 1 bekommt man somit eine Zahl, die alle Bedingungen erfüllt (weil 60 das kleinste gemeinsame Vielfache ist von 2, 3, 4, 5, 6):

$$6(N_1 + 1) + 1 = 6 \cdot (7 \cdot 7 + 1) + 1 = 7 \cdot (6 \cdot 7 + 1) = 7 \cdot 43 = 301\,. \qquad (2)$$

Die zum Schluß im Text stehende Division von 301 durch 7 (siehe oben) wäre dann als Probe dafür anzusehen, ob die erhaltene Zahl wirklich ohne Rest durch 7 teilbar ist.

[6] [Vogel 1954], S. 153.

Der Satz „Und da, sooft ... wurden 50" will dann besagen: weil sich durch Multiplikation einer einstelligen Zahl mit sich selbst kein Vielfaches von 10 ergibt, addiere man zum Produkt, was noch bis zum nächsten Vielfachen von 10 fehlt (zu 49 also 1). Auch der folgende Satz der Vorschrift ist dann sinnvoll: durch Multiplikation mit der „hinter der 7"[7] kommenden 6 wird eine Zahl erzeugt, der an einem Vielfachen von 7 noch eine 1 fehlt – zugleich jene 1, „die du vernachlässigst". Damit ist wohl nicht eine 1 gemeint, „die beim Abzählen nicht vorkam", wie in der Fußnote vermutet wird, sondern jener Rest 1, der beim Abzählen zu je 2, 3, 4, 5 und 6 übrig blieb und bisher in der Rechnung noch nicht berücksichtigt wurde.

Das beschriebene Verfahren läuft mit jeder auf 7 endenden Versuchszahl n_1 in analoger Weise ab (für $n_1 = 17$ erhält man z.B. $721 = 301 + 7 \cdot 60$, d.h. die nächstgrößere Lösung, und jede weitere Vermehrung von n_1 um 10 vergrößert das Ergebnis um 420). Es kann auch „zielbewußt" für beliebige andere Versuchszahlen variiert werden. So würde man etwa für $n_2 = 4$ wie folgt vorzugehen haben:

$$\begin{aligned} & N_2 = 7 \cdot n_2 = 7 \cdot 4 = 28 \equiv 0 \pmod 7 , \\ & N_2 + 2 = 30 \equiv 0 \pmod 6 \ \text{ und} \equiv 0 \pmod 5 , \\ & 2 \cdot (N_2 + 2) = 60 \equiv 0 \pmod{60} , \\ \text{aber } & 2 \cdot (N_2 + 2) + 1 = (2N_2 + 4) + 1 \not\equiv 0 \pmod 7 . \end{aligned}$$

Es gilt nun, dasjenige kleinste Vielfache von $2N_2 + 4 = 60$ zu suchen, das, vermehrt um 1, $\equiv 0 \pmod 7$ wird. Dies ist 5, woraus folgt

$$5 \cdot (2N_2 + 4) + 1 = 5 \cdot (2 \cdot 28 + 4) + 1 = 301 . \tag{3}$$

Die beiden Ansätze unterscheiden sich allerdings in ihrer Tragweite, wenn man sie auf Versuchszahlen mit gleicher Endziffer ausdehnt. Wählt man anstelle von $n_1 = 7$ den Ansatz $n_1 = 10t + 7$, so wird

$$\begin{aligned} N_1 + 1 = 7 \cdot (10t + 7) + 1 &= (7t + 5) \cdot 10 \equiv 0 \pmod{10} , \\ 6(N_1 + 1) + 1 &= (7t + 5) \cdot 60 + 1 = 301 + 420t , \end{aligned} \tag{4}$$

was für alle $t \in \mathrm{N}$ die vorgegebenen Bedingungen erfüllt und die zu (1) gehörende allgemeine Lösung darstellt.[8]

Beginnt man entsprechend mit $n_2 = 10t + 4$, so ergibt sich

$$N_2 + 2 = 7 \cdot (10t + 4) + 2 = (7t + 3) \cdot 10 \equiv 0 \pmod{10} .$$

[7] Könnte die Angabe „hinter der 7" darauf hindeuten, daß dem Schreiber die Anordnung der indisch-arabischen Ziffern in der Reihenfolge 9, 8, 7, ..., 2, 1 vor Augen steht, wie sie durch die Richtung der arabischen Schrift gegeben ist? Vgl. dazu [Curtze 1897], S. XIII und S. 2 (wo Petrus Philomenus de Dacia diese Begründung gibt) sowie [Vogel 1963], S. 9 und S. 45.

[8] Laut [Tropfke 1980], S. 641 geben Bhāskara I und der *Algorismus Ratisbonensis* die Lösung 721, ibn al-Haiṯam, Leonardo von Pisa und Chuquet (um 1500) die allgemeine Lösung.

Nun kann man verschieden fortsetzen. Vervielfacht man mit 6, so ist zwar

$$6(N_2+2) = (7t+3)\cdot 60 \equiv 0 \pmod{60},$$

doch

$$6(N_2+2)+1 = 7\cdot 60t + 181 \not\equiv 0 \pmod{7}.$$

Erst

$$24(N_2+2)+1 = 7\cdot 240t + 721 \equiv 0 \pmod{7}, \tag{5}$$

aber $721 + 1680t$ stellt nur eine Teilfolge der allgemeinen Lösung dar. Es bietet sich aber auch an, $t = 3s$ zu setzen:

$$\begin{array}{rrcllc} & N_2+2 & = & (7s+1)\cdot 30 & & , \\ & 2(N_2+2) & = & (7s+1)\cdot 60 & \equiv 0 \pmod{60} & , \\ \text{doch} & 2(N_2+2)+1 & = & 7\cdot 60s+61 & \not\equiv 0 \pmod{7} & . \\ \text{Erst} & 10(N_2+2)+1 & = & 2100s+301 & \equiv 0 \pmod{7} & . \end{array} \tag{6}$$

Wiederum ergibt sich nur eine Teilfolge, die allerdings die kleinste positive Lösung enthält.

Damit genug der Beispiele für „zielbewußtes Erfassen", das der Autor höher schätzt als das bloße Befolgen einer Methode. Denn an das zuletzt wiedergegebene, mit der Probe endende Zitat schließt er folgende Bemerkungen an:[9]

> Viele Rechnungen werden durchgeführt ohne Begründung durch eine Methode, nur mit ziel⟨bewußtem Erfassen⟩. Und feiner ist jene Rechnung des ziel⟨bewußten Erfassens⟩ gegenüber jener, die mit einer Methode geschieht. Da alle, die methodisch durchgeführt werden, Grund und Anfang der Methode in sich haben[†], werden alle die Rechnungen mit Leichtigkeit durchgeführt, die eine Grundlegung und einen Griff nach den Halben, nach den Dritteln, den Vierteln und so weiter haben. Und so steht es mit der Methode eines derartigen Rechnens.

Ich denke, das Urteil der Herausgeber über die Persönlichkeit des Schreibers müßte man etwas revidieren. Sie schrieben nämlich am Ende ihrer in der Einleitung gegebenen sehr verständnisvollen, ja fast liebevollen Charakterisierung des Schreibers:[10]

> Freilich, wo er über den Schwierigkeitsgrad von Rechenmethoden urteilen möchte, zeigen sich sofort die engen Grenzen seines Verständnisses und seiner Ausdrucksfähigkeit. In dem so selbstbewußt beginnenden Schlußteil der Aufgabe **86** verstrickt sich unser Schreiber nur zu bald in einem nicht mehr auflösbaren Galimathias.

---

[9] [Hunger/Vogel 1963], S. 73.

[†] Widerspricht dem Vorhergehenden. Das ganze ist äußerst unklar. [Anmerkung 3 der Herausgeber]

[10] [Hunger/Vogel 1963], S. 14.

Seine Ausdrucksfähigkeit mag in der Tat begrenzt sein (den griechischen Text kann ich nicht beurteilen). Doch die Intention seiner allgemeinen Bemerkungen scheint, falls die oben gegebene Deutung seiner Überlegungen zutrifft, jetzt besser verständlich zu werden: bloße Rechenschritte, bei denen nur nach Vorschrift zu halbieren, zu dritteln, zu vierteln usw. ist, lassen sich mit Leichtigkeit durchführen, doch feiner ist die Lösung von Aufgaben, bei denen es keine Standardverfahren gibt, die vielmehr nur vom Endziel her begründbare Schritte verlangen. Die durch die knappen, den Grund der vorgenommenen Rechenschritte verschweigenden Formulierungen des Autors bedingte Feststellung der Herausgeber „Das ganze ist äußerst unklar"[11] ist, falls der obige Erklärungsvorschlag akzeptabel erscheint, wohl doch zu revidieren.

Literatur

[Curtze 1897] Curtze, Maximilian (Hrsg.): *Petri Philomeni de Dacia in Algorismum Vulgarem Johannis de Sacrobosco Commentarius*. Kopenhagen 1897.

[Vogel 1954] Vogel, Kurt (Hrsg.): *Die Practica des Algorismus Ratisbonensis. Ein Rechenbuch des Benediktinerklosters St. Emmeram aus der Mitte des 15. Jahrhunderts*. München 1954.

[Hunger/Vogel 1963] Hunger, Herbert und Vogel, Kurt (Hrsg.): *Ein byzantinisches Rechenbuch des 15. Jahrhunderts. 100 Aufgaben aus dem Codex Vindobonensis Phil. Gr. 65. Text, Übersetzung und Kommentar*. Österreichische Akademie der Wissenschaften, Philosophisch-Historische Klasse, Denkschriften, 78. Band, 2. Abhandlung. Wien 1963.

[Vogel 1963] Vogel, Kurt (Hrsg.): *Mohammed ibn Musa Alchwarizmi's Algorismus. Das früheste Lehrbuch zum Rechnen mit indischen Ziffern*. Aalen 1963.

[Tropfke 1980] Tropfke, Johannes: *Geschichte der Elementarmathematik*. 4. Auflage, Band 1: Arithmetik und Algebra. Vollständig neu bearbeitet von Kurt Vogel, Karin Reich und Helmuth Gericke. Berlin, New York 1980.

[11] [Hunger/Vogel 1963], S. 73, Anm. 3.

La version latine médiévale de l'Algèbre d'Abū Kāmil

J. Sesiano

1. Introduction

Si quelqu'un devait s'enquérir du manuscrit latin 7377A de la Bibliothèque Nationale de Paris, il devrait le faire auprès de M. H. Busard: c'est grâce à lui qu'ont été mis en lumière quelques-uns des traités qui en font partie. Ledit manuscrit est en effet un recueil de textes, tous écrits par différentes mains du XIVe siècle, mais qui ont en commun le domaine scientifique et l'origine des sources: la première moitié en est formée par des traductions latines de textes mathématiques arabes, tandis que l'autre moitié contient le *Liber mahameleth* de Johannes Hispalensis, qui n'est certes pas une traduction mais s'appuie sur des traités mathématiques arabes d'Espagne ainsi que sur l'ouvrage fondamental d'algèbre du monde musulman tant oriental qu'occidental, celui qu'écrivit vers la fin du IXe siècle l'Egyptien Abū Kāmil. Or, précédant le *Liber mahameleth*, on trouve justement (fol. 71^{v}-97^{r}) une version latine de cette Algèbre d'Abū Kāmil. Elle est toutefois incomplète de la fin: alors que le traité originel compte une partie proprement algébrique, puis des applications de l'algèbre à des problèmes sur les polygones réguliers, et enfin des équations indéterminées et des problèmes de nature récréative, la traduction s'arrête abruptement peu après le début de la troisième partie. Notre étude a pour but de présenter le texte de la première partie et du fragment de la troisième, M. R. Lorch prenant soin, dans ce même ouvrage, de l'édition du texte des applications aux polygones. On se reportera de plus à son article pour des remarques philologiques sur la traduction et les références à la littérature concernant Abū Kāmil.

Outre son intérêt pour le texte d'Abū Kāmil, dont elle présente occasionnellement de meilleures leçons, la traduction latine est notable pour au moins deux raisons. D'abord, elle est fort tardive; il est en effet surprenant de voir que, au XIVe siècle encore, on traduisait un texte arabe d'algèbre. Ensuite, la belle écriture du XIVe siècle qui l'a transcrite est celle du traducteur lui-même, dont on peut suivre les hésitations, ainsi que les améliorations ou les corrections d'interprétations apportées au moment même de la traduction ou lors d'une lecture ultérieure. En effet, à l'exception de deux notes de lecteur (lignes 216-20 et 1528-30) et de quelques corrections et additions dues à un réviseur contemporain du traducteur (principalement au fol. 90^{v}), tout le texte est d'une seule et même main. Il fallait donc choisir un mode d'édition qui permît de reproduire le texte en laissant à leur place les diverses corrections du traducteur. A cette fin, nous avons rejeté l'emploi d'un appareil

critique et adopté un certain nombre de conventions:
– en *italique* apparaîtront les additions; dans le manuscrit, elles sont placées au-dessus de la ligne concernée (en bout de ligne si elles se rapportent à ses derniers mots), et en marge ou au bas des feuillets pour les plus longues;
– entre [crochets] se trouveront les parties supprimées par le traducteur; de la sorte, la version initiale du traducteur apparaîtra lorsque l'on omettra de lire ce qui est en italique tout en incluant le contenu des crochets (nous avons néanmoins négligé de reproduire des ratures de moindre importance ou de lecture incertaine);
– les ⟨parenthèses angulaires⟩ enfermeront les parties omises par le traducteur et qui se trouvent dans le seul manuscrit arabe conservé (en excellente condition) et qui très certainement apparaissaient aussi dans le texte original d'Abū Kāmil;
– les (parenthèses) contiennent nos propres additions, corrections ou remarques au texte; on trouvera aussi entre parenthèses, mais en caractères gras, l'indication du foliotage du manuscrit latin et, précédée de **A**, de celui du manuscrit arabe (dans la reproduction imprimée de ce dernier, la page numérotée $2n$ correspond au verso du folio n du manuscrit);
– les {accolades} comprennent les gloses du traducteur, distinctes des simples modifications ou compléments de la traduction; à l'exception d'une seule (lignes 1153-60), elles sont toujours à l'extérieur du corps du texte (et apparaîtront donc en caractères italiques); nous avons également utilisé les accolades pour les deux notes de lecteur déjà mentionnées;
– les *astérisques* indiquent d'autres traductions possibles d'un mot, ou d'un groupe de mots, qui ont été ajoutées ultérieurement par le traducteur sans rejet de la traduction initiale.

Deux conventions mineures ont en outre dû être adoptées:
– lorsqu'une addition marginale a une correction, ajoutée au-dessus de la ligne ou ailleurs dans la marge, celle-ci sera transcrite dans un caractère (italique) de plus petite taille;
– lorsque le texte a, en un même endroit, des ratures successives, les ratures antérieures et les crochets qui les enferment apparaîtront souvent avec des caractères plus petits afin que l'on puisse aisément distinguer la suite des crochets.

Quant aux corrections effectuées sur les lettres elles-mêmes, donc dans le corps du texte, elles ont été placées à la suite et écrites dans les mêmes caractères. Ainsi, quadrat[o]um (ligne 1107) signifiera que quadratum a remplacé quadrato, le -um étant substitué au -o; similairement, 1[0]80 (ligne 1252) indique que 100 a été corrigé en 180, le second 0 étant transformé en 8. Exceptionnellement, lorsque la correction devait être clairement mise en évidence, nous avons eu recours à des caractères italiques bien que ladite correction ne se trouvât pas au-dessus de la ligne; c'est le cas dans les lignes 936 (*e*), 3132, 3269 (*i*).

Grâce à ce mode d'édition, qui conserve en leur endroit traductions initiales, corrections et additions, nous serons à même de suivre pas à pas le travail de traduction et d'en établir les caractéristiques avec une notable aisance.

Il apparaît ainsi d'abord que quelques parties entre crochets sont des traductions mal placées, le traducteur ayant par inadvertance sauté à un même mot distant de quelques lignes, avant que de se rendre compte de son erreur et de reprendre sa traduction au bon endroit. Or, si certaines traductions du même membre de phrase sont concordantes (p.ex. en 2088-2090), dans d'autres cas la traduction biffée diffère de la traduction conservée, soit par l'ordre des mots, soit par la fonction grammaticale qui leur est attribuée en latin, soit par l'usage de synonymes tels que *et* et *autem* ou *quod* et *quia* (lignes 439-40, 1892-3, 1915-6, 3267-9). Cette présence de deux traductions différentes d'un même membre de phrase dont l'une est barrée démontre sans équivoque que la main est celle du traducteur.

Si de telles erreurs de répétition apparaissaient d'emblée au traducteur et étaient donc immédiatement corrigées, la rectification d'erreurs de traduction dues à une lecture fautive demandait souvent la compréhension du contenu mathématique du texte. Notre interprète qui, durant son travail de traduction et durant une lecture ultérieure, s'est efforcé de suivre le raisonnement des problèmes, y a en général remédié. En revanche, il a souvent laissé des erreurs dans des passages qui n'ont point directement rapport avec les calculs mais qui introduisent, concluent ou généralisent des sections de l'ouvrage: visiblement, notre traducteur, s'il suit les calculs d'Abū Kāmil, manque de la vue globale que seules une étude prolongée et une assimilation de l'ensemble eussent pu lui apporter. On en trouvera plusieurs exemples parmi les erreurs suivantes, qui sont toutes dues à une lecture initiale fautive du texte arabe:

68: *exponemus eam* (sc. *utramque*) (نبيّن عليهما) pour *exponemus eius causam* (نبيّن علّتهما).

343-4: *in huiusmodi questione* (بمثل ذلك في مسئلة) pour *huius statuemus exemplum in questione* (نمثّل ذلك في مسئلة).

444: *docuimus* (علّمنا) pour *fecimus* (عملنا).

528-9: *ad terminationem *vel continentiam* ipsarum* (الى حدّها) pour *ad unam ipsarum* (الى احدها).

535: *qui multiplicat considerationem *vel addit in consideratione** (من زاد النظر) pour *qui considerare vult* (من اراد النظر).

812: *totum quod additur medietati* (كلّ ما زاد على نصف) pour *totum quod est maius vel minus* (كلّ ما زاد او نقص).

836 (& 2392?): confusion entre *radix* (جذر) et *capere* (اخذ).

1417: *et* $\frac{1}{2}$ (ونصف) pour *et (tu) dimidia* (ونصّف).

1871: *ante ipsa* (قبلهما) pour *ergo tertia ipsorum* (فثلثها).

2250 (& 2403; 1693?): *60* (ستّين) pour *2 res* (شيين).

2311 & 2312: *10 diminuta re* (العشرة الّا شيًا) pour *10 res* (العشرة الاشيا).
2396: *primus* (الاوّل) pour *parvus* (الاقلّ).
2467: *quasdam* (بعضها) pour (*earum*) *medietatem* (نصفها).
2584: *adde* (زد) pour *reduc* (ردّ).
2858: *prohice(re)* (القى) pour *2000* (الفى).
2963: *adde autem illa* (وزدهما) pour *et dragma* (ودرهمًا).
3195: *questio* (? المسئلة) pour *3* (الثلثة).
3206-7: *redibit quia* (فيعيد لان, sic) pour *et erunt equales* (فيعتدلان).
3212: *sicut ipsa est* (بمثله) pour *tribus* (بثلثة).
3213: *secundum mensuram equalitatis* (على قدر عدل) pour *secundum mensuram numeri* (على قدر عدد).
3329: *quantitas* (قدر) pour *ergo DZ* (فدز).
3370: *manifestum est* (تبيّن) pour *manifestum erit* ou *exponemus* (نبيّن).
3501: *omnia illa* (?جمعين) pour *50* (خمسين).

Des erreurs de traduction reflétant une lecture initiale erronée ou incomplète apparaissent aussi lors de la transcription des nombres. Elles sont aisément explicables lorsque l'on a à l'esprit que, dans le traité d'Abū Kāmil comme dans beaucoup de traités d'algèbre arabes, les nombres sont écrits en mots. Si ces erreurs ont été rapidement rectifiées, notre autographe du traducteur en a gardé les traces. Il apparaît que les causes d'erreur sont les suivantes:
- la similitude d'écriture entre 7, سبعة , et 9, تسعة, ou, de manière analogue, entre 70 et 90 (voir les lignes 1574, 1842, 1939 (bis), 2695, 2839 seqq, 2856, 2859, 2881); parfois, la similitude entre 6, ست, et 3, ثلث, (2100, 2833, 2862);
- la position du nombre des unités, qui précède les dizaines dans les numéraux en arabe; voir les lignes 75, 1460, 1631, 1645, 1806, 1814, 1992, 2119, 2311, 2826, 2843-4 & 2853 (unius et 30); de même, les centaines étant nommées en premier, il arrive qu'elles soient tout d'abord transcrites en chiffres seules, sans les dizaines et les unités qui les accompagnent, voire suivies des unités seules en vertu du cas précédent; c'est ce qu'on remarque dans les lignes 1135 (100 changé en 110), 1252 (100 en 180), 2224 (102 en 122), 2233, 2239, 2394, 2397, 2519, 2520, 2753, 3503 seqq.

On trouvera encore quelques exemples de séparations de nombres qui ont fidèlement gardé l'empreinte du texte arabe dans les lignes 1460 (مائة درهم وعشرين درهمًا), 2223 (مائة درهم ودرهم), 2340-1 (مائة درهم ودرهمين), 2883 (جذر مائة وثمانية اموال).

Abū Kāmil fait en outre un usage fréquent des fractions. Or, ces fractions sont, elles aussi, exprimées en mots dans le texte arabe et en chiffres dans notre texte. Pour les fractions aliquotes simples, dont les dénominateurs vont de 2 à 10, on trouve la forme moderne, parfois avec la désinence casuelle au numérateur (v. lignes 45, 550), rarement le mot (sauf pour medietas; autres cas isolés: lignes 1096, 1125, 2905). Quant aux fractions générales du type

$\frac{A}{B}$, qui sont exprimées en arabe par l'expression «A parties de B parties», notre traducteur les a écrites systématiquement sous la forme moderne; mais on retrouve des traces de la désignation arabe, soit parce que le numérateur a initialement été inscrit sur la ligne (v. 2765, 2832, 2833; 2851, 2858), soit parce qu'une fois le mot «parties» a été traduit par inadvertance (2843). Très caractéristique de notre traduction est la notation $\frac{A\ B}{C\ D}$ pour transcrire la somme $\frac{A}{C}+\frac{B}{CD}$ (ligne 1853, $\frac{4\ 1}{7\ 7}$ = اربعة اسباع وسبع سبع). La forme $\frac{0\ B}{C\ D}$ (ou bien $\frac{e\ B}{C\ D}$ avec la désinence casuelle) se réduira donc à la fraction $\frac{B}{CD}$ (ou $\frac{B}{D}\cdot\frac{1}{C}$, ce qui explique la désinence casuelle); voir les exemples des lignes 1096, 1293-4, 1805, 1845-1847, 1853, 2071-2072, 2122-2125, 2500-2512, 2906-2918, 3093-3094, 3464-3466, 3512-3514, de même que celui de la forme $\frac{e\ e\ B}{C\ D\ F}$ en 3093 (sc. quarta octave octave = ربع ثمن الثمن). L'usage de l'écriture $\frac{A\ B}{C\ D}$ pour $\frac{B}{D}+\frac{A}{CD}$ en 2996-2997 ($\frac{1\ 2}{2\ 8}$ = ربع ونصف ثمن), incompatible avec la précédente si l'on fait exception des cas ambigus des lignes 941 et 2070, est en fait dû à une intervention de la seconde main, que l'on reconnaît ici à son écriture pâteuse. Remarquons encore, pour clore le sujet des fractions, que le *et* est parfois omis entre la partie entière et la partie fractionnaire (433, 480, 481, 1426, 2372, 3049 (sic, e corr.)), et que notre traducteur n'a pas conservé la décomposition du texte original dans deux cas, les sommes $\frac{13}{26}+\frac{23}{676}$ et $\frac{11}{26}+\frac{23}{676}$ d'Abū Kāmil apparaissant respectivement comme $\frac{361}{676}$ et $\frac{309}{676}$ aux lignes 1675-6.

En plus des nombres fractionnaires, Abū Kāmil fait usage dès le Livre III de l'Algèbre (commençant à la ligne 2264) de nombres irrationnels, sous la forme d'irrationalités quadratiques. Or, un problème sérieux va se poser pour représenter les expressions contenant des radicaux, par suite de l'absence d'un mode clair de désignation de l'étendue du signe radical. Cette difficulté n'est pas propre à la traduction latine mais remonte au texte même d'Abū Kāmil, qui n'a pu lui trouver de remède. Ainsi, son habitude de marquer la terminaison du radical en insérant, à la suite du radicande, l'expression «sa racine étant prise» (مأخوذ جذره) ne fait réellement que transposer la difficulté; on arrive en outre à cette incongruité que, lorsque une racine doit être multipliée, c'est le radicande qui deviendra complément direct du verbe «multiplier» (v. lignes 3075-7, qui reflètent fidèlement l'original arabe). Notre texte a gardé la trace des soucis causés par ces irrationalités au traducteur. Cela commence assez tôt, avec l'étude de la sommation et de la soustraction de racines: on voit le traducteur peiner pour mettre en latin l'expression $\sqrt{12 \pm 2\sqrt{20}}$ (lignes 1039-55). Cela se poursuit avec le Livre III, qui étend l'usage des racines aux coefficients et aux solutions d'équations. On observe que la bonne volonté de notre traducteur se détériore progressivement, un changement encore amplifié par son embarras devant la traduction du terme منقوص, *diminutus*, dont le placement en tête de phrase fait que la partie soustraite est mentionnée avant le tout dont elle est enlevée; dans la traduction, un ablatif absolu se trouve conséquemment devoir prendre la fonction grammaticale d'un sujet (v. p.ex.

lignes 2320, 2343; mais: 2452-3, 2520), ce qui cause mille tourments au traducteur (2565 seqq). La situation n'est pas forcément meilleure lorsque sous le signe radical se trouvent plusieurs termes positifs, à cause de l'absence susmentionnée de délimitation du signe radical; on remarquera des efforts ultérieurs du traducteur pour rendre la version compréhensible, que ce soit par un changement de cas des noms (2345) ou par l'addition de désinences casuelles aux nombres (2913, 2969 seqq). Toutefois, certaines places n'ont pas été corrigées (cf. 3111 seqq —mais v. 3122 et (solution de facilité?) 3124, 3127). Notre traducteur n'est certes pas toujours à l'abri de reproches, mais répétons qu'il n'est nullement aidé par le texte d'Abū Kāmil, qui lui rend même parfois la vie encore plus difficile, comme par sa fréquente utilisation du terme جذر, racine, non seulement pour désigner le radical, mais aussi pour l'inconnue comme synonyme de شيء (p.ex. en 2333; cf. 569-70).

Délimiter l'expression sur laquelle porte une opération est, en l'absence d'un symbolisme, un problème général, qui ne se pose pas uniquement pour la racine. Une illustration s'en trouve dans notre texte pour le cas d'une division. Ayant établi l'équation de son problème, $x^2 + \frac{100}{x^2} = (\frac{100}{x^3})^2$, Abū Kāmil la transforme en $\frac{x^4+100}{x^2} = \frac{10000}{x^6}$, puis en $\frac{x^8+100x^4}{x^6} = \frac{10000}{x^6}$, ce qui lui donne immédiatement l'équation pour x^4. Seulement, le mode purement verbal de rendre ces calculs est ambigu, et la traduction donne au premier membre dans la première transformation de l'équation la forme $x^4 + \frac{100}{x^2}$ (v. lignes 2936-7). Dès lors, les calculs subséquents perdent tout sens, et le tout a été ultérieurement raturé, une addition (2947-8) transformant l'équation de départ en l'équation de résolution grâce à une multiplication par x^6.

Le traducteur n'était pas seulement en butte aux difficultés inhérentes à l'expression purement verbale des calculs algébriques. Son exemplaire arabe n'avait en effet pas la qualité exceptionnelle de celui qui nous a conservé le texte original de l'Algèbre, car il était en plusieurs endroits lacunaire. Certes, ces lacunes étaient en majorité de simples fautes de copiste par *homœoteleuton*, comme il apparaîtra de la reconstitution du texte dans la version éditée. Mais il y avait aussi au moins deux lacunes importantes. Passons sur l'omission de l'introduction à l'ouvrage (fol. 1^v-2^v,11 du texte arabe): elle est sans doute volontaire, notre traducteur désirant entrer dans le vif du sujet et ne goûtant guère la traduction de passages non mathématiques. Par contre, deux autres lacunes tombent insidieusement au milieu de problèmes; la première recouvre le texte arabe du fol. 26^r,12 au fol. 26^v,13, la seconde s'étend du fol. 50^v,1 au fol. 54^r,11.

Notre traducteur a pris à cœur de remédier aux lacunes qu'il remarquait, tout comme il s'est efforcé de clarifier le sens du texte là où cela lui semblait désirable (on trouvera quantité d'exemples de ce second cas, répartis tout au long de la traduction). Pour tenter de combler les petites lacunes, il a régulièrement inséré des corrections ou des gloses (incidemment, nous déduisons de leur formulation, différente de ce qu'on trouve aux endroits concernés du texte arabe, qu'il n'avait pas d'autre manuscrit à disposition

pour compléter son exemplaire). Tenant compte du texte interprété et du sens désiré, le traducteur effectue ses corrections
– en introduisant des compléments au texte sans modifier le texte traduit, comme dans les lignes 1816-7, 2063-4, 2314, 2670-1, 2708-9, 2970;
– en biffant le ou les mots incompréhensibles et en adaptant et complétant le texte, comme cela est fait dans les lignes 247-9, 1771-3, 1797-8, 2177-8, 2411-2, 2503-4, 2748-50, 2840-2, 3448-50;
– quand le changement doit être plus important, les compléments deviennent, en principe, des gloses du traducteur: 1153-60, 1306-20, 1334-44, 2084-5, 2141-2, 2150, 2420-1, 2782-93, 3313-5.

Plus généralement, les gloses qu'introduit le traducteur ont pour but non de remédier à une lacune du texte originel, mais de lui ajouter quelque chose qui n'y était point à l'origine. Il s'agira alors d'expliquer, de reformuler ou de commenter un passage, comme cela apparaît des lignes 31, 113-4, 167-9 (& seqq), 197 (del.), 244-5, 280-1, 284-6, 379-81, 383-4, 386, 959-61, 1302-4, 1381-9, 1403-7 (vérification d'un calcul), 1799, 2136-40, 2321-2, 2449, 2655(!). On y voit même le traducteur prodiguer des encouragements au lecteur (?!, v. lignes 2522, 2586).

Il est approprié d'introduire ici une précision. Certaines additions sont, nous l'avons dit, d'une deuxième main, que l'on reconnaît à son écriture pâteuse, à sa manière arabe (occidentale) de représenter les chiffres, ainsi qu'au placement (que l'on retrouve chez Léonard de Pise) de la fraction avant l'entier qu'elle accompagne. On doit à cette seconde main, semble-t-il, l'écriture corrigée des fractions des lignes 2996 et 2997 (cas déjà mentionné), ainsi que les additions ou modifications apportées aux lignes 2989-91, 3006, 3013, 3017, 3020, 3022, 3025-6, 3029-32, 3037, 3040, 3085, 3132 et, apparemment, des changements comme ceux que l'on trouve dans les lignes 3185 et 3199-200; on reconnaît la trace de cette même main ultérieurement (voir l'étude de M. Lorch).

Avant cette remarque sur la deuxième main, nous avions établi une distinction entre compléments et gloses, bien que cette différence ne puisse toujours être déterminée par le contenu de l'addition (comparer 3313-5 avec 3354-6). Si nous avons posé cette distinction, c'est à cause de l'importance capitale de ces gloses pour l'histoire de notre texte. D'abord, il faut bien se rendre compte qu'elles remontent au traducteur lui-même, et ne sont pas des gloses de son exemplaire qu'il aurait traduites (cela doit aussi être le cas pour la seule glose placée dans le texte, lignes 1153-60), ce qui explique pourquoi des gloses ont été remaniées ultérieurement (v. 2138-40). On voit aussi qu'il en a écrit certaines quelque temps après avoir traduit (dans la glose des lignes 1306-20, il renvoie à un calcul supposé être connu alors que ledit calcul vient en fait ultérieurement, en 1422-30).

Il nous paraissait souhaitable d'insister sur l'identité du traducteur et du glossateur, que la conformité des écritures seule ne nous autorisait pas à postuler. La raison en est que ces gloses nous permettent d'avoir quelque information d'une part sur le manuscrit utilisé par le traducteur et d'autre part sur le traducteur lui-même.

Nous avons relevé que son exemplaire contenait deux importantes lacunes. Or, le traducteur avait un autre moyen de le remarquer que la coupure du raisonnement: dans les deux cas, il nous mentionne l'existence d'une note en hébreu indiquant la défectuosité du manuscrit arabe (lignes 1304, 2782). Il ne semble pas qu'il y ait eu beaucoup d'autres annotations —du moins que le traducteur aurait utilisées—, sauf peut-être en rapport avec le nom de monnaie فلس, employé par Abū Kāmil, tout comme *dinar*, pour désigner une inconnue auxiliaire. En effet, le traducteur le rend d'abord —correctement— par *moneta* (ligne 2012), qu'il corrige immédiatement après en *blada*, utilisé dans les lignes suivantes (2013-15); ce *blada* (blé comme mesure de compte? transcription de *baladi*, dans le sens de (moneta) *parvi pretii*?) étant peu intelligible, le traducteur l'a remplacé ultérieurement par *bizantius*, qui cette fois était unanimement compris. Toutefois, dans un autre passage (lignes 2155-60), ce même terme فلس a été rendu, ou plus vraisemblablement transcrit de la marge de l'exemplaire, par *zoz* (*zuz* étant le nom d'une monnaie valant un quart de chékel), lequel a été remplacé à nouveau ultérieurement par *bizantius*. Enfin, فلس est traduit par *moneta* dans les lignes 2203-8. Remarquons aussi que, lors de la première apparition de فلس et de ses diverses tentatives de le rendre, notre traducteur a inscrit dans la marge de la feuille, en écriture hébraïque, *dun*, reproduction maladroite de ce qu'il avait sans doute sous les yeux. On peut enfin mentionner que le pluriel *dinarim* témoigne aussi d'une influence hébraïque.

En résumé, nous avons appris que l'exemplaire du traducteur avait au moins deux lacunes importantes, plusieurs lacunes de moindre importance (généralement causées par homœoteleuton), et qu'il contenait quelques indications en hébreu. La connaissance de ses lacunes permet déjà d'affirmer que ce n'est pas l'exemplaire qui a été utilisé pour la traduction de l'Algèbre d'Abū Kāmil en hébreu par Mordekhai Finzi, une piètre traduction du XVe siècle —elle-même tout aussi piètrement éditée et traduite cinq siècles plus tard.

Ces gloses, qui nous permirent de jeter un coup d'œil furtif par-dessus l'épaule du traducteur, nous permettent aussi de lui attribuer un nom, car elles sont en général signées: habituellement *g.*, ou *G.*, une fois *W.* (961) et une autre fois *Guill.* (1343). Une détermination plus précise de l'identité de ce Guillelmus/Willelmus ne peut être que conjecturée. Dans son *Histoire des sciences mathématiques en Italie*, G. Libri rapporte que deux mathématiciens italiens des XIVe et XVIe siècles, R. Canacci et F. Ghaligai, font mention d'un traducteur d'algèbre de l'arabe dans une langue que Ghaligai appelle *nostra lingua* —on a dès lors assumé qu'il devait s'agir de l'italien—, et un fragment qu'ils lui attribuent consiste en une liste de mots transcrits de l'arabe. Apparemment, il n'y a là rien de commun avec notre traducteur: la liste en question ne se trouve pas dans notre texte, et celui-ci est en latin. Mais remarquons d'abord que ladite liste ne se trouvant pas chez Abū Kāmil, on ne s'attendrait pas à la trouver dans notre traduction. Remarquons ensuite que, pour un Italien des années 1500, et s'agissant d'un texte originellement en arabe, *nostra lingua* pouvait tout aussi bien faire référence au latin. Remarquons enfin que le nom du traducteur qu'ils mentionnent, Guglielmo de Lunis, siérait fort bien à notre traducteur, tout comme la qualification de traducteur d'un texte algébrique de l'arabe. Dès lors, il nous semble au moins justifié de supposer que ce Guglielmo de Lunis et notre Guillelmus pourraient bien être une seule et même personne.

Faire une évaluation des capacités du traducteur présentera probablement plus d'intérêt maintenant que nous connaissons son nom et lui avons peut-être trouvé une identité. Aussi allons-nous tenter d'en dresser un portrait tant comme mathématicien que comme traducteur.

On peut d'abord dire qu'il est un mathématicien formé à la géométrie d'Euclide. Régulièrement, il refait les figures: outre que la position en est changée (en général tournée d'un quart de tour et inversée), le mode de désignation est parfois remanié, dans le texte et sur la figure. Il est en tout cas clair qu'il a étudié et compris les raisonnements sur les figures.

Mais comprendre les démonstrations n'implique pas que l'on soit capable d'en composer. Or, Guillaume a désiré par deux fois ajouter une démonstration manquante. Dans le premier cas, le texte qu'il a traduit renvoie (ligne 1334) à une figure qui se trouvait dans la première des deux grandes lacunes et qui illustrait la démonstration du théorème, manquant lui aussi, que l'on énoncerait, *modo Guillelmi*, comme suit: *Omnium duorum numerorum quorum unus dividitur per alterum erit quod exit de divisione sicut illud quod exit de divisione multiplicationis divisi in se per multiplicationem divisi in dividentem.* En reconstituer la démonstration n'est pas vraiment difficile, et Guillaume l'a établie sur une autre figure (lignes 1334-44); mais ceci montre au moins qu'il avait suffisamment bien assimilé la manière de raisonner de l'auteur qu'il traduisait.

Fort de ce succès, Guillaume va s'essayer peu après à l'établissement d'une seconde démonstration, qui concerne un calcul algébrique cette fois (1381-9). La démonstration en question est devenue pour nous lacunaire: l'usure due au temps a causé la rognure du bas de la page et nous en a masqué l'intégralité. Ce qui en reste permet toutefois d'affirmer que ces lacunes, n'autorisant pas un jugement global, ont servi la cause du traducteur.

Dans une autre situation, Guillaume s'est attaché à refaire une démonstration d'Abū Kāmil, qui est en fait la première que l'on trouve dans le Livre III (fol. 62 du texte arabe). On dirait que tous les déboires que Guillaume avait essuyés lors qu'il se débattait avec les expressions à racines lui ont donné des ailes quand il s'est retrouvé en face d'une démonstration géométrique. En effet, on trouve dans le texte latin une démonstration fort claire, en deux figures, qui est bien supérieure au raisonnement étriqué (et sans doute corrompu) que présente Abū Kāmil et qu'il illustre à l'aide d'une seule figure.

Des diverses corrections, il apparaît que le traducteur a vérifié la majorité des calculs des problèmes, et que, alors qu'il avait quelque peu renoncé une main secourable est intervenue. Des gloses témoignent aussi de cette révision des calculs par le traducteur, tout comme sa reconstruction de l'équation de départ après la seconde grande lacune —en conséquence de quoi les effets de cette lacune se trouvent singulièrement atténués pour le lecteur.

Mais lorsqu'il n'arrivait pas à vérifier les résolutions, il ne s'est pas toujours présenté une seconde main pour venir à sa rescousse. Outre que des calculs comprenant des racines restent défectueux, des erreurs apparaissent aussi dans la fin de la première partie, quand Abū Kāmil fait appel à quelques identités pour servir d'auxiliaires à la résolution de problèmes. Ce sont les suivantes. Si l'on a que $a - m\sqrt{a} = b + m\sqrt{b}$ (avec m positif), alors ces expressions sont aussi égales à $\sqrt{a} \cdot \sqrt{b}$, et l'on a $m = \sqrt{a} - \sqrt{b}$ (lignes 3181-91 resp. 3204-15). De même, si a est partagé en deux parties, x et $a-x$, alors, puisque $\frac{a}{a-x} - \frac{x}{a-x} = 1 = \frac{a}{x} - \frac{a-x}{x}$ (lignes 3295-3304), on aura $\frac{a}{x} + \frac{a}{a-x} = \frac{a-x}{x} + \frac{x}{a-x} + 2$ (3304-7); on a en outre que $\frac{a}{x} + \frac{a}{a-x} = \frac{a}{x} \cdot \frac{a}{a-x}$ (3347-50), d'où il suit que, si $\frac{a}{x} + \frac{a}{a-x} = c$, alors $\frac{ta}{x} \cdot \frac{ta}{a-x} = t^2 c = t(\frac{ta}{x} + \frac{ta}{a-x})$ (lignes 3436 seqq). Présentées ainsi, ces relations ne posent aucune difficulté pour le lecteur moderne. Il n'en va pas de même lorsqu'elles sont exprimées d'une manière purement verbale: la description s'en étend sur plusieurs lignes, et la moindre inattention ou erreur d'interprétation rend le tout incompréhensible. Notre traducteur n'a pu éviter ce piège, que seule une bonne assimilation du contenu de l'ouvrage lui eût permis d'éviter.

Allant de pair avec les difficultés de compréhension du texte mathématique, ou en découlant comme conséquence, la traduction apparaît moins soignée à la fin qu'au début. Le traducteur, qui auparavant revenait sur un mot lui semblant peu convenable, qui pesait soigneusement le temps et le mode du verbe pour distinguer l'hypothèse de la conséquence, s'en soucie moins, de même qu'il prête moins garde à l'accord du verbe ayant une expression algébrique comme sujet. Il fait aussi un usage croissant d'abréviations, l'économie des terminaisons de substantifs lui évitant d'avoir à se prononcer sur le sens mathématique. Ceci apparaîtra fort bien du texte édité par M. Lorch —notre version n'a en effet conservé le défaut des terminaisons que dans des cas litigieux (lignes 3066, 3164).

Ce déclin se remarque même dans la traduction de parties qui ne sont pas purement mathématiques, dans lesquelles Abū Kāmil annonce ce qui va être fait ou caractérise en traits généraux ce qui a été présenté, parties où pourtant notre traducteur était déjà régulièrement pris en défaut. On croira même tomber dans une parodie de traduction dans l'introduction à la troisième partie (v. appendice). La fidélité de Guillaume au mot-à-mot, qui lui permettait jusqu'alors de ne pas commettre de bévue trop criante, mettra ses faiblesses en évidence. Il est regrettable que tous ses louables efforts antérieurs s'estompent ainsi, recouverts qu'ils sont par ce que lui-même, las de traîner plus avant sa peine, a jeté comme un linceul sur son travail inachevé.

2. Texte latin

(71^{v}) Primum quod necessarium est aspicienti in hoc libro ⟨meo⟩ est 3 species quas memoratus est Mahammed filius Moysi Algorismi in libro suo. Que sunt: radices, et census, et numerus. Radix autem est omne in se multiplicatum ex uno, et quod est supra ipsum ex unis et fractis suis, et quod est sub ipso ex fractis. Sed census est quod agregatur ex ductu eius quod est in hac radice ex unis, et unis et fractis, ⟨et fractis⟩, in se ipso. Numerus vero **(A3^{r})** est stans per se, super quo non cadit nomen radicis vel census, sed proportionatur ad id quod est in ipso de unis.

Ex istis etiam tribus speciebus quelibet [singulis] *singillatim* duabus aliis proportionatur (vel: proportionari – sc. equari – potest). Quod est ut dicas: census equantur radicibus; item, census equantur numero; et radices equantur numero.

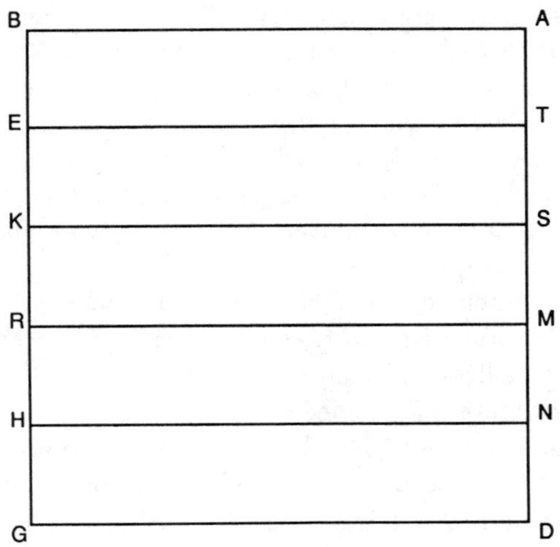

Census autem qui radicibus equa⟨n⟩tur est (sc.: sunt) sicut si dicas: Census equatur 5 radicibus suis. Cuius expositio est quod census equalis est quincuplo radicum (sic, hic et infra) eius. Et radix census erit semper sicut numerus radicum quibus census equalis existit. Que est in hac questione 5, et census est 25. Qui est sicut 5 radices eius. Quare autem sit radix census sicut numerus radicum, hic exponimus. *Et illud est* ut ponamus censum superficiem quadratam, super quam sint A, B, G, D *et* cuius latera *sint* AB, BG, GD, DA. Cuius quadrati quodlibet latus [multi-] ductum in unum ex numero efficit superficiem cuius longitudo est latus **vel: radix** quadrati (sic, pro: ductum in unum ex numero longitudinis superficiei huius est radix huius superficiei). Quare latere AB ducto in unum, quod sit linea BE, provenit superficies AE, que est radix quadrati AG. Et superficies AG sicut quinarius radicum **vel: quincuplum radicum** ipsius existit. Igitur ipsa est quincupla ad superficiem AE. Divisa igitur superficie AG lineis equidistantibus in quinque partes **superficies** equales, scilicet superficie[bus]s **(A3^{v})** AE, TK, SR, MH, NG, erunt linee BE, EK, KR, RH, HG equales {*glosa. cum dicte superficies sint eiusdem altitudinis*}. Linea autem BE [1] u*num* est. Igitur linea BG erit 5. Que est radix census, et census est 25. Qui est superficies AG. Et hoc est quod voluimus exponere.

Et sicut ⟨si⟩ diceres: Medietas census [prop-] equatur 10 radicibus. Totus igitur census est 20 radices, et radix census est 20, et census 400. Item sicut tuum dicere: 5 census equantur [equantur] 20 radicibus. Igitur census equatur 4 radicibus. Ergo radix census est 4, et census est 16. Et sic vel plus vel minus censu ad unum reducatur censum, sic etiam fiat in eo quod [agrega-] iungitur **equatur** eis ex radicibus.

Census autem qui numero equa⟨n⟩tur est (sc.: sunt) ut dicas: Census equatur 16. Igitur ipse est 16, et radix eius est 4. Et sicut dictio tua: 5 [radices equantur 40] census equantur 45 dragmis. Igitur census unus est $\frac{\text{ta}}{5}$ de 45. Que est 9, et radix eius est 3. Et sicut tuum dicere: Tertia census equatur 2[0]7. Igitur census unus est 81, et radix eius 9. Et sic *vel* plus vel minus censu ad unum reducetur censum, et sic fiet in eo quod agregatum **id est: coniunctum, vel equale** est ei[s] ex [radicibus] numeris.

Radices autem que numero equantur sunt sicut si dicas: Radix equatur 4. Igitur ipsa est 4, et census eius 16. Et *sicut* si iterum [s-] dicas: 5 radices equantur 30. Igitur radix equatur 6, et census 36. Et similiter item si dicas: Medietas radicis equatur [r-] 10. Igitur radix equatur 20, et census 400. Et ita vel plus vel minus radice ad unam reduces radicem, ita reducendo quod ei equatur ex numero.

Inveni etiam **(A4^{r})** istas 3 species, scilicet radices, census, et numerum, binas coniunctim (vel: coniungi et binas) proportionari **vel: equari** tertie. Quod est ut dicas: Census et radices equantur numero; item, census et numerus equantur radicibus; item, radices et numerus equantur censibus.

[Verbi gratia] Census autem et radices qui numero equantur sunt ut dicas, verbi gratia: Census et 10 radices equantur 39 dragmis. Cuius expositio est quod aliquis census est super quem si addatur sicut decuplum radicis eius, agregabitur **vel: deveniet** numerus eius (sc.: hoc totum) in 39 dragmas. In qua questione due sunt vie, quarum una te ducit ad scientiam radicis census, reliqua vero ad scientiam census radicis (sic, pro: ad scientiam census). Quarum utramque [adducamus] *afferemus*, et exponemus eam (sic, pro: eius causam) figura geometrica quam intelligent qui librum Euclidis, **(72^{r})** scilicet Elementorum, intellexerunt (vel: aspexerunt).

Viam autem que te ducit ad scientiam radicis census iam narravit Mahamed filius Moysi Algorismi in libro suo. Que est ut dividas **vel: dimidies** radices *per medium, id est: in duas partes equales*, que in hac questione [sunt] [*erunt*] *exibunt* 5. Et ducas ipsas in se [ipsas], et efficient 25. Que addes (ad) 39, et erunt [4] 64. Cuius numeri accipies radicem, que erit 8. De quibus minue medietatem radicum, scilicet 5, et remanebunt 3. Qui sunt radix census, et census est 9. Verum via te ducens ad censum sciendum est quod ducas 10 radices in se, et [erunt] *provenient* 100. Que [ite-] multiplica in 39, quibus equantur census et radices, et erunt 3900. Divide[s] iterum 100 per medium, et exib[i]unt 50. Que duc **(A4^{v})** in se, et provenient 2500. Que adde ad 3900, et erunt 6400. Cuius numeri accipe radicem, et exibunt 80. Que prohice de 50, que sunt medietas de 100, et de 39, quibus equantur census et radices, que simul sunt 89, et remanebunt 9. Que sunt census.

Et similiter si dixerit: 2 census, vel 3, vel plus, vel minus: reduc ad unum censum, et reduc quod est cum eo ex radicibus et numero sicut reduxisti censum. Verbi gratia si dixerit: 2 census et 10 radices equantur 48 dragmis. Cuius expositio est quod, cum duo agregabuntur census et ipsis agregatis agregabitur decuplum radicis census, provenient ad 48, necesse est tibi ut reducas duos census ad unum ⟨censum⟩. Et iam scis quod census duorum censuum est medietas. Reduc ergo totum quod habes ad medietatem *(medi)-*um** ipsius; et erit sicut (sc.: quasi) diceret: Census et quincuplum radicis eius equantur **vel: faciunt** 24. ⟨Cuius expositio est quod aliquis census est super quem si addatur sicut quincuplum radicis eius, provenient ad 24.⟩ Porta faciens te exire ad radicem est quod dividas radices per medium, et erunt 2 et $\frac{1}{2}$. Que duc in se ipsa, et provenient 6 et $\frac{1}{4}$. Que agrega ad 24, et erunt 30 et $\frac{a}{4}$. Cuius numeri accipe radicem, et exibunt 5 et $\frac{1}{2}$. Ex quibus minuens medietatem radicum, que est 2 et $\frac{1}{2}$, habebis 3. Que sunt radix census, et census est 9. Porta autem te [ducens] faciens exire ad censum sciendum est quod ducas 5 radices in se, et provenient 25. **(A5^{r})** Que duc in 24, et provenient 600. Divide iterum 25 per medium, et exibunt 12 et $\frac{1}{2}$. Que extende in se, et provenient 156 et $\frac{a}{4}$. Que adde ad 600, et erunt 756 et $\frac{1}{4}$. Quorum accipe radicem, et exibunt 27 et $\frac{1}{2}$. Que prohice de 12 et $\frac{1}{2}$, que sunt medietas de 25, et de 24, quibus equantur census et radices, que simul sunt 36 et $\frac{1}{2}$, et remanebunt 9. Que sunt census, radix autem eius est 3.

Si autem dixerit: Medietas census et 5 radices equantur 28. Et eius expositio est quod est aliquis census cuius medietati si additur sicut quincuplum radicis ipsius census, provenient ad 28. Tu autem vis, *id est: tibi necesse est*, ut censum reintegres ⟨ut proveniat census integer⟩; quod erit si duplaveris totum quod est in *hac* questione. Et duplatum erit census {*g. quod est duplum medietatis*} et 10 {*que sunt duplum de 5*} radices, que 56 {*duplo de 28*} equantur. Porta autem te faciens exire ad sciendum radicem est ut dividas radices in duo[s] equalia, et exibunt 5. Que in se multiplica, et provenient 25. Que adde ad 56, et erunt 81. Quorum accipiens radicem habebis 9. Ex quibus minue medietatem radicum, que est 5, et remanebunt 4. Que sunt radix census, et census est 16. Sed via te ducens ad censum sciendum est quod ducas **extendas** ⟨10⟩ radices in se, et provenient 100. Que multiplicans in 56 habebis 5600. Divide item 100 **(A5^{v})** in duo equalia, et exibunt 50. Quibus in se multiplicatis provenient 2500. Que addens ad 5600 habebis 8100. Quorum accipe radicem, et exibunt 90. Prohiciens igitur ea de 50, que sunt medietas de 100, et ⟨de⟩ 56, quibus equantur census et radices, que simul sunt 106, remanebunt 16. Que sunt census, et eius radix est 4. Sic autem facies **(72^{v})** in omni eo quod ad manum t[u-]ibi venerit ex censibus et radicibus et numero cu[i]m equantur, cum adiutorio Dei.

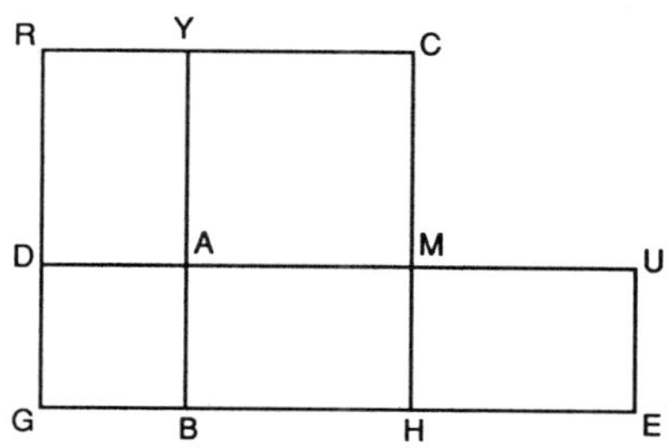

Verum causa in census et 10 radic[ibus]es que 39 equantur: Via que te ducit ad scientiam radicis est quod quod (sic) faciamus **ponamus** censum superficiem quadratam, super quam sint A, B, G ⟨,D⟩. Cui addamus radices que sunt cum eo, scilicet 10 radices, que sint superficies ABE*U*. Manifestum est [igitur] quod linea BE est 10 ex numeris: [propter quod latus] *quia radix* superficiei ABGD, qu[od]e est linea AB, multiplicata in unum *ex numeris* producit un[u]am [igitur] radicem superficiei ABGD, et (sc.: ergo) cum multiplicatur in 10 producit 10 radices eiusdem; *multiplicatur autem in linea BE, et producit 10 radices ipsius, cum superficies ABEU posita sit 10 radices;* igitur linea BE [igitur] est 10. Superficies autem EGD*U* est 39, cum ipsa sit census et 10 radices. Que producitur ex multiplicatione linee EG in lineam GD. Et linea GD est sicut linea GB. Ergo multiplicatio linee EG in lineam GB est 39. Divide igitur lineam BE, que est 10, in duo equalia in puncto H. [Que] *Cui* divis[a]e in duo equalia *in puncto H* addita est [ei] linea BG in lo⟨n⟩gitudine. Est igitur multiplicatio linee EG in lineam GB et *linee* HB in se sicut multiplicatio linee HG in se, sicut dixit Euclides in secundo tractatu libri sui. Multiplicatio autem linee EG in lineam GB est 39, et multiplicatio linee HB in se est 25. Igitur multiplicatio linee HG in se est 64. Ergo linea HG est 8. **(A6^{r})**

Et linea HB est 5. Ergo linea GB residua est 3. Que est radix census, et census est 9.

Si autem vis quod exponam tibi quod dixi ita ut oculis appareat: Super lineam GH faciam superficiem quadratam, que sit superficies HCRG. Et producam lineam AB (sc.: BA) in rectitudine ad punctum Y. Et erit linea GH sicut linea GR, et linea BG sicut linea GD. Remanebit linea BH sicut linea DR. Et erit superficies AH sicut superficies [[B]AR] *AR*. Et superficies HA ⟨est⟩ sicut superficies M[H]E. Superficies igitur ME *est* sicut superficies AR. Ergo superficies MB, BD, DY tres *sunt* 39. Superficies vero AC *est* 25, quoniam ipsa provenit ex ductu linee HB in se, que est 5. Ergo superficies GC tota est 64. Et linea GH est radix eius. Igitur ipsa est 8. Linea autem HB est 5. Ergo linea BG residua est 3. Et hoc est quod voluimus exponere.

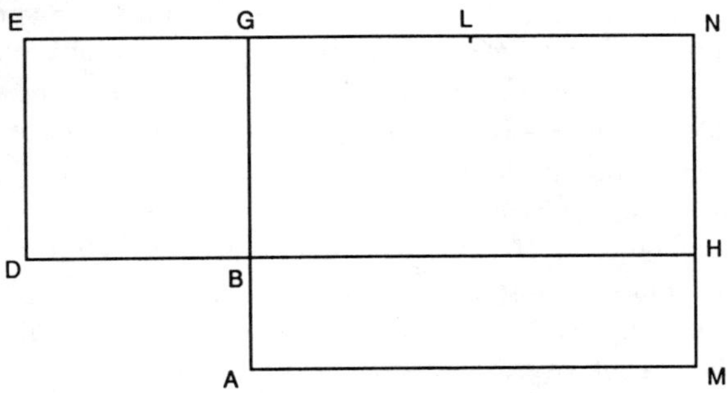

Porta autem te faciens exire ad sciendum censum monstratur ut ponamus censum lineam AB. Cui addamus 10 radices ipsius, que sint linea BG. Et erit linea AG 39. Volumus autem scire quanta sit linea AB. Faciamus igitur super lineam BG superficiem quadratam, que sit superficies DE[B]GB. Ipsa igitur erit 100 census, id est centies sicut linea AB {*g. una enim radix in unam radicem producit unum censum, ergo in 10 producit 10 census; igitur 10 radices in 10 producunt 100 census*}: multiplicata enim linea BG in unum, *quod est sicut decima ipsius linee BG*, de unis, *id est: de lateribus 4 equalibus*, que sunt in linea AB, *id est: censu*, quia linea BG est 10 radices [ipsius] linee AB, prodicit (sic) decuplum ipsius linee, *id est: census*; igitur, multiplicata in se, *id est: decuplum lateris census*, 100 census producit, *id est: centuplum ipsius linee AB* (sic, pro: que sit superficies DEGB. Ipsa igitur erit centies sicut linea AB multiplicata in unam de unitatibus que sunt in linea AB: quia linea BG est 10 radices de AB, et 10 radices alicuius rei multiplicate in se producunt centies sicut ipsam rem). Faciamus igitur (sc.: autem) superficiem AH *sicut* quadratam superficiem, *id est: equalem quadrato*, BE. Que, *scilicet: HA*, est *sicut* multiplicatio linee AB in *unam de* unitatibus que sunt in eo, *scilicet: in AB*, **(A6^{v})** centum vicibus. ⟨Igitur linea AM est 100.⟩ Et compleamus superficiem AN; et erit 3900, propterea quia linea AG est 39 et AM est 100, que, *scilicet: linee*, ipsam superficiem AN continent. Superficies autem AH est sicut superficies BE. Ergo superficies DN est [39] *3900*. Que ex ductu linee NE in EG producitur, cum linea EG sit sicut linea ED. Et linea GN est 100, cum ipsa sit sicut linea AM. Dividamus igitur lineam GN in duo equalia in puncto L. Cui [d-] iam *sic* divise addita est linea GE in longitudine *sua*. Igitur multiplicatio totius linee NE in EG additam, cum multiplicatione linee GL in se ipsam, est sicut multiplicatio linee LE in se ipsam, sicut dixit Euclides in secundo tractatu libri sui. Multiplicatio autem linee NE in **(73^{r})** lineam EG est 3900, et multiplicatio linee GL in se est 2500. Que simul iuncta efficiunt 6400. Que sunt sicut multiplicatio linee LE in se ipsam. Igitur multiplicatio linee LE in se est 6400. Cuius radix est 80[0]. Que est linea LE. Linea autem G[H]E *est* sicut linea GB. Igitur linea LG et linea GB sunt 80. Quando ergo proiecerimus lineam LG et lineam BG, que sunt 80, de lineis AG, *id est:* [*g.*] *39*, et GL, *id est: 50*, que sunt 89 {[*g. quia*]}, remanebit linea AB 9, que est census. Et hoc est quod voluimus exponere.

Census autem et numeri (sc.: numerus) qui radicibus proportionantur **vel: equantur** sunt sicut si dicas *verbi gratia*: Census et 21 dragme equantur 10 radicibus. Quod est **Id est**: Census aliquis est cui si addas 21 dra⟨g⟩mas **(A7^{r})** erit quod ag⟨re⟩gabitur equale 10 radicibus illius census. In hac questione etiam due sunt porte, quarum quelibet faciet te exire ad additionem et diminutionem, *id est: cum additione vel diminutione ad propositum.* Una ipsarum te faciet exire ad sciendam radicem census, *scilicet: cum additione vel diminutione*, et alia te faciet exire ad scientiam census, *vel cum additione vel cum diminutione.*

Porta autem te faciens exire ad sciendam radicem census est quod dividas radices in duo equa(lia), et erunt 5. Que multiplica in se, et provenient 25. Ex quibus minue 21, que fuerunt cum censu, et remanebunt 4. Quorum accipe radicem, et exibunt 2. Minue igitur ea de medietate radicum, que sunt (sic) *5*, et remanebunt 3. Que sunt radix census, et census est 9. Et, si volueris, adde duo medietati radicum, que [sunt] est 5, et erunt 7. Que sunt radix census, et census est 49. {*Nota super hec quod in illa questione possunt concurrere duo casus, quia scilicet census potest esse 3 radices et numerus additus 7, sicut 9 est 3 radices et 21 7^{es} 3^{a}; vel census potest esse 7 radices et numerus additus 3, sicut 49 est 7 radices et 21 3, ita quod census potest extendere numerum vel e contra iterum.* (Adnotatio lectoris)} Scias etiam quia, cum diviseris radices per medium in hac [questione] porta **vel: hoc [vel] capitulo** et multiplicaveris eas in se, et erit quod proveniet minus dragm[arum numero]*is* que sunt cum censu, questio est impossibilis **vel: contradicibilis**; cum autem erit equale dragmis, radix census erit sicut medietas radicum, non minus neque plus (sc.: absque additione et diminutione). Nos autem exponemus totum quod diximus in hac questione figura geometrica (sc.: figuris geometricis), cum adiutorio [Dei re-] Dei.

Sed capitulum quod te ducit ad sciendum censum est quod multiplices 10 radices in se, et provenient 100. Que multiplica in 21, que sunt cum censu, et provenient 2100. Deinde divide [in duo equalia] 100 [per medium], [*id est*] *per 2* [*equa(lia)*], et exibunt 50. Que percute in se ipsa, et provenient 2500. Ex quibus prohice 2100, et remanebunt 400. Quorum accipiens radicem habebis 20. Que minue de 50, que sunt medietas de 100, et remanebunt 30. Ex quibus prohiciens 21, que sunt cum censu, [re-] habebis re*si*duum, scilicet 9. Que sunt census, et eius radix est 3. Et, si vol[u]e*s*, addes 20 ad 50, et erunt 70. Ex quibus prohicies 21, que sunt cum censu, et remanebunt 49. Que sunt census, et eius radix est 7. Omne autem quod ad manum tuam perveniet ex censibus, sive plus sive minus *censu fuerit*, ad unum semper reducas censum sicut exposuimus in capitulo primo **id est: porta prima**.

Causam autem quod census et 21 dra⟨g⟩m[arum] *e* [que] radicibus 10 equantur in [hoc] capitulo quod te ducit ad sciendum radicem [est] cum fuerit census maius (sic, hic et infra) dragmis que sunt cum eo {*g. et tunc invenitur cum diminutione*}, vel *quod* si multiplicatio medietatis radicum ⟨in se⟩ fuerit equalis dragmis que sunt cum ce⟨n⟩su, erit tunc census sicut dragme, dicemus [postea. Sed pri[mo]*us* dicamus ipsam cum fuerit census minus dragmis, *quod fit cum additione*, et est *[sunt]*] *cum adiutorio Dei complete. Nunc* autem (sic, pro: Verum causa in census et 21 dragme que 10 radicibus equantur: In hoc capitulo quod te ducit ad sciendum radicem, census potest esse maior dragmis que sunt cum eo vel ⟨minor eis, **(A7ᵛ)** sed census non est sicut dragme que sunt cum eo;⟩ cum ⟨tamen⟩ fuerit multiplicatio medietatis radicum in se equalis dragmis que sunt cum censu, erit tunc census sicut dragme que sunt cum eo. Dicemus autem totum hoc et exponemus ipsum, cum adiutorio Dei.

Sed prius dicamus ipsum cum fuerit census minor dragmis. Et est ut) ponamus [censum superficiem] dra⟨g⟩mas que sunt cum censu, que sunt 21, [*et sit*] plus ipso. Et faciamus censum superficiem quadratam, super quam sint A, B, G, D, et addamus ei 21, que sunt cum eo, que sint superficies ABLE. Et superficies ABLE est maior quadrato ABDG, sicut posuimus ipsam. Ergo linea BL longior est linea BD. Et superficies ED est 10 radices quadrati **(73ᵛ)** ABGD. Igitur linea DL est 10. *Et* superficies EB producta *est* ex linea LB in BA. Linea autem BA est sicut linea BD. Igitur multiplicatio linee LB in BD est superficies BE. Superficies vero BE est 21. Ergo multiplicatio LB in BD est 21. Dividam lineam LD in duo equalia in puncto H. Linea ergo LD iam divisa est in duo equalia in puncto H et ⟨in⟩ duo inequalia in puncto B. Erit igitur multiplicatio [B]LB in BD et HB in se sicut HD in se, sicut dixit Euclides in secundo tractatu libri sui. Linea autem HD in se ducta producit 25, cum ipsa sit 5. Et linea LB in BD ducta est 21. Remanet ergo HB in se ducta 4. Qu*orum* radix, scilicet 2, est linea BH. Sed linea HD est 5. Remanebit *ergo* linea BD 3. Que [sunt] est radix census, et census est 9.

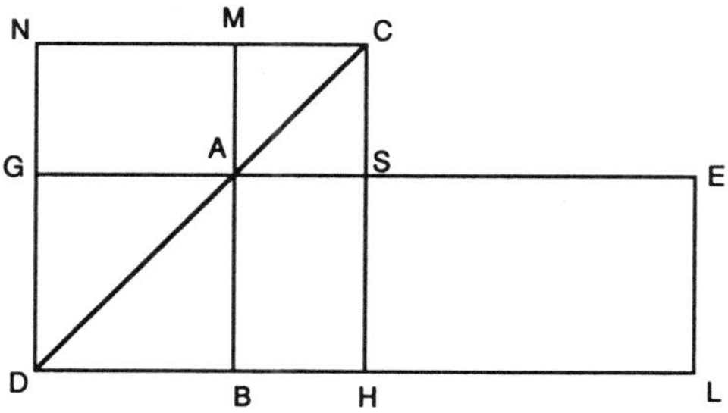

Et si volueris scire [ex] (vel: Et si volueris quod tibi exponamus) quod diximus scientia subiecta oculis: Faciemus super HD superficiem quadratam, que sit superficies [CD]*CNDH.* Que erit 25, (**A8ʳ**) quia linea HD est 5. *Et producamus in rectitudine lineam BA usque ad M. Erit ergo* superficies [autem] HG [est] sicut superficies HE, cum linea LH sit equalis linee HD. Et superficies AH *est* sicut superficies AN {*g. cum sint supplementa*}. Erunt ergo tres superficies AH, AD, AN 21, cum sint sicut superficies EB. Remanet ergo superficies AC 4. Et est quadratum, propterea quod linea CN est equalis linee CH et linea HS equalis linee NM {*quia linea SH est equalis GD, et GD est equalis DB. [Quia] Et DB equalis est NM. [Ergo linea est] Ergo linea SH equalis est linee NM. g.*}; remanet ergo linea MC sicut linea CS. Erit igitur *ex inde* linea MC 2. Que est sicut HB. Sed tota linea HB (sic, pro: HD) est 5. Remanet igitur linea BD 3. Que est radix ⟨census, et census est 9⟩. Et hoc est quod voluimus exponere. Et expositum est ex (sc.: in) hac questione quod, cum medietas radicum ducitur in se, quod provenit est plus dragmis que sunt cum censu secundum quantitatem quadrati AC.

Amplius ponamus dragmas que sunt cum censu, que sunt 21 dragme, minus censu. Et faciamus censum superficiem quadratam, que sit superficies ABGD. Et addamus ei 21, que sunt cum eo, et sint superficies ABEU. Est igitur superficies A[G]D maior superficie AU, sicut posuimus. Sic ergo (sic, pro: sicut posuimus ipsam. Ergo) linea [G]DB maior est linea BU. Superficies autem E[G]D tota est 10 radices superficiei A[G]D. Ergo linea [G]DU est 10. Linea vero BU in B[G]D *ducta* producit 21. Divide ergo lineam [G]DU in duo equalia in puncto H. Linea ergo [*D*G]*DU* divisa est iam in duo equa(lia) in puncto H et in duo inequalia in puncto B. Multiplicatio igitur BU in B[G]D et HB in se est sicut HU in se, sicut dixit Euclides in secundo tractatu libri sui. Multiplicatio autem [HG]*HU* in se producit 25, et multiplicatio B[G]D in BU est 21. Remanet igitur multiplicatio linee BH in se 4. Quorum radix est 2. Que est linea BH. Adde igitur eam ad lineam H[G]D, que est 5, et erit tota linea B[G]D 7. Que est radix census, et census est 49. Iam ergo manifestum est quod, cum posuerimus censum minus dragmis que sunt cum eo, egredietur solutio (vel: questio) cum diminutione; sed cum posuerimus ipsum plus dragmis, exibit cum additione.

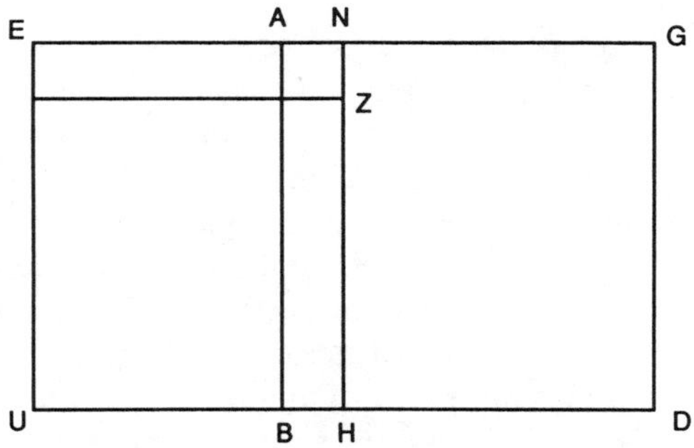

Si autem volueris ⟨etiam⟩ quod tibi exponamus quod diximus expositione que sit oculis manifesta: **(A8^{v})** Faciemus super lineam UH quadratum, super quod sint U, Z. Et producamus HZ ⟨in rectitudine⟩ usque ad N. Erit ergo superficies BZ sicut superficies ZE: propterea quia linea HZ est equalis linee EN, cum utraque sit medietas totius linee UD; et linea [[M]BD] *BH* est equalis linee ZN, propterea quod tota BD est equalis toti HN, *quia ipsa (sc.: BD) est equalis BA, et BA est sicut HN*, et ZH equalis HD, *quia utraque est medietas linee BD* (sic, pro: UD), remanet igitur BH equalis NZ; ergo superficies BZ, ZE sunt equales, quia equis continentur lineis. Igitur 2 superficies ZA et BE sunt 25, cum ipse sint sicut quadratum UZ. Superficies [*quadrata*] autem BE iam fuit 21. Igitur superficies ZA residua est 4, et est quadrata, sicut ostensum est, [quia linea AB est sicut linea BG AD] *quia linea AN est equalis linee BH, cui ostensum est esse equalem lineam ZN*. Cuius radix est AN. Que est equalis BH. Igitur BH est [duo] 2. Qu[e]am addes ad lineam HD, que est 5, et erit tota BD 7. Que [est] sunt radix census, et census est 49. Unde manifestum est iterum in hac questione quod, cum medietas radicum ducitur in se, quod provenit addit super dragmis que sunt cum censu per quantitatem quadrati linee BH, quod est AZ. Et [il-] hoc est quod voluimus exponere. Iam ergo exposuimus quod, cum posuerimus censum minus dragmis que sunt cum censu, exibit questio cum diminutione, et cum posuerimus ipsum plus dragmis, exibit cum additione.

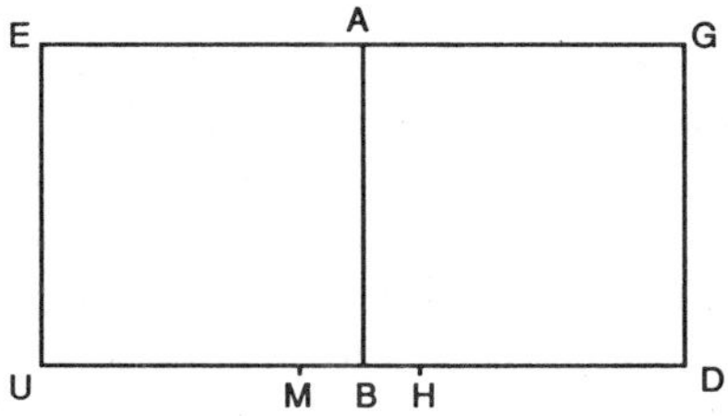

(**74**^{r}) Amplius manifestum [erit] *est* in duobus censibus qu[od]*os* diximus (sic, pro: Amplius manifestum est ex eo quod exposuimus in duabus figuris), scilicet [ratio] *figura* diminutionis et ⟨figura⟩ additionis, quod, cum multiplicabitur medietas radicum in se (**A9**^{r}) [in] in hac specie *harum* trium specierum questionis (sc.: aequationum) et erit quod producetur minus dragmis que sunt cum censu, questio est falsa et impossibilis, secundum quod iam exposuimus in duabus figuris predictis quod multiplicatio medietatis radicum in se addit super dragmis que sunt cum censu. Exponetur etiam quod, si multiplicatio medietatis radicum ⟨in se⟩ erit sicut dragme que sunt cum censu, erit census sicut dragme, et radix census sicut medietas radicum. [Et] In huiusmodi questione *scilicet quod* ponamus quod (sic, pro: Huius statuemus exemplum in questione. Quam ponamus:) census et 25 dragme equantur 10 radicibus. Faciamus censum superficiem quadratam, super quam sint A, B, G, D. Cui addamus 25 dragmas que sunt cum eo, et sint superficies ABEU. Et erit superficies tota GU 10 radices superficiei ABGD. Ergo linea DU erit 10. Cum ergo ipsam divi[seri]demus in duo equalia, impossibile est quin dividatur in puncto B, (scilicet:) aut alio puncto ab ipso versus D, vel versus U. Et [sit] *dividatur* prius versus D, si est possibile, in puncto H. Linea ergo DU iam divisa est in duo equalia in puncto H et in duo inequalia in puncto B. Igitur multiplicatio DB in BU et BH in se est sicut multiplicatio UH in se, sicut dixit Euclides in tractatu secundo libri sui. Sed multiplicatio HU in se est 25, *quia ipsa est 5, quia tota UD est 10.* Ergo multiplicatio DB in BU et BH in se est 25. Sed multiplicatio DB in BU [et quadratum DB] est superficies BE, propterea quia linea DB est equalis BA; et superficies BE est 25. Ergo multiplicatio DB in BU est 25. Quod est impossibile. Non ergo est possibile ut cadat divisio linee [B]DU in duo equalia super puncto H a parte B versus D. Et dico etiam quod non est possibile ut cadat nisi in B (sc.: ut cadat a parte B versus U). Quod si est possibile, cadat in puncto M. Linea igitur DU iam divisa est in duo equalia in puncto M [*[s]et linea BU*] et in duo inequalia in puncto in puncto (sic) B. Ergo multiplicatio DB in BU et BM in se est 25, propterea quod est sicut MU in se, et MU est 5. (**A9**^{v}) Sed multiplicatio DB in BU est

25. Quod est impossibile. Iam ergo expositum est quod medium linee [B]DU non cadit preter*quam in* puncto B. [Linea] Ergo linea BD est 5. Que est radix census, et census est 25. Qui est superficies AD. Et hoc est quod voluimus exponere.

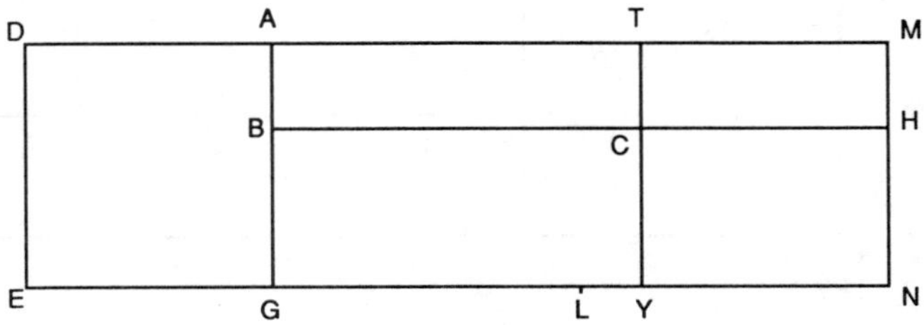

Causa autem illius quod (vel: porte que) te ducit ad sciendum censum est ut ponamus censum lineam AB. Et addamus ei dragmas que sunt cum eo, que sunt 21; sintque linea BG. [Ergo] Erit ergo linea AG 10 radices linee AB. Faciamus etiam super lineam AG superficiem quadratam, et sit quadratum AGDE. Et illud est 100 vicibus sicut linea AB ducta in unum de uni[s]*tatibus* que sunt in eo (sc.: in AB), *[100] id est census 100*, cum linea AG sit 10[0] radices linee AB, 10 autem radices in se producunt 100 census (sic, pro: 10 autem radices alicuius rei in se sunt sicut ipsa res 100 vicibus) {*g. quia una radix in [se] unam producit unum censum, ergo 10 radices in unam radicem producunt 10 census, igitur 10 in 10 100 producunt census*}. Faciamus item superficiem AH equalem quadrato AGDE, sit*que* unum latus eius linea AB. Erit igitur reliquum *latus* [10], scilicet BH, 100 {*g. quia AE est centies sicut linea AB, sicut dictum est*}. Deinde compleamus superficiem AN. Erit ergo superficies BN 2100, propterea quia linea BG est 21 et linea GN {*g. que est equalis linee BH*} 100. Faciamus item superficiem MY equalem superficiei AE. Sed superficies AE equalis est superficiei AH, ergo superficies MY equalis [sunt] superficiei AH. [Abiecta] *Sublata* comuni superficie [A]*T*H, remanet superficies CN equalis superficiei CA. Addita ergo comuni superficie BY, erit *tota* superficies AY equalis ⟨toti⟩ superficiei BN. Que est 2100. Ergo superficies AY est 2100. Ipsa autem producitur ex GY in YN, cum YN sit equalis (**A10r**) YT, quia superficies YM est quadrata. Dividamus ergo lineam GN in duo equalia in puncto L. Divisa est igitur linea NG in duo equalia in puncto L et in duo inequalia in puncto Y. Ergo multiplicatio GY in YN cum quadrato LY est sicut multiplicatio LN in se. Sed LN in se ⟨est⟩ 2500, quia ipsa est 50, et multiplicatio GY in YN est 2100. Remanet ergo quadratum linee LY 400, ergo linea LY 20. Sed LN est 50, sicut dictum est. Remanet ergo YN 30. Sed ipsa est sicut YT, et YT est sicut AG. Ergo linea AG est 30. Sed linea BG est 21. Ergo AB residua est 9, que est census. Et hoc est quod voluimus exponere.

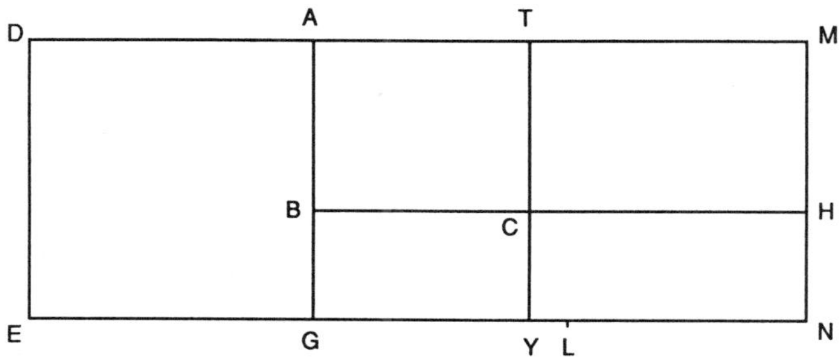

(74^{v}) *Et quia* iam exposuimus causam diminutionis in isto capitulo, exponemus [igitur] etiam causam additionis, cum adiutorio Dei. Quod est ut ponamus divisionem linee GN in duo equalia ex parte Y versus N (in puncto L). Et hoc erit cum ponemus censum plus dragmis que sunt cum eo, quemadmodum exposuimus in causa capituli quod te ducit ad scientiam radicis. Ergo multiplicatio GY in YN, et quadratum YL, est sicut multiplicatio LG in se. Sed LG in se est 2500, et *ex* linea YN in YG *ducta* est 2100. Remanet ergo quadratum LY 400. Quare est linea YL 20. Sed LN est 50. Ergo tota linea YN est 70. Sed NY est sicut YT, et TY sicut AG. Ergo AG est 70. Sed BG est 21. Remanet ergo linea AB 49, que est census. Et hoc est quod voluimus exponere.

Et planum est ex eo quod diximus et ordinavimus quod *si*, cum multiplicabis radices que equantur censui et numero in hac specie trium specierum ⟨compositarum⟩ in se ipsas, et *illius* quod provenerit item medietatem duces in se, erit (productum) sicut dra⟨g⟩me que sunt cum censu ducte in id quod provenit ex ductu radicum in se, census **($A10^{\mathrm{v}}$)** erit sicut dragme que sunt cum eo. Si autem erit, cum multiplicaveris dragmas in id quod provenit ex ductu radicum in se, plus medietate illius quod provenit ex ductu [medietati-] radicum in se ducta in se ipsam, questio est impossibilis. Iam ergo exposuimus ⟨totum⟩ quod est in hac porta.

Radices autem et numerus que equantur censibus sunt sicut si dicas: 3 radices et 4 de numeris equantur censui. In hac etiam questione sunt 2 porte, una quarum facit te exire ad *sciendum* radicem census, secunda vero ad scientiam census. Quas nos sumus [expposituri] [adducturi] *allaturi*, et expposituri ⟨causas earum⟩ figuris geometricis, cum adiutorio Dei.

Verum porta te faciens exire ad sciendum radicem census est quod dividas radices in duo equa(lia). Et exibit in hac questione 1 et $\frac{1}{2}$. Quod in se multiplica, et provenient 2 et $\frac{1}{4}$. Que agrega ad 4, *que sunt cum radicibus*, et erunt 6 et $\frac{1}{4}$. Quorum accipe radicem, et exibunt 2 et $\frac{1}{2}$. Que agrega medietati radicum, que est 1 $\frac{1}{2}$; habebis 4, que sunt [census] radix census, census autem est 16.

Porta vero faciens te exire ad sciendum censum est quod multiplices 3 radices in se, et provenient 9. Item multiplica hec 9 in 4 dragmas, que *sunt* cum radicibus, et provenient 36. Deinde divi*de* 9 in duo equalia, et exibunt 4 et $\frac{1}{2}$. Que multiplica in se, et provenient 20 *et* $\frac{1}{4}$. [Quorum radicem accipe, et e-] Que agrega ad 36, et erunt 56 et $\frac{1}{4}$. Quorum accipe radicem, et exibunt 7 et $\frac{1}{2}$. Agrega igitur ea ad 4 et $\frac{1}{2}$, que sunt medietas de 9, et ad 4 dragmas, que sunt cum radicibus, ⟨que simul sunt 8 et $\frac{1}{2}$,⟩ et erunt 16. Que sunt census, radix vero ⟨est⟩ 4. Totum autem quod erit plus cens[i-]u aut minus reduc ad unum censum, quemadmodum docuimus (sc.: fecimus) hoc in questione prima.

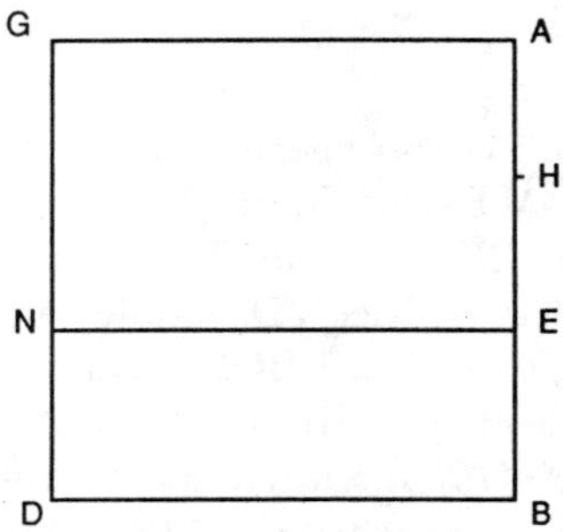

Sed causa questionis huius, scilicet quod 3 radices et 4 de numeris equantur censui, in porta que te ducit ad scientiam radicis, est quod faciamus **ponamus** censum superficiem quadratam, super quam sint A, B, G, D. Et illa erit 3 radices et 4 [de numeris] dragme. Deinde abscide de superficie AD superficiem AN, et pone eam 3 radices superficiei AD. Et remanebit superficies ED 4 dragme. Divide ergo lineam AE in duo equa(lia) in puncto H. Et quia superficies AN est 3 radices superficiei AD, manifestum est ex his que ordinavimus et exposuimus quod linea AE est 3 ex numero et quia superficies **(A11r)** ED ⟨residua⟩ est 4 ex numero (recte sic: Deinde abscide de superficie AD superficiem AN, et pone eam 3 radices superficiei AD. Manifestum est igitur ex his que ordinavimus et exposuimus quod linea AE est 3 ex numero et quia superficies ED residua est 4 ex numero. Divide ergo lineam AE in duo equalia in puncto H). Linea ergo AE divisa est iam in duo equa(lia) in puncto H, et adiuncta est ei in longitudine linea EB. Ergo multiplicatio A[E]*B* in BE et EH in se est sicut HB in se, sicut dixit Euclides in secundo tractatu libri sui. Sed multiplicatio AB in BE est 4, propterea quia AB est sicut BD. Et multiplicatio **(75r)** EH in se est

2 et $\frac{1}{4}$. Ergo multiplicatio HB in se est 6 et $\frac{1}{4}$. Quare linea HB est 2 et $\frac{1}{2}$. Sed linea H[E]A est *1 et* $\frac{1}{2}$. Ergo tota linea AB est 4. Que est radix census, census autem est 16.

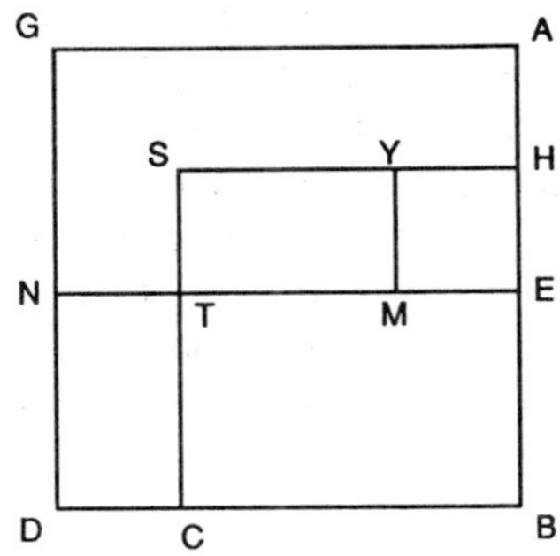

Si autem fecerimus super lineam HB superficiem quadratam H[T]*C*, et faciemus lineam MT sicut lineam ND, et produxerimus lineam MY equidistantem linee BH (sc.: linee EH), erit manifestum quod superficies MS erit sicut superficies TD: propterea quod linea AB est sicut linea DB, et linea HB est sicut BC; remanet linea CD sicut linea AH; et linea AH ⟨est⟩ sicut linea HE, et HE sicut YM; erit YM sicut CD; et linea MT *est* sic⟨ut⟩ ND; erit igitur superficies [TD] YT sicut superficies TD. Posita autem superficie [H]EC comuni, erit tota superficies ED sicut due superficies YT et TB. Sed superficies ED est 4. Ergo due superficies YT et TB sunt 4. Manifestum est etiam quod superficies HM est quadrata: propterea quod linea HB est sicut linea [TH] TE et linea MT sicut linea EB; remanet ergo linea [HB] EH sicut linea [D]EM. Item HE est 1 et $\frac{1}{2}$. Ergo superficies HM est 2 et $\frac{1}{4}$. Igitur tota superficies HC est 6 et $\frac{1}{4}$. Quare linea HB e⟨st⟩ 2 $\frac{1}{2}$. Linea autem AH est [2 $\frac{1}{2}$] 1 et $\frac{1}{2}$. Ergo tota linea AB est 4. Que est radix census, et census [ergo] est 16.

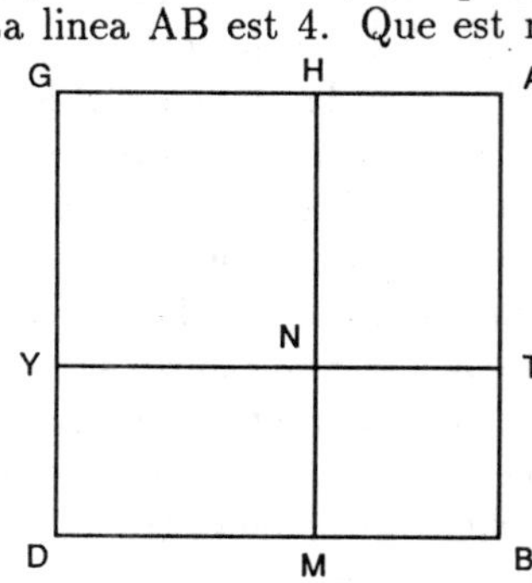

Est etiam ad hoc idem alia figura. Que est [quod] *ut* ponamus censum superficiem quadratam, super quam sint A, B, G, D. Et ipsa est 3 radices et 4 ex numero. Et abscidamus de linea AG lineam AH, ponentes ipsam 1 et $\frac{1}{2}$. Et [faciamus] *producamus* lineam HM equidistantem linee AB. Erit ergo superficies A[B]M radix 1 et [$\frac{1}{2}$] media. Item ponamus lineam YD 1 et $\frac{1}{2}$, et ducamus lineam YT equidistantem linee DB. Erit igitur superficies YB radix et medietas radicis. Planum est ergo quod tres superficies AM et **(A11$^{\mathbf{v}}$)** ND et [DB] NB sunt tres radices. Remanet ergo superficies HY 4 ex numeris et sicut superficies NB. Superficies autem NB est 2 et $\frac{1}{4}$. Ergo est superficies HY 6 et $\frac{1}{4}$. Quare linea [GH] GH est 2 et $\frac{1}{2}$. Sed linea HA est 1 et $\frac{1}{2}$. Ergo tota linea AG est 4. Que est radix census, et census est 16. Et hoc est quod voluimus exponere.

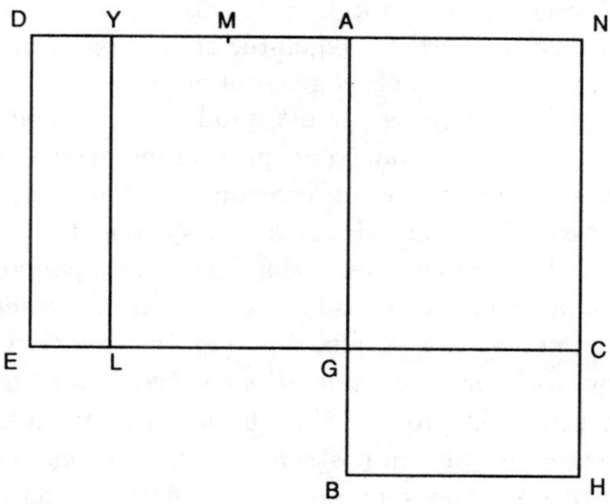

Causa autem porte que te [du-] facit exire ad sciendum censum in hac questione est quod ponamus censum lineam AB. Linea autem (sc.: igitur) AB est 3 radices et 4 ex numero. Et ponamus lineam GB 4 ex numero. Remanebit igitur linea AG 3 radices linee AB. Item faciemus super lineam AG superficiem quadratam, super quam sint A, G, D, E. Que est 9 vicibus quantum est linea AB ducta in unum. Deinde faciamus superficiem AH equalem quadrato AE, super lineam AB. Manifestum est igitur quod linea AN est 9. Que est sicut linea GC. Igitur GC est 9. Sed linea BG est 4. Igitur superficies GH est 36. Abscide item de linea AD sicut est linea AN, et sit linea AY. Et producas lineam YL equidistantem linee DE. Erit superficies YG sicut superficies AC. Iam autem fuit superficies A[H]E sicut superficies AH. Remanet ergo superficies YE sicut superficies CB. Sed superficies CB est 36. Igitur superficies YE est 36. Linea autem YA est 9, cum ipsa sit sicut A[G]N. Divide ergo eam in duo equa(lia) in puncto **(A12$^{\mathbf{r}}$)** M. Linea igitur YA iam divisa est in duo equa(lia) in puncto M, et addita est ei in longitudine linea YD. Erit itaque multiplicatio AD in [BH] DY et YM in se sicut MD in se, sicut dixit Euclides in tractatu secundo. Sed multiplicatio AD in DY est ⟨sicut⟩ superficies YE, que *est* 36. Multiplicatio item YM in se est 20 et $\frac{1}{4}$. Quibus agregatis erunt 56 et $\frac{1}{4}$. Ergo multiplicatio linee MD in se est 56 et $\frac{1}{4}$. Quare linea DM est 7 et $\frac{1}{2}$. Sed linea AM est 4 et $\frac{1}{2}$. ⟨Ergo linea AD est 12. Que est sicut linea AG.⟩ Et linea GB est 4. Ergo tota linea AB est 16. ⟨Que est census.⟩ Et hoc ⟨est⟩ quod voluimus exponere.

(75^{v}) Iste sunt 6 species quas memoravimus et ordinavimus in libro nostro isto et exposuimus equationem (sc.: solutionem) ipsarum et causam. Ex quibus 3 sunt divise (sc.: simplices, vel separate), que sunt: census equantur radicibus; item, census equantur numero; item, radices equantur numero. Tres vero composite, que sunt: census et radices equantur numero; item, census et numerus equantur radicibus; item, radices et numerus equantur censui (sic, pro: censibus). Et [si] plurimum computationis el geber et el mukabale, *id est: restaurationis et oppositionis*, impossibile est quod non te faciat exire ad terminationem **vel continentiam** (sic, pro: ad unam) ipsarum. Et ego ponam cuilibet speciei 6 specierum istarum questionem que te ducet ad computationem (sic, pro: ad quam (sc.: speciem) te ducet computatio) el geber et el mucabele. Sed prius incipiam exponere multiplicationem rerum quarundam in quasdam, et rerum et numerorum in se vel in res et numeros alios, et alia sine quorum scientia impossibile est esse illum qui multiplicat considerationem **vel: addit in consideratione** in libro nostro (sic, pro: et alia quorum scientia necessaria est illi qui considerare vult librum nostrum). Et ego exponam tibi quomodo multiplicantur res, que sunt radices, quedam in quasdam quando erunt separate, vel erunt cum numeris, vel **(A12^{v})** diminute a numeris, vel numerus diminutus erit ab ipsis, et quomodo agregabuntur quedam quibusdam⟨, et quomodo minuentur quedam de quibusdam⟩.

Quando erunt res addite *ambe* [super] numero*is* (sic), multiplicatio quart[e]a addit[e]a erit; que quarta multiplicatio tunc erit rerum quarundam in quasdam ⟨et multiplicatio numerorum quorundam in quosdam⟩: Propterea quia omnes duo numeri *cum* multiplicantur in duo[bu]s numeros, impossibile est ipsorum multiplicationem esse *aliter* quam 4 vicibus, *id est: pluribus vel paucioribus*; ⟨quod est ut ducamus duos numeros primos in unumquemque duorum numerorum aliorum; ergo multiplicatio quadrifariam cadet.⟩ Et quando erunt res diminute ambe a numero*is*, multiplicatio $\frac{a}{4}$ erit addita; et est multiplicatio rerum quarundam in quasdam. Si autem una ipsarum rerum *est* addita, alia vero diminuta, tunc multiplicatio $\frac{a}{4}$ erit diminuta; que est multiplicatio rerum quarundam in quasdam. Sed quando erunt numeri diminuti *vel additi* ambo *rebus*, multiplicatio quarta erit addita; que erit unius numeri in alium. *Et* quando unus numerorum additus *est* rebus, alter vero diminutus a rebus, quarta multiplicatio erit diminuta; que erit multiplicatio unius duorum numerorum in alium. Et quando erunt res addite numero et numerus diminutus a rebus, multiplicatio quarta erit diminuta; que est multiplicatio rerum additarum in numerum diminutum. Et hoc est quod ordinavimus et exposuimus in multiplicatione $\frac{a}{4}$, [illa] que, *scilicet:* $\frac{a}{4}$, est secundum viam ill[am quam]*orum quos* vidimus computantes in multiplicatione [incipere] *operari per ordinem* (sic, pro: Et hoc quod ordinavimus et exposuimus in multiplicatione $\frac{a}{4}$ est secundum viam a qua vidimus computantes in multiplicatione incipere). Possibile tamen est quod sit multiplicatio quarta preter illam quam diximus; summa tamen modi est quod multiplicatio diminuti in diminutum est addita, diminuti vero in additum est diminuta, et additi in additum est addita. Scias etiam quod res ducte in res producunt census, et res in numerum res sive radices, quia res et radix duo nomina sunt cadentia super unum modum **significatum.*

[Unde] Si *igitur* dixerimus tibi: Quantum est **producit** res *ducta* in 10 dragmas? **(A13^{r})** Dic: 10 res. Et porta illius est quod ponamus rem in gradu unius (sc., hic et mutatis mutandis infra: quod substituamus 1 pro re), et ducamus unum in 10; et producentur 10 ⟨. Que sunt 10⟩ res.

Et si dixit: Quantum [est] sunt 2 res, *scilicet: ducte*, in 10 dragmas? Dic: 20 res. Et porta eius *est* quod ponamus duas res in gradu **vel: loco, vel: ordine** 2. Et ducamus 2 in 10, et erunt 20. Que sunt 20 res. Et sic quod addimus ex rebus: Pone [quaml-] unamquamque rem in gradu unius **vel: pro uno**; deinde multiplica ⟨hunc⟩ numerum in numerum (datum); et quod producitur est res, propterea quia omnis numerus ex numeris cum multiplicatur in aliquam speciem specierum producit illam speciem secundum unitates que producuntur (vel: cum multiplicatur in aliquam speciem specierum est omnis unitas illius speciei in gradu unius). Verbi gratia, si erit 1 res producetur 1; si 2, 2 res (sic pro: si est res, ponetur 1; si sunt 2 res, ponentur 2). Et similiter de ce[s]nsibus et aliis; [verbi gratia] *scilicet*, cum *aliqua species* multiplicatur in numerum, erit quod proveniet (sic, pro: deinde multiplica (numerum) in numerum, et quod proveniet est) de illa specie. Et ego ponam illi figuram (sic, pro: figuras) ad exponendum quod dixi expositione oculo manifesta.

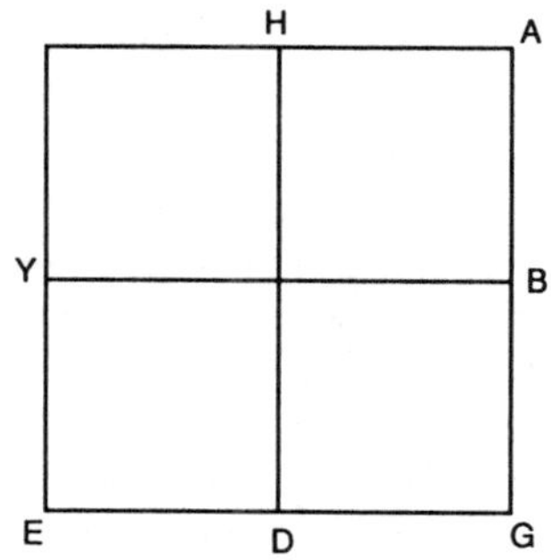

Et si item dixerimus tibi: Quantum est res in rem? Dic: censum. Et porta illi *[vel: expositio] vel: regula ad illud* est quod ponamus rem pro 1. Deinde ducamus 1 in 1, et proveniet 1. Qui est census.

Et si diximus tibi: Quantum sunt 2 res in duas res? Dic: 4 census. Et porta ad illud est quod ponamus 2 res pro 2. Deinde multiplicemus 2 in 2, et erunt 4. Qui sunt census, sicut tibi exposui. Et similiter 3 res in 2 *res*: Duc 3 in 2, et erunt 6, census. Sic $\frac{1}{2}$ rei in $\frac{1}{2}$ rei, erit $\frac{a}{4}$ census. **(76^r^)** Et sic quod agregatur **additur** ex rebus, vel minuitur: Fac **vel: pone** [pro] una*m*quamque rem pro uno et unamquamque fractionem ⟨rei⟩ pro fractione simili unius; deinde multiplica unum in aliud, et quod provenerit est census vel fractio census. Et ego ponam tibi figuram qua faciam te intelligere rationem census et radicis (sc.: rei). [Et] Sed faciamus *nunc* figuram multiplicationis 2 rerum in 2 res. Ponamus lineam AG 2 res et lineam GE 2 res. Et [iam sit *[fuit]* ducta] *ducatur* linea AG in lineam GE, et proveniat superficies AE. Dico quod superficies AE est 4 census. Cuius demonstratio est quod dividamus lineam AG per (vel.: in) numerum rerum que sunt in ea; sintque partes eius [GD, DE. Et producamus a puncto [B]D] AB, BG. Deinde dividamus similiter lineam GE **(A13^v^)** lineis GD, DE. Et producamus a puncto B lineam BY equidistantem GE, et lineam *DH* a puncto D equidistantem [GE] AG. Iam ergo invenimus in superficie AE 4 census (sic, pro: quadrata) equales, qui sunt [superficies] quadrata HB et BD et DY et YH. Quorum unumquodque est census. Et una[m]queque linearum AB, BG, GD, DH (sic, pro: DE) est ducta in 1, que est res (sc.: Et unaqueque linearum AB, BG, GD, DE ducta in 1 est res). Et hec est ratio census et rei in superficie AE 4 censuum. Et hoc est quod voluimus exponere.

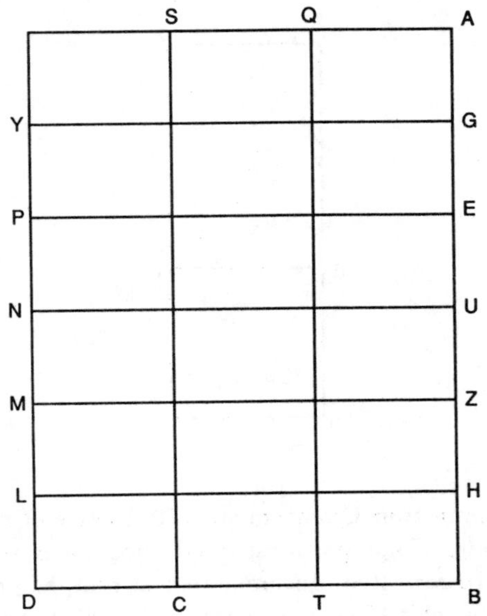

Et si dixerimus tibi: Quantum [e-] sunt 3 res in 6 dragmas? Pone 3 res loco *vel: pro* 3. Deinde multiplica 3 in 6 et provenient 18, *res*. Cuius exemplum est quod ponamus lineam AB 6 ex numeris et lineam BD 3 res. Et ducamus lineam AB in BD, et proveniat superficies AD. Dico quod superficies AD est 18 res. Cuius demonstratio est ut dividamus lineam AB, que est 6, per (vel: in) id quod est in ea de unitatibus; et exibunt 6 divisiones, que sunt AG, GE, EU, [Z]UZ, ZH, HB. Et dividamus lineam BD per (vel: in) numerum rerum [que sunt in] eius, et exibunt divisiones ipsius BT, TC, CD. Deinde pro[ducam]traham a punctis G, E, *U*, Z, H lineas equidistantes linee BD, que sint HL, ZM, UN, EP, GY. Item protrahamus a punctis C, T duas lineas equidistantes linee AB, que sint TQ, CS. Fient igitur in superficie AD 18 superficies equales, sicut vides in figura, quarum unaqueque est equalis superficiei HT, que provenit ex multiplicatione rei, que est linea BT, in unum, qu[e]od **(A14^{r})** est linea HB. Superficies autem HT est res. Igitur tota superficies AD est 18 res. Et hec est causa multiplicationis numeri in res. Et hoc est quod voluimus exponere.

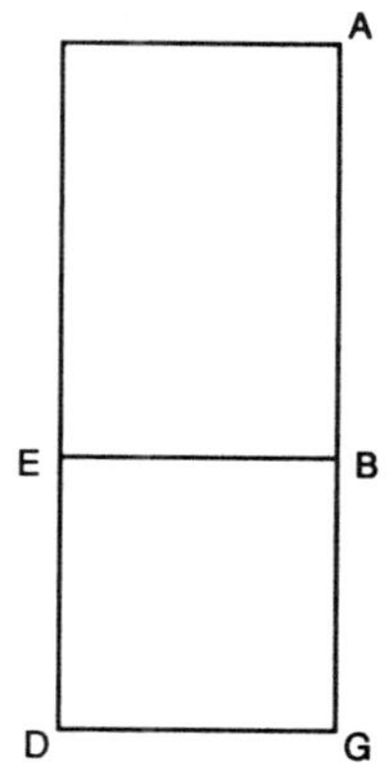

Et si dixerimus tibi: Quantum sunt 10 dragme et res in rem? Dic: 10 res et censum. Cuius porta est quod ducamus rem in 10 dragmas, et provenient 10 res. Item ducamus rem in rem, et proveniet census. Que agregans habebis 10 res et censum. Quod tibi exponemus in hac figura. Quod est ut ponamus 10 dragmas et rem lineam AG, 10 scilicet partem eius AB, rem vero partem eius aliam BG. Deinde ponamus lineam GD rem, et ducamus **percutiamus** lineam AG in lineam GD; et proveniet superficies AD. Dico ergo quod superficies AD est 10 res et census. Cuius demonstratio est quia linea BE est sicut linea GD, et linea GD est sicut linea GB. Ergo linea BE est sicut linea BG. Sed ipsa est res. Ergo linea BE est res. Et linea AB est 10. Superficies autem AE producitur ex multiplicatione linee AB in lineam BE. Ergo ipsa est 10 res. Et superficies BD est census, propterea quod ipsa est quadratum, productum ex multiplicatione BG, que est res, in lineam AD (sic, pro: GD), que similiter est res. Ergo totum agregatum quod est superficies AD est 10 res et census. Et hoc est quod voluimus exponere.

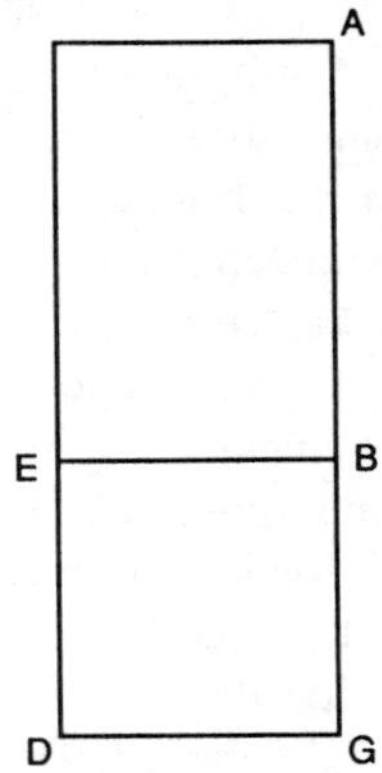

Et si dixerimus tibi: Quantum sunt 10 dragme diminuta re in rem? Dic: 10 res diminuto censu. C[uius]*ui* porta est quod ducamus 10 dragmas in rem, et provenient 10 res. Item ducamus rem in rem, et proveniet census. Quem prohiciamus de 10 **(A14^{v})** rebus, et remanebunt 10 res diminuto censu. Quod tibi exponam in hac figura. Et est quod ponamus 10 dragmas lineam **(76^{v})** AG, et ponamus BG rem. Remanebit ergo linea AB 10 dragme re diminuta. Et ponamus lineam BE rem. Multiplicatio linee AB, que est 10 dragme diminuta re, in lineam BE, que est res, erit superficies AE. Dico ergo quod superficies AE est 10 res diminuto censu. Cuius [p-] demonstratio est quod ponamus lineam GD etiam similiter rem. Et compleamus superficiem A[E]D rectorum angulorum. Superficies *igitur* AD *est* 10 res, propterea quod linea AG est 10 et linea GD est res. [Et] Quadratum autem BD est census, propterea quod ipsum fit ex ductu linee BG, que est res, in lineam GD illi equalem. Remanet igitur superficies AE 10 res diminuto censu. Et hoc ⟨est⟩ quod voluimus exponere.

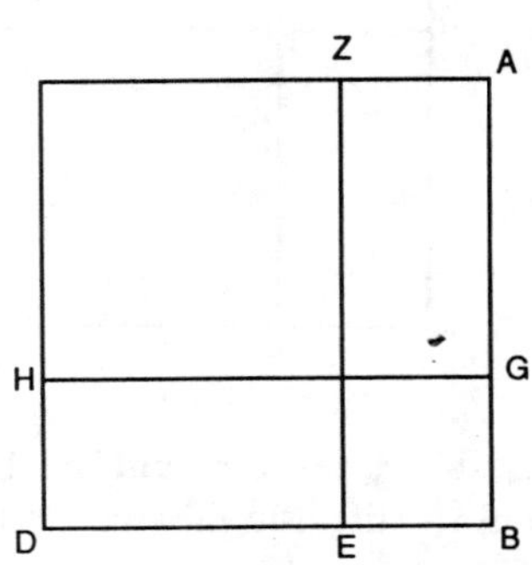

Et si dicemus tibi: Quantum sunt 10 dragme et res in 10 dragmas et rem? Dic: 100 dragme et census et 20 res. Cuius porta est quod ducamus *multiplicemus* 10 dragmas in 10 dragmas, et provenient 100 dragme. Item ducamus rem in 10 dragmas, et provenient 10 res addite. Item multiplicemus rem in 10 dragmas, et provenient 10 dragme (sic, pro: 10 res) addite. Demum multiplicemus rem in rem, et proveniet census additus. Que **Et hec** (omnia) sunt 100 dragme et 20 res et census. Quod tibi exponemus in hac figura. Quod est ut faciamus lineam AB 10 dragmas [et lineam BG] et rem, AG *[videlicet]* ***videlicet*** 10 et GB rem. Et faciamus lineam BD 10 dragmas et rem, lineam sci*licet* DE 10 et BE rem. Deinde ducamus lineam AB in lineam BD; et proveniet superficies AD. Dico ergo quod superficies AD est 100 dragme et census et 20 radices **res*. Cuius [probatio] demonstratio est quod protrahamus **(A15^{r})** a puncto G lineam GH equidistantem linee BD. Item producamus a puncto E lineam equidistantem linee AB, que sit linea [H]EZ. Igitur superficies GE est quadrata ⟨et superficies ZH est quadrata⟩, sicut iam dixit Euclides (in primo tractatu libri sui), et superficies GZ est sicut superficies EH. Et superficies ZH est 100 dragme, propterea quod ipsa provenit ex multiplicatione linee AG, que est 10, in lineam ED, que est 10. Et superficies ZG est 10 res, propterea quod ipsa est ex multiplicatione GB, que est res, in lineam AG, que est 10. Et superficies HE etiam est 10 res, propterea quod ipsa est ex linea BE, que est res, in lineam ED, que est 10. Superficies vero EG quadrata est census, cum ipsa sit ex ductu linee GB, que est res, in lineam BE, rem. Agregate igitur 4 superficies, que sunt superficies ZH, EG, EH, ZG, erunt 100 et census et 20 res. Sed superficies iste 4 sunt sicut superficies AD. Ergo superficies AD est 100 dragme et census et 20 res. Et hoc est quod voluimus exponere.

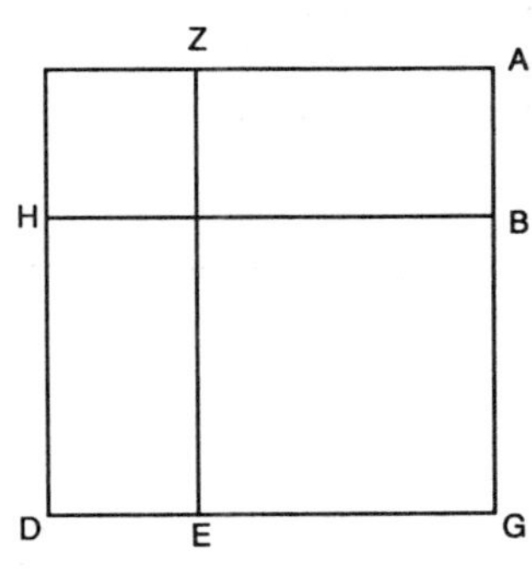

Et si dicemus tibi: Quantum sunt 10 dragme diminuta re in 10 dragmas *re* diminuta? Dic: 100 dragme et census diminutis 20 rebus. Cuius porta est ut ducamus 10 dragmas in 10 dragmas, et provenient 100. Item ducamus rem diminutam in 10 dragmas, et fient 10 res diminute. Deinde ducamus similiter rem diminutam in 10, et erunt 10 res diminute. Demum multiplicemus rem diminutam in rem diminutam, et proveniet census additus. Quibus *omnibus* agregatis provenient 100 dragme et census 20 rebus diminutis. Quod [no-] tibi exponam in ⟨hac⟩ figura. Quod est ut faciamus lineam AB 10 dragmas **(A15^{v})** diminuta re, AG *videlicet* 10 et GB rem. Et faciamus lineam D[B]E 10 dra⟨g⟩mas diminuta re, *scilicet* DG 10 et GE rem. Et percutiamus lineam AB, que est 10 re dimi*nu*ta, in lineam DE, que est etiam 10 re diminuta, et proveniet quadratum ZH. Dico ergo quod [quadratum] *superficies* ZH est 100 dragme et census 20 rebus diminutis. Cuius demonstratio est quod compleamus superficiem AD. Et protrahamus a puncto B lineam equidistantem linee GD, que sit linea BH. Protrahamus etiam a puncto [H] E lineam equidistantem linee AG, que sit linea EZ. Superficies igitur AD est 100, propterea quod linea AG est 10 et linea GD est 10. Sed superficies ZG est 10 res, propterea quod linea AG est 10 et linea GE est res. Superficies etiam *GH* similiter est 10 res, propterea quod linea DG est 10 et linea GB est res. Et superficies quadrata BE est census, propterea quod ipsa est ex multiplicatione BG, que est [res] *res*, in lineam GE, que est etiam res. Remanet ergo superficies HE 10 res diminut[is]*o* censu. Et superficies A[G]E est 10 res. Igitur superficies ZG et superficies EH sunt 20 res diminuto censu. Sed superficies AD est 100. Remanet ergo superficies ZH 100 dragme et census diminutis 20 rebus. Et hoc est quod voluimus exponere.

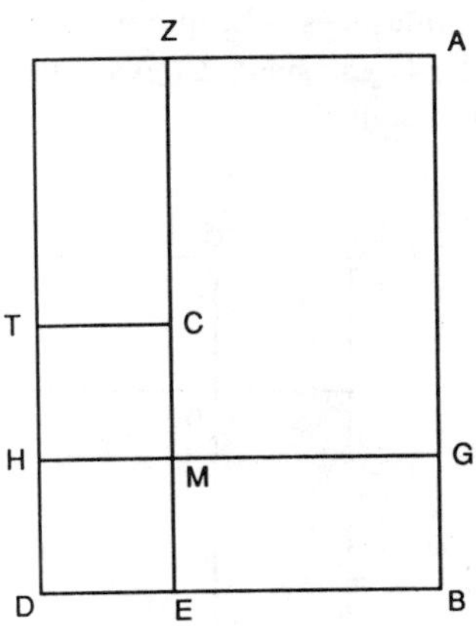

(77^{r}) Et si dicemus tibi: Quantum sunt 10 dragme et res in 10 dragmas diminuta re? Dic: 100 dragme diminuto censu. Cuius porta est quod [fa-] ducamus 10 dragmas in 10 dragmas, et provenient 100 dragme. Item ducamus rem additam in 10, et provenient 10 res addite. Item ducamus rem diminutam in 10, et provenient 10 res diminute. Et abiciamus [*ponamus*] additas cum diminutis; **(A16^{r})** remanebunt 100 dragme. Item ducamus rem additam in rem diminutam, et proveniet census diminutus. Prohice ergo eum de 100 dragmis, et remanebunt 100 dragme diminuto censu. Quod tibi exponam in hac figura. Quod est [quod] *ut* ponamus lineam AB 10 et rem, AG videlicet 10, GB vero rem. Et ponamus lineam BE 10 diminuta re, lineam videlicet BD 10, lineam autem ED rem. Et ducamus lineam AB, que est 10 dragme et re[m]s, in lineam BE, que est 10 diminuta re, et proveni*et* superficies AE. Dico itaque quod superficies AE est 100 diminuto censu. Cuius demonstratio est quod compleamus superficiem AD. Igitur superficies AD est 100 et 10 res; propterea quod superficies AH est 100: que est quadratum (vel: ipsa provenit ex multiplicatione) linee AG in GH, et AG est 10, et GH est sicut BD, et BD est 10, ergo GH est 10; superficies autem GD est 10 res: propterea quod ipsa provenit ex multiplicatione GB, que est res, in BD, que *est* 10; ergo tota superficies AD est 100 dragme et 10 radices. Et superficies etiam ZH similiter est 10[0] res, propterea quod ipsa provenit ex multiplicatione GH in GZ (sic, pro: MZ -vel GH- in MH). Igitur superficies ZH est sicut superficies HB. Abscidamus ergo de superficie ZH sicut *est* superficies MD quadrata[m], que sit superficies CH. Superficies CH etiam quadrata est census. Remanet superficies ZT equalis superficiei MB. Et faciamus superficiem ZG comunem. Superficies *ergo* AM et superficies ZT est (sic) [100 dragme] sicut superficies AE. Et superficies AM et superficies ZT sunt 100 dragme diminuto censu, quia superficies (AH est 100 dragme et superficies) CH est sicut superficies HE (sic, pro: CH est census). Ergo superficies AE, que est sicut due superficies AM et ZT, propterea quod EG est sicut ZT, est 100 diminuto censu. Et hoc est quod voluimus exponere.

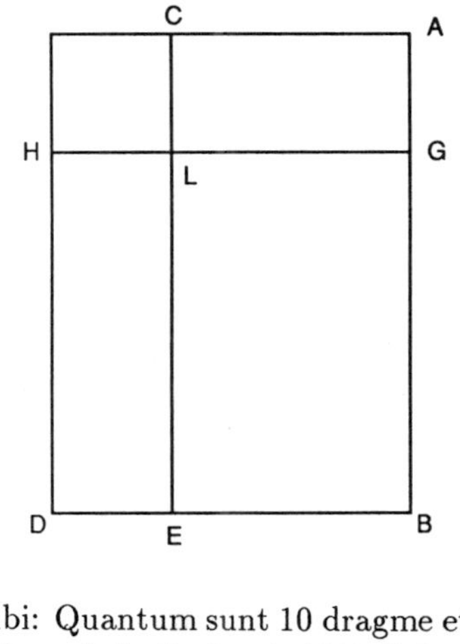

Et si dicemus tibi: Quantum sunt 10 dragme et res in rem diminutis 10 dragmis? Dic: Censu[m]*s* diminutis 100. Cuius porta est quod ducamus rem in rem, et proveni*et* census. Item ducamus rem in **(A16^{v})** 10

dragmas additas, et provenient 10 res addite. Item ducamus rem in 10 dragmas diminutas, et provenient 10 [dragme] *res* diminute. Prohice ergo res additas cum diminutis, et remanebit census. Demum multiplica 10 dragmas additas in 10 dragmas diminutas, et provenient 100 dragme diminute. Prohiciamus ergo eas a censu, et remanebit census diminutis 100 dragmis. Quod tibi exponemus in hac figura. Quod est ut faciamus lineam AB 10 dragmas et rem, AG videlicet 10 et GB rem. Et faciamus lineam BE rem diminutis 10 dragmis, BD vero rem et DE 10. Duc ergo *lineam* AB, que est 10 dragme et res, in lineam BE, que est res diminutis 10 dragmis; erit superficies AE. Dico igitur superficiem AE esse censum diminutis 100 dragmis. Cuius demonstratio est quod compleamus superficiem AD. Que est 10 res et census, propterea quod [superficies] *linea* AB est 10 dragme et res et linea BD est res. Producamus*-*trahamus** ergo (sc.: autem) lineam GH equidistantem linee BD. Et erit superficies GD quadrata census, propterea quod ipsa provenit ex multiplicatione GB, que est res, in BD, que similiter est res. Sed superficies AH est sicut superficies HE, propterea quod linea GH est sicut linea HD et linea AG est sicut linea DE. [[[Fuerunt] [sunt] s-] Erunt] *Sunt* ergo due superficies A[E]H et GE *census, quia sunt equales quadrato GD, quod est census.* Sed superficies CH est 100: propterea quod linea AG est sicut linea CL, sed linea AG est 10, ergo linea CL est 10; linea autem LH est sicut linea DE, sed DE est 10, igitur LH est 10; ergo superficies CH, ab eis contenta, est 100. Remanet itaque superficies AE census diminutis 100 dragmis. Et hoc est quod voluimus exponere.

Et si dicemus tibi: Quantum sunt 10 dragme et $\frac{2}{3}$ rei in 3 dragmas diminutis 6 rebus? Dic: 30 dragme diminutis 4 censibus et 58 rebus. Cuius porta est quod multiplices 10 (**A17**$^{\mathbf{r}}$) dragmas in 3 dragmas, et provenient 30 dragme; et multiplices $\frac{2}{3}$ rei in 3 dragmas, et provenient 2 res addite; deinde duc 6 res diminutas in 10 dragmas, et provenient 60 res diminute, de quibus prohice 2 res additas, et remanebunt (**77**$^{\mathbf{v}}$) [58] 30 dragme diminutis 58 rebus; demum duc 6 res diminutas in $\frac{2}{3}$ rei [ab-] residuas (additas), et provenient 4 census diminuti, quos prohice de 30 dragmis ⟨diminutis 58 radicibus⟩, et remanebunt 30 dragme diminutis 4 censibus et 58 rebus **radicibus.*

Et, in eo quod [divisi] *exposui* tibi de multiplicatione radicum (sc.: rerum) et numerorum quorundam in quosdam, est illud quod sufficit tibi ad intelligendum et ad operandum in [eo] *omni* quod perveniet ad manum tuam de tota **vel: omni** ista specie.

Cum volueris radicem numeri noti duplicare [*id est: ducere*] vel census ignoti, et intellige per duplicare multiplicare *in 2* radice[s]m sua[s]m, et cum volueris ducere 2 radices numeri noti vel census ignoti in numerum alium vel censum alium (sic, pro: in radicem alterius numeri vel alterius census): Pone ⟨2⟩ radices loco 2^{orum}, deinde multiplica 2 in

2, deinde in numerum notum et (sc.: vel) ⟨censum⟩ ignotum; erit *radix eius* [*quadratum 2 radicum*] quod proveniet 2 radices numeri noti vel quadrati (sic, hic et infra, pro: census) ignoti. [Item] *Et* cum volueris 3 radices [*scilicet: ducere, numeri noti etc.*]: Duc 3 in 3, deinde in numerum notum vel quadratum ignotum; et erit *radix eius* quod proveniet 3 radices numeri noti vel quadrati ignoti. Item si volueris medietatem radicis ipsius: Duc $\frac{1}{2}$ in $\frac{1}{2}$, et proveniet $\frac{1}{4}$, deinde in quadratum notum vel ignotum ⟨et erit radix eius quod exibit medietas radicis numeri noti vel census ignoti⟩. Et similiter totum quod additur medietati (sic, pro: totum quod est maius vel minus) secundum hanc ordinem.

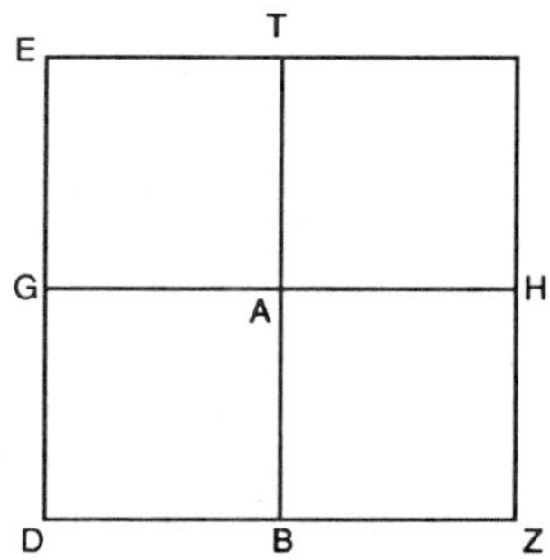

Verbi gratia, quia, cum voluerimus duplicare radicem *de* 16: Ducemus 2 in 2, et provenient 4. Deinde ducemus 4 in 16, et provenient 64. Quorum, *id est: 64*, radix est 8, que sunt duplum radicis de 16. Quod exponam tibi in hac figura. Quod est quod faciamus 16 superficiem quadratam, super quam sint A, B, G, D. Cuius **(A17^{v})** quadrati radix est linea AB. Quam cum voluerimus duplicare, producemus lineam DG secundum rectitudinem ad E, et faciamus [GD sicut] GE sicut GD; igitur GE est radix superficiei ABGD (et DE est duplum radicis superficiei ABGD). Cum autem voluerimus scire cuius numeri sit radix ED: Faciemus super ipsam quadratum EZ. Ergo linea ED est radix superficiei EZ. Dico itaque quod superficies EZ est quadruplum superficiei GB. Cuius demonstratio est quod protrahamus lineam GA secundum rectitudinem ad punctum H, et protrahamus lineam BA etiam secundum rectitudinem usque ad punctum T. Iam ergo invenisti in superficie EZ quatuor quadrata equalia, que sunt quadrata HB et HT et TG et GB. ⟨Manifestum est igitur quod quadratum EZ est quadruplum quadrati GB.⟩ Sed quadratum GB est 16. Igitur quadratum EZ est 64. ⟨Ergo linea DE est radix de 64, que est 8.⟩ Et hoc ⟨est⟩ quod voluimus exponere.

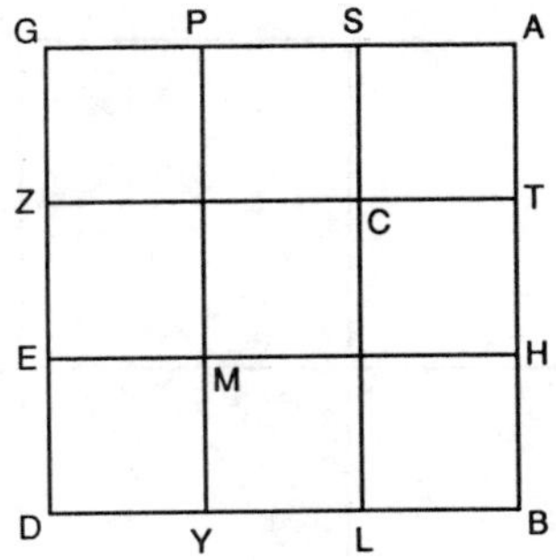

Et si volueris capere medietatem radicis de 9: Duc $\frac{1}{2}$ in $\frac{1}{2}$, et proveniet $\frac{1}{4}$. Quam duc in 9, et provenient 2 et $\frac{1}{4}$. Quorum radix est 1 et $\frac{1}{2}$. Que est medietas radicis de 9.

Et cum volueris ⟨capere⟩ [radicem] $\frac{2}{3}$ radicis de 9: Duc $\frac{2}{3}$ in $\frac{2}{3}$, et provenient $\frac{4}{9}$. Quas duc [in] in 9, et provenient 4. Quorum radix est 2. Que sunt $\frac{2}{3}$ radicis de 9. Et ego exponam tibi hoc in hac figura. Quod est ut faciamus 9 superficiem quadratam, super quam sint A, B, G, D. Cuius radix est linea GD, scilicet radix (de) 9. Cum ergo volemus capere $\frac{2}{3}$ eius, dividemus lineam GD in tria equalia, que sint GZ, ZE, ED. Linea ergo DZ est $\frac{2}{3}$ linee GD. Que est $\frac{2}{3}$ radicis quadrati GB. Cum autem voluerimus scire ipsa in numero: Faciemus super lineam ZD quadratum ZC(**A18^r^**)LD. Dico igitur quod quadratum ZCLD est $\frac{4}{9}$ de superficie *quadrata* GB, que sunt 4 de numero. Cuius demonstratio est quod protrahamus lineam ZC secundum rectitudinem ad punctum T. Et producamus a puncto E lineam equidistantem linee DB, que sit linea [*E*H] EH. Et protrahamus lineam LC secundum rectitudinem ad punctum S. Deinde dividamus lineam LD in duo equa(lia) [super] in puncto Y. [A quo] *A quo* protrahamus lineam equidistantem linee AB, sitque linea YP. Iam igitur facte sunt in superficie GB 9 superficies equales, que sunt EY, Z[P]M, ML, MC, ZP, PC, CA, CH, HL. Superficies autem ZL est 4 superficies ex istis 9. Ergo ipsa est $\frac{4}{9}$ superficiei GB. Que est 9. Ergo superficies ZL est 4. Igitur linea ZD est radix (de) 4, que *est* 2. Et hoc est quod voluimus exponere. [Totum autem] *Et in toto* quod veniet ad manum tuam de hac specie operabis secundum hoc exemplum, cum adiutorio Dei.

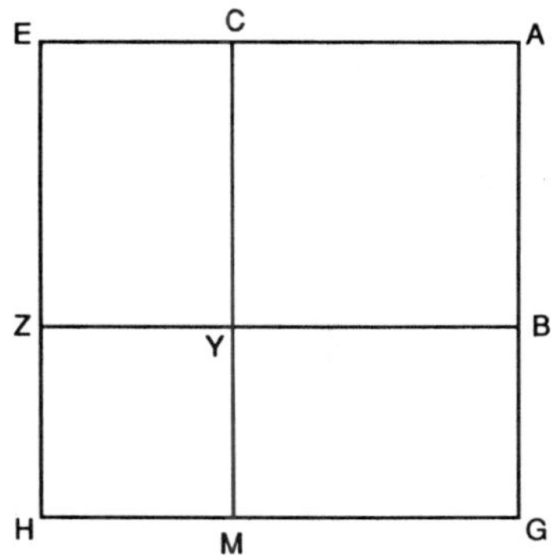

Sed cum volueris multiplicare radicem de 9 in radicem *de* 4: Duc 9 in 4, et provenient 36. Quorum radicem accipe. Que est 6. Et illud est quod provenit ex ductu radicis *de* 9 in radicem de 4. Quod ego exponam tibi in hac figura. Quod est ut faciamus lineam AB radicem de 9 et lineam BG radicem de 4. Cum ergo voluerimus multiplicare lineam AB in lineam BG: Faciemus super lineam AG quadratum AH. Et protrahemus a puncto B lineam equidistantem linee AE, et GH, que sit linea BZ. Quelibet ergo linearum A[G]*E* et GH est radix de 9 et radix de 4. [Et] Ponamus autem lineam MH radicem de 4. Et protrahamus ⟨a puncto M⟩ lineam equidistantem linee AG, et HE; sitque linea MC. Ergo superficies AY est 9, **(A18^{v})** et linea CY est radix eius. ⟨Sed superficies YH est 4 et linea ZY est radix de 4, propterea quod linea ZY est sicut linea YM et linea YM sicut linea BG.⟩ Sed YB est sicut YC *et YZ sicut YM*. Ergo proportio MY ad Y[Z]*C* est sicut proportio ZY ad YB. **(78^{r})** Sed proportio MY ad YC est sicut proportio superficiei YH ad superficiem [YE] *ZC*, et proportio ZY ad YB est sicut proportio superficiei ZC ad superficiem YA. Igitur proportio superficiei ZM ad superficiem ZC est sicut proportio superficiei ZC ad superficiem YA. Ergo multiplicatio numeri qui est in superficie ZM [ad] in numerum qui est in superficie YA est sicut multiplicatio numeri qui est in superficie ZC in se, sicut exposuit Euclides ⟨in sexto tractatu libri sui⟩, dicens: Quando erunt 3 numeri proportionales, erit multiplicatio primi numeri in tertium sicut multiplicatio secundi in se. Multiplicemus ergo quod est in superficie YH de unitatibus, quod est 4, in unitates que sunt in superficie Y[B]A, que sunt 9, et provenient 36. Ergo multiplicatio unitatum que sunt in superficie ZC in se est 36. Ergo superficies ZC est radix (de) 36. Que est 6. Et hoc est quod provenit ex ductu radicis de 9 in radicem de 4, propterea quod linea C[Z]*Y* est radix de 9 et YZ est radix de 4. Et hoc est quod voluimus exponere.

Et si dicemus tibi: Quantum est radix *de* 10, *scilicet: ducta*, in radicem *de* 3? Duc 10 in 3, et provenient 30. Dic ergo: Radix de 30.

Et si dicemus: Quantum sunt 2 radices de 10 in medietatem radicis *de* 5? Scias prius cuius numeri sint radix 2 radices *de* 10 sicut ostendi tibi; et invenies quod sunt radix de 40. Item [intellige] ***animadverte*** cuius numeri sit radix medietas radicis de 5; et invenies sicut tibi exposui quod est radix unius et $\frac{e}{4}$. Est ergo quasi diceremus: Quantum est radix de 40 in radicem *de* 1 et $\frac{1}{4}$. Duc ergo 1 et $\frac{1}{4}$ in 40, et provenient 50. Dic igitur: **(A19^r)** Radix *de* 50. Et ipsa est sicut *2* radi[x]ce*s* *de* 10 in medietatem radicis de 5.

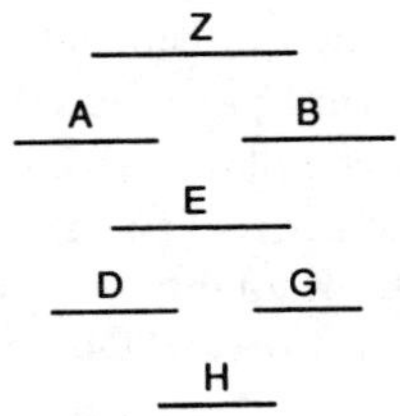

Et ego ponam huic porte formam (vel: figuram) comunem **universalem.* Omn[is]em numeru[s]m multiplicare in numerum et deinde eius quod provenerit accipere radicem est sicut multiplicare radicem unius (eorum) in radicem alterius. Cuius exemplum est quod sit B numerus et A numerus. Radix autem B sit G et radix A sit D. Et ducatur B in A, et proveniat Z; sitque radix Z E. Et multiplicetur G in D, et proveniat H. Dico ergo quod H est sicut E. Cuius demonstratio est: Quia G ducitur in se et producit B, et ducitur in D et producit H, ergo est proportio G ad D sicut B ad H. Et similiter: D ducitur in se et producit[ur] A, et ducitur in G et producit H, ergo [*convertendo*] proportio H ad A[d] est sicut proportio G ad *D*. Iam autem fuit proportio B ad H sicut G ad D. Ergo est proportio B ad H sicut H ad A. Igitur multiplicatio B, primi, in A, tertium, est sicut multiplicatio H, [tert-] secundi, in se. Sed multiplicatio B in A est Z. Igitur multiplicatio H in se est Z. Sed multiplicatio E in se est Z. ⟨Igitur multiplicatio H in se est sicut multiplicatio E in se.⟩ Igitur H est sicut [Z] E. Et hoc est quod voluimus exponere. Et in eo quod tibi exposui in (sc: de) hac specie est sufficientia.

Scias quod cum tu multiplicas partem in partientem, redit partitum **divisum.* Cuius exemplum est quod dividas 10 per **in** 2, et exibunt 5. Ergo cum multiplicabis 5 in 2, provenient **redibunt** 10, qu[i]e [est] *sunt* numerus partitus **divisus.*

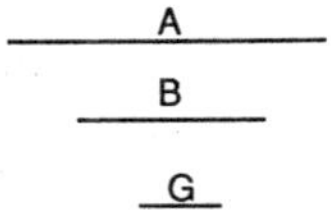

Et ego ponam huic numero (sic, pro: porte) figuram comunem **universalem**, cum adiutorio Dei. Cum quivis numerus dividitur per aliquem numerum, multiplicatio numeri qui exit de [*per*] divisione [*vel: -nem*] in numerum per quem dividitur producit numerum qui fuit divisus. Cuius exemplum est quod numerus dividendus sit A, numerus autem per quem dividitur sit B, numerus qui de divisione (**A19^{v}**) exit sit G. Dico itaque quod multiplicatio B [G] in G [*B*] producit A. Cuius demonstratio est: Quoniam A dividitur per B et exit G, ergo est proportio unius **unitatis** ad G sicut proportio B ad A. Ergo est multiplicatio unitatis, prime, in A, quartum, [est] sicut multiplicatio G, secundi, in B, tertium. Sed multiplicatio unitatis in A producit A. Ergo multiplicatio G in B producit A. Et hoc est quod voluimus exponere.

Et si dixerimus tibi: Divide radicem de 9 per radicem de 4. Divide 9 per 4, et exibunt 2 et $\frac{1}{4}$. Quorum radicem accipe. Que est 1 et $\frac{1}{2}$. Et hoc est radix de 9 divisa per radicem de 4.

Et si dixerimus tibi: Divide radicem de 10 per radicem de 2. Divide 10 per 2, et exibunt 5. Dic ergo exire radicem de 5.

Et si dixerimus tibi: Divide 2 radices de 20 per [tertiam] *3* radic[i]*es* de 6. Scias ergo prius cuius numeri sunt (sic) radix 2 radices de 20 sicut exposui tibi; et invenies quod sunt radix (de) 80. Deinde scias cuius numeri sint radix [$\frac{1}{3}$ radicis] *3 radices* de 6; et invenies quod sunt radix de 54. Est ergo quasi diceremus: Divide radicem de 80 per radicem de 54. Divide igitur 80 per 54, et exibit 1 et [$\frac{4}{9}$ *et*] $\frac{4}{9}\frac{1}{3}$. Et hoc (sic, pro: Et radix huius) est quod [producitur] *producitur ex eo quod* exit de divisione 2 radicum de 20 per 3 radices de 6. Et sic facies in omni eo quod ad te perveniet de hac specie secundum hanc viam.

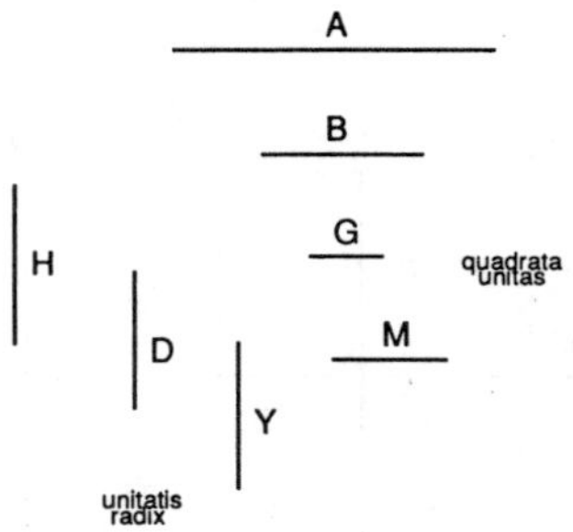

Et ego exponam tibi hanc speciem in figura comuni **universali.* **(78^{v})** Radix numeri qui exit de divisione cuiusvis numeri per alium est sicut illud quod exit de divisione radicis eiusdem, *scilicet: divisi*, per radicem *ipsiusmet* alterius [ipsius], *scilicet: dividentis.* Cuius exemplum est quod numerus A dividatur per numerum B, et exeat numerus G. Radix autem numeri B sit D, et radix numeri A sit H. Dividatur autem H per D, et exeat Y. Dico igitur quod Y est radix *de* G. Cuius demonstratio est: Quia A dividitur per B et exit G, ergo quotiens est G in A, totiens est unitas in B. Ergo est proportio G ad A sicut unitatis ad B. ⟨Similiter: H dividitur per D et exit Y, ergo quotiens est Y in H, totiens est unitas in D. Ergo est proportio Y ad H sicut unitatis ad D.⟩ **(A20^{r})** Et multiplicetur [1] *unitas* in se, et proveniat 1, *unitas*; item multiplicatur D in se, et provenit B, et H in se, et provenit A; et ducatur Y in se, et proveniat M. Ergo est proportio unitatis ad B sicut M ad A {*quia sicut est unitatis ad D, sic est Y ad H; ergo similiter in quadratis unitatis et D et Y et H, que quadrata sunt 1 et M et B* (sc.: B et M) *et A. W.*}. Sed iam fuit proportio G ad A sicut unitatis ad B. Ergo est proportio M ad A sicut G ad A, [sed] [ergo] *et* M et G sunt equalia. Sed Y est radix M. Ergo Y est radix G. Et hoc est quod voluimus exponere.

Si autem vis agregare radicem alicuius numeri radici alterius numeri ut scias cuius numeri ille due radices agregate sint radix: Scias hoc non esse tibi possibile in omnibus numeris, sed in quadratis tantum qui habent radicem, et (in) duobus numeris quorum uno per alterum diviso numerus exeat habens radicem, vel multiplicato uno in alterum numerus producatur habens radicem; in aliis vero numeris ab istis non est possibile illorum radices coniungi ut coniunctum sit radix alicuius numeri. Similiter etiam accidit in subtractione radicis unius numeri a radice alterius. Sed incipiamus [*ab agregatione*] in duobus quadratis; sintque 9 et 4.

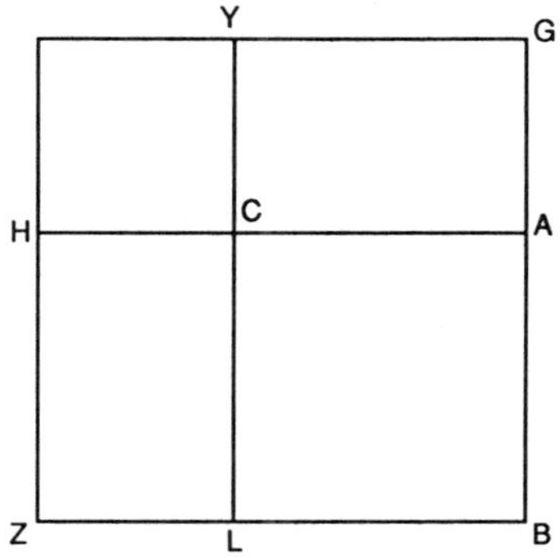

Cum ergo voluerimus agregare radicem de 9 ad radicem de 4 ut sciamus cuius numeri sit radix agregatum ex eis: Agregabimus 9 ad 4, et erunt 13. Servabimus ergo 13. Deinde ducemus 9 in 4, et provenient 36. Quorum accipiemus 2 radices, que erunt 12. Et addemus eas ad 13 que servavimus, et fient 25. Quorum radix est radix de 9 et radix de 4 (agregate), et illa est 5. Quod tibi exponemus in hac figura. Quod est ut faciamus lineam AB radicem de 9 et lineam AG radicem de 4. Cum ergo (**A20**^{v}) libuerit **volemus** scire de linea GB cuius numeri sit radix: Faciemus super lineam GB superficiem quadratam; sitque GZ, linea GB radice ipsius superficiei. Et faciemus super lineam AB quadratum ABCL. Et protrahemus lineam AC secundum rectitudinem ad punctum H, et CL (sc.: LC) secundum rectitudinem ad punctum Y. Igitur superficies CB est 9, propterea quod linea AB est radix de 9; et superficies YH est 4, quia A[H]G, radix de 4, est sicut YC. Superficies autem AY est 6, quia producitur ex multiplicatione radicis *de* 9 [in ra-], que est linea AC, in radicem de 4, que est linea AG; superficies vero CZ etiam *similiter* est 6. Ergo tota superficies GZ est 25. Igitur linea GB est radix (de) 25, que est 5. Et hoc est quod voluimus exponere.

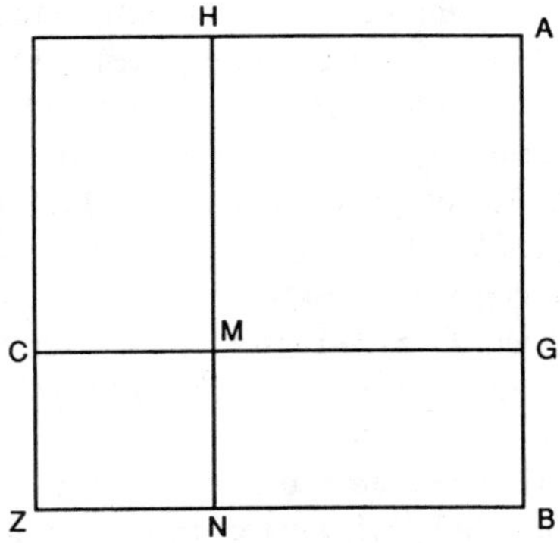

Si vis subtrahere radicem de 4 de radice de 9 ut scias cuius numeri sit radix residuum radicis de 9 subtracta radice de 4: Agrega 9 ad 4, et erunt 13. Que serva. Deinde duc 9 in 4, et provenient 36. Quorum 2 radices, que sunt 12, prohice **subtrahe** de 13 que servasti, et remanebit 1. Cuius radix est 1. Et illud est radix de 9 subtracta radice de 4. Quod tibi exponemus in hac figura. Quod est ut faciamus lineam AB radicem de 9 et lineam AG radicem de 4. Cum autem subtrahemus lineam AG de linea AB, remanebit linea BG. Quam si volueris scire *quanta sit et* cuius numeri sit radix: **(A21^r^)** Faciemus super lineam AB superficiem quadratam AZ. Que erit 9. Item faciemus super lineam AG quadratum AM. Quod erit 4. Et protrahamus lineam HM secundum rectitudinem *ad punctum N* et lineam GM [secundum rect-] in directum ad punctum C. Manifestum est quod quadratum MZ provenit ex ductu GB in se. Que est 1, *unum*, et superficies H[B]C est 2, quia superficies A[C]Z est 9, propter(ea) quod ipsa provenit ex ductu linee AB *in se*, que est radix 9 (sic, pro: Manifestum est quod quadratum MZ provenit ex ductu GB in se. Et superficies HC est 2: quia superficies AC est 6, quia provenit ex ductu AG, que est radix de 4, in CG, que est radix de 9); sed superficies AM est 4; ergo remanet superficies HC 2. Et similiter superficies GN est 2. Remane[n]t ergo quadratum MZ unum. Sed linea MN est sicut linea GB. Ergo linea GB est radix unius, que est unum (sc.: Sed linea MN est eius radix. Que est unum. Et linea MN est sicut linea GB. Ergo linea GB est unum). Et hoc est quod voluimus exponere.

Et si volueris agregare radicem de 18 ad radicem de 8 ut sint radix numeri unius, quia hoc est possibile: *propterea* quia, si diviseris 18 per 8, exibunt 2 et $\frac{1}{4}$, quorum (sic, hic et bis infra, pro: et illis est radix, et eorum) radix est 1 et $\frac{1}{2}$; vel si diviseris 8 per 18 exibunt $\frac{4}{9}$, quorum radix [est] sunt $\frac{2}{3}$; et si duces 18 in 8, provenient 144, quorum radix sunt 12. In quibuslibet enim duobus numeris quorum radices possibile est agregari ut sint radix alicuius numeri concurrunt hec tria iudicia que narravi. Et si aliquod istorum trium concurrit inter aliquos duos numeros, concurrunt inter eos omnia tria. Quod si vis scire [cuius] numerum cuius 2 radices predicte sunt radix, fac sicut narravi tibi. Quod est [quod] ut a*g*greges 18 ad 8, et erunt 26. Que serva. Deinde multiplica 8 in 18, et provenient 144. Quorum 2 radices, que sunt 24, [minue de] *adde ad* 26 que servasti, et erunt 50. Quo⟨rum⟩ radix est radix 18 a*g*gregata ad radicem de 8. **(79^r^)** Si autem volueris subtrahere radicem de 8 de radice de 18: Minue 24 **(A21^v^)** de 26 que servasti, et remanebunt 2. Quorum radix est radix de 18 subtracta radice de 8.

Et si volueris agregare radicem de 10 ad radicem de 2, hoc non est possibile: non *enim* erit aggregatum radix numeri unius, propterea quia, cum diviseris 10 per 2, exibunt 5, que radicem non habent; et si diviseris 2 per 10, exibit $\frac{\text{ta}}{5}$, que similiter non habet radicem; si autem multiplicaveris 2 in 10, provenient 20, que nec etiam radicem habent. Verum si agregaveris eas secundum hoc quod nar⟨r⟩avimus tibi, exi[bit]*bunt 2 radices de 20, que sunt* radix 80, addite [super] ad 12; quorum radix, *scilicet: radicis 80 et 12*, est radix *de* 10 et radix de 2 agregate. Sed, si volueris subtrahere ⟨hanc de illa⟩ et feceris sicut narravi tibi, exibunt [2 radices de 20] 12 preter 2 radices de 20; quorum radix, scilicet residui, est radix *de* 10 subtracta radice de 2. Et questio in hoc est melior responsione. Unde conveniens est, cum nominatur tibi aliquid de hac specie, quod sit responsio tua in eo quod queritur sicut est questio. Melius est enim quod dicas quod "est radix 10 et radix 2" quam quod dicas "est radix [*radicis de*] 12 radice 2 radic[is]um *de* 20 agregata sibi" (sc.: "est 12 et 2 radices de 20, agregati accepta radice", vel: "est radix de 12 et 2 radicum de 20"); et similiter etiam tuum dicere, in subtractione radicis unius numeri de radice alterius, "est radix de 10 subtracta radice de 2" melius est quam tuum dicere: "[est] radix [*superflui radicis super*] de 12 [preter 2 radices de 20] [super] [2 radicibus de 20 subtractis ab ipsis 12] *radice 2 radicum de 20 subtracta*" est radix 10 [et] radi[x]ce 2 [agregate] diminuta; due autem radices de 20 sunt radix de 80. Et in hoc quod tibi exposui est sufficientia tibi, [si Deus] cum adiutorio Dei.

He sunt 6 questiones quas [doc-] dixi tibi quod ponam [cuilibet earum] *unam in unaquaque 6* specierum ad quas te ducet computatio restaurationis **vel: que te ducent ad computationem restaurationis* (sic, pro: ad quam (sc. speciem) te ducet computatio restaurationis (sc. per algebram)):

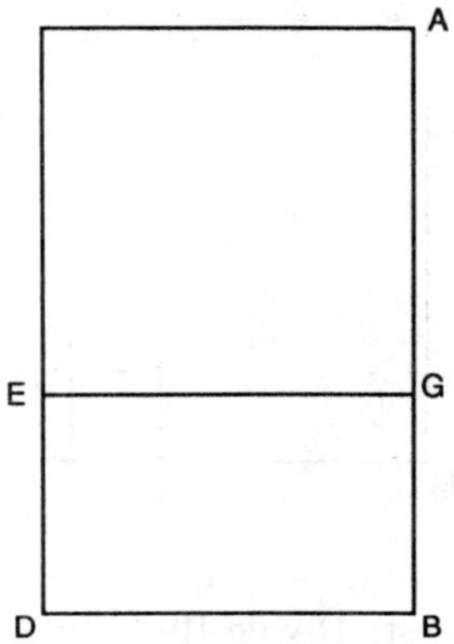

Quarum prima est sicut tuum dicere: Divide 10 in duas partes, et multiplica unam ipsarum in aliam; deinde multiplica partem alteram (sic, pro: maiorem) in se, *et* erit productum *partis ducte* in se sicut multiplicatio unius in alteram duarum partium semel et semis. Cuius exemplum est quod faciamus partem maiorem rem, aliam vero 10 excepta re. Multiplica ergo rem in 10 re diminuta, et erunt 10 res diminuto censu. Deinde multiplica (**A22^{r}**) partem maiorem in se, *id est: (rem) in rem*; et proveniet census equalis 10 rebus diminuto censu, semel et semis. Duc igitur 10 res diminuto censu in 1 et $\frac{1}{2}$; et producentur 15 res diminuto censu et medio, que censui equantur. Restaura **reintegra** ergo 15 per censum et dimidium ut sint 15 res integre, et adde censum et dimidium ad censum **censui**; et erunt 2 census et semis, equales 15 rebus. Ergo unus census equatur 6 radicibus. Ergo radix equatur 6. [Et] Que sunt maior pars, minor vero est 4, que sunt residuum de 10. Iam ergo te duxit hec questio ad primam harum 6 specierum, que est quod census equantur radicibus. Et ego exponam tibi hanc questionem in hac figura. Que est ut faciamus lineam AB 10, lineam autem AG maiorem partem, et lineam GB minorem. Et ducamus lineam AG in BG, et producatur superficies GD, cum fecerimus lineam GE equalem AG. Et ducamus lineam AG in se, et proveniat superficies quadrata AE. E[rit]st igitur superficies *quadrata* AE sicut superficies GD semel et semis. Ergo est linea AG sicut linea GB semel et semis. Igitur ⟨tota⟩ linea AB est sicut linea GB bis et semis. Linea autem AB est 10. Ergo linea GB est 4. Ergo linea AG est 6. Et hoc est quod voluimus exponere.

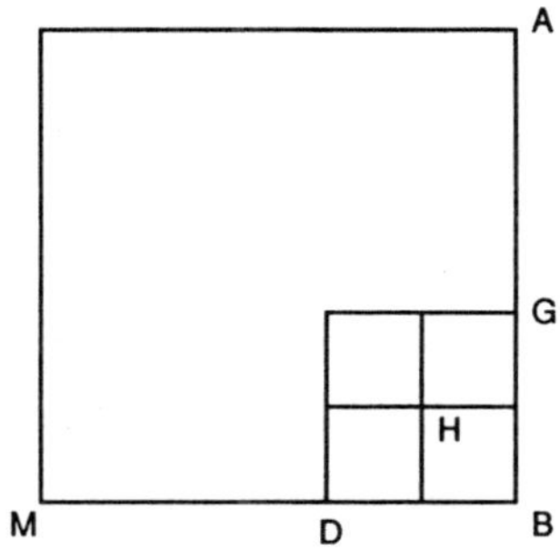

Questio secunda est: Divide 10 in duas partes, et multiplica unam duarum *partium* in se; deinde multiplica 10 in se, sit[*fiet*]que multiplicatio 10 in se sicut multiplicatio unius duarum partium in se [6] secties et 4ª vicis unius. Cuius exemplum est quod facias unam duarum partium rem; et ducas eam in se, et fiet census. Deinde ducas 10 in se, et provenient 100. Sintque hec 100 equalia censui secties et 4ª vicis unius. Duc ergo censum (**A22ᵛ**) in 6 et $\frac{1}{4}$, et fient 6 census et $\frac{1}{4}$ census, qui equantur 100. [Ergo] Reduc igitur eos ad censum unum, qui (de 6 censibus et $\frac{1}{4}$ census) est $\frac{4}{25}$, que sunt $\frac{0}{5}\frac{4}{5}$. Accipe ergo *de 100* $\frac{4}{5}$ quinte unius (sc.: eorum). Que sunt 16. Qu[i]e equantur censui. [Cuius] Quorum radix est 4, que sunt pars quesita. Iam ergo f[e]acit te exire hec questio ad secundam 6 specierum, que est quod census equantur numero. Et eius quod diximus expositio in figura est quod faciamus lineam AB 10 et GB partem quesitam. Deinde ducamus lineam AB, que est 10, in se, et fient 100; que si[n]t superficies *quadrata* AM. Item multiplicemus lineam GB, que est pars quesita, in se, et proveniat quadratum GD. Quadratum ergo AB (sic, pro: AM), quod est 100, est sexcuplum et $\frac{1}{4}$ quadrati GD. Dividamus itaque quadratum GD in quatuor equalia, quorum quodlibet [sit] erit sicut quadratum BH. Ergo erit quadratum AM [25 vicibus equalis] *sicut* quadrat[o]um BH 25 vicibus. Sed quadratum AM est 100. Ergo quadratum BH est 4. Ergo quadratum GD est 16. Quare linea GB, que est eius radix, est 4. Que est pars quesita. Et hoc est quod voluimus exponere.

Questio 3. Divide 10 in duas partes, deinde divide maiorem partem per minorem, et exeat de divisione [5] *4*. Cuius exemplum est quod facias minorem partem rem, maiorem vero 10 re diminuta. Et dividas 10 re diminuta per rem, et exeant 4. Iam ergo (sc.: autem) docui

te quod, quando multiplicas id quod exit de divisione in dividentem, producitur *redit* divisum. Multiplica ergo rem in 4, et producentur 4 res, que equantur 10 [ex-] re diminuta. **(A23^r) (79^v)** Restaura ergo 10 [per rem] *cum re*, et [appone] *adde* rem 4 rebus; et erunt 5 res, equales 10 dragmis. Ergo res equatur 2. Iam ergo te fecit exire questio hec ad tertiam specierum 6 predictarum, que est quod res equantur [ce-] dragmis. Et ego exponam tibi hoc in hac figura. Quod est ut faciamus lineam AB 10, et GB minorem eius partem, AG vero maiorem. Cum autem dividitur linea AB (sic, pro: AG) per lineam GB, ex[eunt]it 4. Ergo linea AG quadrupla est linee GB. Ergo tota iinea AB est quincupla eius. Ergo linea GB est quinta linee AB. Et AB est 10. Ergo GB est 2. Et hoc est quod voluimus exponere.

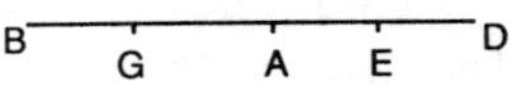

Questio 4^a. Dividuntur 10 in duas partes et ducitur minor in 9 et maior in se, et [[erit] sunt] equalia producuntur. Cuius exemplum est quod faciamus maiorem partem rem, minorem vero 10 subtracta re. Et ducamus rem in se, et fiet census. Et multiplicemus 10 diminuta re [in] in 9, et provenient 90 dragme diminutis 9 rebus, que sunt equales censui. Restaura ergo *eas* cum 9 rebus, et appone illas ad censum; et erunt census et [90] *9* res, equales 90 dragmis. Porta te ducens ad sciendum rem est quod dimidies radices, et erunt 4 et $\frac{1}{2}$. Quas duc in se, et provenient 20 et $\frac{1}{4}$. Que adde (ad) 90, et erunt 1[0]10 et $\frac{1}{4}$. Quorum radicem accipe, et exibunt 10 dragme et $\frac{1}{2}$. Ex quibus prohice medietatem radicum, que est 4 et $\frac{1}{2}$, et remanebunt 6. Que sunt res, et ipsa sunt pars maior. Sed porta te ducens ad sciendum censum est quod ducas 9 res in se, et provenient 81. Que duc in 90, et provenient 7290. Deinde dimidia 81, et exibunt 40 et $\frac{1}{2}$. Que duc in se, et provenient 1640 et $\frac{1}{4}$. Que adde **(A23^v)** ad 7290, et erunt 8930 et $\frac{1}{4}$. Quorum radicem accipe, et exibunt 94 et $\frac{1}{2}$. Que prohice de 40 et $\frac{1}{2}$, que sunt medietas de 81, et de 90, que censui equantur et radicibus, que [sunt] simul sunt 130 et $\frac{1}{2}$, et remanebunt 36. Que sunt census. Iam fecit te exire hec questio ad quartam speciem 6 predictarum, que est quod census et radices equantur numero. Et ego exponam tibi hoc in hac figura. Qu[e]od est ut faciamus lineam AB 10, et lineam AG maiorem partem ipsius, lineam vero GB minorem. Et sit [G]AG in se sicut GB in 9. [Sed AG in 9 et GB in 9 est 9[9]0], sitque linea AD 9 (sic, pro: Ergo AG in se est sicut GB in 9. Sed AG in 9 et GB in 9 est 90 ⟨, et GB in 9 est sicut AG in se. Ergo AG in se et in 9 est 90. Faciamus igitur 9 lineam AD. Ergo AG in se et in AD est 90; et sunt sicut DG in GA. Ergo DG in GA est 90⟩. Et DA est 9.) {*glo-* Ergo quanta est GB in AD, tanta est AG in se. Ergo est proportio BG ad GA, secundam, sicut GA, secunde, ad AD, tertiam. Ergo, componendo, est proportio BG et GA, *prime,* ad GA, secundam, sicut proportio [A]GA et AD, tertie, ad AD. Ergo quanta est multiplicatio BA, prime, in AD, quarta⟨m⟩, tanta est multiplicatio GA, secunde, in GD, tertiam. Sed multiplicatio AB in AD, quartam, est 90, quia AB est 10 et AD est 9. Ergo multiplicatio A[D]G, secunde, in GD, tertiam, est 90. *-sa*}. Et postquam ita est, dividamus lineam AD, que est 9, in duo equalia in puncto E. Linea ergo AD iam divisa est in 2 equalia in puncto E et addita est ei in rectitudine linea AG. Ergo multiplicatio totius linee DG in lineam AG, cum quadrato linee AE, est equalis quadrato EG. Sed multiplicatio AG in [A]D*G* iam fuit 90, et quadratum linee AE est 20 et $\frac{1}{4}$, cum ipsa *sit* 4 et $\frac{1}{2}$. Ergo quadratum linee EG est 110 et $\frac{1}{4}$. Ergo linea EG est 10 et $\frac{1}{2}$. Sed AE est 4 et $\frac{1}{2}$. Ergo linea AG remanet 6. Que est maior pars. Igitur linea GB est 4. *Que·est* minor pars [linee]. Et hoc est quod voluimus exponere.

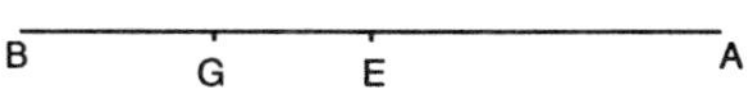

Questio 5. [Dividuntur] Divide 10 in duas partes et multiplica unam illarum in alteram, et sint **proveniant** 21. Cuius exemplum est quod facias unam duarum partium rem et alteram 10 diminuta re. Et multiplices rem in 10 diminuta re, et provenient 10 res diminuto censu, que equantur 21. Restaura 10 radice*s per* censum, et adde eum ad 21; et erunt census (**A24^r^**) et 21, equalia 10 radicibus. Porta te ducens ad scientiam rei est ut dimidies res, et exibunt 5. Que duc in se, et provenient 25. Ex quibus prohice 21, et remanebunt 4. Quorum radicem accipe. Que est 2. Minue igitur eam de 5, que sunt medietas radicum, et remanebunt 3. Que sunt una pars; reliqua vero est residuum de 10, quod est 7. Sed, si vis, adde 2 ad 5, et erunt 7. Que sunt una duarum partium de 10, reliqua autem est residuum, scilicet 3. Porta vero te ducens ad censum est quod ducas 10 radices in se, et provenient 100. Que duc in 21, ⟨que sunt cum censu,⟩ et provenient 2100. Deinde dimidia 100, et exibunt 50. Que duc in se, et provenient 2500. Ex quibus minue 2100, et remanebunt 400. Quorum radicem accipe. Que est 20. Que minue de 50, que sunt medietas de 100, et remanebunt 30. De quibus minue 21, que fuerunt cum censu, et remanebunt 9. Que sunt census. Sed [per] *si cum* additione[m] *vis invenire*: Adde 20 ad 50, et erunt 70; de quibus minue 21, et remanebunt 49. ⟨Que sunt census.⟩ Iam [fecit te] traxit te hec questio ad quintam specierum 6 dictarum, que est quod census et numerus equantur radicibus. Et ego exponam tibi hoc i[n i]sta figura. Que (sc.: Quod) est ut faciamus lineam AB 10; et dividamus eam in duas partes in puncto G; sitque multiplicatio AG in GB 21. Dividamus ergo lineam AB in duo equa(lia)

in puncto E. Est ergo multiplicatio AG in GB, et quadratum EG, sicut quadratum EB. Sed quadratum EB est 25, et multiplicatio AG in GB est 21. Remanet ergo quadratum EG 4. Ergo linea EG est 2. Sed EB est 5. Ergo GB est 3, et AG est 7. Et hoc est quod voluimus exponere.

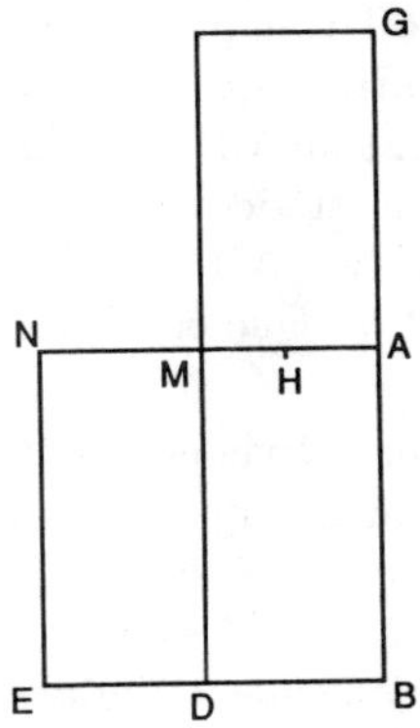

(**A24^{v}**) Questio 6^{a}. Census est, super quem **vel: cui** si addi*deris* 8 dragmas, deinde duces [in se] agregatum in 4 dragmas, proveniet equale (**80^{r}**) ipsi censui ducto in se. Cuius exemplum est quod faciamus censum rem, et addamus ei 8 dragmas; et erunt res et 8 dragme. Que multiplica*bimus* in 4, et [erunt] *provenient* 4[0] res et 32 dragme. Multiplica*bi*mus etiam censum, qui est res, in se, et proveniet census, qui equatur 32 dragmis et 4 rebus. Porta te ducens ad sciendum rem est quod dimidies res, et exibunt 2. Que multiplica in se, et provenient 4. Que adde ad 32, et erunt 36. Quorum radicem accipe. Que est 6. Et adde eis medietatem radicum, que est 2, et erunt 8. Que sunt census qu[i]em pos[itus fuit]*uimus* re[s]m. Sed p[roptereal]orta te ducens ad censum qui equatur radicibus et numero est quod ducas 4 radices in se, et provenient 16. Que percute in dragmas, que sunt 32, et provenient [5] 512. Deinde dimidia 16, et exibunt 8. Que duc in se, et provenient 64. Que adde ad 512, et erunt 576. Quorum radicem accipe, et exibunt 24. Que adde ad 8, que sunt medietas de 16, et *ad* 32, que sunt cum radicibus, ⟨que simul sunt 40,⟩ et erunt 64. Que sunt census qui equatur radicibus et numero. Cuius radicem accipe. Que est 8. Qui est census quesitus. Iam traxit te hec questio ad speciem 6^{tam} specierum 6 predictarum, que est quod numerus et radices equantur censui (sic, pro: censibus). Et ego exponam tibi hoc in figura. Quod est ut faciamus [lineam AB] censum lineam AB, et 8 dragmas lineam AG, et faciamus lineam BD 4. Multiplicemus igitur lineam BG in BD, et proveniat superficies GD. Deinde ducamus lineam AB, que est (**A25^{r}**)

census, in se, et proveniat quadratum AE. Erit igitur superficies GD sicut quadratum AE. Ergo, superficie AD abiecta comuni, remanebit superficies ME sicut superficies MG. Que est 32, propterea quia linea GA est 8 et linea AM, que est sicut [B]DB, est 4. Ergo superficies ME est 32. Sed ipsa provenit ex ductu AN, que est sicut NE, in NM. Ergo multiplicatio linee AN in NM est 32. ⟨Et AM est 4.⟩ Dividamus igitur lineam AM in duo equa(lia) in puncto H. Et erit multiplicatio linee AN in NM cum quadrato MH sicut quadratum NH. Sed multiplicatio AN in NM est 32; et quadratum MH est 4, quia MH est duo, quia ipsa est medietas linee AM, que est sicut DB, que est 4. Ergo quadratum NH est 36. Ergo HN est 6. Sed AH est 2. Ergo [Ergo] AN est 8. Que est sicut AB. Ergo AB est 8. Que est census. Et hoc est quod voluimus exponere.

Et si dixerimus tibi: Divi[de]*sisti* 10 in duas partes et multiplicasti unamquamque partem in se, et minuisti minus de maiori, et remanserunt [[10]80] *80*. Eius exemplum est quod faciamus partem minorem rem, maiorem vero 10 excepta re. Quam multiplicemus [unamquamque] in se, et provenient 100 dragme et census diminutis [10] *20* rebus. Deinde multiplicemus rem in se, et proveniet census. Quem minue de 100 et censu diminutis 20 rebus, et remanebunt 100 dragme diminutis 20 rebus, que equantur 80. Restaura igitur 100 dragmas cum 20 rebus, et adde eas [cens-] *ad* 80, et erunt 20 res et 80 dragme, equales 100 dragmis. Minue ergo 80 de 100, et remanebunt 20 dragme, que equantur 20 rebus. Ergo res equatur uni. Que est pars minor. Ergo m[in]aior est 9, que sunt residuum de 10. Et si faciemus maiorem partem rem, minorem vero 10 diminuta re, minuemus [censum] de censu *censum* et 100 dragm[i]*as* diminutis 20 rebus; remanebunt [100 dragme] *20 res* diminutis [20 rebus] *100 dragmis*, que equantur 80 dragmis. Restaura ergo [[cu-] 100 cum] 20 re[bus]s *cum 100 dragmis*, **(A25^{v})** et adde eas 80 dragmis, et erunt 1[0]80, que equantur 20 rebus. Ergo res equatur 9. Que est maior pars, residua vero *pars de 10, que est* minor, [que] est 1. Et, si volueris, divide[re] 10 in duas partes preter consuetam divisionem cogitantium, *id est: alia regula*, dividere 10; et erit tibi leve opus ad sciendum partem maiorem et minorem, et facies cadere *eandem regulam* in plurimo illius quod tibi nominatur de questionibus [in ill-] ponderatis cum divisione radicum (sic, pro: et erit tibi leve opus et discernere poteris partem maiorem a parte minore, et levabit te labore mediandi radices (sc. solvendi aequationes secundi gradus compositas) in multis questionibus que tibi nominantur). Quod est [*Que sunt*] [quod] *ut* faciamus maiorem duarum partium 5 et rem, minorem vero 5 diminuta re. Et ducatur [una] utraque earum in se, et prohiciatur illud quod provenit ex minore de eo quod provenit ex maiore earum, et remanebunt 20 res, que equantur 80 dragmis. Ergo res equatur 4. Que agregentur ad [maiorem] 5 ⟨, vel minuantur ex eis⟩; et erunt *9, que* *sunt* maior pars, reliqua vero unum.

Et si dicemus tibi: Divide 10 in duas partes, et divide unam*quamque* per aliam, *id est: unam per aliam et aliam per aliam*, et exeant 4 et $\frac{1}{4}$. Exemplum eius est quod facias unam duarum partium rem, reliquam vero 10 diminuta re. Et ducas rem in se, et proveniet census. Item ducas 10 re diminuta in se, et producentur 100 et census diminutis 20 rebus. Que agrega, et erunt 100 dragme et 2 census diminutis 20 rebus. Que reserva. Deinde multiplica unam duarum partium in alteram, id est, rem in 10 diminuta re, et provenient 10 res diminuto censu. Quas multiplica in *id* quod exit de divisione, s*cilicet* in 4 et $\frac{1}{4}$, et provenient 42 res et dimidia res diminutis 4 censibus et $\frac{1}{4}$ census, que equantur ⟨ei quod exit de multiplicatione utriusque partis in se, scilicet⟩ 100 dragmis et duobus censibus diminutis 20 rebus: quia *in* omn[es]*ibus* duo*bus* numeri*s* quorum [unus] *uterque* dividitur per alterum er[i]unt multiplicationes utriususque (sic) duorum numerorum in se agregate sicut multiplicatio unius duorum numerorum in alterum, deinde in id quod exit de divisione utriusque duorum numerorum per alium, *scilicet: primi per secundum et secundi per* ⟨*p*⟩*rimum.* Et ego exponam tibi causam huius post complementum questionis. Restaura ergo 100 dragmas et 2 census ⟨diminutis 20 rebus⟩ cum 20 radicibus, et adde illas *ad* 42 res et $\frac{1}{2}$ (**A26ʳ**) rei diminutis 4 censibus et $\frac{1}{4}$; et erunt 62 res et $\frac{1}{2}$ rei diminutis 4 censibus et $\frac{1}{4}$ census, equales 100 dragmis et 2 censibus. Restaura item 62 res et $\frac{1}{2}$ rei cum 4 censibus et $\frac{1}{4}$, et adde illos 100 dragmis et 2 [et] censibus; (**80ᵛ**) et erunt 100 dragme et 6 census et $\frac{1}{4}$ census, que equantur [42] *62* rebus et $\frac{1}{2}$ rei. Reduc ergo ⟨eos⟩ ad censum unum. Et iam scivisti quod census unus de 6 *censibus* et $\frac{1}{4}$ census est [$\frac{4}{5}$. Accipe igitur] $\frac{0\;4}{5\;5}$. Accipe igitur totius quod habes $\frac{0\;4}{5\;5}$, et exibunt census et 16 dragme [que equantur] equales 10 rebus. Dimidia ergo res, et exibunt 5. Quas duc in se, et provenient 25. Ex quibus minue 16, et remanebunt 9. Quorum radicem, que ⟨est⟩ 3, minue de medietate radicum, et remanebunt 2. Que sunt una duarum partium, alia vero est 8. [*va-* Et si vis, divide 10 in duas partes in eo quod exit a divisione que non est ex computa*tione* cogitantium quod divid[as]ant 10 (sic, pro: Et si vis, divide 10 in duas partes secundum id quod ⟨te docui⟩ ex divisione que non est ex consuetudine cogitantium ut dividant 10); et erit tibi leve opus {*G. id est, aliter quam sit divisio [ipsorum] 10 in rem et 10 re diminuta, et est: Dividas 10 in 5 et rem et 5 re diminuta*}. Hic deficit. Scriptum est in Ebraico: "Hic deficit". *-cat*].

{*Et potest sic leviter fieri, sicut mihi videtur. Dividas 4 et $\frac{1}{4}$ in 2 partes quarum una ducta in alteram faciat unum [quaru-]. Et hoc iam supradictum est qualiter fieri debeat. Et exibit maior pars 4, minor vero unum* (sic, pro: $\frac{1}{4}$).*Duc ergo 4, que est maior pars, in 5 minus re; et provenient 20 minus 4 rebus, equales 5 et rei. Restaura et oppone,* *et invenies rem equari 3. Que [d-] addas addas* (sic) *ad 5, et erunt 8. Que est maior pars 10. Vel minuas a 5, et remanebunt 2. Que est minor. Vel, si volueris, ducas* $\frac{1}{4}$, *que est minor pars de 4 et* $\frac{1}{4}$, *in 5 et rem, que est maior pars de 10, et proveniet 1 et* $\frac{1}{4}$ *[et res] et* $\frac{1}{4}$ *rei, equales 5 minus re. Restaura ergo 5 cum re, et adde eam ad 1 et* $\frac{1}{4}$ *et rem* (sic) *et* $\frac{1}{4}$ *rei; et erunt 5, equales uni dragme et* $\frac{1}{4}$ *dragme et [2 rebus] rei et* $\frac{1}{4}$ *rei. Prohice ergo dragmam et* $\frac{1}{4}$ *de 5 dragmis, et remane(bu)nt 3 dragme et* $\frac{3}{4}$ *dragme, equale⟨s⟩ [2 rebus] rei et* $\frac{1}{4}$. *Igitur res equatur 3. Que similiter adde ad 5, et erunt 8. Que est maior pars de 10, residua vero est duo. G.}* (. . .) (**A26^v^**) (. . .)

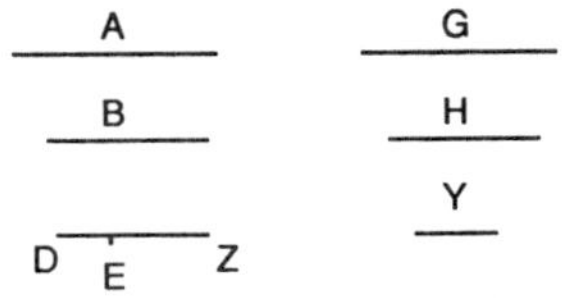

Omnium duorum numerorum si dividatur unusquisque eorum per alterum, erit quod exibit de divisionibus duabus sicut illud quod exibit de divisione multiplicationis uniuscuiusque duorum numero⟨rum⟩ in se, *id est: duorum quadratorum coniunctorum ipsorum*, per multiplicationem unius eorum in alium. Cuius exemplum est quod numerus A dividatur per numerum B, et exeat DE; et dividatur numerus B per numerum A, et exeat EZ; et multiplicetur A in se, et proveniat G; et multiplicetur numerus B in se, et proveniat H; et multiplicetur A in B, et proveniat Y. Dico quod, quando G H (GH in cod., hic et infra pro: G et H (agregati)), *id est: unusquisque eorum*, dividetur per Y, exibit DZ. Cuius demonstratio est: Quia A dividitur per B et exit DE et multiplicatur, *id est: A*, in se et provenit G, et multiplicatur A in B et producitur Y, *ergo*, quando dividitur G per Y, exibit DE, secundum illud quod est in figura que est supra istam. {*Glosa. Quod sic probatur. Quia A ducitur in se et provenit G, et ducitur in B et provenit Y, ergo est proportio G ad Y, productorum, sicut A ad B, producentium. Et quia A dividitur per B et exit DE, ergo est proportio A ad B sicut DE ad unitatem. Sed proportio A ad B fuit sicut G ad Y. Ergo est proportio DE ad unitatem sicut G ad [B]Y. Proportio autem eius quod exit de divisione G per [B]Y [eadem ratio-] ad unitatem similiter est sicut proportio G ad [B]Y. Ergo est proportio DE ad unitatem sicut eius quod exit de divisione G per [B]Y ad unitatem. Ergo DE est quod exit de divisione G per [B]Y. Et hoc est quod demonstrare voluimus. Guill. Hic videtur esse defectus, sed compleri potest sicut diximus.*} Et sic etiam: B dividitur per A et exit EZ, et B multiplicatur in se et producitur H, et multiplicatur in A et fit Y, (**A27^r^**) ergo, quando dividetur H per Y, exit EZ, sicut exposuimus *etiam*. Iam ergo planum est quod, quando dividitur H G per Y, exit DZ. Ergo quando ducemus DZ in Y, proveniet H G. Iam ergo manifestum est quod omnium duorum numerorum quorum unus*quisque* dividitur per alium, erunt *2* quadrata ipsorum simul iuncta sicut multiplicatio unius eorum in alterum ducta in id quod exit de divisionibus duabus duorum numerorum uniuscuiusque eorum per alium. Et hoc est quod voluimus exponere.

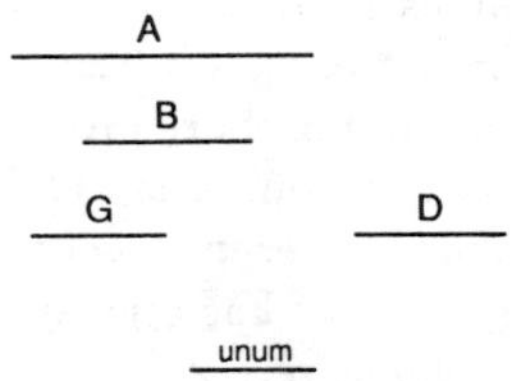

Omnium duorum numerorum quorum unusquisque dividitur per alium, multiplicatio unius duarum partium, *id est: exeuntium de divisione*, in aliam producit unum, *dragmam*, *[tantum]* semper. Cuius exemplum est quod dividamus numerum A per numerum B, et exeat G. Et dividamus numerum B per numerum A, et exeat D. Dico igitur quia multiplicatio G in D est [dragma] semper dragma. Cuius demonstratio est: Quia A dividitur per B et exit G, ergo B est in [G]A secundum unitates G. Ergo unitas est in G secundum proportionem B ad A. Ergo cum [permutaverimus] *converterimus*, erit proportio A ad B sicut proportio G ad unum. Et ita etiam: B dividitur per A et exit D, ergo est [*proportio*] A [ad] *in* B s[icut [*sicut*] proportio]ecundum unitates D. Ergo est proportio unitatis ad D [est] sicut proportio A[B] *ad B* [Ergo convertendo erit proportio BA et B ad A sicut G ad unum]. Sed iam fuit proportio A ad B sicut proportio G ad **(A27ᵛ)** [dragmam] unum. Ergo proportio [unius] G ad 1, *unum*, est sicut proportio unius ad D. Ergo est multiplicatio G in D sicut quadratum unius. Sed [mult-] quadratum unius est unum. Ergo [G] multiplicatio G in D est unum. Et hoc est quod voluimus exponere.

Et, si volueris, intellige (sc.: computa) [super] hoc *secundum* viam aliam. Que est: Si dixerimus tibi: 10 divide in duas partes, et divide unamquamque earum per aliam, et exeant 4 et $\frac{1}{4}$. Fac igitur unam duarum partium rem, aliam vero 10 diminuta re. Et divide 10 excepta re per rem, et exibit de divisione id quod si multiplicetur in rem, provenient 10 diminuta re. Multiplica igitur 4 et $\frac{1}{4}$ in rem, et provenient 4 res et $\frac{1}{4}$ rei. Et prohice ex ipsis 10 dragmas diminuta re, et remanebunt 5 res et $\frac{1}{4}$ *rei* exceptis 10 dragmis. Deinde multiplica rem in rem, et proveniet census. Quem divide per 10 dragmas diminuta re, et exibunt 5 res et $\frac{1}{4}$ rei diminutis 10 dragmis. { *Verbi gratia. A, res, dividatur per B et exeat DE, et B per A et exeat EZ. [Ergo est proportio EZ ad uni-*

tatem sicut unitatis ad DE, sicut probatum est nuper]. Et ducatur DZ in A, et proveniat G H. Deinde auferatur ab eo B, et remaneat G Y. Dico ergo quod [si quadratum de D] B ductum in G Y producit quadratum de A, qui est census . . . quia H (sc. B), *qui est sicut H Y, provenit ex multiplicatione EZ in A. Ergo G Y, residuum, provenit ex ductu DE, residui, in A. Sed A provenit ex ductu eiusdem DE in B. Ergo est proportio B ad A sicut A ad G Y . . .* }. [Quia] Iam *enim* exposuimus quia, quando multiplicabimus quod exit de divisione in divide⟨n⟩tem, redibit divisum sicut fuit prius. Sed multiplicatio 5 rerum et $\frac{1}{4}$ rei diminutis 10 dragmis in 10 dragmas diminuta re producit 62 res et $\frac{1}{2}$ rei diminutis 5 [*6*] censibus et $\frac{1}{4}$ census et diminutis 100 dragmis, que equantur censui. Restaura ergo et appone, et erunt 6 census et $\frac{1}{4}$ census et 100 dragme, equales [4] 62 rebus et $\frac{1}{2}$ rei. Fac ergo sicut exposui tibi, et exibit una duarum partium 8, reliqua vero 2.

[est]E*st* etiam via alia. Que est quod faciamus unam duarum partium rem [rem], reliquam vero 10 excepta re. Deinde divide 10 excepta re per rem, et exeat dinar. Ergo cum multiplicabis dinar in rem, provenient10 diminuta re. Deinde divide rem per 10 diminuta re, et ex[eant]*ibunt* 4 et $\frac{1}{4}$ diminuto dinar. Ergo multiplica 4 et $\frac{1}{4}$ diminuto dinar [*in rem*, et exibunt] *in* 10 diminuta re. Ergo erunt 52 dragme et $\frac{1}{2}$ diminutis 5 rebus et $\frac{1}{4}$ rei ⟨et⟩ diminutis 10 dinarim, equales rei {*Glosa. Quia dinar diminutum in rem diminutam facit 10 addita diminuta re. Que adde ad 42 et $\frac{1}{2}$, et erunt 52 et $\frac{1}{2}$, rem autem diminuta⟨m⟩ adde ad 4 res diminutas et $\frac{1}{4}$ rei, et erunt 5 res diminute et $\frac{1}{4}$ rei. Verum est ergo quod dicit auctor. G.*}. Restaura ergo et appone sicut tibi exposui, et erunt 10 dinarim, equales 52 dragmis (**A28**^{r}) et [$\frac{1}{4}$] $\frac{1}{2}$ diminutis 6 rebus et $\frac{1}{4}$. Ergo dinar equatur 5 dragmis et $\frac{1}{4}$ diminutis $\frac{5}{8}$ rei. Iam autem (sc.: igitur) scivisti quia, quando tu divides 10 dragmas diminuta re per rem, exibunt 5 et $\frac{1}{4}$ diminutis $\frac{5}{8}$ rei. Multiplica igitur 5 [vel] et $\frac{1}{4}$ di(**81**^{r})minutis $\frac{5}{8}$ rei in rem, et provenient 5 res et $\frac{1}{4}$ rei diminutis $\frac{5}{8}$ census, equales 10 dragmis diminuta re. Restaura et appone in ipsis, et erunt 6 res et $\frac{1}{4}$ rei, que equantur 10 dragmis et $\frac{5}{8}$ census. Reduc ergo $\frac{5}{8}$ ⟨census⟩ ad censum unum. Quod est ut addas eis $\frac{3}{5}$ census (sic, pro: $\frac{3}{5}$ earum); adde etiam ad omne quod habes $\frac{3}{5}$ eius. Et erunt census et 16 dragme equales 10 [dragmis] *rebus* [et $\frac{1}{2}$ rei]. Dimidia ergo res, et exibunt 5. Quas duc in se, et provenient 25. De quibus prohice 16, et remanebunt 9. Quorum radicem accipe, que sunt 3; et prohice ea[s]m de medietate radicum, que est 5, et remanebunt 2. Que sunt una duarum partium, reliqua vero est 8.

Est etiam in ipso (vel: in ipsa (sc.: questione)) [*est*] via alia. Iam preposuimus et diximus quia in omnibus duobus numeris, cum dividitur unusquisque eorum per alium, multiplicatio eius quod exit de duabus divisionibus, unius scilicet in aliud, sem*per* producit unum. Quod autem exit de duabus divisionibus in hac questione est 4 $\frac{1}{4}$. Faciamus igitur [una-] quod exit de una divisione rem, quod autem exit de alia 4 et $\frac{1}{4}$ re diminuta. Deinde ducamus unum eorum in aliud, et provenient 4 res et $\frac{1}{4}$ rei diminuto censu, que equantur dragme. Restaura ergo in eis et appone, eritque res, si voluerimus, 4, si autem voluerimus, $\frac{1}{4}$. Deinde redibimus ad principium questionis, et dicemus unam duarum partium rem, reliquam vero 10 *dragmas* diminuta re. Et dividamus 10 dragmas re diminuta per rem, et ex[ibu]eant 4. Ergo *si* multiplicemus 4 in rem, etiam provenient 4 res, equales 10 dragmis diminuta re. Restauremus ergo et apponamus in eis, et er[un]*i*t res 2. Que est una duarum partium, reliqua vero est residuum de 10, qu[e]od est 8. Et, si volueris, dic: 10 ⟨diminuta re⟩ divisimus per rem, et exivit $\frac{1}{4}$ dragme. Multiplica ergo $\frac{1}{4}$ dragme in rem, et provenient $\frac{1}{4}$ rei, que equatur 10 dragmis diminuta re. Restaura *ergo* et oppone *in* ipsis, et erit res 8. Que est una **(A28^{v})** duarum partium, reliqua vero est 2.

Et si dixit: Divide 10 in duas partes, et divide unamquamque earum per aliam; deinde prohice unum duorum, *id est: exeuntium*, de alio, et remaneant $\frac{5}{6}$ dragme. [Ver*bi* gratia] Verbi gratia, faciamus unam duarum partium rem, reliquam vero 10 diminuta re. Et ducamus rem in se, et proveniet census. Deinde ducamus 10 diminuta re in se, et provenient 100 dragme et census diminutis 20 rebus. Prohice ergo unum eorum ex alio, et remanebunt 100 dragme diminutis 20 rebus. Serva igitur numerum hunc (sc.: congregatum hoc). Deinde duc unam duarum partium in aliam, scilicet rem in 10 diminuta re, et provenient 10 res diminuto censu. Que multiplica in residuum duarum partium postquam proieceris unum eorum ex alio, quod est $\frac{5}{6}$ dragme, et provenient 8 res et $\frac{1}{3}$ rei diminutis $\frac{5}{6}$ cens[ibus]us, que equantur 100 dragmis *diminutis 20 rebus. Restaura [ergo et oppone] ergo eas cum 20 rebus, et adde eas ⟨ad⟩ 8 res et $\frac{1}{3}$ ⟨rei diminutis $\frac{5}{6}$ census⟩; et erunt 28 res et $\frac{1}{3}$ rei diminutis $\frac{5}{6}$ census, equales 100 dragmis.* Restaura ergo *eas* cum $\frac{5}{6}$ ⟨census⟩, et adde ⟨eas⟩ 100 dragmis, et erunt 100 dragme et $\frac{5}{6}$ *census*, equales 28 rebus et tertie rei. Re[staura erg-]*integra* item $\frac{5}{6}$ census donec sit census ⟨unus⟩; [con-] re[staurabis]*integrabis* autem ut addas ei $\frac{1}{5}$ eius (sc.: ut addas eis $\frac{1}{5}$ earum); et adde [si-] ei toti quod habes $\frac{1}{5}$ ipsius. Et erunt census et 1[0]20 dragme [et 20 dragme] equales [4] 34 rebus. Dimidia igitur radices, et multiplica eas, *id est: dimidiatas*, in se, et prohice ex eo quod provenit 120 dragmas; et remanebunt 169. Quorum radicem accipe, ⟨que est 13,⟩ et prohice eam de medietate radicum, que est 17, et remanebunt 4. Que sunt una duarum partium, reliqua vero est 6.

Causa autem huius questionis est sicut causa questionis superioris. Sed differentia inter ipsas est quod necessarium fuit in questione prima hoc, scilicet quod in omnibus duobus numeris ⟨quorum unusquisque dividitur per alium⟩ multiplicatione*s* **(A29^{r})** uniuscuiusque eorum in se coniuncte sunt sicut multiplicatio unius *duorum* numerorum in alium, deinde in id quod exit de divisionibus uni⟨us⟩cuiusque eorum per alium; et iam exposuimus hoc (vel: causam huius). In hac questione vero necessarium est quod in omnibus duobus numeris quorum unusquisque dividitur per alium, erit, *cum* multiplicatio unius eorum in se subtrahetur de multiplicatione alterius in se, residuum sicut multiplicatio unius duorum numerorum in alium ducta in superfluum illius quod exit de divisione unius duorum numero⟨rum⟩ per alium subtracto ex eo quod exit de divisione alterius per primum. Ille [tamen] *etiam* {*videtur quod contrarium debeat dicere*} cui grav[e]i*s* [erit invenire] *fuit* causa[m huius] questionis [, per causam] prioris inveniet eam [,] ex causa questionis istius, quoniam hec similis est prime. Et tres modi qui sunt in questione prima intellig[i]*un*tur in hac questione, cum adiutorio Dei (sic, pro: Ille qui de causa prime questionis meditatus erit inveniet causam huius questionis, quoniam hec similis est illi. Et secundum tres modos qui sunt in questione prima operabit in hac questione, cum adiutorio Dei).

Et si dicemus tibi: Divisisti 50 ⟨dragmas⟩ [hominibus] per homines, et contigit cuilibet res; deinde addidisti illis 3 homines, et divisisti item per ipsos 50 dragmas [et $\frac{1}{2}$ et $\frac{1}{4}$], et conti[n]git cuilibet eorum [*quod* minus eo] quod contigit cuilibet primorum *diminutis* 3 dragmis et $\frac{1}{2}$ et $\frac{1}{4}$. [Verbi gratia] *Eius expositio est, *sive solutio**: Multiplica homines primos in diminutum quod est inter illud quod contigit cuilibet primorum et quod contingit cuilibet secundorum; et quod provenerit de multiplicatione divide per superfluum quod est inter homines primos et secundos; et quod exibit duc in homines secundos. Et quod proveniet est census divisus (sc.: 50).

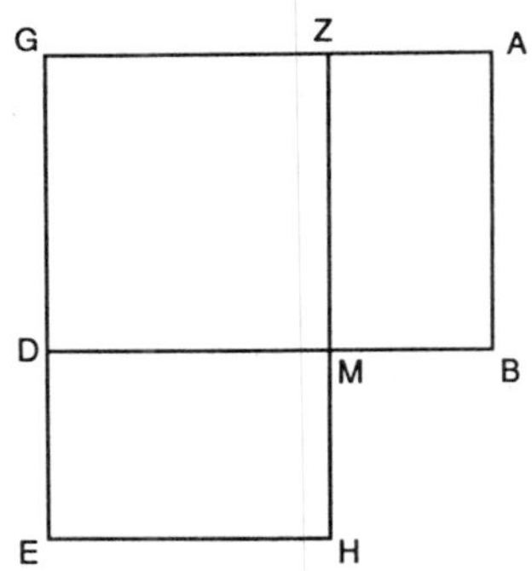

Et ego exponam tibi causam huius porte. Quod est ut faciamus numerum divisum, qui est 50 dragme, ⟨superficiem⟩ ABGD, [residuum autem] homines vero primos lineam GD, et quod contigit cu*i*libet ipsorum lineam AG. Quia linea (**81**^{v}) AG in GD [est] *producit* superficie[s]m GB, cum ergo diviserimus superficiem AGBD per lineam DG, que est homines primi, exibit quod contingit uni linea AG. Deinde addemus [-s] hominibus primis, qui sunt linea GD, 3 homines, qui sint linea DE; *et* erit linea G[B]E homines secundi **ultimi*. Faciamus etiam superficiem GEZH 50, et dividamus eam per lineam GE, qu[od]e est homines ultimi; et exibit linea GZ. Que *est* id *quod* contingit uni (**A29**^{v}) hominum ultimorum. Ergo linea AZ est 3 et $\frac{1}{2}$ et $\frac{1}{4}$, cum ipsa sit superfluum illius quod contingit uni primorum super eo quod contingit uni ultimorum. Multiplica ergo lineam MZ, que est homines primi, quia ipsa est sicut GD, in lineam AZ, que est superfluum *illius* quod contingit uni primorum super eo quod ⟨con⟩tingit uni postremorum. Et producetur superficies ZB. Que est sicut superficies ME: quia superficies AD est sicut superficies GH ⟨postquam unaqueque earum est 50⟩; et superficies GM est comunis; (sublata superficie GM) remanet [igitur] ergo superficies ZB sicut superficies DH. Dividamus etiam superficiem DH per lineam DE, que est superfluum quod est inter homines priores et posteriores; et exibit linea DM. Que est sicut linea GZ, que est id quod contingit uni posteriorum hominum. Que ducta in lineam GE, que est homines ultimi, producit superficiem GH, que est 50. Et hoc est quod voluimus exponere.

Et postquam expositum est hoc ⟨et stabilita est causa huius porte⟩, faciamus homines priores rem. Et ducamus eam in diminutum quod est inter hoc quod contingit uni priorum et quod contingit unius (sic) posteriorum, quod est 3 et $\frac{1}{2}$ et $\frac{1}{4}$. Et quod provenerit (sc. 3 res et $\frac{1}{2}$ et $\frac{1}{4}$ rei) dividamus per superfluum quod est inter homines priores et posteriores, quod est 3. Et exibit res et $\frac{1}{4}$. Quam ducamus in homines ultimos, qui sunt res et 3, et provenient census et $\frac{1}{4}$ et 3 res et $\frac{3}{4}$ rei, equales 50 dragmis. Reduc ergo totum quod habes ad censum unum; et erunt census 1 et 3 radices equales 40. {*Quia $\frac{1}{4}$ census unius et $\frac{3}{4}$ radicis unius eiusdem sunt equales 5*^{e} *numeri totius cui omnia predicta dicta sunt adequari.* (Nota lectoris)} Dimidia igitur radices, et duc [medieta-] eas in se, et adde quod provenerit ad 40; et erunt 42 et $\frac{1}{4}$. Quorum radicem, que est 6 et $\frac{1}{2}$, [minue] accipe (sic, pro: Quorum radicem accipe. Que est 6 et $\frac{1}{2}$;) et minue ex ipsis medietatem radicum, que est 1 et $\frac{1}{2}$; et remanebunt 5. Qu[i]e sunt homines primi.

Et si dixerimus tibi: Divisisti 10 dragmas per homines, et conti[n]git uni*cuique* eorum res; deinde addidisti (**A30**^{r}) illis *hominibus* 4, [dein-] et postea divi*si*sti per ipsos 30 dragmas, et contigit unicuique posteriorum minus eo quod conti[n]git uni primorum 4 dragmis. Eius exemplum est quod ducas homines priores in diminutum quod est inter hoc quod conti[n]git uni priorum et uni posteriorum; deinde addas ei quod provenerit superfluum quod est inter [homines per quos dividitur] divisa; et quod [prov-] agregabitur divide per superfluum quod est inter homines priores et posteriores; et quod exibit duc *in* (homines) posteriores. Et quod agregabitur est numerus [numerus] maior qui dividitur, et est in hac questione 33 (sic, pro: 30).

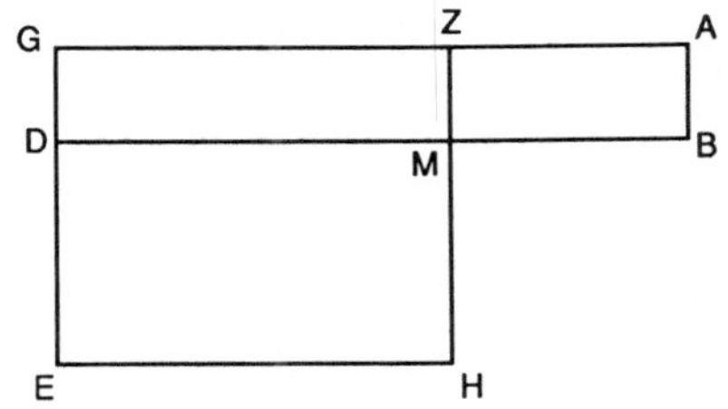

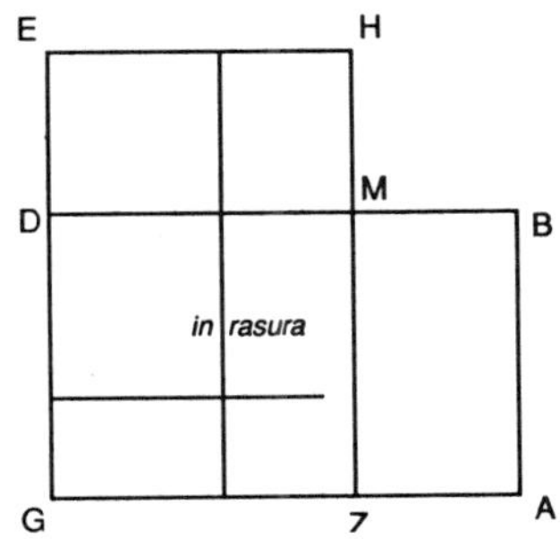

Et ego ponam causam huius porte. Quod est ut faciamus numerum minorem divisum, qui est 10, superficiem ABGD, et homines primos lineam GD; quod autem contingit uni primorum [erit] sit AG. Deinde addamus hominibus primis, qui sunt linea GD, 4 homines, qui sint linea DE; erit tota linea GE homines [posteriores] postremi. Et postmodum faciamus superficiem [GH]GE⟨H⟩Z 30 dragmas. Quam dividamus per lineam GE, et exibit linea GZ, que est quod contingit uni postremorum. Remanet ergo AZ 4 dragme, que sunt superfluum illius quod contigit uni primorum super id quod contigit uni postremorum. Multiplicemus igitur homines priores, qui sunt linea MZ, quia ipsa est sicut DG, in lineam AZ, et producetur superficies ZB. Cui adde superfluum quod est inter 30 et 10, que sunt 20, et [est] *erit* superficies DH *agregatum*: quia [erit] superficies DH *est* addens super superficiem ZB 20, que sunt superfluum inter duos numeros divisos, quia superficies AD est 10 et superficies GH est 30; que addens est super superficiem AD 20; *igitur* reiecta *_vel: sublata_* comuni superficie GM, remanet superficies DH addens super superficiem ZB 20. Divide item superficiem DH per lineam DE, que est 4; et exibit linea DM, que est sicut linea GZ, et hoc est quod contingit uni postremorum. Duc[itur]ta igitur GZ in GE, producitur superficies GH, que est 30. Et hoc est quod voluimus exponere.

Et postquam exposuimus causam huius porte, **(A30^{v})** faciamus homines primos rem. Et ducamus eam in diminutum quod est inter eos (sc.: inter contingentia), quod est 4 dragme, et provenient 4 res. Quibus addamus 20 dragmas, que sunt superfluum quod est inter duos numeros divisos, et erunt 4 res et 20 dragme. Qua⟨s⟩ dividamus per superfluum quod est inter homines priores [et] et posteriores, quod est 4, et exib[i]unt res et 5 dragme. Quas ducamus in homines ultimos, qui sunt res et 4, et provenient census et [7] *9* res et 20 dragme, que equantur censui **pecunie** maiori diviso (sic), que est 30. Minue ergo ex eis 20; et remanebunt census et [10 dragme] *9 res*, que equantur 10. [Minue 2] Dimidia igitur radices, et exibunt 4 et $\frac{1}{2}$; et multiplica eas in se, et provenient 20 et $\frac{1}{4}$. Quibus adde 10, et erunt 30 et $\frac{1}{4}$. Quorum radicem ac⟨c⟩ipe. Que est 5 et $\frac{1}{2}$. Ex quibus minue medietatem radicum, que est 4 et $\frac{1}{2}$, et remanebit 1. Quod est homines primi.

Et si dixit: 10 ⟨dragmas⟩ divisisti per homines, et contigit uni res; deinde addidisti illis 4 homines, et divi*si*sti [illis] per illos 60 dragmas, et contigit uni *postremorum* plus eo quod contigit uni primorum 5 dragmis. Eius regula est quod ducas homines postremos in superfluum quod est inter illud quod contigit uni [primorum] hominum postremorum et illud quod contigit uni primorum; et *ei* quod habebis addes **(82ʳ)** numerum minorem divisum; et agregatum minue de numero maiori diviso, et residuum divide per superfluum quod est inter homines primos et postremos; et quod exibit duc in homines primos. Et quod proveniet est numerus [divisus] minor divisus [qui est 10].

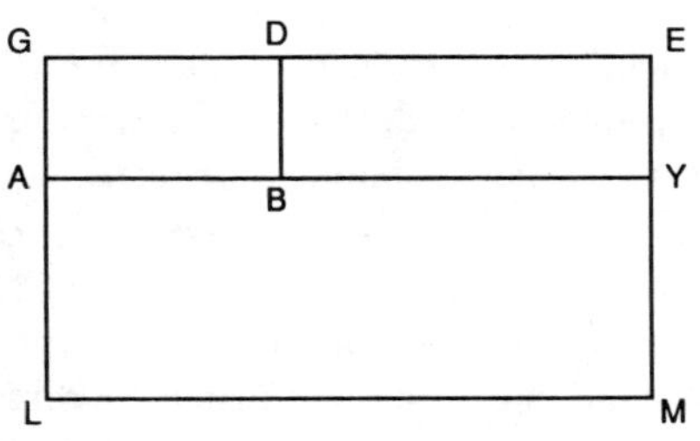

Et ego ponam causam huius capituli. Que est quod faciamus numerum minorem divisum, qui est 10, superficiem **(A31ʳ)** ABGD, et homines primos lineam GD, et quod contigit [cuilibet] uni hominum primorum lineam AG. Deinde addamus super homines primos, qui sunt linea GD, 4 homines, qui sint linea DE. Erit igitur linea GE homines postremi. Et faciamus superficiem GEML 60. Quam dividamus per lineam GE, qu[od]e est homines postremi, et exibit linea GL. Que est id quod conti[n]git uni postremorum hominum. Linea ergo AL est 5, propterea quod *ipsa est* superfluum quod est inter hoc quod conti[n]git uni hominum postremorum et quod contigit uni primorum. Multiplicemus ergo homines postremos, qui sunt linea AY, quia ipsa est sicut linea GE, in [se] AL, que est superfluum *inter id quod contigit uni primorum et postremorum*, et proveniet superficies AM. Deinde adiunge ei superficiem GB, que est numerus minor divisus, et est 10. Et agregatum prohice de tota superficie GM, que est numerus maior divisus, qui est 60. Remanebit superficies DY. Quam divide per lineam DE, que est 4, que est superfluum inter homines primos et postremos, et exibit linea DB. Que est sicut linea AG, et illa est quod conti[n]git uni primorum. Duc etiam illud in lineam G[E [*D*]]D, que est homines p[ostremi]rimi, et proveniet superficies GB 10. Et hoc est quod voluimus exponere.

Et postquam exposui causam huius porte, facia*mus* homines postremos rem et 4. Et multiplicabimus e[os]a in superfluum quod est inter contingentia, quod est 5, et provenient 5 res et 20 dragme. Quibus addemus numerum minorem divisum, qui est 10, et erunt 5 res et 30 dragme. Quibus diminutis a numero maiori diviso, qui est 60 dragme, remanent 30 (**A31^{v}**) dragme diminutis 5 rebus. Quas dividemus per superfluum quod est inter homines primos et postremos, quod est 4, et exibunt [4] *7* et $\frac{1}{2}$ diminuta re et $\frac{1}{4}$ rei. Que [equantur] ducemus in homines primos, qui sunt res, et provenient 7 res et $\frac{1}{2}$ rei diminuto censu et $\frac{1}{4}$ census, que equantur numero minori diviso, qui est 10. Reduca*mus* ergo totum quod habemus ad censum; et erunt 6 res diminuto censu, que equantur 8. Resta*ura* ergo *6* res cum censu[m], et adde ipsum ad 8; et erunt census et 8 dragme, *e*quales 6 rebus. Dimidia igitur res, et exibunt 3. Quas duc in se, et provenient 9. Ex quibus prohice [9] 8, et remanebit 1. Cuius radicem accipe; que est 1. Quam minue de medietate radicum, que est 3, et remanebunt 2. Que sunt homines primi. Et, si volueris, adde 1 ad medietatem radicum, que est 3, et erunt 4. Que sunt (etiam) homines ⟨primi⟩.

Et si dixit: Divisi 60 dragmas per homines et contigit uni res; deinde addidi illis 3 homines, et iterum divisi per illos 20 dragmas, et contigit uni minus eo quod contingit primo (sc.: uni primorum) [6] 26. Eius exemplum est quod ducas homines primos in diminutum quod est inter eos (sc.: inter contingentia); et ex eo quod provenerit prohice superfluum quod est inter duos numeros divisos; et quod remanet divide per superfluum quod est inter homines primos et postremos; et quod exibit multiplica in homines postremos. Et productum erit numerus minor divisus.

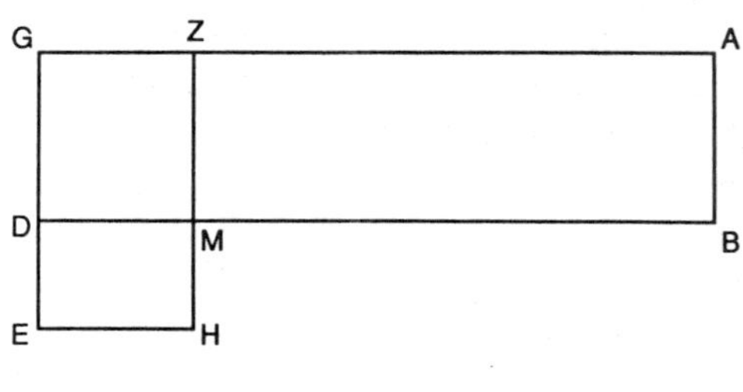

Et ego ponam causam huius porte. Que est ut faciamus numerum maiorem divisum, qui est 60 dragme, superficiem ABGD, et homines primos lineam GD, et quod contigit unicuique primorum hominum lineam [G]AG. Deinde addamus primis hominibus, qui sunt linea GD, 4 (sic, pro: 3) homines, qui sint linea DE. Erit linea GE homines postremi. Et faciamus superficiem GEZ*H* 20. Et dividamus eam per lineam GE, que est homines postremi, et exibit linea GZ. Et illud est quod contigit uni postremorum. Ergo linea AZ est [6] 26, quia ipsa est superfluum quod est inter illud quod contigit uni primorum (**A32^{r}**) et uni postremorum. Ducamus ergo homines primos, qui sunt *sicut* linea ZM, in lineam AZ; erit superficies ZB. Abiciamus ergo ex ipsa superfluum quod est inter duos numeros divisos, quod est 40, et remanebit superficies D[E]H. Manifestum est igitur quod [linea] superficies ZB est plus [26] superficie *D*[E]H 40: quia superficies A[B]D est [4] 60 et superficies G[D]H 20; ergo ipsa, *scilicet: AD*, est maior ea, *scilicet: GH*, 40; proiecta igitur superficie ZD comuni, remanet ZB maior DH 40. [Et] Divide ergo superficiem DH per lineam DE, *que est superfluum inter homines primos et postremos, et* que est 3; exibit linea DM. *Et ipsa est sicut GZ*, que est id quod contigit uni postremorum. Multiplica igitur lineam GZ [in lineam BE] in lineam GE, que est homines postremi, et proveniet superficies GH. Que est numerus minor divisus, et ipse est 20. [et]

Et postquam exposita est causa huius porte, faciamus homines primos rem. Et percutiamus eam in diminutum quod est inter *contingentia* eos, quod est 26, et provenient 26 res. Ex quibus prohice superfluum quod est inter duos numeros divisos, quod est 40, et remanebunt 26 res diminutis 40 dragmis. Quas divide per superfluum quod est inter homines primos et postremos, (**82^{v}**) quod est 3, et exibunt 8 res et $\frac{2}{3}$ rei diminutis 13 dragmis et $\frac{1}{3}$ dragme. Quas duc in homines postremos, qui sunt res et 3, et [ex-] provenient 8 census et $\frac{2}{3}$ census et 12 res et $\frac{2}{3}$ rei diminutis 40 dragmis, que equantur numero minori ⟨diviso⟩. Quod est 20 dragme. Restaura *ea* cum 40 dragmis, et adde eas ad 20 dragmas; et erunt 8 census et $\frac{2}{3}$ census et [12] *12* (**A32^{v}**) res et $\frac{2}{3}$ rei, equales 60 dragmis. Reduc ergo totum quod habes ad censum [unum]. Iam autem scivi[sti]mus quod census de 8 censibus et $\frac{2}{3}$ census est $\frac{3}{26}$. Accipe igitur totius quod habes $\frac{3}{26}$, et exibunt census et res et $\frac{12}{26}$ rei, ⟨que equantur⟩ 6 dragmis et $\frac{24}{26}$ dragme. Dimidia igitur radicem (sic), et exibunt [19] $\frac{19}{26}$. Qua⟨s⟩ multiplica in se, et provenient $\frac{361}{676}$. Quas adde 6 dragmis et $\frac{24}{26}$, et erunt 7 et $\frac{309}{676}$. Quorum radicem accipe, et exibunt 2 et $\frac{19}{26}$. Ex quibus minue medietatem radicum, que est $\frac{19}{26}$, et remanebunt 2. Que sunt homines primi.

In istis autem questionibus 4 et similibus istis sunt 2 vie, que sunt similes duabus viis quas fecimus in [capi-] porta *de* 10 qu[as]e *est*: Divisi *10* in duas divisiones quarum [quel-] unaqueque divi[ditur]*sa est* per aliam et exivit 4 et $\frac{1}{4}$. Et ego exponam tibi *in* una[m], per quam addisces *in* alia[m] (sc.: Et ego exponam tibi unam que te docebit (vel: e qua inferes) alteram).

Quod est si dicemus tibi: Divisi 20 dragmas per homines, et contigit uni res; deinde addidi illis 2 homines, et postea divisi illis 60 dragmas, et conti[n]git uni plus eo quod contigit uni primorum 5 dragmis. Cuius exemplum est quod faciamus homines primos rem. Per quos divide*as* (sic) 20, et conting[it]*at* uni dinar. Cum ergo multiplicabis dinar in rem, provenient 20 dragme. Deinde adde hominibus primis, qui sunt res, 2; et erunt res et 2, que sunt homines postremi. [Per quos] Divid[as]*e ergo* 60 dragmas per rem et 2, et exibunt dinar et 5. Ergo cum multiplicabis **(A33^{r})** dinar et 5 in rem et 2, provenient [2] 60 dragme. Multiplica *itaque dinar* et 5 in rem et 2, et provenient 30 dragme et 5 res et *2* dinarim, que equantur 60 dragmis [diminutis 5 rebus]. Prohice ergo 30 dragmas et 5 res de 60, et remanebunt [6]30 dragme diminutis 5 rebus, que equantur *ad 2* dinarim. Ergo dinar equatur 15 dragmis diminutis 2 rebus et $\frac{1}{2}$. Deinde redeas ad principium questionis, et dic: Divisi 20 dragmas per rem, et exiverunt 15 dragme diminutis 2 rebus et $\frac{1}{2}$. [Et] Du[xi]c [*Dic ergo*] *ergo* rem in [[5][25]25] *15* diminut[a re]*is 2 rebus* et $\frac{1}{2}$, et proven[erunt]*ient* 15 res diminutis duobus censibus et $\frac{1}{2}$, que equantur 20 dragmis. Reduc [igitur] *itaque* totum quod habes ad censum. Iam autem scivimus quod census de 2 censibus et $\frac{1}{2}$ est $\frac{2}{5}$. Accipe igitur totius quod habes $\frac{2}{5}$, et exibunt 6 [*10*] res diminuto censu, equales 8 dragmis. Restaura ergo 6 res cum censu, et adde illum 8 dragmis, et erunt census et 8 dragme, equales 6 rebus. Dimidia itaque radices, et exibunt 3. Qu[e]*as* duc in se, et provenient 9. Ex quibus prohice 8, et remanebit 1. Cuius radicem accipe. Que est 1. Et minue ipsum de medietate radicum, que est 3, et remanebunt 2. Que sunt homines primi. Et, si vis, adde unum ad medietatem radicum, que est 3, et [pro-] erunt 4. Que sunt homines postremi (sic, pro: Que sunt (etiam) homines primi). [Intellige] *Animadverte* ergo quod tibi dixi, et intelliges et prosperabis, cum adiutorio Dei.

Et si dicemus tibi: 10 dragmas divisi per homines, et contigit uni res; deinde addidi illis 6 homines, et postea divisi per eos 40 dragmas, et contigit uni *postremorum* sicut id quod contigit uni primorum. Eius exemplum est quod duca[mus]*s* numerum minorem divisum in superfluum quod est inter homines primos et postremos; et quod provenerit dividas per superfluum quod est inter numeros divisos. Et quod exierit est homines primi. Duc ergo numerum minorem divisum, qui est 10, in superfluum quod est inter primos et ultimos, quod est 6, et provenient 60. Que divide per superfluum quod est inter duos numeros divisos, **(A33^{v})** quod est 30, et exibunt 2. Que sunt homines primi. Si autem volueris, divides superfluum quod est inter 2 numeros divisos, quod est 30, per superfluum quod est inter homines primos et postremos, quod est 6, et exibunt 5. Per quem divide numerum minorem divisum, et illud est 10, et exibunt 2. Que sunt homines ⟨primi⟩. Et causa huius questionis **in hac questione** est manifesta et plana.

Et in ipsa *questione* est via alia, et est: Quia iam scivisti quod proportio hominum primorum ad homines postremos ⟨est⟩ sicut proportio eius quod contingit uni primorum **vel: hominibus primis** (sic recte) ad illud quod contingit uni postremorum **vel: hominibus postremis**; iam etiam scivisti quod id quod contingit hominibus primis est 10 et quod contingit postremis est 40; sed proportio 10 ad 40 est $\frac{1}{4}$; ergo homines primi sunt $\frac{1}{4}$ hominum postremorum. Faciamus igitur homines primos rem, et postremos rem et 6 [*dragme*] *dragmas*. [Et] Iam *igitur* scivimus quod res et 6 *dragme* sunt quadruplum rei. Accipe itaque quartam rei et 6 dragmarum. Que est $\frac{1}{4}$ rei et dragma et $\frac{1}{2}$, et ipsa equantur rei. Prohice igitur $\frac{1}{4}$ rei de re, et remanebunt $\frac{3}{4}$ rei, equales uni dragme et $\frac{1}{2}$ dragme. Ergo res equatur 2^{bus}. Que sunt homines primi.

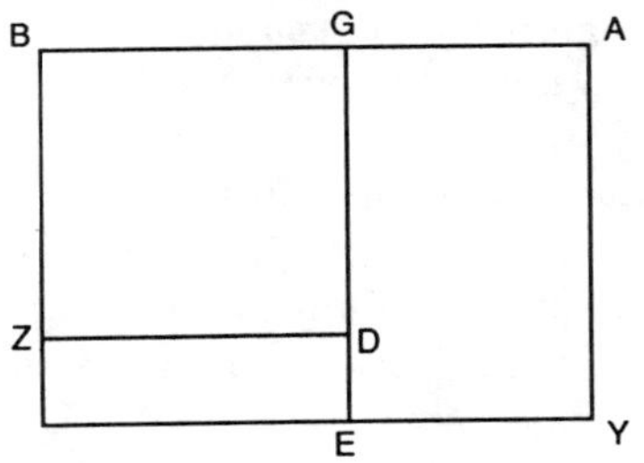

Et si dicemus tibi: 10 [diversorum pretiorum] pannos emerunt 2 homines, quos*dam* [*aliquantos*] *de 10* maiori pretio, alios vero minori, emerunt tamen illos 10 pannos 72 [denariis] *dragmis*, de quibus 72 dra⟨g⟩mis unusquisque solvit medietatem, [id est] *scilicet* 36; et accepit unus aliquot pan⟨n⟩os ad unum forum *unumquemque*, secundus vero accepit residuos ad aliud **(83^{r})** forum unumquemque, fuit autem uniuscuiusque panni *unius hominis* pretium addens super pretium panni unius*cuiusque* alterius viri 3 dragmas. Eius exemplum *est* quod faciamus pannos lineam AB, quod autem accepit unus [ex ip-] eorum ex pannis lineam AG et quod alius lineam GB. Et faciamus pretium cuiusque panni de linea GB lineam GD. [Omne] Totum ergo pretium pannorum linee GB est superficies DB. Quod est 36, sicut positum est, ergo superficies DB est 36. Et faciamus pretium uniuscuiusque panni de linea AG lineam GE. Ergo totum pretium pannorum linee AG est superficies AE. Ergo ipsa similiter etiam *est* 36. Linea autem DE est 3, quia ipsa est superfluum quod est inter duas partes **divisiones**, (sc.) *quod est inter pretium unius panni de pannis unius hominis et pretium unius panni de pan⟨n⟩is alterius viri.* Postea faciamus lineam GB rem. Que est **(A34^{r})** sicut linea DZ, et linea DE *est* 3. Ergo superficies EZ est 3 res. Tota itaque superficies BY est 72 et 3 res. Sed linea AB est 10. Ergo linea AY est 7 et $\frac{1}{5}$ et $\frac{3}{10}$ rei. Et ipsa est sicut linea GE. Sed linea DE est 3. Remanet ergo linea GD 4 et $\frac{1}{5}$ et $\frac{3}{10}$ rei. Duc ergo eam in lineam GB, que est res, et provenient 4 res et $\frac{1}{5}$ rei et $\frac{3}{10}$ census, equales 36 dragmis. Restaura (sc.: reintegra) $\frac{3}{10}$ census ut sit census unus (vel: integer). Quod est ut ducas eas in 3 et $\frac{1}{3}$, et ducas similiter totum quod habes in 3 et $\frac{1}{3}$. Et provenie⟨n⟩t census et 14 res equales 120 dragmis. Fac **restaura** (sc.: solve) sicut dixi tibi, et exibit res 6. Que sunt numerus pannorum linee GB, alterius vero est residuum de 10 pannis, quod est 4, [q-] *et* illud est linea AG. Et illud est quod voluimus exponere.

Huic autem questioni et [huiusmodi] huic consimili via est sicut *[illud]* expo[sui]*nam* tibi (sic, pro: Huic autem questioni et huic consimilibus est via ⟨altera⟩ sicut exposui). Et est quod faciamus *id* quod accepit unus ipsorum *hominum* [rem] ex 10 pannis rem, et quod alius 10 re diminuta. Iam autem scivimus quod, *cum* multiplicamus rem in pretium panni unius*cuiusque pannorum GB* et [3] in 3, proveniunt 36 et 3 res, sed cum multiplicamus 10 diminuta re in pretium cuiuslibet panni et in 3, proveniunt 36. Cum autem (sc.: igitur) multiplicamus ⟨omnes⟩ 10 pannos in pretium cuiusque panni pan⟨n⟩orum [qu[od]e est] *qui sunt res* et in 3, provenient 72 et 3 res. Cum ergo diviserimus 72 et 3 res per 10, exib[un]it pretium cuiuslibet pan⟨n⟩i de pannis qui sunt 10 excepta re 7 et $\frac{1}{5}$ et $\frac{3}{10}$ rei. Ergo erit pretium cuiuslibet panni pannorum qui sunt res 4 (**A34^{v}**) et $\frac{1}{5}$ et $\frac{3}{10}$ rei. Postea fac sicut dixi tibi ⟨in his que precesserunt⟩, et exibit res 6 panni. Qui sunt id quod accepit unus ipsorum, alius vero accepit residuum de 10 pannis, quod est 4 panni. Et hec questio est sicut ⟨si⟩ dicas: Divisisti 10 in duas partes et multiplicasti unam duarum partium in rem, et provenerunt 36, [deinde] *et* multiplicasti aliam in rem et 3, et provenerunt 36. Fac itaque de ipsis sicut premisimus et exposuimus superius.

Et si dixit: Divide 10 in duas partes et multiplica unam duarum partium in 6, et quod provenerit divide per partem reliquam; deinde illius quod exiverit tertiam [duc] accipe, et adde illam ei quod provenit ex ductu unius duarum partium in 6, et [[s-] fuerunt] *fiet* congregatum 56. Eius exemplum est quod faciamus unam duarum partium rem, aliam vero 10 diminuta re. Et ducamus re[s]m in 6, et provenient 6 res. Quas cum dividemus per 10 re diminuta, erit tertia eius quod exibit 56 diminutis 6 rebus [quia ⟨dixit⟩: "quando addemus ad 6 res, erit *congregatum* 56"]. *Totum ergo* quod exibit de divisione *erit* 168 diminutis 18 rebus {*g. que sunt triplum de 56 et 6 rebus diminutis*}. Ducamus ergo 168 diminutis 18 rebus in 10 diminuta re, et provenient 168[8]0 et 18 census diminutis 348 rebus, que equantur 6 rebus. Restaura ergo 348 res (sc.: Restaura ergo 1680 et 18 census cum 348 rebus), et adde illas ad 6 res; et erunt [168[0]] *1680* dragme et 18 census, equales 354 rebus. Reduc itaque totum quod habes ad censum. Iam autem scivisti quia census 18 censu*um* est $\frac{0\ 1}{9\ 2}$. Accipe igitur totius quod habes $\frac{0\ 1}{9\ 2}$, et exibunt census et [3] *93* dragme et $\frac{1}{3}$ dragme, equales 19 rebus et $\frac{2}{3}$ rei. Restaura **Fac** ergo in illis sicut exposui tibi, et exibit res 8. Que sunt una duarum partium, reliqua vero est 2. Intellige ergo.

Et si dicemus tibi: Est census de quo prohicies $\frac{1}{3}$ eius et 2 dragmas, postea duces residuum in se; et redib[un]it census et (insuper) 24 dragme. Eius exemplum est quod faciamus censum rem. Et prohiciamus ex ea $\frac{1}{3}$ ipsius et 2 dragmas, **(A35ʳ)** et remanebunt $\frac{2}{3}$ ⟨rei⟩ diminutis 2 dragmis. Qua*s* ducemus in se, et provenient $\frac{4}{9}$ census et 4 dragme diminutis 2 rebus et $\frac{2}{3}$ rei, equales rei et [4] 24 dragmis. Restaura igitur (sic, pro: ⟨Prohice ergo 4 dragmas de 24 dragmis;⟩ et restaura) $\frac{4}{9}$ (census) cum 2 rebus et $\frac{2}{3}$, et adde eas dragmis et rei, *et prohice hinc inde 4 dragmas comunes*; et erunt 2[4]0 dragme et 3 res et $\frac{2}{3}$ rei, equales $\frac{4}{9}$ census. Restaura (vel: reintegra) item $\frac{4}{9}$ censu[m]*s* ut sit census integer; quod est [quod] *ut* ducas eas in 2 et $\frac{1}{4}$; et ducas totum quod habes in 2 et $\frac{1}{4}$, et provenient 45 dragme et 8 res et $\frac{1}{4}$ rei equales censui. Fac ergo sicut exposui tibi, et erunt 12. Que sunt census (sc. quesitus).

Est et⟨iam⟩ alia via. Que est quod faciamus censum censum diminutis 24 dragmis. Et prohiciamus ex eo tertiam eius et 2 dragmas; et remanebunt $\frac{2}{3}$ census diminutis 18 dragmis, equales radici *illius census: quia census diminutis 24 dragmis $\frac{3}{2}$ sunt $\frac{2}{3}$ census diminutis 16 dragmis; que sunt equales rei et 2 dragmis; prohice item 2 dragmas de $\frac{2}{3}$ census diminutis 16, et remanebunt $\frac{2}{3}$ census diminutis 18, equales rei.* Restaura ergo et appone sicut exposui tibi, et exib[un]it census 36. Ex quibus prohice 24, [sicut] *quia* iam fecisti censum censum diminutis 24, et remanebunt [[2] 16] *12*. Que sunt census quesitus.

Et si dicemus tibi: Est census cuius 3 radices et 4 radices residui *sunt* 20 dragme. Eius exemplum est quod facias ex censu *tuo* censum. Deinde dicas: [4] *3 radices et 4* radices *alicuius* census [diminutis 3] [*additis [cum]*]*cum 3* radicibus *ab eo* (sc.: 3 radices census et 4 radices ipsius census diminutis ab eo 3 radicibus eius) equantur 20 dragmis. Prohice 3 radices de 20, et remanebunt 20 dragme diminutis 3 radicibus, equales 4 radicibus census diminutis 3 radicibus *[scilicet reiecta] ab ipso. Ergo $\frac{1}{4}$ eius*, qu[od]e est radix census diminutis **(83ᵛ)** diminutis (sic) [4] $\frac{3}{4}$ radici[bu]s (sic, pro: 3 radicibus), *[scilicet] ab ipso censu*, [quod] est 5 dragme diminutis $\frac{3}{4}$ radicis. Quas duc in se, et provenie⟨n⟩t $\frac{1}{2}$ census et $\frac{0\ 1}{8\ 2}$ census et 25 dragme diminutis [9] *7* radicibus et $\frac{1}{2}$ radicis, equales censui diminutis 3 radicibus. Restaura ergo eam (sic) cum [10] *7* radicibus et $\frac{1}{2}$ radicis, et adde ea⟨s⟩ ad censum diminutis 3 radicibus; et erunt census et 4 radices et $\frac{1}{2}$, equales medietati census et $\frac{0\ 1}{8\ 2}$ census et 25 dragmis. Prohice ergo medietatem census et $\frac{0\ 1}{8\ 2}$ census ex cens[i-]u, et remanebunt $\frac{3}{8}$ census et $\frac{0\ 1}{8\ 2}$ census et 4 radices et $\frac{1}{2}$ radicis, equales

25 dragmis. Restaura **Reintegra** *igitur* censum tuum; (**A35**$^{\mathbf{v}}$) et tua reintegratio *in eo* e[orum]rit quod addas ei sicut *ipsum* est et $\frac{2}{7}$ eius, et erit census integer. Adde ergo *ei* toti quod habes sicut ipsum est et sicut $\frac{2}{7}$ ipsius; et erunt census et 10 res et $\frac{2}{7}$ rei equales 57 et $\frac{1}{7}$. Dimidia ergo radices, et exibunt 5 et $\frac{1}{7}$. Que duc in se, et provenient 26 et $\frac{3}{7}$ et $\frac{0}{7}\frac{1}{7}$. Quibus adde 57 et $\frac{1}{7}$, et erunt 83 et $\frac{4 \text{ et } 1}{7 \;\; 7}$ (sic). Quorum radicem accipe. Que est 9 et $\frac{1}{7}$. Ex quibus prohice $\frac{1}{2}$ radicum, que est 5 et $\frac{1}{7}$, et remanebunt 4. Que sunt radix census, census autem est 16.

Et, si volueris, dic: 3 radices et et (sic) 4 radices census diminutis 3 radicibus *ab eo* equantur 20 dragmis. Et 4 radices census diminut[a]is [$\frac{1}{3}$] 3 radicibus [census] equantur 20 dragmis *diminutis 3 radicibus.* Et 4 radices census diminutis 3 radicibus que est (sic, pro: diminutis 3 radicibus sunt) radix 16 censuum diminutis 48 [Erg-] ⟨radicibus. Dic ergo: 3 radices et radix 16 censuum diminutis 48⟩ radicibus equantur 20 dragmis. Prohice 3 radices de 20, et remanebunt 20 diminutis 3 radicibus, equales radici 16 *censuum* diminutis 48 radicibus. Duc ergo 20 diminutis *3 radicibus in se, et oppone productum cum 16 censibus diminutis* 48 radicibus *in se* (*in se* delendum), et fac sicut dixi tibi, et exibit census 16.

Et, si volueris, fac residuum de censu, postquam proiecisti ex ipso 3 radices census, censum. Cuius 4 radices accipe. Que sunt 4 radices. Et scitum est quod, quando addes has 4 radices ad 3 radices census (sc. quesiti), erunt 20 dragme. Et manifestum est quia 3 radices census (primi) sunt 20 dragme *diminutis 4 radicibus.* [Ante ipsa] *Igitur* radix census [que] est (sic, pro: Ergo tertia ipsorum est radix census; que est) 6 dragme et $\frac{2}{3}$ diminuta radice et $\frac{1}{3}$. Que duc in se, et provenie⟨n⟩t census et $\frac{7}{9}$ census et 44 dragme et $\frac{4}{9}$ dragme diminutis 17 rebus et $\frac{7}{9}$ rei. (**A36**$^{\mathbf{r}}$) Que sunt census quesitus a nobis. Prohice igitur ex eo 3 radices ⟨eius⟩, que sunt 20 dragme diminutis 4 radicibus, et remanebunt 24 dragme et $\frac{4}{9}$ dragme et census et $\frac{7}{9}$ census diminutis 13 radicibus et $\frac{7}{9}$ radicis, equales censui. *Fac ergo* sicut exposui tibi; exibit res 2. Que est radix residui ex censu postquam proiecisti ex eo 3 radices eius. Duc igitur ea in se, et provenient 4. Que [est] sunt residuum census. Accipe ergo 4 radices de 4, et erunt 8. Quas prohice de 20, et remanebunt 12. Que sunt 3 radices census. Et tertia ipsorum est radix census. Que est 4, census autem est 16. Que sunt census quesitus.

Et si dixerimus tibi: De censu proi[ecisti]ce $\frac{1}{3}$ ⟨eius⟩, et postea residuum duc in 3 radices census [census] *primi*, et redibit census primus. Iam ergo scivisti quia, quando prohicies ex censu $\frac{1}{3}$ eius, residuum census erit $\frac{2}{3}$ eius, quas [cum] duces in 1 *dragmam* et $\frac{1}{2}$ ut proveniat census [unus] integer. Ergo dragma et $\frac{1}{2}$ est 3 radices [eius] census. Igitur radix census est $\frac{1}{2}$ dra⟨g⟩me, census autem est $\frac{1}{4}$.

Et si dixerimus tibi: De censu prohice $\frac{1}{3}$ ipsius, postea duc residuum in 3 radices residui ex censu, et redibit census. Iam sci[visti]*s* quod dragma et $\frac{1}{2}$ est 3 radices residui de censu, que sunt $\frac{2}{3}$, et quia radix *de* $\frac{2}{3}$ [et quod radix *de* $\frac{2}{3}$] est $\frac{1}{2}$ dragme; ergo $\frac{2}{3}$ sunt $\frac{1}{4}$. Igitur census integer est $\frac{3}{8}$ dragme.

Et si dixerimus tibi: 3 radices census et 2 radices residui equantur ei, *scilicet: census.* Iam sci[visti]*s* quod 2 radices residui *ex eo* equantur residuo ex ipso. Ergo residuum est 4. Que cum 3 radicibus equantur censui. Fac [ig-] itaque sicut dixi tibi, et proveniet census 16. [ex-]

Et si dixerimus tibi: 3 radices census et 4 radic[i]es residui ex eo sunt sicut census et 4 dragme. Iam scivisti quod residuum ex censu, postquam proiecisti 3 radices eius, est sicut 4 radices eius diminutis 4 dragmis. Voca igitur *nomen* residu[um]i ex censu postquam proieceris 3 radices eius censum. Deinde dic: [Census est] 4 radices [eius] *census* diminutis **(A36^{v})** 4 dragmis equantur censui. Igitur census est 4. Deinde [di-] prohice ex censu (sic, pro: Deinde dic: Ex censu proiecerimus) 3 radices ipsius, et remanebunt 4 dragme. Ergo census est 16 dragme. Que sunt census quesitus.

Et si dixit: Divisisti 10 in duas partes et divisisti unam duarum partium per aliam; et fuit quod exivit de divisione cum dividente **eo per quem fuit divisum** 5 et $\frac{1}{2}$. Eius exemplum est quod faciamus unam duarum partium rem et aliam 10 diminuta re. Iam scivisti quod, quando dividemus 10 re diminuta per rem, [erit] quod exibit de divisione, additum rei, erit 5 et $\frac{1}{2}$. Cum autem proiecerimus rem de 5 et $\frac{1}{2}$, erit residuum quod conti[n]git parti, *id est: exit de divisione*, quod est 5 et $\frac{1}{2}$ diminuta re. [Iam autem scivimus] Amplius dicemus: Divisimus 10 diminuta re per rem, et exiverunt 5 et $\frac{1}{2}$ diminuta re. Et iam scivimus quod, quando multiplicamus quod exit de divisione **(84^{r})** in id per quod dividitur, redit divisum. Multiplicemus igitur 5 et $\frac{1}{2}$ diminuta re in rem, et opponemus cum eo quod provenerit de multiplicatione ad 10 diminuta re, et exibit res 4.

Et si dixit: Divi[de]*sisti* 10 in duas partes, et [multiplicasti] *divisisti* unam duarum partium per aliam, et postea quod exivit de divisione addidisti diviso; deinde multiplicasti agregatum [per] in dividentem, et provenerunt 30. Exemplum est quod faciamus unam duarum partium rem *et* aliam 10 diminuta re. Cum ergo dividemus 10 diminuta re per rem, iam scivimus quod, cum multiplicabimus quod exibit de divisione *in rem*, provenient 10 ⟨diminuta re. Et cum multiplicabimus 10 diminuta re in rem, provenient 10⟩ res diminuto censu. Cui addamus 10 diminuta re, et erunt 10 dragme et 9 res diminuto censu, equales 3[6]0 dragmis. Restaura et appone sicut exposui tibi, et exibit res; que est una duarum partium, per quam dividitur, que *est* 4 (sc.: et exibit res 4; que sunt una duarum partium, per quam dividitur).

Et si dixit: Divide 10 in duas partes, et divide unam duarum partium per aliam; deinde ducas quod exibit de divisione in divisum, et provenient 9 dragme. Exemplum: Quod facias unam duarum partium rem et aliam 10 excepta re. **(A37^{r})** Cum autem divi[s]des 10 excepta re per rem et quod exibit de divisione duces in rem, provenient 10 diminuta re; sed quod exivit, quando multiplicabimus ⟨illud⟩ in 10 diminuta re, provenient 9 [*7*], *sicut positum est.* Congrega *ergo* ea, et erunt 1[7]9 diminuta re. Planum autem (sc.: igitur) est ex eo quod diximus quia quod exit de divisione 10 diminuta re per rem, cum multiplicabimus illud in 10 dragmas, [equa[n]bi-] reintegrabunt (sic) 19 dragmas diminuta re. Dividamus igitur 19 diminuta re per 10, et exibit dragma et $\frac{9}{10}$ dragme diminuta $\frac{1}{10}$ rei. Et illud est quod exit ex divisione. Deinde redibimus, et dicemus: Divisimus 10 dragmas diminuta re per rem, et exivit dragma et $\frac{9}{10}$ dragme *diminuta* $\frac{1}{10}$ *rei.* Postea fac sicut exposuimus tibi, et exibit res 4; que s[unt]unt una duarum partium, per ⟨quam⟩ divisum fuit.

Et si dixerit tibi: Divide 10 in duas partes, et divide unam (duarum) partium per aliam; deinde duc quod exit de divisione in se, et postea in dividentem, et provenient 32 dragme. Exemplum: Quod faciamus unam duarum partium rem, reliquam vero 10 diminuta re. Quando dividemus 10 diminuta re per rem *et* quod exibit de divisione multiplicabimus in rem, provenient 10 diminuta re, et hoc est divisum; cum autem ducemus ipsum in 10 diminuta re [*et in id*], quod exi[b]vit [*id est: in se*] de divisione, *scilicet: 10 re diminuta per rem*, erunt 32 (sic, pro: cum autem ducemus 10 diminuta re in id quod exivit de divisione, erunt 32). Quia omnis numerus qui ducitur in se et postea ⟨in⟩ alium numerum est sicut multiplicatio numeri primi in numerum alium, deinde [agregatum] *quod provenit* in numerum primum. Et hoc manifestum est. Et postquam ita est (vel: Et postquam manifestum est hoc), [sicut] *est quasi* diceret: 10 divisi in duas partes, et divisi unam duarum partium per aliam, et postea du[c]xi quod exivit de divisione in divisum, et fuerunt 32. Et iam exposuimus hoc in questione superiori.

Et si dixit: 10 divisi in duas partes, et divisi unam duarum partium per aliam; deinde duxi quod exivit de divisione in superfluum quod est inter 2 partes, et fuit quod provenit 24 dragme. Exemplum: Quod faciamus unam duarum partium rem, reliquam vero 10 re diminuta. Cum igitur divi[demus]*serimus* 10 diminuta re per **(A37^{v})** rem, scimus quia, cum ducemus id quod exit de divisione in [residuum] *superfluum* quod est inter divis[iones]*a*, et illud est 10 diminut[a]is *2* rebus, provenient 24 dragme. Iam autem scivimus quia multiplicatio eius quod exit de divisione in rem est 10 diminuta re, et in 2 res 20 diminutis 2 rebus. Que agregabimus ad 24, et erunt 44 diminutis 2 rebus. Manifestum est autem (sc.: igitur) ex eo quod diximus quod multiplicatio eius quod exit ex divisione in 10 est 44 diminutis 2 rebus. Dividamus ergo 44 diminutis 2 rebus per 10, et exibunt 4 dragme et $\frac{2}{5}$ dragme diminut[is]a $\frac{1}{5}$ rei. Et hoc est quod exit de divisione. Deinde redibimus, et dicemus: Divisimus 10 diminuta re per rem, et exiverunt 4 dragme et [$\frac{1}{5}$] $\frac{2}{5}$

dragme diminuta $\frac{1}{5}$ rei. Et faciemus sicut dixi, et exibit res 2. Que est una duarum partium, per *quam* dividitur, reliqua vero, que dividitur, est 8.

Et si dice[mus]tur: 10 divisi in duas partes, et divisi unamquamque duarum partium per aliam, et agregavi ea; deinde multiplicavi agregatum in unam duarum partium, et fuit quod provenit 34 dragme. Exemplum: Quod faciamus unam duarum partium rem, reliquam vero 10 diminuta re. [Et quia] Divisa sunt *ergo* 10 diminuta re per rem et res per 10 re diminuta, et agregata sunt que de $\mathit{2}^{bus}$ division[e]*ibus* exierunt, deinde agregatum ductum est in 10 diminuta re, et [exiveru(n)t] provenerunt 34 dragme. Iam itaque (sc.: autem) [scivi-] scivimus quod, cum multiplicas quod exit tibi de divisione rei per 10 diminuta re in 10 diminuta re, provenit res. Prohice ⟨ergo⟩ rem de [4] 34, et remanebunt 34 diminuta re. Planum est etiam (sc.: ergo) ex eo quod diximus et exposuimus quod, quando [ducimus] *dividimus* **(84^{v})** 10 diminuta re [in] *per* rem, de*inde* ducimus quod [provenit] *exit nobis* de divisione in 10 diminuta re, provenient 34 diminuta re. Cum igitur (sc.: autem) ducemus quod exit de divisione 10 diminuta re per rem *in rem*, provenient 10 diminuta re. Que addemus ad 34 diminuta re; erunt 44 diminutis 2 rebus. Et hoc est sicut quod *provenit de multiplicatione eius quod* exit de divisione 10 diminuta re per rem [cum] *in* 10. Divide *ergo* 44 diminutis 2 rebus per 10 [diminuta re], et exibunt 4 dragme et $\frac{2}{5}$ dragme diminuta $\frac{1}{5}$ rei. Deinde revertamus, et dicamus: Divisimus 10 dragmas diminuta re per rem, et exiverunt 4 dragme et $\frac{2}{5}$ dragme diminuta $\frac{1}{5}$ rei. Iam autem scivimus quia, quando multiplicamus quod **(A38^{r})** exit nobis de divisione per (sic, pro: in) dividentem, redit census noster divisus. Ducamus 4 dragmas et $\frac{2}{5}$ dragme diminuta $\frac{1}{5}$ rei in rem, et provenient 4 res et $\frac{1}{5}$ (sic) rei diminuta $\frac{1}{5}$ census, equales 10 dragmis diminuta re. Age ergo in eis sicut dixi tibi, et exibit res 2. Que [sunt] est una duarum partium, reliqua vero est 8.

Sed, si volueris, fac unam duarum partium rem, reliquam vero 10 diminuta re. Deinde dicemus: Divisimus rem per 10 diminuta re, et exivit dinar; et divisimus 10 diminuta re per rem, et exivit [moneta] *[blada] bizantius.* Deinde agregavimus dinar et [bladam] *bizantium.* Iam autem scivimus quia, quando multiplicamus dinar et [bladam] *bizantium* [per] in rem, provenient 34, quando vero multiplicabimus [bladam] *bizantium* in rem, provenient 10 dragme diminuta re. Prohice ergo 10 diminuta re de 34, et remanebunt 24 et res. Et iam etiam scivimus quia, si multiplicabimus dinar in rem, provenient 24 dragme et res, et quoniam (sc.: quando) ducemus dinar in 10 diminuta re, proveniet res. Addamus ergo eam ad 24 *et rem*, et erunt 24 et 2 res. Iam vero exposuimus quod, quando ducimus dinar in 10, provenient 24 dragme et 2 res. Divida[g]mus ergo 24 et 2 res per 10, et exibunt 2 dragme et $\frac{2}{5}$ *dragme* et $\frac{1}{5}$ rei. Que sunt dinar. Deinde revertamus, et dicamus: Divisimus rem per 10 diminuta re, et exiverunt 2 dragme et $\frac{2}{5}$ dragme et $\frac{1}{5}$ rei. Et iam sci[s]vimus quia, cum ducimus quod exit nobis de divisione in id per quod divisimus, redit census noster divisus. Ducamus igitur 2 dragmas et $\frac{2}{5}$ *dragme* et $\frac{1}{5}$ rei in 10 diminuta re, et provenient 24 dragme diminutis $\frac{2}{5}$ rei et diminuta $\frac{1}{5}$ census, equales rei. Age ergo in eis secundum id quod exposui tibi, et exibit res 8. Que est una duarum partium, reliqua vero est 2.

Et si dixit: Divide 10 in duas partes, et divide maiorem per minorem; deinde agrega [ei] *id* quod exit de divisione ad 10, et postea multiplica a*g*gregatum in partem minorem, et provenient 46 dragme. Exemplum: Quod faciamus minorem partem rem, maiorem vero 10 diminuta re. Iam scivimus quia, cum [ducimus] *dividimus* 10 diminuta re per rem et aggregamus quod **(A38^{v})** exit de divisione ad 10 et aggregatum multiplicamus in rem, proveniunt 46. Manifestum est etiam quia multiplicatio nostra de 10 in rem est 10 res. *Quas* prohiciemus de 46, et remanebunt 46 diminutis 10 rebus. Que *sunt* sicut multiplicatio illius quod exit de divisione 10 diminuta re per rem in rem. Sed multiplicatio eius quod exit de divisione 10 diminuta re *per rem* in rem [est *[et]*] 10 diminuta re est 10 diminuta re, que equantur 46 diminutis 10 rebus (sic, pro: Sed multiplicatio eius quod exit de divisione 10 diminuta re per rem in rem est 10 diminuta re. Erunt ergo 10 diminuta re equalia 46 diminutis 10 rebus). Agemus in ipsis secundum id quod exposui tibi, et exibit res 4.

Et si dixerimus tibi: Divide 10 in duas partes, et divide partem maiorem per partem minorem; deinde adde quod exit de divisione ad 10, postea duc [agregatum] coniunctum in partem maiorem, et provenient [6[0]9] *69* dragme. Exemplum: Faciamus partem minorem rem et maiorem 10 re diminuta. *Et* quando divides 10 diminuta re per rem et addes id quod exit de divisione ad 10 et postea multiplicabis coniunctum in 10 diminuta re, provenient [2] 69. Duc ergo 10 dragmas in 10 diminuta re, et provenient 100 diminutis *10* rebus. Que prohice ex 69 dragmis, et remanebunt 10 res diminutis 31 dragmis. Que sunt sicut multiplicatio illius quod exit de divisione 10 diminuta re per rem in 10 diminuta re. Sed multiplicatio*nem* eius quod exit de divisione 10 diminuta re per rem in 10 diminuta re addemus (sic, pro: Sed multiplicatio eius quod exit de divisione 10 diminuta re per rem in rem est 10 diminuta re. Que addemus) ad 10 res diminutis 31 dragmis, et erunt [69] 9 res diminutis 21. Et hoc est ⟨sicut⟩ multiplicatio eius quod exit de divisione 10 re diminuta per rem in 10. Dividamus igitur 9 res diminutis 21 per 10, et exibunt $\frac{9}{10}$ rei diminutis 2 dragmis et $\frac{1}{10}$. *Et ducamus ea* (sc.: et exibunt $\frac{9}{10}$ rei diminutis 2 dragmis et $\frac{1}{10}$. ⟨Et hoc est quod exit de divisione 10 diminuta re per rem. Ducamus igitur $\frac{9}{10}$ rei diminutis 2 dragmis et $\frac{1}{10}$⟩) in rem; et provenient $\frac{9}{10}$ census diminutis 2 rebus et $\frac{1}{10}$ rei, que equantur 10 diminuta re. Et agemus in eis **(A39^{r})** et reintegrabimus censum nostrum, et erit census equalis 11 dragmis et $\frac{1}{9}$ drag(**85**$^{\mathbf{r}}$)me et rei et $\frac{2}{9}$. Dimidia igitur re[s]m et $\frac{2}{9}$, et exibit $\frac{1}{2}$ et $\frac{1}{9}$. Qu[e]as duc in se, et provenient $\frac{1}{4}$ et $\frac{1\ 1}{9\ 9}$. Quas addemus ad 11 dragmas et $\frac{1}{9}$, et erunt 11 dragme et $\frac{1}{4}$ et $\frac{1}{6}$ et $\frac{e\ 1}{9\ 2}$ et $\frac{e\ 1}{9\ 9}$. *Quorum* accipias radicem, et exibunt [9] 3 et $\frac{1}{3}$ et $\frac{0\ 1}{9\ 2}$. Quibus adde medietatem radicum, que est medietas et $\frac{1}{9}$, et erunt 4. Que sunt res, que est pars minor.

Si autem volueris, facias unam duarum partium rem, sitque minor, maiorem vero 10 diminuta re. Quam dividemus per rem, et exeat dinar; addamusque dinar ad 10, et erunt 10 dragme et dinar. Cum autem duces 10 dragmas et dinar in 10 diminuta re, provenient 69 dragme. Multiplicatio vero 10 dragmarum in 10 dragmas diminuta re est 100 dragme diminutis 10 rebus. Quas prohice ex 69, et remanebunt 10 res diminutis 31 dragmis. Iam etiam (sc.: ergo) manifestum est quia multiplicatio dinar in 10 dragmas diminuta re producit 10 res diminutis 31. [Ergo] ⟨ Et multiplicatio dinar in rem producit 10 dragmas diminuta re. Que addamus ad 10 res diminutis 31, et⟩ erunt {*g. Sed multiplicatio dinar in rem diminutam producit rem diminutis 10. Ergo remanent*} 9 res diminutis 21. Iam autem (sc.: igitur) scivimus quod multiplicatio dinar in 10 dragmas [s-] producit 9 res diminutis 21. Divide igitur 9 res diminutis 21 per 10, et exibunt $\frac{9}{10}$ rei diminutis 2 dragmis et $\frac{1}{10}$. Que sunt dinar. Et quasi diceret: Divisimus 10 dragmas re diminuta per rem, et exiverunt $\frac{9}{10}$ rei diminutis 2 dragmis et $\frac{1}{10}$. [Que sunt dinar.] Postea fac sicut dixi tibi, et exibit res 4. Que sunt pars minor, maior vero est 6.

Et si dixit: Divide 10 in duas partes, et divide partem maiorem per minorem et minorem per maiorem, et aggrega ea, et adde aggregatum ad 10; (**A39**$^{\mathbf{v}}$) deinde duc aggregatum in partem maiorem, et provenient [30] 73 dragme. [E-] Exemplum. Faciamus partem maiorem rem, minorem vero 10 excepta ⟨re⟩. Cum ergo divi[si]*de*mus 10 diminuta re per rem et rem per 10 re diminuta et addemus quod exit de 2 divisionibus ad 10, deinde ducemus aggregatum in rem, provenient [76] 73. Et iam scivimus ⟨quod⟩, cum multiplicamus 10 in rem, proveniunt 10 res. Prohice ergo eas ex 73, et remanebunt 73 diminutis 10 rebus. Et [hoc] *illud* est sicut multiplicatio eius quod exit de divisione rei per 10 diminuta re et de divisione 10 diminuta re per rem in rem. Sed multiplicatio eius quod exit de divisione 10 diminuta re per rem in rem *est 10* diminuta re. Minue ergo ea ex 73 diminutis 10 rebus, et remane[nt]bunt 63 diminutis 9 rebus. Et est sicut multiplicatio illius quod exit de divisione rei per 10 diminuta re in rem. Sed cum ducemus quod exit de divisione rei per 10 diminuta re in 10 diminuta re producetur res. Quam addemus ad 63 diminutis 9 rebus, et erunt 63 diminutis 8 rebus. Et est sicut multiplicatio eius quod exit de divisione rei per 10 dragmas diminuta re in 10. Dividamus igitur 63 diminutis 8 rebus per 10, et exibunt 6 dragme et $\frac{3}{10}$ dragme diminutis $\frac{4}{5}$ rei. Et manifestum est ex eo quod diximus quia, cum dividemus rem per 10 diminuta re, exibunt 6 dragme et $\frac{3}{10}$ dragme diminutis $\frac{4}{5}$ rei. Iam autem scivimus quod, cum ducemus quod exit nobis de divisione in id per quod dividitur, redibit census noster divisus. Multiplicemus ergo 6 dragmas et $\frac{3}{10}$ dragme diminutis $\frac{4}{5}$ rei in 10 diminuta re, et provenient 63 dragme et $\frac{4}{5}$ census diminutis 14 rebus et $\frac{3}{10}$ *rei*, equales rei [divis-]. Oppone ergo in illis, et erunt [5] *15* res et $\frac{3}{10}$ rei, equales $\frac{4}{5}$ census et [60] *63* dragmis. Reintegra igitur censum, et erit census et 78 dragme et [$\frac{e}{8}$-] $\frac{1}{2}$ et $\frac{1}{4}$, equales 19 rebus et $\frac{1}{8}$ rei. Dimidia igitur radices, et exibunt 9 et $\frac{1}{2}$ et $\frac{e}{8}\frac{1}{2}$. **(A40^{r})** Que duc in se, et provenient 91 et $\frac{1}{4}$ et $\frac{1}{8}$ et $\frac{e}{8}\frac{1}{2}$ et [$\frac{1}{186}$] $\mathit{\frac{1}{256}}$. Ex quibus prohice 78 et $\frac{1}{2}$ et $\frac{1}{4}$, et remanebunt 12 et $\frac{1}{2}$ et $\frac{1}{8}$ et [$\frac{1}{186}$] $\mathit{\frac{e}{8}\frac{1}{2}}$ *et* $\mathit{\frac{1}{256}}$. Quorum radicem accipe, et exib[i]unt 3 et $\frac{1}{2}$ et $\frac{e}{8}\frac{1}{2}$. Que minue ex medieta*te* radicum, que est 9 et $\frac{1}{2}$ et $\frac{e}{8}\frac{1}{2}$, et remanebunt 6. Que sunt res, que est pars maior.

Et si dixit: Divide 10 in duas partes, et divide unamquamque duarum partium per aliam; deinde accipe differentiam que est inter illa que de divisione exeunt, et duc eam in unam partium, et provenient 5 dragme. Exemplum. Faciamus unam duarum partium rem, residuam vero 10 diminuta re. Iam scivimus quia, cum dividimus rem per 10 diminuta re, et 10 diminuta re per rem, et accipimus differentiam eorum *que exeunt de divisione*, et ducimus eam in rem, provenient 5 dragme. [*Et*] Cum autem ducemus integrum quod exit [de] de duabus divisionibus (sc.: complementum ad summam (vide glossam)) in rem, [erit sicut id quod exit de divisione] provenient 20 dragme diminutis 2 rebus {*id est, duplum minoris exeuntis, quia tunc rei⟨n⟩tegratur quando additur [ei quantum erat minus g.] ad differentiam que est inter maius et minus. Verbi gratia: 7 ad 5 differentiam, que est 2, reintegra[n]t duplum de 5, id est 10, ut fiant 12, que sunt 7 et 5*}. Quia integrum illud quod exit de duabus divisionibus in rem est sicut (sic, pro: 2 vicibus) illud quod {*scilicet: duplatum g.*} exit de divisione 10 diminuta re per rem ⟨in rem⟩. Addamus ergo illud ad 5, et erunt [5 dragme] 25 dragme diminutis 2 dragmis (sic, pro: 2 rebus). Iam *autem* (sc.: igitur) manifestum est quod multiplicatio eius quod exit de divisione uniuscuiusque partium per aliam [si agregabitur] *agregati* in rem *erit* 25 dragme diminutis 2 rebus. Quasi diceret: Divisi 10 in duas partes, et divisi unam⟨quamque⟩ duarum partium per aliam; deinde **(85^{v})** multiplicavi ⟨ea⟩ in unam duarum partium, et provenerunt 25 dragme diminutis 2 rebus; et est 25 dragme diminuta sicut (sic, pro: 2 vicibus) una {*duplata g.*} duarum partium. Fac ergo in eis sicut exposuimus superius.

Si autem volueris, pone unam duarum partium rem, aliam vero 10 diminuta re. Deinde dicemus: Divisimus rem per 10 diminuta re et exivit dinar, et divisimus [per] 10 diminuta re per rem, et exivit [zoz] *bizantius*. Et multiplicavimus superfluum quod est inter dinar et [zuz] *bizantium*, quod est dinar [pa-] diminuto [zoz] *bizantio*, in rem, et provenerunt 5 dragme, secundum hanc questionem. Et iam scivimus quod *id* quod est inter dinar et 2 [zozim] *bizantios* (sic, pro: et bizantium) qui exiverunt de divisionibus duabus et inter dinar diminuto zoz **bizantio** *est* 2 zozim **bizantii.* Et multiplicatio duorum zo[n]zim in rem [[sunt] *[est]* producit] *producit* (**A40^{v}**) 20 dragmas diminutis 2 rebus. Quibus addemus 5 dragmas, et erunt 25 dragme diminutis 2 rebus. Quasi diceret: 10 divisi in duas partes, et divisi unamquamque duarum partium per aliam, et aggregavi ea; deinde multiplicavi hoc in unam duarum partium, et provenerunt 25 dragme diminutis 2 rebus. Age ergo in eis sicut exposui tibi.

Et si dixit: 10 divisi in duas partes, et divisi maius per minus, et addidi quod exivit de divisione ad maius; et divisi minus per maius, et addidi quod exivit de divisione ad minus; deinde multiplicavi unum eorum in aliud, et fuerunt 35. Exemplum. Faciamus partem maiorem rem et minorem 10 diminuta re. Et dividamus rem per 10 re diminuta, et quod exit addamus ad rem; et erit **sit** (recte: erit) res *et* quod [erit] exit de divisione rei per 10 diminuta re. Deinde dividamus 10 diminuta re per rem, et quod exit addamus ad 10 diminuta re; et erit 10 dragme diminuta re et quod exit de divisione 10 diminuta re per rem. Quod multiplicemus in rem et quod exit de divisione rei per 10 diminuta re. *Et erunt 4 multiplicationes, scilicet:* [et] multiplicatio rei [et quod exit de divisione rei per] *in* 10 diminuta re (sc.: Quod multiplicemus in rem et quod exit de divisione rei per 10 diminuta re. Et multiplicatio rei et eius quod exit de divisione rei per 10 diminuta re ⟨in 10 diminuta re et quod exit de divisione 10 diminuta re per rem est sicut multiplicatio rei in 10 diminuta re⟩), et multiplicatio eius quod exit de divisione 10 diminuta re per rem in rem, et multiplicatio eius quod exit de divisione rei per 10 diminuta re in 10 dragmas diminuta re, et multiplicatio eius quod exit de divisione rei per 10 diminuta re in (id) quod exit de divisione 10 diminuta re per rem. Et iam scivisti quia multiplicatio rei in 10 diminuta re est 10 res diminuto censu; et quia multiplicatio eius quod exit de divisione ⟨rei per 10 diminuta re in 10 diminuta re est res; et quia multiplicatio eius quod exit de divisione⟩ 10 diminuta re per rem in rem est 10 [dragme] diminuta re; et quia multiplicatio eius quod exit d[i]*e* divisione rei per 10 diminuta re in id quod exit de divisione 10 diminuta re per rem est dragma, quia iam exposuimus quia omnium duorum numerorum cum dividitur unusquisque per alium [est] multiplicatio exeuntium inter se producit unum semper. Aggregemus ergo multiplicationem totam, et est [10 res diminuto censu *et res* et] 10 res diminuto censu et res et 10 dragme (**A41^{r}**) diminuta re et dragma. Erit *ergo* totum hoc 11 dragme et 10 res diminuto censu, que equantur 35 dragmis. Age in eis secundum [ordinem] algebram sicut dixi tibi, et exibit res 6. Que [sunt] est pars maior, reliqua vero ⟨est⟩ 4.

Et, si volueris, fac[ias] maiorem partem rem, minorem vero 10 diminuta re. Et divide rem per 10 diminuta re, et exeat dinar. Quod adde ad rem, et erit res et dinar. Divide [ergo] *etiam* 10 diminuta re per rem, et exeat moneta. Quam adde ad 10 diminuta re, et erit 10 et moneta diminuta re. Qu[am]e duc in rem et dinar. [Et provenient] Quod est *ut ducas* re[s]m in 10 dragmas diminuta re, et erunt 10 dragme (sic, pro: res) diminuto censu; duc etiam dinar in 10 dragmas diminuta re, et erit res; et duc monetam in rem, et provenient 10 dragme diminuta re; duc etiam dinar in [hoc] *monetam*, et provenie[n]t dragma. *Que* (omnia) agregemus; et erunt 11 dragme et 10 res diminuto censu, que equantur 35 dragmis. Fac ergo in illis sicut dixi tibi.

Et si dix[er]imus tibi: Divisimus 10 in du[as partes]*o*, et divisimus maius per minus, et addidimus quod exivit de divisione [per] *ad* 10; et divisimus min[orem]*us* per maius, et addidimus quod exivit de divisione ad 10; et du[ca]x*i*mus unum eorum in aliud, et proven[ient]erunt [112] *122* dragme et $\frac{2}{3}$. Exemplum. Faciamus quod exit de divisione maioris per minus rem ma[iorem]*gnam*. Quam addamus ad 10, et erunt 10 et res magna. Et faciamus quod exit de divisione minoris per maius rem parvam. Et addamus eam ad 10, et erunt 10 et res parva. Que ducamus in 10 et rem magnam: et ducamus 10 in 10, et provenient 100; deinde ducamus rem magnam in 10, et provenient 10 res magne; et ducamus rem parvam in 10, et provenient 10 res parve; et ducamus rem magnam in rem parvam, et proveniet dragma. **(A41^v^)** Que agrega, et erunt 10[0]1 dragme [et dragma] et 10 res magne et 10 res parve, que equantur 1[0]*22* dragmis et $\frac{2}{3}$. Prohice igitur 101 dragmas ex 122 et $\frac{2}{3}$, et remanebunt 21 dragme et $\frac{2}{3}$, equales 10 rebus magnis et 10 parvis. Faciam[is]us ergo ex rebus magnis et parvis rem magnam et rem parvam 2 dragmas et (sic, pro: Ergo erunt res parva et res magna 2 dragme et) [$\frac{2}{3}$] $\frac{1}{6}$. Quasi diceret: 10 divisi in duo, et divisi unumquodque eorum per aliud, et exiverunt 2 *dragme* et [$\frac{1}{3}$] $\frac{1}{6}$. Iam autem exposuimus hoc.

Et si diceremus tibi: 10 divisi in duo, et divisi maius per **(86^r^)** minus, et addidi quod exivit de divisione ad 10; deinde divisi *minus* per maius, et minui quod exivit de divisione de 10; deinde multiplicavi unum eorum in aliud, et provenerunt [100] 107 et $\frac{1}{3}$. Exemplum. Faciamus quod exit de divisione maioris per minus rem magnam, et addamus eam ad 10, et erunt 10 et res magna. Et faciamus quod exit de divisione minoris per maius rem parvam, quam diminuamus ex 10, et remanebunt 10 diminuta re parva. Qu[od]e ducamus in 10 et rem magnam, et provenient 99 dragme et 10 res magne diminutis 10 rebus parvis, que equantur 10[0]*7* dragmis et $\frac{1}{3}$. Age ergo in eis secundum algebram, et remanebunt 10 res magne diminutis 10 rebus parvis, equales 8 *dragmis* et $\frac{1}{3}$. *Igitur* res magn[e]a diminuta re parva [s-] equantur (sic) $\frac{5}{6}$ dragme. Quasi diceret: 10 divisi in duo, et divisi unumquodque illorum duorum per aliud, et proieci minus in illis (sc.: minus eorum) de maiori, et remanserunt $\frac{5}{6}$ dragme. Fac igitur sicut exposui tibi.

Et si dicemus tibi: Divide 10 dragmas [inter] *per* 2 et radicem 3. Exemplum. Faciamus id quod contigit uni rem. Quasi diceremus:

Divisimus 10 dragmas per 2 dragmas et radicem 3, et exivit res. Iam autem scivimus quod, cum multiplicas quod exit de divisione in dividentem, redit census divisus. Multiplica ergo rem in **(A42^{r})** 2 et radicem *de* 3, et provenient [60] *2 res* et radix 3 censuum, equales 10 dragmis. Prohice 2 res de 10, et remanebunt 10 diminutis 2 rebus, equales *radici* 3 cens[ibus]uum. Duc 10 diminutis 2 rebus in se, et provenient 100 dragme et 4 census diminutis 40 rebus, equales 3 censibus. Fac ergo secundum algebram, et erit census ⟨et⟩ 100 dragme, que equantur 40 radicibus. Dimidia igitur radices, et erunt 20. Que duc in se, et provenient 400. Ex quibus prohice 100 que sunt cum censu, et remanebunt 300. Accipe igitur radicem [diminuti] *eorum et minue* ⟨eam⟩ de 20, et *residuum* est quod contigit uni quando divisa sunt 10 per 2 et radicem 3, *trium. Et est* quod contigit uni 20 diminuta radice 300 (sc.: Radix eorum diminuta ex 20 est quod contingit uni. Quando ergo dividimus 10 per 2 et radicem de 3, est quod contingit uni 20 diminuta radice 300).

Explicit liber 2, cum adiutorio Adiutoris.

3 Et inveni questiones multas de computatione reintegrationis *ad* qu[e]as te duc[un]it aliud quam 6 species nominate (sic, pro: que te ducent ⟨ad⟩ aliud quam 6 species nominatas) in libro meo isto. Et ego exponam tibi quasdam que te docebunt viam hanc (sc.: speciem hanc) computationis reintegrationis, cum Dei adiutorio.

Si dicemus tibi: Multiplica censum in se et radicem 10 dragmarum, et proveni[e]ant 9, *noncuplum,* census. Exemplum. Faciamus censum rem. Et ducamus eam in rem et radicem 10 dragmarum, et provenie[n]t (sic) census et radix 10 censuum, equales 9 rebus. Prohice radicem 10 censuum de 9 rebus, et remanebunt 9 res diminuta radice 10 censuum, [equales] que equantur censui. Ergo res equatur 9 dragmis diminuta *radice* 10 dragmarum. Que est census.

Et si dicemus tibi: Duc radicem octupli alicuius census in radicem tripli eius; deinde adde ei quod provenit [10] 20 dragmas, et erit sicut census in se. Exemplum. Faciamus censum rem, et ducamus radicem octupli eius in radicem tripli ⟨eius⟩, et provenient (sic) radix 24 censuum. Cui adde 20, et erunt 20 et radix 24 censuum, que equantur censui ⟨quia dixit:⟩ "et fuit *id quod fuit* agregatum sicut census in se", **(A42^{v})** census autem est [radix] *res.* Dimidia ergo radicem (sc.: Dimidia ergo ⟨res, que sunt⟩ radix) 24 censu⟨u⟩m, et exibit radix 6. Quam duc in se, et provenie⟨n⟩t 6. Que adde ad dragmas, que sunt 20, et erunt 26 dragme. Quarum radicem accipe, et adde ei radicem 6. Et *id* quod habebis est census; quod est radix 26 et radix 6.

Et si dicemus tibi: Ducas radicem sexcupli alicuius census in radicem quincupli eius; deinde adde ad id quod provenit ex multiplicatione decuplum eius (sc.: census) et 20 dragmas, et erit sicut census in se. Exemplum. Faciamus censum rem, et ducamus radicem sexcupli eius in *radicem* quincupl[um]i ipsius, et proveniet radix 30 censuum; deinde adde ei 10 res et 20 dragmas, [et erunt 10 res] et erunt 10 res et 20 dragme et radix 30 censuum, equales censui. Dimidia igitur [radices] *res* et radicem 30 censu[m]um, et exibunt 5 et radix 7 [*censuum*] et $\frac{1}{2}$ [*census*]. Que duc in se, et provenient 32 dragme et $\frac{1}{2}$ et radix 750 dragmarum. Que adde ad 20 dragmas, et erunt 52 dragme et $\frac{1}{2}$ et radix 750 dragmarum. Et radix huius *congre⟨g⟩ati* addit[a]is ei 5 et radice 7 et $\frac{1}{2}$ est census.

Questio ⟨ ⟩. Si dicemus tibi: 10 divisi in 2 et duxi unum eorum in aliud; deinde divisi quod provenit ex multiplicatione per superfluum quod fuit inter duas partes, et exivit radix 6. Exemplum. Faciamus unam duarum partium rem, reliquam vero 10 diminuta re. Et ducamus unam *partium* earum (sc.: earum, *partium,*) in aliam, et provenient 10 res diminuto censu. Qu[as[od]]e dividamus per superfluum quod est inter duas partes, quod est 10 diminutis 2 rebus, et exibit radix 6. Iam autem scivimus quia, quando multiplicabimus quod exit nobis de divisione in dividentem, redit census noster divisus. Multiplicemus igitur radicem (**86^v^**) 6 in 10 diminutis 2 rebus, et proveniet radix 600 dragmarum diminuta radice 24 censuum, equalis 10 rebus diminuto censu. Reintegra radicem 600 dragmarum [ad] *[cum] per* radicem de [4] 24 cens[uum]ibus, et adde illam ad 10 [diminuta re] (sic, pro: ad 10 res) diminuto censu; deinde rein*te*gra 10 *res* diminut[a]o [re] *censu* (sic, pro: reintegra – vel restaura – 10 res) cum censu, et adde illum ad radicem 600; et *erunt census et 600 radix* (sc.: et adde illum ad radicem 600 ⟨; et erunt census et radix 600⟩), equali*a* 10 rebus (**A43^r^**) et radici 24 censuum. Dimidia itaque 10 [diminuto censu] *res* et radicem 24 censuum, et exibunt 5 et radix 6. Que duc in se, [*et agrega*] et provenient 31 dragme et radix 600. Ex quibus prohice[m-] radicem 600 dragmarum que fuit cum censu, et remanebit 31. Quorum radicem accipe. *Et* diminuta *ea* ex 5 et radice 6, quod *remanet* est una duarum partium {(sc.:) *et minue eam ex 5 et radice 6, et residuum est una duarum partium 10 g.*}, reliqua vero est residuum de 10, et est 5 et radix 31 diminuta radice 6.

Et si dicemus tibi: Divisi 10 in duas partes, et multiplicavi unam [in aliam] duarum partium in se et aliam in radicem 8; deinde proieci quod [agregatum] *productum* fuit *ex* multiplicatione unius duarum partium in radicem 8 ex eo quod provenit ex multiplicatione (alterius) in se, et remanserunt 40 dragme. Exemplum. Faciamus unam duarum partium rem, reliquam vero 10 diminuta re. Et ducamus 10 diminuta re in se, et erunt 100 dragme et census diminutis 20 rebus. Deinde multiplica rem in radicem de 8, et proveniet radix 8 censuum. Quam prohice ex 100 dragmis et censu diminutis 20 rebus, et remanebunt 100 dragme et census 20 [radicibus] *rebus diminutis* et diminuta radice 8 censuum, que equantur 40 dragmis. Restaura ergo 100 et censum cum 20 [radicibus] *rebus* et radice 8 censuum, et adde ⟨eas⟩ ad 40 dragmas. Et habebis 100 dragmas et censum, que equantur 40 dragmis et 20 rebus et [rei] *radici* 8 censuum. Prohice ergo 40 dragmis (sic) ex 100 dragmis; et remanebunt 60 dragme et census, que equantur 20 rebus et radici[bus] 8 censuum. Dimidia ergo 20 res et radicem 8 censuum, et exibunt 10 et radix 2 censuum (sic). Quas duc in se, et provenient 100 dragme et 2 dragme (sic, pro: 102 dragme) et radix [8 censuum] 800. Ex quibus prohice 60 dragmas que sunt cum censu, et remanebunt 42 dragme et radix 800. Quorum radice diminuta ex 10 et radice 2, *residuum* est pars ducta in radicem 8; et [postquam] reliqua est residuum de 10, qu[e]od est 42 dragm[e]*arum* et radi[x]*cis* (**A43**$^{\mathbf{v}}$) 800 accepta radice [eius] *et* diminuta ex ea radice duorum.

Et, si volueris, duc rem in se, et proveniet census. Et multiplica partem alteram, que est 10 diminuta re, in radicem de 8, et provenient (sic) *radix* 800 diminuta radice 8[00] *censuum*. Prohice igitur radicem 800 dragmarum diminuta radice 8 censuum ex censu[m], et remanebit census et radix 8 censuum ⟨diminuta radice 800 dragmarum, que equantur 40 dragmis. Restaura ergo censum et radicem 8 censuum⟩ accipe radicem (sic, pro: cum radice) 800 dragmarum, et adde illam ad 40 dragmas; et erunt 40 dragme et radix 800 dragmarum, que equantur censui et radici 8 censuum. Dimidia igitur radicem 8 censuum, et exibit radix 2. Quam duc in se, et erunt 2. Que adde ad 40 dragmas et radicem 800 dragmarum, et erunt 42 dragme et radix 800 dragmarum. Radix eorum diminuta ex ea radice 2^{orum} est pars ducta in se.

Questio ⟨ ⟩. Et si dicemus tibi: Divisi 10 in duo, et duxi unum eorum in [aliud] radicem 10, reliquum vero in se, et fuerunt equalia. Exemplum. Faciamus unam duarum partium rem, reliquam vero 10 diminuta re. Et ducamus rem in se, et proveniet census. Deinde ducamus 10 diminuta re in radicem de 10, et provenie[n]t radix 1000 dragmarum diminuta radice 10 censuum ⟨equalis censui⟩. Restaura *igitur* radicem 1000 dragmarum cum radice 10 censuum, ⟨et⟩ adde illam ad censum; et erunt census et radix 10 censuum, equales radici 1000 dragmarum. Dimidia radicem 10 censuum, et exibit radix 2 et $\frac{1}{2}$. Quam duc in se, et provenient 2 et $\frac{1}{2}$. Que adde ad radicem 1000, et erunt 2 dragme et $\frac{1}{2}$ et radix 1000 dragmarum. Cuius radix diminuta ex ea radice duorum ⟨et $\frac{1}{2}$⟩ est pars ducta in se, et alia est [residuum] *id* quod remanet de 10, quod est 2 et $\frac{1}{2}$ et radix 1000 accepta radice eius, *scilicet: radicis 1000 et 2* $\frac{1}{2}$, *et* diminuta ex 10 et radice 2^{orum} et $\frac{1}{2}$.

Et si dicemus tibi: Censui addidimus 10 dragmas, deinde duximus quod agregatum fuit in radicem 5, et fuit quod provenit (**A44**$^{\mathbf{r}}$) sicut census in se. Exemplum. Faciamus censum rem. Et addamus ei 10 dragmas, et erunt 10 dragme et res. Quas ducamus in radicem 5 dragmarum, et provenie⟨n⟩t radix 500 dragmarum et radix 5 censuum, *que* equantur censui. Dimidia igitur radicem 5 censuum, et exibit radix unius et $\frac{1}{4}$. Quam duc in se, et proveniet 1 et $\frac{1}{4}$. Que adde ad radicem 500 dragmarum, et erunt 1 et $\frac{1}{4}$ et radix 500 dragmarum. Radix eius, additis ei 1 et $\frac{1}{4}$ (sic, pro: addita ei radice 1 et $\frac{1}{4}$), est census.

Questio ⟨ ⟩. Et si dicemus tibi: Differentia inter duos census est 5 dragme, et multiplicavi maiorem in 10, *decuplum*, eius; deinde accepi radicem eius quod [agregatum] *productum* fuit, et exivit sicut minor in se. Exemplum. Faciamus unum duorum censuum rem, reliquum vero rem diminutis 5. Et ducamus maiorem, qui est res, in 10, *decuplum*, sui, et erunt 10 [res] census. Radix ergo 10 censuum equatur 25 dragmis et censui diminutis 10 radicibus. Restaura ergo 25 et (**87$^{\mathbf{r}}$**) et (sic) censum cum 10 rebus, et adde eas ad radicem 10 censuum; et erunt 10 res [que sunt res] et radix 10 censuum, equales censui et 25 dragmis. Dimidia ergo 10 res et radicem 10 censuum, et exibunt 5 et radix 2 et $\frac{1}{2}$. Que duc in se, et provenient 27 et $\frac{1}{2}$ [Accipe radicem 250] *et radix 250.* Ex quibus prohice dragmas que sunt 25 (sc.: 25 dragmas que sunt cum censu); residuum *est* 2 et $\frac{1}{2}$ et radix 2[0]50. Radix eius, *scilicet: 2 et* $\frac{1}{2}$ *et radicis 250*, addita ad 5 et radicem 2^{orum} *et* $\frac{1}{2}$ est census magnus, census autem primus (sc.: parvus) est diminutis ab eo *toto* 5, et est 2 dragme et $\frac{1}{2}$ et radix 2[0]50 accepta radice eius *et* addita ad radicem 2 dragmarum et $\frac{1}{2}$.

Et si dicemus tibi: Duc censum in duplum eius, deinde accipe radicem eius quod provenit; cui adde 2 dragmas, et postea duc eum in censum, et erit 30 dragme. Exemplum. Faciamus censum rem. Et ducamus eam in duplum eius, et provenient 2 census. Deinde duc 2 dragmas [in] cum radice[m] 2 censuum in rem, et provenient [60] *2 res* et radix 2 censuum *census*, que equantur 30 dragmis. Re[staura]*duc* quicquid habes ad censum. Quod est ut ducas radicem 2^{orum} censuum [census] (sic) in radicem $\frac{1}{2}$ (**A44$^{\mathbf{v}}$**) dragme ut sit census; et ratio [eius] quod hoc sit ita *est* propterea *quod* census est radix census census, et radix census census (sic, pro: radix 2 censuum census) cum multiplicatur in radicem $\frac{1}{2}$ dragme [exit] *provenit* radix census census, et radix census census est census. Multiplica ergo totum quod habes in radicem $\frac{1}{2}$ dragme. *Et* radix $\frac{1}{2}$ dragme [*ducta in radicem 2 censuum*] [erit radix 2 censuum] *ducta in 2 res producit radicem 2 censuum*: (sic, pro: ⟨Multiplica ergo⟩ radicem $\frac{1}{2}$ dragme ⟨in radicem 2 censuum census, et proveniet census. Deinde multiplica 2 res in radicem $\frac{1}{2}$ dragme⟩, et proveniet radix 2 censuum:) propterea quod, cum multiplicas 2 res in se, provenient 4 census; deinde duc 4 census in $\frac{1}{2}$ dragme, et provenient 2 census; radix huius est radix [est] $\frac{1}{2}$ dragme in 2 res. Deinde duc 30 dragmas in radicem $\frac{1}{2}$ dragme, [*et* erit 4 census et 50 dragme (sic, pro: et proveniet radix 450 ⟨. Deinde agrega (totum) hoc; et erunt census et radix 2 censuum, que equantur radici 450⟩)] {*provenient* (sic) *radix*

450. Erunt ergo census et radix 2 censuum equales radici 450. g.}. Dimidia igitur radicem 2 censuum, et exibit radix $\frac{1}{2}$. Quam duc in se, et proveniet $\frac{1}{2}$. Quam adde ad radicem 450, et erunt $\frac{1}{2}$ dragme et radix 450. *De* radice *quorum* diminuta [ex eo] radice $\frac{1}{2}$ dragme, *remanebit res*, que est census quesitus.

Si dicemus tibi: Divide 10 in 2 partes, et divide unamquamque duarum partium per aliam; et agrega ea, et erit radix 5 dragmarum. Exemplum. Faciamus unam duarum partium rem, reliquam vero 10 diminuta re. Et ducamus unamquamque in se, et agregemus, et erunt 100 dragme et 2 census diminutis 20 rebus. Custodi ergo hunc numerum (sc.: Serva ergo hoc congregatum). Deinde duc unam duarum partium in aliam, videlicet rem in 10 diminuta re, et erunt 10 res diminuto censu. Qu[e]as duc in radicem 5, et [erunt] *provenie[n]t* radix 500 censu[s]um diminuta radice 5 censuum census, equales 100 dragmis et 2 censibus diminutis 20 radicibus. Restaura radicem 500 *[-pli]* cens[us]*uum* cum radice 5 censuum census, et adde illas (sic) ad 100 dragmas et 2 census diminutis 20 rebus. Et restaura 100 dragmas et 2 census diminutis 20 rebus cum 20 rebus, et adde illas ad radicem 500 census (sic). Et erunt 20 res et radix 500 cens[us]*uum*, equales 100 dragmis et 2 censibus **(A45ʳ)** et radici 5 censuum *census*. Reduc igitur totum quod habes ad censum. Quod est [quod] *ut* ducas 2 census et radicem 5 censuum census in radicem 5 dragmarum diminutis 2 dragmis; et [quia indigemus] *quod cogit nos* multiplicare hec in radicem 5 dragmarum diminutis 2 dragmis *est* quia, quando dividis dragmam per duas res (sic, pro: per duo) et radicem 5, contingit uni radix 5 diminutis 2 dragmis; et iam exposuimus hoc. Multiplica igitur totum quod habes in radicem 5 dragmarum diminutis duabus dragmis; et provenient census et radix 50000 diminutis 200 dragmis, equales 10 rebus *{proiecto comuniter addito et diminuto. Intellige. g.}*. Dimidia ergo radices, et exibunt 5. Quas duc in se, et provenient 25 dragme. Ex quibus prohice radicem 50000 dragmarum diminutis 200 dragmis, et remanebunt 225 dragme diminuta radice 50000. Radix eius subtracta de 5 est una duarum partium, et addita ad 5 est pars altera.

Et, si volueris, computa ea secundum viam aliam. Qu[od]e est quod dividas radicem 5 in dua⟨s⟩ partes, et erit (sc.: et sit) multiplicatio unius earum in aliam dragma 1. Fac unam duarum partium rem, reliquam vero *radicem de* 5 diminuta re. Et duc unam earum in aliam, et [erit] *proveniet* radix 5 censuum diminuto censu, equalis dragme. Restaura radicem 5 censuum cum censu, et adde illum ad dragmam. Et erunt census et dragma, equales radici 5 censuum. Dimidia ergo radicem [cum] 5 cens[ibus]uum, et exibit radix unius et $\frac{1}{4}$. Quam duc in se, et proveniet 1 et $\frac{1}{4}$. Ex quo prohice dragmam, et remanebit $\frac{1}{4}$. Cuius radicem accipe, et exibit $\frac{1}{2}$. Prohice ergo ipsam de radice unius et $\frac{1}{4}$, et remanebit radix unius et $\frac{1}{4}$ diminuta $\frac{1}{2}$ dragme. ⟨Que est res quesita.⟩ Et iam exposuimus tibi quod rerum que sunt [in] *una de* tribus speciebus dictis (sic, pro: que sunt in tribus speciebus compositis) quando duces [quasdam] *partem earum* (sic, pro: medietatem) in se [quod] productum ex multiplicatione est numerus; et radi[ces]x censu[s]um est res; quando *ergo* accipis medietatem eius et ducis eam in se, provenit numerus. Amplius revertere ad principium questionis. Et fac unam duarum partium rem, reliquam vero 10 diminuta re. Divide 10 diminuta re per rem, et exibit, *secundum* illud quod exposuimus, *radix* un[um]ius et $\frac{1}{4}$ diminuta $\frac{1}{2}$ dragme. Que (sc.: Quam) duc in rem, et proveniet radix census **(A45^{v})** et $\frac{1}{4}$ census diminuta $\frac{1}{2}$ rei, que equantur (sic) 10 diminuta re. Restaura igitur radicem census et $\frac{1}{4}$ census cum medietate rei, et adde illam ad 10 **(87^{v})** diminuta re. Erunt 10 diminuta $\frac{1}{2}$ rei, equales radici census et $\frac{1}{4}$ census. Duc igitur 10 diminuta rei $\frac{1}{2}$ in se, et provenient 100 dragme [diminuta] *et* $\frac{1}{4}$ census diminutis 10 rebus, que equantur censui et $\frac{1}{4}$. Restaura igitur [censum] 100 dragmas et $\frac{1}{4}$ census cum 10 [diminuta re] *rebus, et* adde illa*s* ad censum et $\frac{1}{4}$. Et [restaura] *prohice comuniter* $\frac{1}{4}$ census cum $\frac{1}{4}$ census, et [erit] *remanebunt* census [et 20 dragme] et *10 radices, equales 100 dragmis. Igitur res equatur 125 dragmarum radici* diminutis 5 dragmis. Et illud est una duarum partium, pars vero alia *est residuum* de 10; que est 15 dragme diminuta radice 125 dragmarum.

Et, si volueris, computa per viam aliam. Que est ut facias unam duarum partium rem, aliam vero 10 diminuta re. Et dividas 10 dragmas diminuta re per rem, et exeat dinar. Ergo cum duces dinar in rem, provenient 10 diminuta re. Deinde divide rem per 10 diminuta re, et exibit radix 5 diminuto dinar. Ergo cum duces radicem 5 diminuto dinar in 10 diminuta re, proveniet res. Multiplica igitur radicem 5 dragmarum diminuto dinar in 10 diminuta re, et provenient 10 dragme diminuta re et radi[ce]x 500 dragmarum *diminuta radice* 5 censuum et diminutis 10 dinar⟨im⟩, que equantur rei. Restaura ergo ea cum 10 dinarim, et adde ea ad rem. Et erunt res et 10 dinar⟨im⟩, equales 10 dragmis diminuta re et radic[e]i 500 dragmarum diminuta radice 5 censuum. Prohice rem ex ipsis, et remanebunt 10 dragme diminutis 2 rebus et radi[ce]x 500 dragmarum diminuta radice 5 censuum, equales 10 dinar⟨im⟩. Ergo dinar equatur dragme et radici 5 dragmarum diminuta $\frac{1}{5}$ *rei* et diminuta radice $\frac{0\;1}{10\;2}$ census. Amplius redeas, et dicas: Divisi 10 dragmas diminuta re per rem, et exivit dragma et radix 5 dragmarum diminuta $\frac{1}{5}$ rei et diminuta radice $\frac{0\;1}{10\;2}$ census. *Igitur* cum duces ea in rem, provenient 10 dragme ⟨diminuta re⟩. *Igitur* multiplicatio **vel: productum** [dragme] *eius, que est res* et radi[cis]x (sic, pro: Duc igitur dragmam et radicem ⟨5 dragmarum diminuta $\frac{1}{5}$ rei et diminuta radice $\frac{0\;1}{10\;2}$ census in rem; et provenient res et radix⟩) 5 censuum **(A46^{r})** diminuta $\frac{1}{5}$ census et diminuta radice $\frac{0\;1}{10\;2}$ census *census,* equatur (sc.: que equantur) 10 dragmis diminuta re. Restaura ergo ea cum $\frac{1}{5}$ census et radice $\frac{e\;1}{10\;2}$ census census, et adde illa ad 10 dragmas diminuta re; et reintegra 10 dragmas cum re, et adde illam ad rem et radicem 5 censuum. Et erunt 2 res et radix 5 censuum, equales 10 dragmis et $\frac{e}{5}$ census et radici $\frac{0}{10}\frac{1}{2}$ census census. Reintegra censum tuum ut sit census integer; et tua reintegratio erit quod ducas eum in radicem 500 dragmarum diminutis 20 dragmis. Multiplica ergo totum quod habes in radicem 500 dragmarum diminutis 20 dragmis, et provenient 10 res equales censui et radici 50000 dragmarum diminutis 200 dragmis. Dimidia ergo res, et exibunt 5. Que duc in se, et provenient 25. Ex quibus prohice radicem 50000 diminutis 200 dragmis, et remanebunt [205] 225 dragme diminuta radice 50000. Cuius radix [diminuta] *sublata* de 5 est una duarum partium. Et radix 2[0]25 [*225*] dragmarum diminuta radice 50000 [que] est radix 125 diminutis 10 dragmis {*Operare, etiam invenies. g.*}. Minue ergo ea de 5, et remanebunt 15 dragme diminuta radice 125. Que sunt una duarum partium, reliqua vero est residuum de 10, et est radix 125 dragmarum diminutis 5 dragmis.

Et si dicemus tibi: 10 divisi in duas partes, et divisi unamquamque earum per aliam; deinde duxi unamquamque duarum partium que exiverunt de divisione uniuscuiusque partium per aliam in se, et provenerunt 3 dragme. Et iam scivisti quia multiplicatio unius duarum partium que exiverunt de divisione uniuscuiusque *duarum* partium (de) 10 [in] per aliam in aliam est dragma semper. Igitur multiplicatio eius in aliam bis producit 2 dragmas. Addamus ergo 2 dragmas ad 3, et erunt 5. *Et* manifestum est ex eo quod dixit Euclides (in secundo tractatu libri sui) quod multiplicatio unius duarum partium que exeunt de divisione cuiusque earum ex 10 per aliam in se *est* 5 dragme (sc.: quod multiplicatio duarum partium (agregatarum) que exiverunt de divisione uniuscuiusque duarum partium de 10 per aliam in se est 5 dragme). (**A46**[v]) Igitur ipsa (sc. duo exeuntia de divisione) sunt radix 5 dragmarum. Quasi diceret: 10 divisi in duas partes, et divisi unamquamque earum per aliam, et exivit radix 5. Et iam exposui hanc questionem.

Et si dicemus tibi: 10 divisi in duas partes, et divisi unamquamque earum per alteram; deinde proieci quod exivit de divisione minoris per maiorem ex eo quod exivit de divisione maioris per minorem, et remanserunt $\frac{5}{6}$ dragme. Exemplum. Faciamus quod exit de divisione minoris per maius rem. Et addamus eam ad $\frac{5}{6}$ dragme, et erunt res et $\frac{5}{6}$ dragme. Et hoc est quod exit de divisione maioris per minus. Multiplicemus ergo ipsum in rem, et provenient census et $\frac{5}{6}$ *rei*, equales dragme. Et iam exposuimus hanc questionem in his que precesserunt de libro nostro, preter hanc viam **alia via*. Et, si volueris, utere in ea via alia.

Si dicemus tibi: 10 divisi in duas partes, et divisi unamquamque earum per aliam; deinde multiplicavi unamquamque duarum partium que exiverunt de divisione cuiusque earum per aliam in se; et proieci minus de maiori, et remanserunt (**88**[r]) 2 dragme. Exemplum. Faciamus quod exit de divisione minoris per maius rem. Et multiplicemus eam in se, et [erit] *proveniet* census. Quem addemus ad 2 dragmas, et erunt census et 2 dragme. Manifestum est etiam ex eo quod diximus quod multiplicatio eius quod exit de divisione maioris *eorum* per minus in se est census et 2 dragme. Ducamus ergo censum in censum et 2 dragmas, et provenie⟨n⟩t census census et 2 census, equales dragme; quia iam exposuimus quod unum eorum in aliud est [2] dragm[e]a, et multiplicatio quadrati unius eorum in quadratum alterius est sicut dragma in se, multiplicatio autem dragme in se est dragma. Dimidia igitur 2 census, et exibit unus. Quem duc in se, et proveniet 1. Quem adde ad dragmam, et erunt 2. [Radix 2] *De quorum radice* sublat[o]a [ex ea] unitate *erit* accepta radi[ce]*x* residui [est] quod exit de divisione minoris (**A47^{r}**) per maius. Deinde redibimus ad principium questionis, et dicemus: Divisimus 10 dragmas diminuta re per rem, et exivit [radix] *de radice* 2^{orum} sublato [ex ea] uno [accepta] rad*ix* residui. *Quam* multiplicabimus in rem. Cum autem volueris hoc: Duc 10 diminuta re in se, et provenient 100 dragme et census diminutis 20 rebus; multiplica rem in se, et proveniet census; et duc radicem 2 diminuta **sublata** ex e[o]a un[o]*itate et residui* accepta radice in se, proveniet radix 2 sublato ex ea 1; deinde dicemus: Divisimus 100 dragmas et censum diminutis 20 rebus per censum, et exivit radix 2 sublato ex ea 1; ducamus ipsam in censum, et proveniet radix *2* cens[us]*uum* census diminuto **sublato** ex e[o]a censu, equalis 100 dragmis et censui diminutis 20 radicibus. Faciemus secundum algebram in eo. Et est ut restauremus 100 dragmas et censum cum 20 radicibus, et addamus eas ad radicem *2* cens[us]*uum* census; et restauremus radicem *2* cens[us]*uum* census cum censu[m], et addamus eum ad 100 dragmas ⟨et censum⟩; et prohiciamus radicem *2* [ex] censu*um* census ex 100 dragmis et 2 censibus. Et remanebunt 100 dragme et 2 census diminuta radice [radicis] *2* cens[us]*uum* census, que equantur 20 radicibus. [Adde] *Reduc* totum quod habes ad censum; quod est quod ducas totum quod habes in unum et radicem $\frac{1}{2}$ unius {*Intellige. g.*}. Et ducas 2 census diminuta radice *2* cens[us]*uum* census in dragma[s]m et radicem $\frac{1}{2}$ dragme, et proveniet census; deinde duc 20 res in dragmam *et radicem $\frac{1}{2}$ dragme*, et provenient 20 res et radix 200 censuum; deinde ducas 100 dragmas in dragma[s]m et radicem $\frac{1}{2}$ dragme, et provenient 100 dragme et radix 5000 dragmarum. Quas agrega ad censum et 100 dragmas et radicem 5000 dragmarum (sic, pro: Que agrega, et erunt census et 100 dragme et radix 5000 dragmarum), *que* equantur 20 rebus et radici 200 dragmarum (sic, pro: censuum). Dimidia igitur res (sc.: 20 res et radicem 200 censuum), et exibunt 10 et radix 50. Quas duc in se, et provenient 150 dragme et radix 20000. Sub[latis]*trahe* ex eis 100 dragmas et radicem 5000[0] dragmarum, *et* remane[nt]*bunt* 50 dragme et radix 5000 dragmarum. Radice eorum sublat[is]a (**A47^{v}**) a[b *ea*] 10 et radice 50: et hec [*residuum*] est pars maior, minor vero est residuum de 10.

Et, si volueris, computa [in] hoc via alia, et est: Scitum est quia quadratum uniuscuiusque (duarum) partium 10 quando divisa fuerit unaqueque illarum per aliam et prohiciemus (sc.: proiecerimus) maius de minori (sic, pro: minus de maiori) erit residuum 2 dragme: *et* (delendum) quia omnis numerus qui dividitur per numerum, quando multiplicatur quod exit de divisione in se, erit quod agregabitur *_proveniet_* ex eo, *scilicet: quadratum eius*, sicut quod exit de divisione multiplicationis numeri divisi in se per multiplicationem numeri dividentis in se. Faciamus *igitur* quod exit de divisione quadrati maioris per quadratum minus (sc.: minoris) 2 dragmas et rem, et quod exit de divisione quadrati minoris per quadratum maius (sc.: maioris) rem. Et iam exposuimus quia unum eorum in aliu[m]d est dragma. Ducamus ergo rem in 2 dragmas et rem, et provenient 2 res et census. Faciamus igitur secundum algebram in eis unum (sic, pro: Opponamus igitur ea cum uno); exibit res radix 2^{orum} diminuto [*subtractis*] ex ea 1. Deinde redibimus ad 10, et faciemus unam duarum partium rem, reliquam vero 10 diminuta re; quadratum autem unius earum ⟨erit⟩ census, et quadratum alterius 100 dragme et census diminutis 20 radicibus. Deinde dividamus 100 dragmas et censum diminutis 20 radicibus per censum, et exibit radix 2 sublato ex ea 1. Quam duc in censum, et proveniet radix *2* cens[us]*uum* census [de-] subtracto ex ea censu, equa[tur]lis 100 dragmis et censui diminutis 20 radicibus. Age ergo secundum algebram in eo sicut exposui tibi in questione ⟨prima⟩ (sc. in prima parte questionis huius), cum adiutorio Dei.

Et si dicemus: Proice ex censu 2 radices eius et 10 dragmas; deinde duc residuum in se, et proveniet octuplum eius (sc.: census). Exemplum. Facias censum tuum censum, et prohice ex eo 2 radices eius et 10 dragmas; et remanebit census diminutis 2 radicibus et diminutis 10 dragmis, equalis radici 8 censuum, quia hoc dixit: "Ducamus residuum in se, et provenient (sic) octuplum census", et residuum est radix 8 censuum. Age secundum algebram in eis, et erunt 2 radices et 10 dragme et radix 8 censuum equales censui. Et dimidia 2 radices et radicem 8 censuum, et exibunt unus et radix 2. Qu[as]e duc in se, **(A48^{r})** et provenient 3 et radix 8. Que agrega ad 10, et erunt 13 et radix 8. Radix eius agregata **(88^{v})** ad 1 et radicem 2^{orum}, illud quod erit est radix census. Duc illud in se, et quod provenerit est census.

Et si dicemus tibi: *2* radices census et radix medietatis eius et radix tertie eius equantur censui, quantum est census? Exemplum. Facias censum tuum censum, et dic: Radix *_res_* (sic, pro: 2 radices/res) et radix medietatis census et radix tertie census equa⟨n⟩tur censui. Res *ergo* equatur 2^{bus} et radi[x]*ci* $\frac{1}{2}$ et radi[x]*ci* $\frac{1}{3}$. [Que] *Et hoc* est radix census, et census est 4 et $\frac{1}{2}$ et $\frac{1}{3}$ *et radix 8* et radix 5 et $\frac{1}{3}$ et radix $\frac{2}{3}$.

Et si dicemus tibi: Census radi[x]*ces 2* et radix medietatis eius et radix tertie eius est (sic) 20 dragme, quantus est cen (sic) census?

Exemplum. Facias censum tuum censum, et accipias 2 radices eius et radi[x]*cem* $\frac{1}{2}$ ⟨eius⟩ et radi[x]*cem* $\frac{1}{3}$ eius [est 20]; *et* erunt 2 res et radix medietatis census et radix $\frac{1}{3}$ census, equa[ntur]*les* 20 dragmis. Prohice 2 res de 20 dragmis, et remanebunt 20 dragme diminutis 2 rebus, equales radici $\frac{1}{2}$ census et radici $\frac{1}{3}$ census *ductis in se* (haec addita delenda). Duc 20 dragmas diminutis 2 rebus in se, et provenient 400 dragme et 4 census diminutis 8*0* radicibus, equales ⟨radici⟩ $\frac{1}{2}$ census et ⟨radici⟩ $\frac{1}{3}$ census ⟨ductis in se, que sunt $\frac{1}{2}$ census et $\frac{1}{3}$ census et radix $\frac{2}{3}$ census census⟩. Fac ergo secundum algebram in ipsis, et erunt 80 res equales 400 dragmis et 3 cens[us]*ibus* et $\frac{1}{6}$ census *minus radice $\frac{2}{3}$ census census.* Reduc totum quod habes ad censum unum, et fac sicut dixi tibi, et intellige {*et erit radix 1200 subtracta de 40 radix census. g.*}

Et si dicemus tibi: Adde censui 4 radices eius et radicem medietatis eius et radicem $\frac{2}{3}$ eius, et [p-] erunt 10 dragme, quantus est census? Exemplum. Faciamus censum censum, et addamus ei 4 radices eius et radicem $\frac{1}{2}$ eius et radicem $\frac{2}{3}$ eius; et erunt census et 4 res et radix medietatis census et radix $\frac{2}{3}$ census, que equantur 10 dragmis. Fac sicut exposui tibi, et invenies, cum adiutorio Dei.

Et si dicemus tibi: Adde censui radicem eius et radicem medietatis eius; deinde duc quod agregatur in se, et provenient (sic) **(A48^{v})** 5^{plum} census. Hic census et radix *et* radi[cis]*x* $\frac{1}{2}$ census equantur radici 5 censuum. Res ⟨igitur⟩ equatur radici 5 dragmarum diminuto (sic) dragma et diminuta radice $\frac{1}{2}$ dragme. Et hoc est radix census, et census est 6 dragme et $\frac{1}{2}$ et radix 2^{orum} diminuta radice 20 et diminuta ⟨radice⟩ 10.

Et si dicemus tibi: Adde censui radicem eius et radicem $\frac{1}{2}$ eius, deinde duc additum in se, et provenient 20 dragme. Hic census et res et radix $\frac{1}{2}$ census [erit] *sunt sicut radix 20. Accipe ergo medietatem rei et radicis $\frac{1}{2}$ ⟨census⟩. Que est* (sc.: Hic census et res et radix $\frac{1}{2}$ census ⟨equantur radici 20 dragmarum. Dimidia ergo rem et radicem $\frac{1}{2}$ census,⟩ et erunt) $\frac{1}{2}$ et radix $\frac{1}{8}$. Duc ea[m]*s* in se, et provenient $\frac{3}{8}$ dragme et radix $\frac{1}{8}$ dragme. Adde illa ad radicem 20 dragmarum, et erunt $\frac{3}{8}$ dragme et radix 20 dragmarum et radix $\frac{1}{8}$ dragme. Accipe radicem eorum, et prohice ex ipsis $\frac{1}{2}$ et radicem $\frac{1}{8}$; et residuum est radix census.

Et si dicemus tibi: Censui adde radicem medietatis eius; deinde duc additum in se, et proveniet quadruplum census. Exemplum. Fac censum tuum rem, et adde ei radicem medietatis eius, et [prov-] erunt res et radix $\frac{1}{2}$ rei. Que duc in se, et provenie⟨n⟩t census et $\frac{1}{2}$ rei et radix 2 cuborum, equales 4 rebus. Et scias quia census provenit ex ductu rei in se, et cubus ex ductu rei in censum, et census census ex ductu census in censu⟨m⟩, et cubus cubi ex ductu cubi in cubum. Census et $\frac{1}{2}$ rei et radix 2 cuborum equantur 4 rebus. Prohice [*va-* 2 cubos, et duc 3 res et $\frac{1}{2}$ rei diminuto censu in se, et proveniet census *-cat*] *comuniter censum, et remanebit $\frac{1}{2}$ rei et radix 2 cuborum, equales 4* *rebus diminuto censu. Prohice etiam* (sc.: Ergo census et $\frac{1}{2}$ rei et radix 2 cuborum equantur 4 rebus. Prohice censum et) dimidiam [etiam] rem ex 4 rebus, et remanebunt 3 res et $\frac{1}{2}$ rei diminuto censu [et $\frac{1}{2}$ rei diminuto censu], equales radici 2 cuborum. Duc ergo 3 res et $\frac{1}{2}$ rei diminuto censu in se, et provenient census census et 12 census et $\frac{1}{4}$ census diminutis 7 cubis, equales 2 cubis. Restaura igitur ea **(A49^{r})** cum 7 cubis, et adde illos ad 2 cubos; et erunt census census et 12 census et $\frac{1}{4}$ census equales [7] *9* cubis. Reduc totum quod habes ad censum; et erit census et 12 dragme et $\frac{1}{4}$ dragme, equales 9 rebus. Dimidia igitur res, et exibunt 4 et $\frac{1}{2}$. Quas duc in se, et erunt 20 et $\frac{1}{4}$. Prohice ex ipsis 12 et $\frac{1}{4}$, et remanebunt 8. Radix eorum sublata ex 4 et $\frac{1}{2}$, *quod* remanebit erit census, et est 4 et $\frac{1}{2}$ dragme *diminuta* radi[x]*ce* 8.

Et, si volueris, fac censum tuum censum, et adde ei radicem medietatis ipsius, et erunt census et radix medietatis census, equales *radici* 4 cens[ibus]*uum*, [*et*] quia di[x]*c*is: "Quando ducimus *e[um]a* in se, [erit] *proveniet* quadruplum census". Est *ergo* census et radix $\frac{1}{2}$ census, equales radici 4^{pli} censu[um]*s*. *Et hoc est 2 res. Prohice radicem* $\frac{1}{2}$ *census ex radice 4 censuum*; remanebit radix 4 censuum diminuta radice *medietatis* census, equalis censui. Igitur radix census [et] *est* radix 4 diminuta radice $\frac{1}{2}$ *census equalis* $\frac{1}{2}$ *dragme, et census est 4 et* $\frac{1}{2}$ *diminuta radice* 8 (sic, pro: Igitur radix census est radix 4 diminuta radice ⟨$\frac{1}{2}$, que est 2 dragme diminuta radice $\frac{1}{2}$ dragme, et census est 4 et $\frac{1}{2}$ diminuta radice⟩ 8).

Et, si volueris, fac censum tuum 2 census. Quibus adde radicem $\frac{1}{2}$ ⟨eius⟩, et erunt 2 census et [radix] *res*. Que duc in se, et provenient *census et* 4 census census et 4 cubi, qui equantur 8 censibus. Prohice [ex] censum de 8 censibus, et remanebunt 4 census census et 4 cubi equales 7 censibus. Redu*c* igitur totum quod habes ad [censum] unum censum census; et exibit census census et cubus equales censui et $\frac{3}{4}$ census. Reduc etiam totum quod habes ad censum [census]; quod est ut dividas totum quod habes per censum [census]: [et] id autem quo indiges ad dividendum totum quod habes per censum est quia, cum divides censum census per censum, [et] *exibit census, et cum* divides cubum per censum, exibit res, et divides censum et $\frac{3}{4}$ census **(89^{r})** per censum, et exibit dragma et $\frac{3}{4}$. *Ergo* census [erit] [census] et res equa[les]*ntur* dragme et $\frac{3}{4}$ dragme. Dimidia igitur radices, et erunt $\frac{1}{2}$. Quam duc in se, **(A49^{v})** et proveniet $\frac{1}{4}$. Quam adde ad dragmam et $\frac{3}{4}$, et erunt 2 dragme. Radix 2 dragmarum sublata ex ea $\frac{1}{2}$, illud est radix $\frac{1}{2}$ census. Quam duc in se, et provenient 2 dragme et $\frac{1}{4}$ diminuta radice 2; et hoc est $\frac{1}{2}$ census. Duplica ergo illud, et erunt 4 dragme et $\frac{1}{2}$ dragme diminuta radice 8; et hic est census.

Et si dicemus tibi: 2 cens[ibus]uum *equalium* addimus uni radicem 3, alii vero radicem 2 dragmarum; deinde ducimus unum ipsorum in alium, et provenient 20 dragme. Exemplum. Faciamus unumquemque 2 censuum rem, et addamus uni radicem 3, alii vero radicem 2 dragmarum. Deinde ducamus unum in aliud, quod est res et radix 3 in rem et radicem 2 dragmarum; et provenient census et radix 6 dragmarum et radix 3 censuum et radix 2 censuum, equales 20 dragmis. Prohice radicem 6 dragmarum ex 20 dragmis, et remanebunt 20 dragme diminuta radice 6 dragmarum, equales censui et radici 2 censuum et radici 3 censuum. Dimidia igitur res (sc.: radicem 2 censuum et radicem 3 censuum), et exibit radix $\frac{3}{4}$ et radix $\frac{1}{2}$. Duc ea[m] in se, et provenient dragma et $\frac{1}{4}$ et radix dragme et $\frac{1}{2}$. Que adde ad 20 dragmas diminuta radice 6 dragmarum, et erunt 21 dragme et $\frac{1}{4}$ et radix dragme *et* $\frac{1}{2}$ diminuta *radice* 6 dragmarum. [Quod est 21 et $\frac{1}{4}$ diminuta radice 1 et $\frac{1}{2}$ (perperam cancellata). Radice eius] *De quorum radice* subtracta [ab eo] radice $\frac{3}{4}$ et radice $\frac{1}{2}$, residuum est unus duorum censuum, alius vero est sicut ipse.

Et si dicemus tibi: Adde censui 7 dragmas; deinde duc agregatum in radicem 3^{pli} census, et proveniet decuplum census [et est radix 3 rerum; erunt]. *Fac igitur censum rem. Et erit radix ex 3 rebus, ducta in rem et 7, equalis 10 rebus. Et sunt* (sc.: et proveniet decuplum census. ⟨Exemplum. Fac censum tuum rem. Cui adde 7 dragmas, et erunt res et 7 dragme. Quas duc in radicem tripli census,⟩ que est radix 3 rerum; et erunt) radix 3 cuborum et radix [1[00]74] 147 rerum, equales decuplo [*eius*]*census*, quod est 10 res. Divide totum quod habes **(A50^{r})** per radicem 3 rerum, et exibit (sic) res et radix 49 dragmarum, equales *radici de* 33 rebus et $\frac{1}{3}$ rei. Duc rem et radicem 49 dragmarum in se, et provenient census et 49 dragme et radix 196 census eius (sic, pro: censuum), equales 33 rebus et $\frac{1}{3}$ rei. Prohice ergo radicem 196 censuum ex 33 rebus et $\frac{1}{3}$ rei, et remanebunt 33 res et $\frac{1}{3}$ rei diminuta radice 196 censuum, equales *censui et radici* 49 dragm[is]*arum* (sic, pro: equales censui et 49 dragmis). Et, si volueris, dimidia res diminuta radice 196 censu[s]*um*, et [e-] *fac* sicut diximus tibi. Et, si volueris, accipe radicem 196 censuum, que est 14 res, et minue eas ex 33 rebus ⟨et $\frac{1}{3}$ rei⟩. Et remanebunt 19 res et $\frac{1}{3}$ rei, equales censui et 49 dragmis. Et dimidia res, et exibit 9 et $\frac{2}{3}$. Quas duc in se, et provenient 93 et [4] $\frac{4}{9}$. Minue ex ipsis dragmas, que sunt 49, et remanebunt 44 et $\frac{4}{9}$. Quorum accipe radicem, et exibunt 6 et $\frac{2}{3}$. Minue eas ex medietate radicum **rerum**, que est 9 et $\frac{2}{3}$, et remanebunt 3, que sunt census. Et, si volueris, [computa] adde 6 et $\frac{2}{3}$ ad medietatem radicum, que est 9 et $\frac{2}{3}$, et erunt 16 et $\frac{1}{3}$, que sunt census. Et, si volueris, quando applicabis in questione ad rem et radicem 49 dragmarum: Accipe radicem 49, que est 7. Et duc rem et 7 in se. Et fac secundum algebram in producto *cum* 33 rebus et $\frac{1}{3}$ rei (sc.: Et oppone cum producto 33 res et $\frac{1}{3}$ rei). Et fac sic⟨ut⟩ dixi tibi, et exibit census 3 vel 16 et $\frac{1}{3}$, *et* id *est* quod vis (sic, pro: utrum volueris).

Et, si volueris, computa hoc per viam aliam. Que est ut facias censum tuum rem; et addas ei 7 dragmas, et e[rit]*runt* res et 7 dragme. Iam scivisti quia, quando ducis [eum] *eas* in radicem 3 rerum, provenient [2-] 10 res. Divide *ergo* 10 res per radicem 3 rerum; exibit radix 33 rerum et $\frac{1}{3}$ rei, (**A50**^{v}) equalis rei et 7 *dragmis. Fac sicut dictum est.*

{*"Deficit hic", in Ebraico. Et ego addo id quod subscribitur ad complendum intellectum littere. Et forte multa deficiunt. Sed hoc sufficit ad intelligendum quod invenio in libro. g.*}

{*Si volueris scire quid exeat de divisione radicis [facta] 10 per 2 et radicem 3 facta: Pone quod exeat res. Ergo res, cum ducetur in 2 et radicem 3, producit radicem de 10. Duc ergo eam in ipsa, et provenient 2 res et radix 3 censuum, equales radici de 10. Minue ergo 2 res de radice 10, et remanebit radix 10 diminutis 2 rebus, equalis radici 3 censuum. Duc ergo radicem 10 diminutis 2 rebus in se, et provenient 10 et 4 census diminuta radice 160 censuum, equales 3 censibus. Restaura, et erunt 3 census et radix 160 censuum, equales 10 dragmis et 4 censibus. Prohice ergo 3 census g.* } [censuum] ex 4 censibus; et remanebit census et 10 dragme, equales radici 160 censuum. [Et] Dimidia radicem 160 censuum, et exibit radix 40. Quam duc in se, et provenient 40. Ex quibus prohice 10, et remanebunt 30. Radi[x]c*e* 30 sublata de *radice* 40, residuum est res, et est quod contigit uni, que est radix 40 diminuta radice 30. Iam ⟨ergo⟩ expositum est quia, quando dividitur radix 10 dragmarum per 2 et radicem 3, quod contingit uni est radix 40 dragmarum diminuta radice 30 dragmarum. Manifestum est etiam quia, quando dividitur radix 10 censuum per 2 et radicem 3, exit uni radix 40 censuum diminuta radice 30 censuum: propterea quia *in* omni[s] numer[us]*o* qui dividitur per numerum, numerus divisus quando ducitur in numerum numerorum (sc.: in aliquem numerum, qualiscumque erit), deinde dividitur productum ex multiplicatione per numerum primum [ex-], *id est*, dividentem, exiens (**A54**^{v}) de divisione est primus (sic, pro: exiens ⟨est sicut exiens⟩ de divisione prima) quando multiplicabis eum in numerum in quem multiplicasti numerum divisum; nos autem, cum diviserimus radicem 10 dragmarum per 2 dragmas et radicem 3, exibit *radix* 40 dragm[e]*a*rum diminuta radice 30 dragmarum; ergo, cum diviserimus radicem 10 censuum per 2 et radicem 3, erit quod exibit radix 40 censuum diminuta radice 30 censuum ⟨; quia radix 10 dragmarum, que est numerus divisus, quando multiplicamus eam in rem, provenit radix 10 censuum; multiplica igitur radicem 40 dragmarum diminuta radice 30 dragmarum in rem, et proveniet radix 40 censuum diminuta radice 30 censuum⟩. Iam *igitur* expositum est quia, quando dividemus radicem 10 censuum per 2 et radicem 3, exibit uni radix 40 censuum diminuta radice 30 censuum. Ergo radix 40 censuum diminuta radice 30 censuum equatur rei diminutis 10, (**89^{v}**) quia dixisti: "contigit uni sicut census diminutis 10", et census est res. Duc rem diminutis 10 in se, et provenie⟨n⟩t census et 100 dragme diminutis 20 rebus, equales radici 40 censuum diminuta radice 30 censuum ducta (sic) in se, et est 70 census diminuta radice 4800 censu[s]*um* census. Prohice censum ⟨ex 70 censibus⟩, et remanebunt 69 census. Restaura 100 dragmas cum 20 rebus, et adde illas ad [9] 69 ⟨census⟩, et restaura 69 census cum r[e]a*dice* 4800 censu[s]*um* census, ⟨et adde illam ad 100 dragmas,⟩*et erunt 100 dragme et radix 4800 censuum census*, equales 69 censibus et 20 rebus. Prohice 69 census de 100 dragmis et radice 4800 cens[us]*uum* census; remanebunt 20 res, equales 100 dragmis et radi[x]*ci* 4800 cens[us]*uum* census diminutis 69 censibus. Reduc igitur totum quod habes ad censum census (sic). Quod est [quod] *ut* ducas totum quod habes in dragmam et [30] $\frac{30}{39}$ dragme et radicem 3 dragmarum et [267] *237* (sic) $\frac{237}{152[0]1}$ dragme; quia, quando [divides] dividitur dragma per radicem 4800 diminutis 6(**A55^{r}**)9, [erit] *est* id quod continget uni dragma et $\frac{30}{39}$ dragme et radix 3 dragmarum et $\frac{237}{1521}$ dragme, manifestum est autem ex eo quod diximus quia, quando nos dividimus unum *censum* per radicem de 4800 censuum ⟨census⟩ diminutis 69 censibus, erit quod contingit censui dragma et $\frac{30}{3[7]9}$ dragme et radix 3 dragmarum et $\frac{237}{1521}$ dragme; cum autem ducemus *100 dragmas et* radicem 4800 census (sic) census diminutis 69 censibus in dragmam et $\frac{30}{3[7]9}$ dragme et radicem 3 dragmarum et $\frac{237}{1521}$ [sed $\frac{30}{39}$ dragme et radix*cem* (sic) 3 dragmarum et $\frac{237}{1521}$] unius *dragme*, [sunt] provenient 176 dragme et [36 partes] $\frac{36}{3[7]9}$ dragme et radix [unius et] 3[0]1558 dragmarum et $\frac{[2]\mathit{1182}}{1521}$ dragme *et census [census]* (sic, pro: cum ergo ducemus radicem 4800 censuum census diminutis 69 censibus in dragmam et $\frac{30}{39}$ dragme et radicem 3 dragmarum et $\frac{237}{1521}$ dragme ⟨, proveniet census. Deinde duc 100 dragmas in dragmam⟩ et $\frac{30}{39}$ dragme et radicem 3 dragmarum et $\frac{237}{1521}$ unius, et provenient 176 dragme et $\frac{36}{39}$ unius et radix 31558 dragmarum et $\frac{282}{1521}$ dragme). Deinde duc 20 res in dragmam et $\frac{30}{39}$ dragme et radicem 3 dragmarum et $\frac{237}{1521}$ unius, et provenient 35 res et $\frac{15}{\mathit{39}}$ rei et radix 1262 censuum et $\frac{4[4]\mathit{98}}{1521}$ census. Ergo erunt census et ^{176} (sic) dragme et $\frac{36}{39}$ dragme (**A55^{v}**) et radix [unius et] 3[0]1558 dragmarum et $\frac{[280]\mathit{1182}}{1521}$ (recte: $\frac{282}{1521}$) dragme equales 3[6]5 ⟨rebus⟩ et $\frac{15}{39}$ [census] rei et radici 12[2]62 censuum et $\frac{498}{1521}$ census. Dimidia igitur 35 res et $\frac{15}{39}$ rei et radicem 1262 censuum et $\frac{498}{1521}$ census; erunt 1[9*[7]*]7 et $\frac{27}{39}$ et radix 315 et $\frac{885}{1521}$. Duc ea[m] in se, et provenient [68-] 628 dragme et $\frac{912}{1521}$ dragme et radix [3[5]95130] *395130* dragmarum [Prohice ergo] *et* $\frac{[1]\mathit{2057670}}{\mathit{2313441}}$ [ex] dragme. Prohice ex ipsis dragm*as* que sunt cum censu, que sunt 1[9]76 dragme et $\frac{36}{39}$ dragme et radi[x]*cem* 31558 et $\frac{80}{1521}$ (sic, hic et infra, pro: $\frac{282}{1521}$) dragme. Et remanebunt 451 dragme et $\frac{1029}{1521}$ (**A56^{r}**) dragme et radix 395130 dragmarum et [$\frac{2057[3]678}{2134}$] $\frac{2057670}{2313441}$ dragme, diminutis (sic, pro: diminuta) radice 31558 dragmarum et $\frac{80}{1521}$ dragme. Accipe radicem eius, et [quod addisti] *adde* ei ⟨quod provenerit⟩ 1[6]*7* dragmas et $\frac{27}{39}$ dragme et radix*cem* (sic) 315 dragmarum et $\frac{885}{1521}$ dragme. *Et* quod [autem contingit] *applicabit* tibi est census.

Et si dixit tibi: 2 cessus (sic) sunt, quorum unus est triplum alterius; et addisti [cuilibet] *unicuique* eorum radicem ⟨eius⟩; deinde duxisti unum ipsorum in alium, et fu[erunt]*it* 10^{plum} maioris census. Exemplum. Faciamus unum de 2 censibus rem, alium vero 3 res. Et addamus rei radicem eius; et erit res et radix rei. Deinde addamus ad 3 res radicem earum; et erunt 3 res et radix 3 rerum. Ducamus rem et radicem *rei* in 3 res et radicem 3 rerum, et provenient 3 [res] census et radix 3 censuum et radix 9 cuborum et radix 3 cuborum, equales decupl[i]o census maioris, quod est 3⟨0⟩ res. Divide totum quod habes *per rem*, et exibunt 3 res et radix 3 dragmarum et radix 9 rerum et radix 3 rerum, equales 30 dragmis. Prohice ergo 3 res et radicem 3 dragmarum ex 30, et remanebunt 30 dragme diminutis 3 rebus et diminuta radice 3 dragmarum, que equantur radici 9 rerum et radici 3 rerum. Duc 30 dragmas diminutis 3 rebus et diminuta radice 3 dragmarum in se, et provenient [7]903 dragme et [fac] 9 census et radix 108 censuum diminutis **(A56^{v})** 180 rebus et diminuta radice 10800, equales [16] *12 rebus* et radici 10[0]8 [dragmarum et 8] cens[ibus]*uum*. Prohice radicem 108 censuum cum radice 108 censuum, et remanebunt 903 dragme et 9 census diminutis 180 rebus et diminut[is]a radice 10800 dragmarum, equales 12 rebus. Restaura igitur in ipsis 180 res (sc.: Restaura igitur ea cum 180 rebus), et adde illas ad 12 ⟨res⟩; et provenient 192 res, equales 903 dragmis et 9 censibus diminuta radice 10800 dragmarum. Deinde fac sicut exposui tibi.

Et si dixit tibi: Radix census et radix radicis eius et radix 2 radicum eius et radix 5^{pli} eius sunt 10 dragme. Exemplum. Fac censum tuum censum. Et accipe radicem eius et radicem radicis et radicem 2 radicum ipsius et radi[x]*cem* 5^{pli} eius; et exibit res et radix rei et radix 2 rerum et radix 5 censuum, equales 10 dragmis. Prohice ergo rem et radicem 5 censuum ex 10 dragmis, et remanebunt 10 dragme diminuta re et diminuta radice 5 censuum, equales [rei] radici rei et radici 2 rerum. Duc 10 dragmas diminuta re et diminuta radice 5 censuum in se, et provenient 100 dragme et 6 censu*s* et radix 20 censuum *census* diminutis 20 rebus et diminuta radice *2000* cens[us]*uum*, equales radici rei et radici 2 [radicum] *rerum* ductis in se, que sunt 3 res et radix 8 censuum. Fac *in eis* secundum algebram, et erunt postea 23 res et radix *2000* cens[us]*uum* et radix 8 censuum, equales 100 dragmis et 6 censibus et radici 20 censuum **(90^{r})** census. Reduc ergo totum quod habes ad censum; quod est quod ducas ipsum in $\frac{3}{8}$ dragme diminuta radice $\frac{5}{8}$ octave dragme: et duc 6 census et radicem 20 censuum census in $\frac{3}{8}$ diminuta radice $\frac{e}{8}\frac{5}{8}$ dragme, et proveniet census; et duc 100 dragmas in $\frac{3}{8}$ diminuta radice $\frac{e\;5}{8\;8}$ dragme, et provenient *37* et $\frac{1}{2}$ diminuta radice 781 dragmarum et $\frac{1}{4}$; et duc 23 res in $\frac{3}{8}$ diminutis (sic, pro: diminuta radice) $\frac{e\;5}{8\;8}$, **(A57^r^)** et provenient 8 res et $\frac{5}{8}$ rei diminuta radice 41 censuum et $\frac{1}{4}$ census et $\frac{e\;5}{8\;8}$ census; et ducamus radicem 2000 censuum in $\frac{3}{8}$ diminuta radice $[\frac{5}{8}]$ $\frac{e\;5}{8\;8}$, et proveniet radix 281 censuum et $\frac{1}{4}$ census diminutis 12 rebus et $\frac{1}{2}$ rei; deinde duc radicem 8 censuum in $\frac{3}{8}$ diminuta radice $\frac{e\;5}{8\;8}$, et proveniet radix census et $\frac{1}{8}^{ve}$ census diminuta radice $\frac{[e]\;5}{[8]\;8}$ census. Erunt igitur post hec omnia census et 37 dragme et $\frac{1}{2}$ dragme diminuta radice $[\frac{1}{5}$ census et$]$ 781 dragmarum et $\frac{1}{4}$, equales [8 rebus et $\frac{5}{8}$ rei] radici 281 censuum et $\frac{1}{4}$ census et radici census et $\frac{1}{8}$ census, diminutis 3 rebus et $\frac{7}{8}$ rei et diminuta radice 41 censuum et $\frac{1}{4}$ census et $\frac{0\;5}{8\;8}$ census et diminuta radice $\frac{[e]\;5}{[8]\;8}$ census. Postea fac sicut exposui tibi.

Et si dixit: 3 censuum *non equalium* cum duces minorem in se et medium in se, proveniet sicut maior census in se; et si duces minorem in maiorem, proveniet sicut medius in se; et si duces minorem in medium, provenient 10 dragme. Exemplum. Fac minorem censum rem; et erit medius 10 dragme divise per rem, et erit maior 100 dragme divise per cubum: quia quadratum medii cum dividitur per minorem exit maior, quadratum *autem* medii est 100 dragme divise per censum; cum itaque dividetur per minorem, qui est res, exibit census magnus 100 dragme divise per cubum. Iam autem (sc.: ergo) expositum est nobis quod census minor quando fiet res, erit census medius 10 dragme divise per rem, et census maior 100 dragme divise per cubum. Ducamus igitur censum minorem, qui est res, in se, et proveniet census. Et ducamus censum medium, qui est 10 dragme divise per rem, in se, et provenient 100 dragme divise per censum. Et agregemus ea, et erunt census et 100 dragme divise per censum. [Ducamus etiam censum in [se] censum, et proveniet census census. **(A57^v^)** Et erit multiplicatio census minoris in se et medii in se census census et 100 dragme divise (sc.: , ⟨hec⟩ divisa) per censum. Ducamus ipsum in censum census, et provenient census census census *census* et 100 census census [census]. Iam autem (sc.: igitur) manifestum est quia, ⟨quando⟩ divid[itur]*emus* census (sic) census census *census* et [c-] 100 census census [census] per cubum cubi, exibit quadratus census medii et quadratus census minoris] (haud recte deleta). Deinde ducamus censum maiorem, qui est 100 dragme divise per cubum, in se, et proveniet quadratus census maioris 10000 dragme divise per cubum cubi, equales quadrato census medii et quadrato census minoris, [qui] *et* est census [census census census] et 100 [census census census (recte: census census)] divisi per [cubum cubi. Ergo] *censum. Multiplicemus hec omnia per censum census census, hoc est per cubum cubi; et erunt* 10000 dragme, *que* equantur censui census census census et 100 cens[us]ibus census [census]. Dimidiemus igitur 100 census census [census], et exibunt 50. Que duc in se, et provenient

2500[0]. Que adde ad 10000, et [provenient] *erunt* 12500. Quorum radicem accipe, et prohice *ex* ea[m] 50 dragmas. Et residui accipe radicem radicis [eius], et quod exibit est census minor.

Et si dicemus tibi: Divide 10 dragmas in 3 partes, et duc minorem in se et mediam in se, et proveniet sicut maior in se; et duc mino*rem* in maiorem, et proveniet sicut media in se. Et aufer nunc 10 dragmas; deinde [dicemus: Si] (haud recte deleta) dicemus tibi: 3 censu[s]*um* inequalium cum ducis minorem in se et medium in se, et (delendum) proveniet sicut maior in se, et cum ducis minorem in maiorem, erit sicut medius in se. Exemplum. Faciamus minorem dragmam, et medium rem, et maiorem censum quia, cum duces minorem in maiorem [in maio-], proveniet sicut productum ex medio in se. Deinde duc minorem in se et medium in se, et agrega ea; et erunt census [Adde autem illa] *et dragma*, equa[ntur]*les* censui census, [qui] *et* est sicut maior in se. Fac etiam sicut exposui tibi, et erit census $\frac{1}{2}$ dragme et radix 1 et $\frac{1}{4}$. Qui est [ce-] census maior, medius vero est radix huius, et est $\frac{1}{2}$ dragme et radix 1 et $\frac{1}{4}$ *postquam* acci[piam-]perimus (sic) eius *totius* radicem, et census minor est dragma. Addamus *hos* census 3 *simul*; *et* erunt dragma et $\frac{1}{2}$, et radix 1 et $\frac{1}{4}$, et $\frac{1}{2}^{tis}$ dragme et rad*icis* 1us et $\frac{1}{4}^{te}$ accepta radi[ce]x [eius]. *Et divid[e]amus per ea 10 et* exibit res (sc.: ⟨Deinde redeamus ad 10 dragmas, **(A58^{r})** et dicamus: Divisimus 10 dragmas per dragmam et $\frac{1}{2}$ et radicem 1 et $\frac{1}{4}$ et medietatis dragme et radicis 1 et $\frac{1}{4}$ acceptam radicem,⟩ et exeat res). Et iam scivimus quia, quando ducimus quod exit nobis de divisione [[de] per] *in* dividentem, provenient 10 dragme (vel: redit divisum). Duc igitur rem in dragmam et $\frac{1}{2}$ et radicem [census] 1us et $\frac{1}{4}^{te}$ et *in* $\frac{1}{2}^{tis}$ dragme et *radicis* 1us et $\frac{1}{4}^{te}$ acceptam radicem [eius], *et* [erunt] *provenient* res et $\frac{1}{2}$ rei et radix census et $\frac{1}{4}$ census, et $\frac{1}{2}^{tis}$ census et radi[[x]*cis*]*cis* census census et $\frac{1}{4}$ census census accepta radi[ce]*x* eius; equa⟨n⟩tur 10 dragmis. Prohice igitur rem et $\frac{1}{2}$ rei et radicem census et $\frac{1}{4}$ census ex 10 dragmis, et remanebunt 10 dragme diminuta re et $\frac{1}{2}$ et diminuta radice census et $\frac{1}{4}$ census, equales $\frac{1}{2}^{tis}$ census et radici*s* census census et $\frac{1}{4}$ census census accept[a]e radic[e]*i* eius. Duc etiam 10 dragmas diminuta re et $\frac{1}{2}$ et diminuta radice census et $\frac{1}{4}$ in se, et provenient 100 dragme et 3 census et $\frac{1}{2}$ census et radix 11 censuum census et $\frac{1}{4}$ census census diminutis 30 rebus et diminuta radice 500 censuum, equales $\frac{1}{2}$ census et radici census census et $\frac{1}{4}$ census census [census]. Restaura 100 dragmas cum 30 rebus et cum radice [5] 500 censuum, **(90^{v})** et adde [sicut] ipsa [super] *ad* $\frac{1}{2}$ census et radicem census census ⟨et $\frac{1}{4}$ census census⟩, *et tolle $\frac{1}{2}$ census ab utraque parte et radicem unius census census et $\frac{1}{4}$ census census* (sc.: ⟨et prohice $\frac{1}{2}$ census ex 3 censibus et $\frac{1}{2}$, et prohice radicem census census et $\frac{1}{4}$ census census ex radice 11 censuum census et $\frac{1}{4}$ census census⟩). Et remanebunt 100 dragme et 3 census et radix 5 censuum ⟨census⟩, equales 30 rebus et radici 500 censuum. Et reduc totum quod habes ad censum; quod est ut ducas eum in $\frac{3}{4}$ diminuta radice [$\frac{1}{4}$] et (delendum, hic et infra) $\frac{1}{2}\frac{2}{8}$, et duc totum quod habes in $\frac{3}{4}$ dragme diminuta radice [$\frac{1}{4}$] et $\frac{1}{2}\frac{2}{8}$; et provenient census et 75 dragme diminuta radice 3125 dragmarum, equales 10 rebus. Dimidia igitur res, et exibunt 5. Et duc ea in se, et provenient 25. Ex quibus prohice 75 diminuta radice 3125, et remanebit radix 3125 dragmarum diminutis 50 dragmis. Radix eius subtracta [ex ea] *de* 5, residuum est pars [pars] minor de partibus 3 que omnes tres sunt 10 dragme.

Et cum volueris scire partem maiorem, fac [censum maiorem rem] **(A58^{v})** *eam dragmam, et partem minorem rem,* medi[um]am vero radicem rei; [et] quo*d* [indigere] *cogit* nos [scire] *facere* [census 3 super hoc nomine quod vocavimus eos: quia] *[quia proponetur] ut sit* multiplicatio [*maioris census in*] minoris in maiorem sicut medii in se. Deinde ducimus minorem in se, et provenit census, et medium in se, et provenit *res*. Deinde addimus illa; et sunt census et res, equales maiori in se, quod est dragma. *Ergo* res equatur radici uni⟨us⟩ et $\frac{1}{4}$ diminuta $\frac{1}{2}$ dragme; est census minor, et medius radix huius, et hoc est radix 1 et $\frac{1}{4}$ diminuta $\frac{1}{2}$ accepta radice eius, et maior dragma. Add[ita]*e hos* census 3; erunt $\frac{1}{2}$ dragme, et radix 1 et $\frac{1}{4}$, diminuta $\frac{1}{2}$ *et radice de* $\frac{1}{4}$ *1* accepta radix eius (sc., hic et infra: et radix 1 et $\frac{1}{4}$, ⟨et radix 1 et $\frac{1}{4}$⟩ diminuta $\frac{1}{2}$ accepta radix eius). Deinde fac sicut fecisti in minori parte. Quod est ut redeamus ad 10 dragmas, et dicamus: Divisimus 10 dragmas per $\frac{1}{2}$ dragme et radicem 1 et $\frac{1}{4}$ diminuta $\frac{1}{2}$ *et radice de* $\frac{1}{4}$ *1* et accepta radice eius, exit ut res. Ergo cum ducemus quod exit nobis de divisione per dividentem, provenient 10 dragme. [duca-] *Duca*mus igitur rem in $\frac{1}{2}$ dragme et radicem 1 et $\frac{1}{4}$ diminuta $\frac{1}{2}$ *et radice de* $\frac{1}{4}$ *1* accepta radice eius, et provenient $\frac{1}{2}$ rei, et radix census et $\frac{1}{4}$ census [census], diminuta $\frac{1}{2}$ census [census] *et radix unius census census et* $\frac{1}{4}$ *census census* accepta radice eius, equales 10 dragmis. Prohice igitur $\frac{1}{2}$ rei et radicem census et $\frac{1}{4}$ census de 10 dragmis, et remanebunt 10 dragme diminuta $\frac{1}{2}$ [census] *rei et diminuta radice unius census et* $\frac{1}{4}$ *census equales radici unius census census et* $\frac{1}{4}$ *unius census census diminuta* $\frac{1}{2}$ *census* accepta radice eius. Duc 10 dragmas diminuta $\frac{1}{2}$ rei et diminuta radice census et $\frac{1}{4}$ in se, et provenient 100 dragme et census et $\frac{1}{2}$ census et radix census census et $\frac{1}{4}$ census census diminut[a]is [$\frac{1}{2}$] *10 rebus et diminuta radice 500 censuum; et multiplica radicem unius census* ⟨*census*⟩ *et quarte minus* $\frac{1}{2}$ *census accepta eius radice in se, venie*⟨*n*⟩*t* ⟨*minus*⟩ $\frac{1}{2}$ *census et radix unius census* ⟨*census*⟩ *et quarte* census census (sc.: et provenient 100 dragme et census et $\frac{1}{2}$ census et radix census census et $\frac{1}{4}$ census census ⟨diminutis 10 rebus et diminuta radice 500 censuum, equales radici census census et $\frac{1}{4}$ census census⟩ diminuta $\frac{1}{2}$ census). Restaura igitur 100 dragmas cum 10 rebus et radice 500 censuum, et adde illa(s) ad radicem census *census* et $\frac{1}{4}$ census census diminuta $\frac{1}{2}$ census census (sic); restaura [igitur] *etiam* radicem census census et $\frac{1}{4}$

census (**A59^{r}**) census cum $\frac{1}{2}$ census, et adde illam ad censum et $\frac{1}{2}$; et prohice radicem census census et $\frac{1}{4}$ census *census* in radice[m] (sc.: cum radice) census census et $\frac{1}{4}$ census census. Et remanebit 100 dragme et 2 census, equales 10 rebus et radici 500 cens[us]*uum* [census]. Reduc 2 census ad censum unum. Et iam scivisti quia census unus 2 censuum est medietas. Accipe igitur totius quod habes $\frac{1}{2}$; et exibunt census et 50 dragme, equales 5 rebus et radici 12[0]5 censuum. Dimidia ergo [radices] *res* et radicem [121] *125* censuum, et exibunt 2 et $\frac{1}{2}$ et radix 31 et $\frac{1}{4}$. Que duc in se, et provenient 37 et $\frac{1}{2}$ et radix 781 et $\frac{1}{4}$. Ex quibus prohice 50 dragmas; et remaneb[unt]it radix 781 et $\frac{1}{4}$ diminuta $\frac{1}{2}$ 12. Radi[ce]x eius sublata ex 2 et $\frac{1}{2}$ et radice 31 et $\frac{1}{4}$, residuum est maior pars. Et iam exposuimus quia minor pars est radix 3125 diminutis 50 dragmis accepta radice eius sublata ex 5. Manifestum ⟨est igitur ex eo quod diximus⟩ quia pars media *est* residuum ex 10, et [hoc] est dragma (sic, pro: 2 dragme) et $\frac{1}{2}$, et radix 781 et $\frac{1}{4}$ dragme et $\frac{1}{4}$ (et $\frac{1}{4}$: delenda) diminutis 12 et $\frac{1}{2}$ accepta radice eius, et radix 3125 diminutis 50 accepta radice eius, sublata ex toto hoc radice 31 dragmarum et $\frac{1}{4}$. Et hoc etiam similiter erit preter hanc respo⟨n⟩sionem et ducunt ad viam unam (sc.: Et hoc etiam similiter erit per aliam responsionem, et eodem recidunt); que est [s-] ut dicamus quod pars media ⟨est⟩ 12 et $\frac{1}{2}$ et radix 781 et $\frac{1}{4}$ accepta radice eius, diminuta ex ea radice 31 dragmarum et $\frac{1}{4}$ diminutis 2 dragmis et $\frac{1}{2}$; et, si volueris, dic: 2 dragme et $\frac{1}{2}$, et 12 dragme et $\frac{1}{2}$ et radix 781 et $\frac{1}{4}$ accepta radice eius, sublata ex toto hoc radice 31 et $\frac{1}{4}$.

Et, si volueris, *fac* (sc.: Et si volueris facere) in parte media sicut fecisti in parte minori et maiori: Fac censum medium de 3 censibus inequalibus 2 dragmas, et censum minorem rem, et censum maiorem radicem census et 4 dragm(arum). Duc rem in radicem census et 4 dragmas (sic, pro: dragmarum), (**A59^{v}**) et proveniet radix census census et 4 censuum, equales (sic) 4 dragmis. *Duc 4 dragmas* in se, et provenient 16 dragme, equales censui census et 4 censibus. Census igitur equatur radici 20 diminutis 2 dragmis. Radice (sic) eius est census minor, et medius 2 dragme, census vero maior 2 dragme et radix 20 accepta radice eius. Dividamus 20 per hos census ⟨inequales⟩ 3 quos diximus; et quod facit nos dividere 20 dragmas et non preter 20 est quia fecimus censum medium 2 dragmas; et exibit (sc.: exeat) res. Et duximus rem in 2 dragmas et in radicem 20 diminutis 2 dragmis accepta radice eius et cum 2 dragmis (sic, pro: in 2 dragmas) et radicem 20 accepta radice eius; erunt 2 res et radix 20 censuum census et 2 census accepta radice eius et radix 20 censuum census diminutis 2 censibus accepta radice eius, equales 20 dragmis. Prohice igitur 2 res de 20 dragmis, et remanebunt 20 dragme diminutis 2 rebus, equales 2 censibus et radici 20 censuum census accepta radice eius et radici 20 censuum census diminutis 2 censibus ac(**91^{r}**)cepta radice eius. Duc 20 dragmas diminutis 2 rebus *2 drag* (sic) in se, et provenient 400 dragme et 4 census diminutis 80 rebus, equales 2 censibus et radici 20 censuum census ⟨accepta radice eius⟩*et radici 20 censuum census* diminutis 2 censibus accepta radice eius in se, quod est [80]*8* census et radix 80 censu[s]um census. Et provenient (sc.: Erunt igitur) 8 census et radix 80 censuum census equales 400 dragmis et 4 censibus diminutis 80 rebus. Restaura ergo *400 et* 4 census cum 80 rebus, et adde *sicut* illas ad 8 census et radicem 80 censuum census; et prohice 4 census de 8 censibus. Et remanebunt 400 dragme, equales 4 censibus et 8*0* rebus et radici 80 censuum census. Reduc totum quod habes ad censum; quod est ut ducas totum quod habes in radicem $\frac{e\ 1}{8\ 8}$ et $\frac{e\ e\ 1}{8\ 8\ 4}$ diminuta $\frac{e\ 1}{8\ 2}$ dragme. Et provenient census et radix 125 dragmarum (sic, pro: censuum) diminutis ex ea (sc.: eis) 5 rebus, equales radici 3125 dragmarum sublatis ex ea 15 (sic, pro: 25) dragmis. Dimidia radicem 125 ⟨censuum⟩ [d-] sublatis ex ea 5 rebus, *ergo* erit radix 31 et $\frac{1}{4}$ diminutis 2 et $\frac{1}{2}$. Duc eam in se, et provenient 37 et $\frac{1}{2}$ diminuta radice 781 et $\frac{1}{4}$. Cui adde [3] radicem 3(**A60^r**)125 dragmarum diminutis 25 dragmis; et erunt 12 et $\frac{1}{2}$ et radix 781 et $\frac{1}{4}$. Radix eius subtracta ex ea radice 31 et $\frac{1}{4}$ diminutis 2 et $\frac{1}{2}$, residuum est pars media, quod est 2 dragme et $\frac{1}{2}$ et radix 781 et $\frac{1}{4}$ additis ei 12 et $\frac{1}{2}$ accepta radice eius subtractio (sic, pro: subtracta) ex toto hoc radice 31 et $\frac{1}{4}$.

Et si dicit: Divide 10 in 2 partes, et adde uni duarum partium 2 radices eius, et minue ex alia 2 radices eius, et erunt equales. Exemplum. Faciamus unam 2 partium 5 et rem, aliam vero 5 diminuta re. Et addamus ad 5 dragmas diminuta re 2 radices eius, et erunt 5 diminuta re et radix 20 diminutis 4 rebus. Et prohice ex 5 et re 2 radices eius, et remanebunt 5 et res diminuta radice 20 et 4 rerum, equales 5 diminuta re et radici 20 diminutis 4 rebus. Restaura 5 et re[s]m cum radice 20 et 4 rebus (sic, pro: rerum), et adde illas (sc.: illam) ad 5 diminuta re et radicem 20 diminutis 4 rebus; erunt 5 diminuta re et radix 20 et 4 res (sic, pro: rerum) et radix [10] 20 diminutis 4 rebus, equales 5 et rei. Restaura etiam 5 diminuta re et radicem 20 et 4 res (sic, pro: Restaura etiam 5 et radicem 20 et 4 rerum) et radicem 20 diminutis 4 rebus cum re, et adde illam ad 5 et rem, et prohice 5 [in] cum 5. Et remaneb[un]*i*t res, equalis (sic, pro: Et remanebunt 2 res, equales) radici 20 et 4 rebus (sic, pro: rerum) et radici 20 diminutis 4 rebus. Duc 2 res in 2 res, et provenient 4 census. Duc radicem 20 et 4 res (sic, pro: rerum) et radicem 20 diminutis 4 rebus in se. Si volueris hoc: duc radicem 20 diminutis 4[0] rebus in se, et provenient 20 diminutis 4 rebus; et duc radicem 20 et 4 re[s]*rum* in se, et erunt 20 et 4 res; adde illa, et erunt 40 dragme; deinde duc radicem 20 diminutis 4 rebus in radicem 20 et 4 (**A60^v**) re⟨rum⟩, et provenient (sic, pro: et proveniet radix) 400 diminutis 16 censibus; dupla illud, et erit radix 1600 diminutis [radice] 64 censibus. Radice eius addita ad 40, et hoc est radix 20 et 4 re⟨rum⟩ et radix 20 diminutis 4 rebus in se, producit 40 dragmas et radicem 1600 diminutis 64 censibus, equales 4 censibus. Prohice ergo 40 dragmas de 4 censibus, et remanebunt 4 census diminutis 40 dragmis, equales radici 1600 diminutis 64 censibus. Duc 4 census diminutis 40 dragmis in se, et provenient 16 census census et 1600 dragme diminutis [64] *320* censibus (sc.: et 1600 dragme ⟨diminutis 320 censibus, equales 1600 dragmis⟩ diminutis 64 censibus). Fac secundum algebram in eis, et erunt 16 census census equales 256 censibus. Census census ergo equatur 16 censibus, et census equatur 16 dragmis. Ergo res equatur 4 dragmis. Quos (sic) adde ad 5 et prohice ipsos ex 5, quia tu fecisti unam duarum partium 5 et rem et aliam 5 diminuta re. Ergo erit una duarum partium 9, reliqua vero 1.

Et, si volueris, fac illa secundum hanc viam. Que est quod facias unam duarum partium 5 et rem, aliam vero 5 diminuta re. Et adde ad minorem 2 radices eius, et erunt 5 diminuta re et 2 radices 5 diminuta re; et minue ex alia 2 radices eius, et remanebunt 5 et res diminutis 2 radicibus 5 et rei, equales 5 diminuta re et 2 radicibus 5 diminuta re. Restaura 5 et rem cum ⟨2⟩ radicibus 5 et rei, et adde illas ad 5 diminuta re et 2 radices 5 diminuta re; et erunt 5 diminuta re et 2 radices 5 et rei et 2 radices 5 diminuta re, equales 5 *et* re[bus]*i*. Restaura etiam 5 cum re, et adde eam ad 5 et rem; prohice 5 cum 5, et remanebunt 2 res, equales 2 radicibus 5 diminuta re et 2 radices (sic, pro: 2 radicibus) 5 et rei. **(A61r)** Res una equatur radici 5 et rei et radici 5 diminuta re. Duc rem in rem, et proveniet census. Et duc radicem 5 in rem (sic, pro: et rei) et radicem 5 diminuta re in se, et provenient 10 dragme et radix 100 dragmarum diminutis 4 censibus, equales censui. Prohice 10 dragmas de censu, et remanebit census diminutis 10 dragmis, equalis radici 100 dragmarum diminutis 4 censibus. Duc censum diminutis 10 dragmis in se, et provenient census census et 100 dragme diminutis 20 censibus, equales 100 dragmis diminutis 4 censibus. Oppone per ipsa, et erit res equalis 4 dragmis. Adde ergo eas ad 5 et minue eas a 5, et erit una duarum partium 9, reliqua vero 1. Et, si volueris, cum applicabis ad rem equalem radici 5 et rei et radici 5 diminuta re: Prohice[s] radicem 5 et rei de re, et remanebit res diminuta radice 5 et rei, equalis radici 5 diminuta re. Duc ergo rem diminuta radice 5 et rei in se. Et quando **(91v)** volueris hoc: duc rem in se, et proveniet census; deinde duc diminuta radice 5 et rem (sic, pro: et rei) in rem, et provenient radix 5 censuum et cubus dimin(utus) (recte: et proveniet radix 5 censuum et cubi diminuta); deinde duc etiam similiter diminuta radice 5 et rems (sic, pro: et rei) in se (sic, pro: in rem), et [erunt] provenient radix 5 censuum et cubus diminutus **sublatus** (sic, pro: et proveniet radix 5 censuum et cubi diminuta/sublata); amplius duc radicem 5 et rei in se, et provenient 5 et res addita. Amplius agrega illa, et erunt census et 5 dragme ⟨et res⟩ diminuta radice 20 censuum et 4 cubi (sic, pro: cuborum), equales 5 ⟨dragmis diminuta re. Restaura ergo 5 cum re, et adde illam ad censum et 5 dragmas et rem; restaura etiam censum et 5 et rem cum radice 20 censuum et 4 cuborum, et adde illam ad 5; et prohice 5 cum 5. Et remanebunt census et 2 res, equales radici 20 censuum et 4 cuborum⟩. Duc censum et 2 res in se, et provenient census census et 4 census et 4 cub[os]*i*, equales 20 censibus et 4 cubis. Prohice 4 cubos cum 4 cubis et 4 census cum 4 censibus, et remaneb[un]*it* census census, equalis 16 censibus. *Census ergo* **(A61^{v})** *equatur 16 [drag]* dragmis. [Ergo res equatur] (perperam erasa) 4 dragmis. Quam (sc.: Quas) adde ad 5 et minue a 5, et erit una de 2 partibus 9, reliqua vero erit 1. Et, si volueris, computa illa per viam hanc (aliam), que est: Quia [omnis] omnes 2 numeri qui dividuntur (sic, pro: omnis numerus qui dividitur) in 2 partes et addimus uni eorum (sic, pro: earum) ⟨radicem eius et diminuimus ab alia radicem eius et erunt equales, vel addimus uni earum⟩ *2* radice[m]s[*x*] eius et diminuimus ab alio (sc.: alia) 2 radices eius ⟨et⟩ erunt equales, vel id quod vis ex radicibus postquam (sc.: dum modo) equa[n]tur numerus radicum in augmento vel (sc.: et) in diminutione, et (sc.: tunc) scitum est quia ipse (sc.: partes post additionem et diminutionem) equantur semper cum multiplicatione radicis unius duarum partium in radicem partis alterius, erunt id quod erunt (sc.: quecumque erunt) radices. Exemplum. [Faciamus] *Dividamus* 10 in 2 partes, et faciamus unam duarum partium 9 et aliam 1; et quando addemus uni 2 radices eius et minuemus ex 9 2 radices ⟨eius⟩, et (delendum) remanebunt de 9 3 et contingit unum ad 3, et fient equalia; manifestum *scitum* est etiam quia questio (sic, pro: et manifestum/scitum est quia 3 ⟨provenit⟩) de multiplicatione radicis 9 in radicem un(i)us. Et [alterius] postquam sic est res, fac unam duarum partium censum et aliam 10 diminuto censu. Et duc radicem census in radicem 10 diminuto censu, et proveniet radix 10 censuum *minus censum census*, qu[ia]e [hic census] equatur censui et 2 radicibus. Duc censum et 2 radices in se, et oppone p[re]er ipsa 10 censibus diminuto censu *census*; erit radix census dragma. Et census est ⟨dragma. Et est⟩ una duarum partium, et reliqua est 9.

Et, si volueris, fac illa secundum [hanc] viam aliam. Que est quia *in* omnes*bus* (sic) duo*bus* numeri*s* inequal[es]*ibus si* subtra[cta]*amus a* maiori radicem eius et addamus minori radicem eius, redibit numerus, quia (sic, pro: et addamus minori radicem eius et erunt equales, erit) radix numeri maioris maior radice numeri minoris 1, quia ipse dixit “radix et radix”; et si ⟨dixerit:⟩ “minuamus ex maiori 2 radices eius et addamus minori 2 radices eius, ⟨et⟩ erunt equales”, fuit (sc.: erit) radix maioris maior radice minoris ⟨duobus; et si dixerit: “tres radices et tres radices”, erit radix maioris maior radice minoris⟩ sicut ipsa est (sic, pro: tribus), secundum [propterea] mensuram equalitatis radicum *cum* quibus equantur (sic, pro: secundum mensuram -vel: quantitatem- numeri radicum que (in additione et diminutione) equantur). Et sic erit ex 2 radicibus ⟨2⟩ ex numero, et similiter *id* (sc.: omne) quod fuit magis aut minus quando equantur radices cum additione et diminutione secundum hanc viam, cum adiutorio Creatoris.

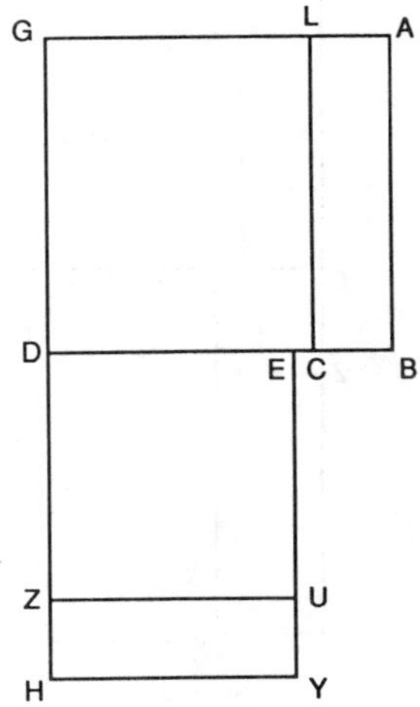

(A62^{r}) Exemplum. Faciamus 2 census *2* quadrata super lineam unam, que sit linea GDZ[DB], et faciamus [quadratum] *censum* maiorem quadratum AGBD et censum minorem ⟨quadratum⟩ EDUZ. Et faciamus *quod* quadratum maius quando au[f]ferentur ab eo 2 radices eius, et [sint] add[ite]entur ad quadratum minus 2 radices eius, [et] sint equal[e]i*a*. Et scitum est quia latus quadrati ABGD longius est latere EDUZ *in* linea EB. Dico quia linea BE *est* 2. Et si non erit 2, erit amplius vel minus ⟨2^{bus}⟩. Faciamus prius *eam* maius (sic) 2^{bus} et faciamus lineam BC 2. Et producamus lineam a C equidistantem linee GD, que sit CL. Erit igitur superficies AC 2 radices quadrati ABGD. Et producamus EU usque ad Y, et ponamus UY similiter 2. Et compleamus superficiem UH. Eritque ipsa similiter 2 radices quadrati EDUZ. Erit igitur, secundum positionem, superficies LCGD equalis superficiei EYDH. Verum linea DC maior est linea DE, que est pars eius. Et linea DG maior est linea DH: quoniam DG est sicut tota DB; sed DH est [tantum] sicut DE et CB, tantum quantum DZ est sicut DE et ZH est 2 sicut CB; quare maior *est* DG quam DH. Ergo superficies GDCL, que a duabus lineis DG et DC maioribus [duabus aliis, scilicet DH et DE, a quibus superficies DEYH continetur] *continetur*, maior est superficie DEYH, a duabus lineis minoribus, videlicet DH et DE, contenta. Quod est impossibile.

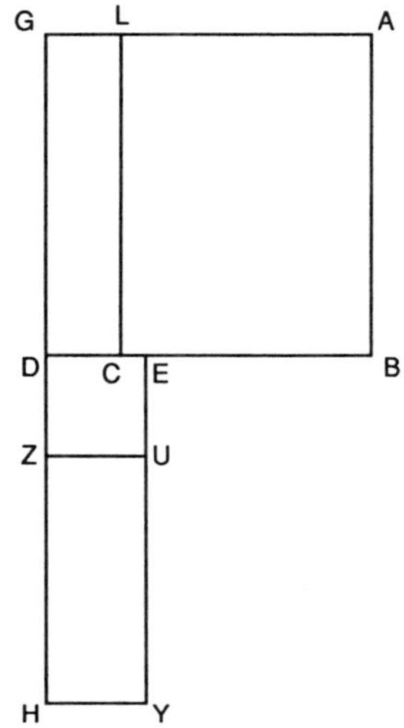

Dico etiam quod linea BE non est minor 2. Quod, si est possibile, sit minor. Et faciamus lineam BC 2, et producamus a puncto C lineam CL equidistantem linee GD; *fiet superficies AC 2 radices quadrati ABGD*. Et producamus EU usque ad Y, facientes UY similiter duo, et compleamus superficiem UH. Erit igitur superficies UH similiter 2 radices quadrati EDUZ. Quare, secundum positionem, erit superficies LCDG equalis superficiei EYDH. Sed linea DE longior est DC. Et DH longior est DG, quoniam ipsa est sicut DE, que est equalis DZ, et CB, que est duo, sicut ZH. Ergo maior est superficies DY, que a longioribus lineis, superficie DL, que a brevioribus. Quod est impossibile. Igitur BE neque est maior 2^{bus} neque minor. Igitur est 2. **(A62^{v})**

Iam igitur manifestum ⟨est⟩ quia, quando prohicies ex censu maiori 2 radices eius et addes minori 2 radices eius, fueri*ntque* equal[is]es, maior *est* radix maioris census radice minoris in duobus. Et cum simili huic forma scies quia, quando prohicies ex maiori 3 radices eius et addes minori 3 radices eius et equabuntur, differentia inter radices est 3 dragme. Et sic est in omnibus quando, diminutis [eis] *aliquot* radicibus et additis *tot*, equantur (sc.: differentia inter radices est 3 dragme, que sunt secundum quantitatem numeri radicum que in additione et diminutione equantur). Et causa in illo ex via restaurationis vera et manifesta est: Quia *in* omnes*ibus* (sic) 2 cens[us]*ibus* quorum unius radix addit super radicem alterius 1 dragma, quando minues [minorem] ex maiori radicem eius et addes maiori (sic, pro: minori) radicem eius, ⟨equabuntur,⟩ secundum id quod exposui tibi. Quod est quod facias radicem unius duorum censuum rem et alterius rem et dragmam. Et duc unumquodque ipsorum in se; et proveniet minor census, maior vero census et 2 res et dragma. **(92^{r})** Et quando addes ad minorem radicem eius, que est res, et minues ex maiori radicem eius, que est res et dragma, remanebit [ex maiori radix eius, que est res et dragma, remanet] ex maiori census et res et contingit minor*i* [ad] censu[m]*s* et re[m]*s*. *Quare* iam facti sunt equales. Et sicut (sc.: similiter) si dixit: Fecimus radicem minoris rem et maioris rem et 2 dragmas, et multiplicavimus unumqu[em]odque ipsorum in se; et fuit minor census, et maior census et 4 res et 4 dragme. Et cum addimus ad minorem 2 radices eius, que sunt 2 res, et minuimus ex alio 2 radices eius, que sunt 2 res et 4 dragme, remanet ex maiori censu*s* et 2 [radic[ibus]es] *res* et contingit minor ad censum et 2 res. Et iam facti sunt equales etiam. Et sic quod erit plus hoc aut minus. Deinde redeamus (sic, pro: Et sic quod erit plus hoc aut minus redibit) ad id quod diximus postquam (sc.: dum modo) equantur radices in additione earum et diminutione.

Et postquam ita est, *sit* radix partis *minoris de* 10 dragm[arum]is res, et erit pars minor census. Deinde faciamus radicem partis maioris, que est residuum de 10, rem **(A63^{r})** et 2 dragmas. Et ducamus eam in se, et provenient census et 4 res et 4 dragme. Que est pars maior. Cui addamus parte[s]m minorem, que est census; et erunt 2 census et 4 radices et 4 dragme, equales 10 dragmis. Opponamus in eis (sc.: Faciamus in eis secundum algebram) sicut exposui*(m)us* tibi, et erit res 1. Que est pars minor (sc.: Que est ⟨radix⟩ partis minoris). Et radix partis maioris est maior ea 2 dragmis. Quare est 3. Et pars maior est 9, minor vero 1.

Et si dixit: 10 divide *in* 2 partes, et divide 10 per unamquamque duarum partium, et exibunt 6 et $\frac{1}{4}$. Prohice ex 6 et $\frac{1}{4}$ 2 semper, et remanebunt 4 et $\frac{1}{4}$. Quasi diceret: 10 divide in 2 partes, et divide unamquam⟨que⟩ 2 duarum (sic) partium per aliam, et exeat 4 et $\frac{1}{4}$. Fac ergo sicut exposui tibi. Et causa in 2 (sc.: in ⟨dictione⟩ "2") vera ex via signi *est*: Quia iam scivisti quia, quando divid[e]is 1 duarum partium per aliam et exibit ex divisione res, quando dividentur 10 per ipsam rem (sc., per ipsam partem per quam divisisti), quod exit ex divisione *est* res et dragma; quia in 10 *est* sicut ipsa pars [cuius pars] *per quam dividitur 10* et sicut quod dividitur per ipsam (vel: quia 10 est sicut pars divisa et pars dividens). Et sic (vel: similiter) [exibit] exibit ex latere alio: Quod est quia, quando dividentur 10 per partem aliam, [quia] quod exit ad⟨d⟩it[ur] super id quod exit ex divisione al(ter)ius p[e]*ar*[divisum quod divisisti per ipsa 10 dragmas]*tis per ipsam partem unam dragmam.* Iam *ergo* manifestum est ex h[o]ac via quia, quando dividentur 10 per unamquamque ex divisionibus (sc.: duabus partibus), quia id *quod exit* addet super id quod exit ex divisione [10 unusquisque] *uniuscuiusque partis de 10* per aliam 2 ⟨semper⟩.

Et planum est *etiam* ex via geometrica. Et est quod dicamus quia omnis numerus qui dividitur in 2 partes deinde dividitur numerus per unam duarum partium, [quia quod exit ex divisione maioris ex (sic, pro: maius est) eo quod exit de divisione nostra partis ⟨per⟩ quam non dividitur per partem per quam dividitur numerus unus (sic, pro: uno) semper] {*erit id quod exit ex divisione totius per unam partium eius addens super id quod exit ex divisione alterius partis numeri per ipsam partem per quam dividitur numerus totus unam dragmam semper. g.*}.

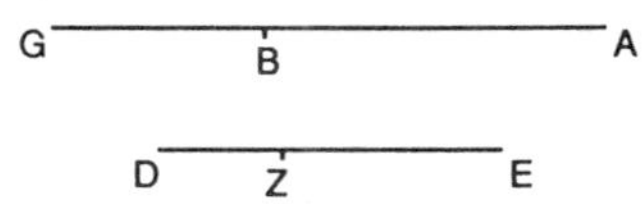

Exemplum. Dividamus numerum AG in duas divisiones in puncto B; que sunt partes AB, BG. Et iam divis[a est]*imus* AG [in duas partes in puncto B] per BG, et exi[bit]*vit* DE. Et dividitur AB per BG, et exi[bi]t ZE; [et DE] *minuaturque EZ*, m[ai]inor, ex DE, *maiori, et remaneat* DZ. Dicam quod DZ est unum. **(A63^{v})** Probatio. Quia AG iam divisa est per BG et exivit D[G]*E*, ergo multiplicatio DE in BG est AG, sicut premisimus et exposuimus. Et AB dividitur per BG et exit ZE, ergo multiplicatio ZE in BG est AB. Iam autem fuit multiplicatio DE in BG AG. Remanet *ergo* multiplicatio DZ in BG sicut [est] BG; [*va- et* quia multiplicatio BG in DE est sicut multiplicatio DE (sic, pro: ZE) in BG et D[E]Z in BG, et multiplicatio D[Z]E in BG est AG, et multiplicatio [D]ZE in BG est AB, remanet multiplicatio DZ in BG est (sic) BG *-cat*]. Et multiplicatio unius in BG est BG. [E-] Ergo multiplicatio unius in BG est sicut multiplicatio DZ in BG. *Ergo* quantitas *DZ* equalis uni est unus (sc.: Ergo DZ est equalis uni). Et hoc est quod voluimus exponere. Et, si volueris: Iam scivisti quia multiplicatio DE in BG est AG; ergo BG ad unum est secundum proportionem GA ad DE. Et multiplicatio ZE in BG est AB; igitur BG ad unum est secundum proportionem BA ad ZE. Ergo proportio totius, AG, ad totum, DE, est sicut partis, BA, ad partem, ZE; remanet igitur etiam sic⟨ut⟩ GB ad DZ. Sed iam fuit A[B]G ad DE, et BA ad ZE, sicut GB ad DZ (sic, pro: ad unum). Ergo GB ad DZ est sicut (proportio) ipsius BG ad unum. Igitur DZ est unum. Et hoc est quod voluimus exponere. Et cum hac forma scies quia, quando dividimus AG per AB, [et] quod exit de divisione addit super id quod exit de divisione GB per BA uno etiam. Iam ⟨ergo⟩ manifestum est quod omnis numerus qui dividitur in 2 partes, deinde dividitur *ipse* numerus per unamquamque duarum partium, quod exit de divisione addit super id quod exit de divisione uniuscuiusque duarum partium per aliam 2 semper.

Et si dixit: 10 divide in 2 partes, et divide 10 per unamquamque duarum partium; deinde duc unam earum (sc.: exeuntium) in aliam, et provenient 6 et $\frac{1}{4}$. Dicam quod, quando multiplicas id quod exit de *duabus* divisionibus un[ius earum in aliam]um, *scilicet: eorum, in aliud,* est sicut si agregas 2 duas partes (sic, pro: est sicut si agregas exeuntia de duabus divisionibus). Et causa in hoc [est] vera et plana ex via innuitionis est. Quia iam exposuimus quia, quando dividimus 10 per unamquamque (duarum) (**A64^r**) [divis-] partium *10*, quod exit de divisione [addit unaqueque partium quod exit de divisione 10 per unam earum super aliam dragma] *10 per unamquamque partium eorum addit super id quod exit de [divi-] divisione uniuscuiusque partis de 10 per aliam in unaquaque divisione 1.* Et planum est ex hoc quod diximus quia *id* quod exit de divisione [10] uniuscuiusque partis *10* per aliam est 4 et $\frac{1}{4}$. Quando nos *ergo* faciemus unam ipsarum (sc., exeuntium) rem et aliam 4 et $\frac{1}{4}$ diminuta re, erit quod exit de divisione [partium eius] (**92^v**) 10 per unamquamque ipsarum *partium* [ex divisis unius earum] *in parte una* res et dragma et alia 5 et $\frac{1}{4}$ diminuta re. Dico quia multiplicatio dragme et rei in 5 et $\frac{1}{4}$ diminuta re [quod] est sicut dragma et res et 5 et $\frac{1}{4}$ diminuta re, et est 6 et $\frac{1}{4}$. Et probatio eius est: Quia iam scivisti quod multiplicatio dragme et rei in 5 et $\frac{1}{4}$ diminuta re est sicut multiplicatio rei in 4 et $\frac{1}{4}$ diminuta re et dragme in 4 et $\frac{1}{4}$ diminuta re et dragme in rem et dragme in dragmam. Et iam exposuimus quia multiplicatio rei in 4 et $\frac{1}{4}$ diminuta re *est* dragma; et multiplicatio dragme in 4 et $\frac{1}{4}$ diminuta re *est 4 et $\frac{1}{4}$ diminuta re*, et dragme in rem *est res*, et dragme in dragmam *est* dragma. Agregemus ea, et erunt 6 et $\frac{1}{4}$. Et manifestum [est] *erit* hoc per viam geometricam. Quod est ut dicamus quia omnibus duobus numeris qui dividuntur (sic, pro: quia omnis numerus qui dividitur) in 2 partes, deinde dividitur numerus per unamquamque partium, multiplicatio eius quod exit de divisionibus unius [earum] *eorum* in ali[am]*ud* est sicut [quod ex⟨it⟩ de divisis] *agregatum ex multiplicandis.*

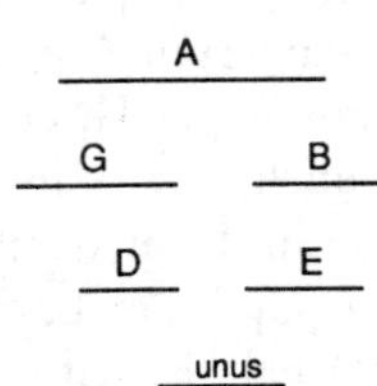

Exemplum. Quia numerus A dividitur in duas partes, que sunt numerus B, et G. Iam divisus est A per B, et exit E, et divid⟨it⟩ur A per G, et exit D. Dico quia multiplicatio E in D est sicut E et D. Cuius demonstratio est: Quia A dividitur per B et exit E, ergo multiplicatio E in B est A. Et similiter multiplicatio D in G est A. Ergo est multiplicatio E in B sicut D in G, ergo proportio G ad B sicut proportio E ad

D. Et iam exposuimus quia multiplicatio D in G est A. Et multiplicatio unius in A est A. Ergo multiplicatio unius in A est sicut multiplicatio D in G, ergo proportio unius ad D sicut proportio G ad A. [Sed] Et A est sicut B et G. Ergo proportio unius ad D est sicut proportio G ad B *et* G. Et proportio G ad B et G est sicut proportio E ad E et D: quia iam exposuimus quod proportio G (**A64**$^{\mathbf{v}}$) ad B est sicut proportio E ad D; cum ergo componemus, erit proportio G ad B *et* G sicut proportio E ad D et E. [Et fuit] *Ergo est* proportio unius ad D sicut proportio E ad E et D. [Sed] Ergo multiplicatio unius in E et D est sicut multiplicatio E in D. Sed multiplicatio unius in E et D est E et D. Ergo multiplicatio E in D est E et D. Et hoc *est* quod voluimus exponere.

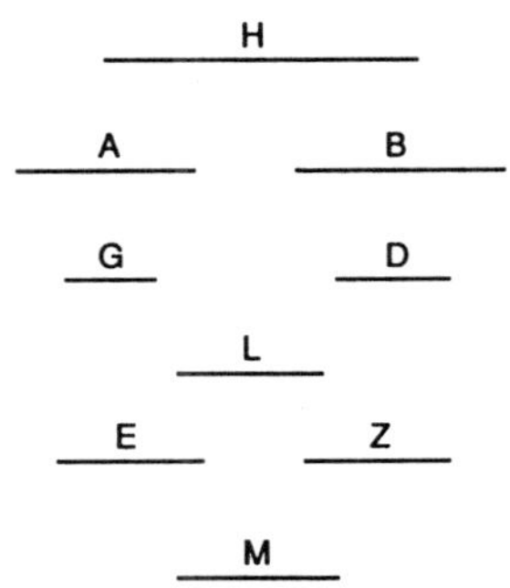

Et dico etiam similiter quia omnes duo numeri quorum unusquisque dividitur per *unum* numerum et ducitur quod exit de divisione duorum numerorum per duos numeros [unus eorum in alium (sic, pro: unum eorum in aliud)], quod provenit ex ductu *unius eorum in alium* (sc.: aliud) est sicut id quod exit de divisione multiplicationis 2 numerorum divisorum unius eorum in alterum per productum ex multiplicatione duorum numerorum divi[sorum]dentium unius eorum in alium. Exemplum. Dividamus numerum A et numerum B per numerum G et numerum D. Dividamus A per G, et exeat E, et dividamus B per D, et exeat Z. Et [iam] duc[i]etur A in B et provenit [G] *H*, et ducitur G *[H]* in D et producitur L, et ducitur E in Z et provenit M. Dico quia, quando dividitur [G] *H* per L, exit M. Cuius probatio est: Quia A dividitur per G ⟨et⟩ exit E, igitur multiplicatio E in G est A. Et multiplicatio E in Z est M. Ergo E ducitur in 2 numeros, qui sunt *Z et* G, et produc[i]untur ex ductu eius M et A. Ergo proportio Z ad G est sicut proportio M ad A. Ergo est multiplicatio Z in A sicut multiplicatio M in G. Faciamus autem D comune in multiplicatione. Et ducamus D in Z, deinde in A, eritque (sic, pro: Et multiplicatio Z in A deinde in D erit) sicut multiplicatio M in G deinde in D [*va-* et faciamus GDLZ comune in multiplicatione *-cat.* Multiplicatio Z in A, deinde in D *-cat*].

Et multiplicatio M in G, deinde in (**A65^r^**) D, est sicut multiplicatio G in D, deinde in M; et multiplicatio G in D est L; *ergo* multiplicatio M in L est sicut multiplicatio G in D, deinde in M. Et iam fuit multiplicatio G in D, deinde in M, est (sic) sicut multiplicatio M in G, deinde in D. [Quod est] *Et multiplicatio* M in G, deinde [B] in D est sicut multiplicatio Z in A, deinde in D. [Et multiplicatio Z in D fuit] *Ergo est* multiplicatio M in L sicut multiplicatio Z in A [et multiplicatio B in A], deinde in D [deinde]. *Et multiplicatio Z in A, deinde in D, est sicut* multiplicatio Z in D, deinde in A; et multiplicatio Z in D *iam* fuit *B; ergo* multiplicatio A in B *est* sicut multiplicatio Z in A, deinde in D. Et iam exposuimus quia multiplicatio Z in A, *deinde in D*, est sicut multiplicatio M in L. *Ergo* multiplicatio M in L est sicut multiplicatio A in B. Et multiplicatio A in B est [G] *H*. Ergo multiplicatio M in L est [G] *H*. Iam *ergo* manifestum est quia quando dividitur [G] *H* per L exit M: quia omnes duo numeri quorum unus *[ipso-]* ducitur in alium, quod provenit ex multiplicatione [est sicut] quando dividitur per unum duorum numerorum exibit numerus alius. Et hoc est quod voluimus exponere.

Iam expositum (vel: probatum) est ex eo quod preposuimus quia, quando dividentur 10 in 2 partes et dividentur 10 per unamquamque duarum partium ⟨et⟩ exibit 6 et $\frac{1}{4}$, quia, quando multiplicas numerum unum (sic, pro: unius) duarum partium *exeuntium* (recte: partium de 10) in alium, deinde in 6 et $\frac{1}{4}$, erit sicut multiplicatio tua *de* 10 in se. Et necesse est ex eo quod premisimus etiam quia, quando dividuntur 10 in 2 partes et dividu⟨n⟩tur 20 dragme per una⟨m⟩quamque duarum partium ⟨et⟩ exibunt 12 et $\frac{1}{2}$, quia [erit], quando dividentur 10 per unamquamque duarum partium *eius* (sc.: de 10), *[quia] est* (sc.: erit) quod exit [est] $\frac{1}{2}^{tas}$ de 12 et $\frac{1}{2}$, que [sunt] *est* 6 et $\frac{1}{4}$; quia 10 est $\frac{1}{2}$ de 20. Et necesse est etiam quia, quando dividuntur 10 in duas partes et dividuntur 5 per unamquam⟨que⟩ duarum partium ⟨et⟩ exibunt 3 et $\frac{1}{8}$, quia erit, quando dividentur (**A65^v^**) 10 per unamquamque duarum partium, quod exit duplum 3 et $\frac{1}{8}$, quod est 6 et $\frac{1}{4}$; quia 10 [est] sunt dupla [de] (**93^r^**) ad 5. Et necesse (est) etiam quod sit, quando dividuntur 10 in 2 partes et dividuntur 20 per unamquamque *ipsarum* duarum partium deinde multiplicas unam ipsarum in aliam [quia exibit [$\frac{1}{5}$] et $\frac{1}{4}$ de], *quod proveniet erit* 25, que [est] *sunt quadruplum de* 6 et $\frac{1}{4}$; quia 10 [est] sunt $\frac{1}{2}$ 20, et $\frac{1}{2}$, quando ducitur in se, [exit $\frac{1}{4}$] *erit productum* sicut $\frac{1}{4}$ eius, *scilicet: dupli,* in se (sc.: deinde multiplicas unum eorum (exeuntium) in aliud et exibunt 25, ⟨quia, quando divides 10 per unamquamque duarum partium deinde duces unum eorum (exeuntium) in aliud, exibit $\frac{1}{4}$ de 25,⟩ que est 6 et $\frac{1}{4}$; quia 10 sunt $\frac{1}{2}$ de 20, et $\frac{1}{2}$ (alicuius rei), quando ducitur in se, exit sicut $\frac{1}{4}$ illius rei in se). Et necesse est ut sit, quando dividis 10 in 2 partes et dividis 30 per unamquamque *ipsarum* duarum partium deinde ducis unam earum *partium exeuntium* in aliam ⟨et⟩ provenient 56 et $\frac{1}{4}$, quia [erit], quando divides 10 per unamquamque *ipsarum* duarum partium et duces unam ipsarum (sc.: exeuntium) in aliam, [quod exibit] proveniet $\frac{1}{9}$ *de* 56 et $\frac{1}{4}$, qu[od]e est 6 et $\frac{1}{4}$; quia 10 est $\frac{1}{3}$ de 30, et $\frac{1}{3}$ (rei), quando ducitur in se, proveniet *sicut* $\frac{1}{9}$ ipsius (rei), *scilicet: 30*, in se. Et necesse est ut sit, quando dividuntur 10 in duas partes et divid[it]ent*ur* per unamquamque *ipsarum* duarum partium 5 deinde duces unam earum *exeuntium* in aliam ⟨et⟩ proveniet dragma et $\frac{1}{2}$ et $\frac{e\ 1}{8\ 2}$, quod erit, quando dividentur 10 per unamquamque duarum partium deinde duces unum eorum in aliud, quia exibit **proveniet** 4^{plum} dragme *et* $\frac{1}{2}$ et $\frac{e\ 1}{8\ 2}$, quod est 6 et $\frac{1}{4}$; quia 10 sunt duplum ad 5, et duplum ⟨rei⟩, quando ducis [in duplum] in se, proveniet quadruplum eius, *scilicet: dimidii* (recte: quadruplum illius ⟨rei in se⟩). Et iam exposuimus quia, quando dividitur numerus in 2 partes et dividitur numerus per unamquamque duarum partium deinde ducis unum eorum in aliud, quia quod provenit est sicut quod [exit de] *agregatur ex illis* 2 partibus (sc.: divisionibus). Et postquam [fuit] *est* hoc sicut exposuimus, manifestum est quia, quando dividuntur 10 in 2 partes et dividuntur 10 per unamquamque duarum partium deinde ducis unam earum (sc., exeuntium) in aliam ⟨et⟩ erit 6 et $\frac{1}{4}$, quia quod exit ex unaquaque duarum [partium per aliam etiam] *divisionum,* (**A66^r^**) *agregatum,* est ⟨etiam⟩ 6 et $\frac{1}{4}$.

Et manifestum est etiam (sc.: manifestum erit) ex eo quod diximus si dicemus tibi: 10 divide in duas partes, et divide 40 per unamquamque 2 partium *decem*; deinde duc unum eorum in aliud, et provenient 100 dragme. Operatio autem (sc.: Quia operatio) in hoc est ut faciamus unam duarum partium 5 et rem et aliam 5 diminuta re. Et ducamus unam earum in aliam, et provenient 25 diminuto censu. Ducamus ipsa in 100 dragmas, et provenient [25 diminutis] 2500 diminutis 100 censibus, equales 40 dragmis in se, que sunt 1600 dragme. Et fac sicut exposui tibi, et erit census equalis 9 dragmis, et res radici eorum (sc.: earum), que est 3. Addamus ea ad 5 et diminuamus ea a 5, et erit una duarum partium 8, et alia 2. Et quod [fec-] coegit nos facere sic est quia iam exposuimus quia omn[es]i*um* duo*rum* numer[i]*orum*, quando dividitur unusquisque eorum per numerum et ducitur quod exit de divisione 2 numerorum per duos numeros un[i]us eorum in alium (sc.: unum eorum in aliud), quia agregatum **productum** ex multiplicatione est sicut *id* quod exit de divisione multiplicationis duorum numerorum divisorum unius eorum [per] *in* alium per agregatum ex multiplicatione 2 numerorum divisorum (sic, pro: dividentium) unius ipsorum in alium; et iam diximus quia una duarum partium ex 10 est 8 et alia est 2; et quando dividuntur 40 per 8 et 40 per 2, erunt 40 et 40 in gradu 2 numerorum, et 8 et 2 duo numeri etiam similiter *se habent.*

[Et] Quod manifestum est (sc.: Et manifestum erit) etiam ex eo quod diximus quia, quando *dicitur tibi:* 10 divid[is]e in 2 partes et divid[is]e [omnia illa] *id est: 50* per unam duarum partium et 40 [et 40] per aliam deinde duc[is] unam ipsarum (sc.: exeuntium) in aliam, et erunt [100] 125 dragme. [[Quia] *Quando*] Faciemus (enim) unam duarum partium ex 10 5 et rem et aliam 5 diminuta re. Ducamus unam ipsarum in aliam, et erunt 25 dragme diminuto censu. Ducamus ea in [100] 125, et provenient 3[0]125 dragme diminutis [1[00]25] *125* censibus, equales 50 in 40, qu[od]e sunt 2000[0]. Fac autem sicut dixi tibi, et erit census equalis 9 dragmis; et res ⟨erit⟩ radix eorum (sc.: earum), que est 3. Quam add[a]emus ad 5 ⟨et minuemus **(A66ᵛ)** ex 5⟩, et erit una duarum partium 8, reliqua vero 2. Et, si vis: Iam scivisti quod, quando dividuntur 10 per unamquamque duarum partium, deinde ducis unam ipsarum (sc.: exeuntium) in aliam, quod exit **provenit** ⟨est⟩ $\frac{e\ 1}{10\ 2}$ de 125, que ⟨est⟩ 6 et $\frac{1}{4}$: quia 10 est $\frac{1}{4}$ de 40 et $\frac{1}{5}$ de 50, sed $\frac{1}{4}$ (alicuius rei) cum ducitur in $\frac{1}{5}$ (rei), proveniet sicut [10 ^{1}]$\frac{e\ 1}{10\ 2}$ illius rei multiplicate in se; et manifestum est ex eo quod diximus etiam ex eo quod premisimus et exposuimus de porta 10 *quod est ut tibi dicatur* "Divide 10 in 2 partes, et divide unamquamque duarum partium per aliam, et exibunt 4 *et* $\frac{1}{4}$": quia addimus ad 4 et $\frac{1}{4}$ 2 semper, et erunt 6 et $\frac{1}{4}$. Deinde facimus unam duarum partium 5 et rem et aliam 5 diminuta re, et ducimus unam ipsarum in aliam, *et* provenient 25 dragme diminut[is 6](o) cens[ibus]*u* [et $\frac{1}{4}$ census]. Ducamus ipsa in 6 et $\frac{1}{4}$, et provenient 156 et $\frac{1}{4}$ diminutis [8] *6* censibus et $\frac{1}{4}$ census, equales 100 dragmis. Oppone ergo in ipsis et restaura et reduc totum quod habes ad censum; et erit census equalis 9 dragmis, et res, que radix eorum (sc.: earum) est, 3. Quam (sc.: Que) adde ad 5 et minue ea ex 5, et erit una duarum partium 8 et reliqua 2.

Et si dixit: 10 divide in 2 partes, et divide 10 per unamquamque duarum partium; deinde duc quod exit de duabus divisionibus in se, et provenient 20 dragme et $\frac{1}{4}$. *Ergo* radix 20 et $\frac{1}{4}$ est quod exit de duabus divisionibus, quod est 4 et $\frac{1}{2}$. Et iam exposui tibi opus in hoc.

Et si dixit: Duximus quod exit de duabus divisionibus in se, et provenient 30. Radix ⟨ergo⟩ 30 est quod exit de duabus divisionibus. Si volueris scire quantum est unaqueque duarum partium, fac sicut exposui tibi. Et est quod facias unam duarum partium ex 10 5 et rem et aliam 5 diminuta re. Et duc unam earum in aliam, et provenient 25 diminuto censu. Que duc in radicem 30. Et si vis hoc: Duc 25 dragmas diminuto censu in se, et provenient 625 dragme et census census diminutis 50 censibus; deinde duc illa **(A67ʳ)** in 30, et provenient 30 census census et 18750 dragme diminutis 1500 censibus. Radix illius totius equatur 100 dragmis. **(93ᵛ)** Duc 100 dragmas in se, et provenient 10000 dragme, equales 30 censibus census et 18750 dragmis diminutis 1500 censibus. Restaura et oppone et reduc totum quod habes ad censum census; et erunt census census et 291 dragme et $\frac{2}{3}$ dragme, equales 50 censibus. Dimidia igitur census, et exibunt 25. Que duc in se, et provenient 625. Ex quibus prohice 291 et $\frac{2}{3}$; et remanebunt 333 et $\frac{1}{3}$. Radi[x]*ce* eorum [d-] subtracta ex [eis] 25 accepta radice residui, addita ad 5, et hoc est una duarum partium, et dimin*uta* ex 5 est pars reliqua.

Et si dixit: 10 divide in 2 partes, et divide 40 per unamquamque duarum partium; deinde duc quod exit *ex* eo in se, et provenient 625. Necesse est ut, quando dividis 10 per unamquamque duarum partium et ducis quod exit ex eo in se, [$\frac{e}{8}\frac{1}{2}$] proveniet $\frac{e}{8}\frac{1}{2}$ de 625, quod est 39 et $\frac{e}{8}\frac{1}{2}$; quia 10 est $\frac{1}{4}$ de 40, et $\frac{1}{4}$ (alicuius), quando ducitur in se, proveniet ex ipso sicut $\frac{e}{8}\frac{1}{2}$ illius quod provenit ex toto in se. Et iam exposuimus operationem in hoc.

Et si vis [facere] *agregare* (vel: colligere, deducere) questiones singulares *multas* ex isto modo questionis secundum istam viam et aliam, lev[e]*is* tibi erit operatio in ipsis secundum id quod premisi tibi ex eo quod exposui tibi in libro ⟨meo⟩ hoc. Unde [laus et] gloria et laus sit soli Creatori.

3. Glossaire

La liste qui suit donne des exemples de la plupart des mots présents dans notre texte, éventuellement des tournures dans lesquelles ils apparaissent. Des renvois systématiques ont été faits pour quelques mots importants, comme *Euclides* ou *restaurare*, et toutes les démonstrations peuvent être localisées en regardant sous *velle*, qui se rencontre régulièrement dans leur conclusion.

Abiicere: 1. ذهب IV: 729. Cfr. prohicere,3.
— 2. سقط IV: 1224, 1648. Cfr. prohicere,1.
— 3. لقى IV: [388] (mut. in sublatus).
abscidere (قطع): 450, 485, 505, 747.
accidere (-): Cfr. similiter,2.
accipere: 1. اخذ : 1744, 1745-6, 1749, 1774, 1784-5; (- radicem:) 75, 82, 98, 104, 117, 123, 212, 234, 432,; (- differentiam:) 2128, 2132; (- fractionem:) 1096, 1673, 1737, 1792, 1805, 2469, 3044. Cfr. capere.
— 2. مأخوذ : 2345, 2371, 2397, 2566, [2569], 2573, 2967 (مأخوذًا = postquam acceperimus), 2970-2983, 3012-3027, 3051-3061, 3072-3086, 3102, 3546. Cfr. radix,2.
addens (اكبر , اكثر; super aliquid ... aliquo/aliquid: من...ب): 1558, 1560, 1562, 1747.
addere: A. (= augere): 1. زاد (على : alicui, ad, super aliquid): 63, 75, 81, 104, 109, 117, 123, 144, 187, 202, 237, 260, 305, 324,; v. 1074, 1199; (في : D.) 1487, 1536, 1582, 1630; (abs., = كثّر) 579, 599 (mut.), 812. Cfr. agregare,1; apponere,1.
— 2. ضاف IV (الى : D.): 132, 163, 294, 347, 372, 3284.
— 3. جمع : 3009, 3012, 3123. Cfr. agregare,3.
— 4. ضمّ : [1797]. Cfr. agregare,5.
— B. (= superare): 1. زاد (على : super aliquid, super aliquo): 3302, 3306, 3340, 3343, [3353].
— 2. اكثر من (كان): 327, 340, 3261, ([3310-1]).
— C. (- comuni(ter):) جعل ... مشتركًا : 390. Cfr. comunis,1.
addiscere (دلّ X; على...ب : per...in): 1683. Cfr. docere,2.
additio (زيادة): 204, 309, 332, 335-6, 403, 1188, 3217, 3279.
additum, n. (ما اجتمع): 2669, 2679. Cfr. agregatum.
additus: 1. زائد (abs.; على : D.): 542, 543, 550, 551, 554, 555, 558-568, 672-675, 703, 727-730, 760 seqq,.
— 2. مزيد : 2297.
— 3. مزاد : 2381, 2395, 2397, 2453, 3102, 3126, 3546.
— 4. زيادة : 729 ((res) addite). Cfr. diminutus,6.

adducere (اتى): [67], [426-7]. Cfr. afferre.
adiungere (زاد; ل , على : D.): 460, 1603.
afferre (اتى): *68*, *427*. Cfr. adducere.
agere: (قبل III; ب : in): 2008, 2028, 2045, 2067, 2165; (agere secundum algebram:) 2198, 2239, 2621, 2630. Cfr. facere,3.
ag(g)regare: 1. زاد (على : ad / D.): 89 (post.; prius: -), 97, 431, 433, 439, 440,; (abs.) 599 (agregatur, mut. in: additur); 2634 (مزاد; v. additus,3), 3349 (sicut si agregas = مثل جمعك). Cfr. addere,A,1.
— 2. بلغ : 64 (agregari, mut. in: devenire); 1542. Cfr. provenire,3.
— 3. جمع : 88, 540, 639, 693, 703, 966 (-), 975, 976, 994, 1017, 1023, 1027,; (الى , مع : ad / D.:) 540, 1974; 1030, 1281 (مجموعين). Cfr. addere,A,3; coniungere, coniunctus; iungi.
— 4. agregari = اجتمع : 6, 202, 1544 (= producitur), [2326] & [2384] (mut. in: produci), 2374, 2605 (mut. in: provenire), 2663. Cfr. agregatum.
— 5. ضمّ : 965. Cfr. addere,A,4.
— 6. فرع II: *3555*.
— 7. عدل III: [40] (corr. in: iungi, equari), 48 (mut. in: coniunctum, equale).
— 8. - : *3472* (يخرج, emend.), *3477* (-).
— *gg*: *1027*, *1030*, 2036, 2094, 2164, 2194.
agregatio (-): *973*.
ag(g)regatum (- / ما اجتمع): 976, 1035, *1557-8*, 1989, 2094-5 / 1200, 1587, 1604, 1923, [1960] (corr. in: quod provenit), 1984-5, 2033, [2049] (corr. in: coniunctum), 2095,; 650 (totum agregatum = جميع), 2036-7 (ما يجتمع); (agregatum ex multiplicatione = ما (الذى) يجتمع من الضرب :) 3492 (mut.), 3494. Cfr. additum; productum,2 (= ما اجتمع).
— *gg*: 1035, *2033*, 2036-7, 2094-5, 2095, 2099.
algebra, n.f. (الجبر): Cfr. agere; facere,3.
Algorismi, Maham(m)ed filius Moysi (محمّد بن موسى الخوارزمى): 2, 72.
aliquantus (بعض): [*1742*]. Cfr. aliquot; quidam.
aliquid (شيء): 1046.
aliquis: 1. اى : 63, 109, 201.
— 2. - numerus: 921 (عددٌ), 2276, 2287; 582, 965 (عدد من الاعداد); 971-2 (عدد ما).
aliquot (بعض): 1745. Cfr. aliquantus; quidam.
alius: 1. آخر : 207, 551, 555, 557, 802, 802, 946, 1064, 1067, 1269, 1283, 1325, 1350,. Cfr. alter; posterior; reliquus; residuus,2.
— 2. -غير : 534 (post.), 587, 970, 2265, 3556.
— 3. باقٍ : 11. Cfr. residuus,1.

— 4. صاحبه : 1963.
— 5. Cfr. versus.
alter (آخر): 555, 900, 965, 968, 969, 973, 1066, 1171, 1172, 1274, 1280, 1282,. Cfr. alius,1.
altitudo: {*31*}.
ambo (جميعًا): *542*, 550, 554.
amplius: 1. ثمّ : 292, 1915, 2470, 2500, 3168, 3169. Cfr. deinde,1.
— 2. وايضًا : 333.
— 3. (adj.) اكثر : 3226.
angulus: Cfr. rectus.
animadvertere: 1. نظر : *891* (corr. ex: intelligere).
— 2. فهم : *1712* (corr. ex: intelligere).
apparere: Cfr. oculus.
applicare (بلغ = pervenire; ad: من , الى): 2770, 2865 (corr. ex: contingere), 3158.
apponere: 1. زاد : [1118] (corr. in: addere), 1132. Cfr. addere,A,1.
— 2. قبل III: Cfr. restaurare,2.
aspiciens (ناظر): 1.
auferre: 1. ترك : 2956.
— 2. نقص : 3222. {1384}.
augmentum (زيادة): 3187. Cfr. additio.
aut (او) : 3277; 350 (-). Cfr. sive; vel,2.
autem: 1. و : 3, 19, 32, 100, 107, 108, 126, 139, 146, 151,; 665 (corr. ex: et). Cfr. et,1.
— 2. ف (errore?): 110, 114, 999, 1030, 1123.
— 3. واما...ف : 50, 162, 199, 371, 423, 496.
— 4. فاما...ف : 14, 42, 61, 71, 209, 242,.
Bini (-): 57.
bis (مرتين): 1085, 2532.
bizantius (فلس): *2013 -2015* (corr. ex: moneta, *blada*); *2155-2160* (corr. & mut. ex: zoz).
blada (فلس): [*2012*-2015] (corr. in: bizantius).
brevis (-): 3249 (brevior; ant. longior).
Cadere: 1. وقع : 8, 360, 363; 362 (ut cadat nisi in: ان يقع خارجًا من); 368 (non cadit preterquam in: لا يقع خارجًا من ... ولا داخلًا من).
— 2. (facit -) سقط IV : 1256.
— 3. عقب VI: 570.
capere (اخذ): 833 (- medietatem), (836), 841 (duas tertias). Cfr. accipere.
capitulum (باب): *222*, 229, 241 (mut. in: porta), 243, 402, 406, 1591, [1680] (corr. in: porta). Cfr. porta.

causa: 1. علّة : 129, 242, 371, 402, 406, 446, 496, 522, 1285, 1465, 1465, 1479, 1479,.
— 2. معنًى : 634.
census: 1. مال (quadratus rei incognitae, sc. x^2): 3, 5, 8, 12, 12, 14, 15, 15, 16-17,; 2682 (def.).
— 2. مال = quantitas (incognita), ut in: 201, 799-800 seqq, 1199 seqq (v. 1223, 1234), 1809 seqq, 2005, 2269.
— 3. عدد : [1121] (?), 1495; 1575 (mut. in: pecunia).
census census (مال مال, sc. x^4): 24*04*-(2413), 2434-2442, 25*07*-2512, 2560, 2576-2586, 2619-20, (2651)-*2653*, 2692-2694, 2714-2721, 2824-2845, 28*98*-2905, 2935-2950, 2964, 2978-(2994), *3022* -3041, 3067-3092, 3175-3177, *3199*-32*01*, 3537-3543; 2683-4 (def.).
census census census (sc. x^6): *2947*.
census census census census (مال مال مال مال, sc. x^8): 293*8*-2948-9.
cogere (ألجأ): *2443*, *3005*, 3488 (quod co(e)git nos). Cfr. cum,B,4; facere,6; indigere; qui,3; quia; ratio,1; tantum,1; unde,2.
cogitans (حاسب): 1255, [1299]. Cfr. computans.
complementum (فراغ): 1285.
complere (تمّ): 181, 384, 663, 711, 738, 771, 3229-30; 3244 (-).
componere (ركب II): 3388. {1155}.
compositus (مقرون): 524 (v. 415, 2466-7).
computans (حاسب): 562. Cfr. cogitans.
computare: 1. حسب : 2454, 2486, 2600, 2776 (v. [2769]). Cfr. intelligere,3.
— 2. عمل : 3181.
computatio (حساب): 527, 530-1, 1059, 2264, 2268, ([1299], v. 1301).
comunis: 1. مشترك : 388-9, 1224, (1513), 1561, 1653 (اسقط/القى...المشترك); 390, 474 (جعل...مشتركًا), 3409 (جعل...مشتركًا في الضرب); 750; *1817*, *2687* (-).
— 2. جامع (mut. in: universalis): 897, 919, 945.
concurrere (concurrit = يكون , يُوجد , موجود): 1023, 1024, 1025.
congregare (جمع): 1939.
congregatum (-): 1793, [*1797-8*], *2297*.
coniunctim: 57 (pro: coniungi = قرن VIII).
coniunctum (- / ما اجتمع): 971 / 2049 (corr. ex: agregatum), 2052-3. Cfr. agregatum.
coniunctus: 1469 (مجموعين, v. agregare,3); *1324* (-).
coniungere (جمع): 971. Cfr. agregare,3.
consideratio (نظر): 535 (errore).
consimilis (شبه IV): 1771. Cfr. similis.
consuetudo (عادة): ([1299], v. 1301).
consuetus (الذى من عادة): 1254-5.

continentia (حدّ): *529* (errore). Cfr. terminatio.
continere (-): 183, 319, 782, 3237, 3238.
contingens (-): 1613, *1661*.
contingere: 1. صاب IV: 1487-1522, 1535 seqq; 1914 (mut.), 2246, 2258, 2259, 2445, 2797, 2799, 2820, 2835, 2838. Cfr. exire,4.
— 2. نهى VIII (ad: الى): 3194, 3269, 3276.
— 3. بلغ : [2865].
contradicibilis (يستحيل): *224* (mut. ex: impossibilis).
conveniens est (ينبغى): 1045. Cfr. necesse.
convertere (خلف III): [*906*] (-); 1362 (corr. ex: permutare).
cubus (كعب, sc. x^3): 2682-2695, 2714-2722, 2753, 2874, 2874, 2925, 2928, 2930, 2943, 3164-3177; 2683 (def.).
cubus cubi (كعب كعب, sc. x^6): 2941-*2948*; 2684 (def.).
cum: A. مع : 86, 211, 223, 231, 236, 238, 246, 258, , 308 (prius; post.: ب),.
— B.1. لانّ : 139, 185, 271, 279, 377, 392,. Cfr. propterea; quia.
— 2. متى , متى ما : 336, 915, 917, 1991, 2025, 2115, 2248, 2488, 2490, 2503, 3018. Cfr. quando,2.
— 3. اذا , ان : 88, 221, 224, 290, 307, 309 (prius), 326, 329, 331 (2^m), 349,.
— 4. فانّما : 2809 (فانّما...لانّ, v. cogere).
custodire (حفظ): 2430. Cfr. servare.
Deinde: 1. ثمّ : 231, 437, 449, 580, 596, 601, 620, 679, 701, 803,. Cfr. amplius,1; demum; item,3; iterum,1; postea,1.
— 2. 384, 502, 596, 789 (و); 1398 (ف).
demonstratio (برهان): 607, 623, 644, 662 & 681 (corr. ex: probatio), 711, 738, 770,. Cfr. probatio.
demum (ثمّ): 674, 702, 762, 791. Cfr. deinde,1.
denarius: [1743] (corr. in: dragma).
Deus: 128, 228, *249*, 256, 403, 428, 857, 920, 1057, 1481, 1713, 2268, 2623, 2661 (cum adiutorio Dei); 2263 (Adiutor); 3218, 3559 (Creator).
devenire (بلغ): *64* (e mut.). Cfr. provenire,3.
dicere: 1. قال : 146, 151, 190, 227, 270,; e.g.: 12, 58, 61-2, (ut dicas = كقولك, نحو قولك); 14-5, 50, 51-2, 87, 200, (sicut si dicas, si dixerit = مثل قولك , نحو قولك , كقولك); 45-6, 1050-1, 1052, 1063, ((sicut) tuum dicere = (نحو/ك) قولك , تقول); 576 (si dixit = ان قال); 85, 108, 3270-1 (si dixerit, si dixit = لو قال); 571, 619 (si dixerimus = ان قيل); 653 (si dixerimus = ان قال); 669, 697 (si dicemus = ان قيل); 893 (est quasi diceremus = كانّه قال لك); 3292 (quasi diceret = فكانّه قال); 878 ((Euclides) dicens = قال); 1431 (dicere = vocare).
— 2. ذكر : 354, 3073. Cfr. exponere,9; memorare,2; narrare,1; nominare,3.
— 3. وصف : 334, 1767, 1783, 1865, 1940, 1975, 1980, 2091, 2113, 2198, 2533, 2557, 2654, 2762, 2774, 2836,; 413 (dicere et ordinare; v. ordinare,1);

1993-4 (dicere et exponere). Cfr. exponere,8; narrare,2.
— 4. خبر IV: 566.
— 5. علم IV: 1058. Cfr. docere,1.
— 6. بان II: 685, 2210. Cfr. preponere,2.
— 7. - : 398.
dictio (قول): 43 (sicut dictio tua = كقولك; cfr. dicere,1).
dictus (-): 1191, (2466). Cfr. predictus.
differentia: 1. فضل : 2128, 2132, 3255. Cfr. superfluum,1.
— 2. فرق : 1466.
— 3. 2382 (differentia inter duos census = مالان بينهما).
dimidiare (نصف II): *73*, 1134, 1140, 1176, 1184, 1206, 1212, 1295, 1418, 1460,. Cfr. dividere,4.
dimidius (نصف): 1073, 1074 (census et dimidius = $1\frac{1}{2}$ census); 1277. Cfr. medius,1.
diminuere: 1. سقط IV (من : ab): 1615; [1905]. Cfr. prohicere,1.
— 2. نقص (من : ex, ab): 2236, 3185, 3487. Cfr. minuere,1.
diminutio (نقصان): 204, 308, 331, 335, 402, 3187-8, 3217, 3279.
diminutum, n.: 1. نقصان : 1491, 1521, 1539, 1568, 1632, 1661.
— 2. استثناء : 566, 567 (2^m).
diminutus: 1. منقوص (ا منقوص من ب = A diminutus ex B, sc. B-A): [2257], 2320, 2343, 2346, 2358, 2369, 2372, 2424, [2520] (corr. in: sublatus), 2572 & 2576 (mut. in: sublatus), 2614, 3059, 3095, [3546] (corr. in: subtractus), 354*7*. Cfr. sublatus,1; subtractus,2.
— 2. ناقص (abs.; من : ab): 539, 540, 549-560, 567 (3^m), 700-703, 728-731, 761-764,; 3167 (mut. in: sublatus). Cfr. sublatus,2.
— 3. اقلّ (من...ب : ab...aliquo): *1489* (corr. ex: minus), 2396.
— 4. الّا (abl. abs.): 653, 654, 657, 659, 660, 662, 667, 697, 698,. Cfr. exceptus; parvus,5.
— 5. غير : 1069, 1173 (post.), 1275, 1275, 1400, 1401 (prius), 1402, 1862, 1916 (prius), 1993, 2080, 2131, 2251,.
— 6. نقصان : 729 ((res) diminute). Cfr. additus,4.
dinar (دينار; pl. dinarim): 1399-1409, 1689-1697, 2012-2023, 2076-2089, 2154-2159, 2201-2208, 2488-2499.
directus (in directum = على استقامه): 1004 (e corr.). Cfr. rectitudo.
diversus esse (فضل V): [1741].
dividendus (مقسوم): 923. Cfr. divisus,1; partitus.
dividens (مقسوم عليه): *948*, 1115, 1390, 1909 (mut.), 1923, 1951, 2005, 2248-49, 2307, 2607, 2806,; 3399 (corr. ex: divisus = مقسوم). Cfr. partiens.
dividere: 1. قسم (على : per): 916 (mut.), 920, 926, 931, 931, 934, 934,; (مقسوم عليه = (numerus) per quem divisimus/dividitur/divisum fuit, e.g.:) 922, 924, 1917-8, 1931, 1947-8, 2025-6; 1486 seqq (divisisti 50 dragmas per homines = قسمت خمسين درهمًا على رجال) —interdum على = D.: [1486], 1686.

— 2. قسم (ب , - : in): 28, (607), (623-4), (626), 841, 1063, 1088, 1105, 1111, 1127,; (بنصفين = in duo equalia, in duo, in 2:) 142, 143, 186, 186, 267, 268, , 2211 (corr. ex: in duas partes), 2228, 2230, 2242, 2299,; (in duo inequalia = بقسمين مختلفين :) 268 passim, v. inequalis.
— 3. قسم (بين : [inter] *per*): 2245.
— 4. نصف II (= dividere per medium, in duas partes equales, in duo equalia): 72-3 (mut.!), 80, 96, 102-3, 115-6, 121-2, 210, 221, 231-2 (mut.!), 430, 437-8. Cfr. dimidiare.
— 5. قطع : 849 (lineam, بنصفين).
— 6. يقع القسم : 350. Cfr. cadere.

divisio: 1. قسمة : 404 (in duo equalia = بنصفين), 947, 1255, [1301], 1323, 1476,.
— 2. قَسم : 921, 946, 1115, 1276, 1283, 1322, 1376, 1390, 1425, 1426,; 1756 (corr. ex: pars; v. pars,2).
— 3. قَسم pro قِسم saepe, ut in: 625, 627, 1681, 3305, 3316, [3352]. Cfr. pars,2.
— 4. (divisio in duo equalia:) نصف : 360.
— 5. تنصيف : 1258 (errore).
divisus, adj. & (divisum) n.: 1. مقسوم : *916* & *918* (corr. ex: partitum), 922-3 (= qui fuit divisus), 933, *947*, 1391, 1495, 1542, 1918 (مال مقسوم), 1923, 1934, 1954,; 1116 (العدد الذى قسمته). Cfr. dividendus; partitus.
— 2. قِسم : *1971*.
— 3. مفرد : 522. Cfr. separatus.
docere: 1. علم IV: (444), [1058], 1114. Cfr. dicere,5.
— 2. دلّ X: 2267 (لتستدلّ = que te docebunt). Cfr. addiscere.
donec (حتّى): 1457.
dragma: 1. درهم (unitates numeri): 44, 62, 65, 201, 202, 223, 225, 242, 244, 257, 291,; (dragma = واحد :) [1367], cfr. unus,3.
— 2. dragme = من العدد , عدد : 330, 449 (e corr.), 1121.
ducere: 1. خرج IV (الى : ad scientiam (alicuius), ad sciendum (aliquid) / ad (aliquid)): 66, 71, 78, 119, 130, 229, 243, 371, 406, 447,; [100] & [496] corr. in: faciens & facit exire; / 530, 1077, 1182, 1209, 2265; 1059 (مخرج). Cfr. exire,3; trahere.
— 2. خرج IV: 488 (lineam).
— 3. ضرب : 78, 81, 101, 102, 120 (corr. in: extendere), 327, 416, 578, 594, 598, 671 (corr. in: multiplicare); (de lineis:) *606*, 622, 642 (mut. in: percutere), 735, 768,. Cfr. extendere; multiplicare,1; percutere.
— 4. ادى II: 3056 (يودّيان الى شىء واحد = ducunt ad viam unam).

ductus: 1. (n.m., -us) ضرب : 6, 158, 185, 417, 419, 420, 666, 692, 884, 1005, 1227, 1793,. Cfr. multiplicatio.
— 2. (adj.) مضروب : 22, 25, 376, 501, 615,. Cfr. multiplicare,1.
— 3. - : 271, 271, 272, 417, (ductus in = في). Cfr. se.
duplare (ضعف II): 112, 3125; 113 (-). Cfr. duplicare,1.
duplicare: 1. ضعف II: 799, 814, 819, 2728. Cfr. duplare.
— 2. pro اضعاف : 800.
duplus, duplum (n.): 1. ضعف : 816, 3444, 3445 (duplum [de] ad), 3467, 3467, [3468]; *3450* (-).
— 2. مثلا : 2399.
Efficere (كان , -): 23, 75, 192.
egredi (خرج): 308. Cfr. exire,1.
emere (-): 1741, 1743.
enim: 1047 (لانّ); *1389* (-).
equalis: 1. مساوٍ (ل : D.): 16, 17, 631 (post.), 667, (963), 3329.
— 2. متساوٍ / (- esse, produci, fieri:) سوى VIII: 29, 31, 613, 631 (prius), 828, 852, *2730* / 1128, 2360, 3105, 3186, 3194, 3210, 3224, (3231), 3252, 3270, 3276.
— 3. مثل : 202, 224, 246, 280, 283, 284, 314, 358, , 1093,. Cfr. sicut,1.
— 4. (- esse:) عدل I / III: 1075, 1118, 1131, 1133, 1175, 1245, 1527, 1528, 1705, / *48*. Cfr. equari,1 & 2.
— 5. non equalis, مختلف : *2920*. Cfr. dividere,2
— 6. 841, 1105 (equalia = اقسام متساوية).
— 7. dividere in duo equalia: cfr. dividere,2 & 4.
equalitas: 3213 (errore).
equari: 1. عدل : 12, 12, 13, 14, 15, 35, 37, 38, 42, 43, 44,; (كان ... يعدل :) 79, 83, 106, 125, 1143,; 93 (mut.). Cfr. proportionari.
— 2. عدل III: *40* & *57-8* (e mut.), 55, 128.
— 3. سوى VIII: 3187, 3189, 3214, 3217, 3255, 3257, 3279.
equatio (قياس) : 521.
equidistans: 1. موازٍ (ل : D.): 469, 486-7, 488, 611, 612, 628, 629-30, 682, 683,.
— 2. وزى III: 506, 773, 847, 850,.
— 3. - : 29.
equus: dividere in duo equa (interdum pro: equalia): 210, [*232*], 300, 430, 452, 460, 510, 511, 849,.
ergo: 1. ف : 91, 141, 148, 149, 157, 159, 160,; 1514 (corr. ex: igitur). Cfr. igitur,1; itaque; item,2; quare,1; tunc,2.
— 2. وإذًا : 853.
esse: 1. - , هي , هو / كان : 2, 3, 4, 4, 5, 5, 6, 7, 9, / 17, 75, 76, 80, 82, 86, 96, 98, 104, 117, , (mut. aut corr. in: provenire:) 78, 1171, 2433, 2458, 2556,

2704, (corr. ex: provenire:) 1711, *2951*; 3190 (erunt id quod erunt = كائنًا ما كانت); 3225 (et si non erit 2, erit amplius = فان لم يكن اثنين فليكن اكثر).
— 2. صار : 19, 164, 195, 325, 416, 492, 695, *778* (prius; e corr.),; 971, 976 (صار II).
— 3. عدل : 36.
— 4. ($\frac{A}{B} = \frac{G}{1}$: قدر الواحد من ج كقدر... = unitas est in G secundum proportionem... , ب يعدّ ا بقدر ما فى ج من الآحاد = B est in A secundum unitates (de) G): 1360-1, 1361, 1364-5; 3332, 3333-4.
— 5. (non esse:) ليس : [1299].
et: 1. و : 3, 3, 4, 5, 5, 6, 6,; (pro additione, e.g.:) 58 seqq, 195 seqq, 865 (post.), 979 (post.), 1047, 2217 seqq, 2460, 3385 seqq.
— 2. (cum fut. / conj. / imperat.:) ف , ut in: 16, 74, 81 (bis), 154, 155, 430, 574, 578, 1064, 1088, 1432, 2201 (bis).
— 3. ثمّ : 1073 (post.), 1582, 2168, 3458.

etiam: 1. ايضًا : 203, 424, 690, 719, 748,; (= etiam similiter:) 663, 716, 744, 9*91*,; (= sic etiam:) 1344. Cfr. ita,2; iterum,2; similiter,3.
— 2. و , - , ف : 10, 40, 56, 221, 341, 361, 374, 403, 476, 483, 568, 708, 713,; *1347*.
— 3. ثمّ : 1204, *2202*.

Euclides (اقليدس): 69, 146, 190, 270, 302, 355, 462, 513, 685, 878, 2533.
exceptus (الّا ; abl. abs.): 1068, [1117], 1239, 1375, 1379, 1398, 1398-9, 1782, 1936, 1936, 2097. Cfr. diminutus,4.
exemplum: 1. مثال : 621, 857, 900, 923, 948-9, 1325, 1357,. Cfr. via,2.
— 2. قياس : 1067, 1091, 1112, 1128, 1171, 1201, 1238, 1270, 1538-9, 1632, 1688, 1717,. Cfr. regula,2; solutio,2.
— 3. 916 (كأنّ = cuius exemplum est quod).
exire: 1. خرج (de operatione, solutione, questione): 309, 330, 331, 625, 627, 916, 921, 925, 926, 932, , (mut. in: provenire:) 3466 & 3512; 1917, 2974 (quod exit = الخارج); 2194 ((duo) exeuntia = ما خرج من القسمين); 2806 (exiens = الذى يخرج. Cfr. agregare,8; egredi; provenire,6.
— 2. - , كان : *74* (-), 80, 98, 103, 105, 116, 122, 123, 212, 232, 430,; 935 (-),.
— 3. (facit - , faciens - :) خرج IV (الى : ad sciendum, scientiam / ad): 101 (e corr.), 115, 162, 205, 207, 209, *425*-(6), 429, 435, / 96, 204, 528, 1098, 1119, 1145, [1190] (corr. in: trahere). Cfr. ducere,1; trahere.
— 4. صاب IV: 1909, *1914* (quod exit de divisione = الذى يصيب القسم), 2802, 2818. Cfr. contingere,1.
ex(s)istere (-): 17, 27.
exponere: 1. بان II: 19-20, 68, 151, 226, 241, 310, 329 (post.), 339, 341, 402, 427, 521,; وذلك ما اردنا ان نبيّن : v. velle,1. Cfr. premittere.

— 2. مبيّن , n. ag. (- انا : exponam): 537, 945, 1682, 2267.
— 3. بان V: 289, 367 (تبيّن = expositum est), 1611 (= exposui), 1660, 2798, 2817, 2928. Cfr. manifestus,1.
— 4. بان X: 2020.
— 5. شرح : 421, *795*, 2540, 3516, 3530, 3558.
— 6. فسر II: 597, 913, 938, 1772, 3286, 3553.
— 7. واضع , n. ag.: 1496. Cfr. ponere,2.
— 8. وصف : 1395, 1408, 1807, 1821, 1829, 1878, 1930, 1946-7, 2029, 2166, 2661, 2965,. Cfr. dicere,3; ordinare,1.
— 9. ذكر : 3263. Cfr. dicere,2.
— 10. صحّ : 3431.
— 11. اتى : [426]. Cfr. afferre.
— 12. خبر IV (ordinare et exponere): 560.
— 13. 532 (ابتدى = incipiam exponere).
— 14. (-): Cfr. premittere.
expositio: 1. معنًى : 15, 63, 88, 109. Cfr. qui,3.
— 2. قياس : *1490* (mut. in: solutio). Cfr. exemplum,2.
— 3. باب : [*593*]. Cfr. porta.
— 4. 1100 (eius quod diximus expositio in figura est = ابيّن ذلك في هذه الصورة). Cfr. oculus.
extendere (ضرب): 103, *120* (mut. ex: ducere). Cfr. ducere,3; multiplicare,1; percutere.
Facere: 1. جعل : 177, 259, 276, 293, 311, 346,; [130], [448], [600] mut. in: ponere. Cfr. ponere,1.
— 2. عمل : 126, 152, 165, 374, 467, 500, 823, 863, 983, 984, 1001, 1002,; 1767 (mut. in: restaurare).
— 3. قبل III (ب : in): *1807* (corr. ex: restaurare), 2210; 2254, 2577-8 (def.), (2612-3), 2652, (2772), 2901, 3134 (قابل = opponere cum; facere secundum algebram - sc. aequationem questionis ad unam sex specierum reducere). Cfr. agere.
— 4. عدل : *93-4* (mut. ex: equari).
— 5. فرع II: [3555]. Cfr. agregare,6.
— 6. انّما : 3073 (et quod facit nos dividere = وانّما قسمنا), [3488]. Cfr. cogere.
— 7. 943 (sic facies = كذلك).
— 8. Cfr. exire,3.
falsus: 339 (falsus et impossibilis: يستحيل). Cfr. contradicibilis.
fieri: 1. فعل : 40, 48.
— 2. - , كان : 666, 700, 979, [*1089*], 1092, 1094, 1102, 1130,; *1793* (corr. ex: esse). Cfr. equalis,2.
— 3. حدث : 630, 851. Cfr. invenire,3.
— 4. جمع VIII: 1346. Cfr. producere,3; provenire,4.
— 5. 2929 (quando fiet = اذا جعلناه).

figura: 1. شكل : 68, 227, *335*, 339-40, 427, 590, 919, 945, 1334,.
— 2. صورة : 483, 603, 604, 631, 640, 657, 676, 705, 732, 765, 817, 838, 861, 980, 998, 1079, 1100, 1121, 1147, 1192, 1219,.
forma: 1. شكل : 897. Cfr. figura,1.
— 2. صفة : 3254 (cum simili huic forma = بهذه الصفة); 3339 (cum hac forma).
forum (-): 1745, 1746.
fractio (كسر): 601, 601 (602: -). Cfr. fractus.
fractus, n. (كسر): 5, 5, 7. Cfr. fractio.
(el) **Geber** (الجبر): 527 (el geber, id est restauratio), 532.
geometricus: 1. هندسى : 68, 227, 427 (figura -).
— 2. الهندسة : 3308, 3370 (via -).
gradus (in gradu = بمنزلة): 573, 577 (mut.), 580 (mut.), 3497. Cfr. locus; ordo,2; pro; sicut,6.
gravis esse (دبر V (= meditari de)): 1478.
Habere: 99 (فيبقى = habebis), 236 (= habebis residuum scilicet; corr. ex: remanere); 117, 121, 123, 234, 433, 639 (habebis = فيكون), 2335 (= فيكون معك); 1586, 2286 ((id) quod habebis = ما بلغ); 968, 969, 970 (كان (ل)); 1036, 1037, 1039 (non habet = ليس ل); 1293, 1416, (habes, habemus = معك; v. totus); *3498* (-).
homo (رجل): 1486-*1774*.
huiusmodi: (344), [1771].
Iam (قد + Perf.): 71, 91, 268, 300, 307, 329, 339, 352, 363,; e.g.: 91 (iam scis = قد علمنا), 268 (iam divisa est = قد قسم); 1830 (iam fecisti = كنت جعلت).
idem: 483 (ad hoc idem = تودّى الى هذا; v. via,9).
igitur: 1. ف : 27-8, 28, 32, 36, 38, 43, 44, 46, 51, 52,; 286 (igitur *ex inde*); [1513] (corr. in: ergo). Cfr. ergo,1.
— 2. [135] (اذًا); 693 (ثمّ).
ignotus (مجهول): 800, 801, 804, 806, 808, 809, 811.
impossibilis: 1. impossibile est: 359, 367, 3239 (خلف لا يمكن). Cfr. possibilis,3.
— 2. impossibile est quin: 350 (ليس يخلو من ان).
— 3. impossibile est (quod non, esse sine, esse aliter quam): 528, 534, 546 (لا بدّ).
— 4. (de questione) يستحيل : 224 (mut. in: contradicibilis), 421. Cfr. falsus.
incipere: 1. بدأ : 973 (incipiamus = نبدأ اوّلاً).
— 2. بدأ VIII: 532 (incipiam exponere), [562].
indigere (انمّا): [2443], 2720 (id quo indiges), [3005]. Cfr. cogere.
inequalis (مختلف): 268, 301, 353, 364, 395 (dividere in duo equalia et in duo

inequalia = قسم بنصفين وبقسمين مختلفين); 2958, 3065 (census inequales = اموال مختلفة), 3205 —v. equalis,5.
innuitio (اشارة): 3351. Cfr. signum.
integer: 1. تامّ : 1819, 1850, 2513. Cfr. unus,4.
— 2. - : 1894.
— 3. اصل : 1888 (ut proveniat (vel: redeat) census integer = حتّى يعود المال الى اصله).
— 4. سواء : 1073 (integrum esse = كان سواء).
integrum, n. (تمام): 2134, 2140.
intelligere: 1. فهم I & II vel IV: 69, [1712] (corr. in: animadvertere); 603 (faciam te intelligere). {2449}, {2586}.
— 2. نظر (فى) : 69-70, [891]. Cfr. animadvertere,1.
— 3. حسب : 1372.
— 4. pro معنًى : 800.
— 5. عمل : 797 (intelligere et operare); 1481 (intelligere; v. 1484).
— 6. - : 1713, 1808, 2655.
invenire: 1. وجد : 56, 891, 892, 938, 939, 2264.
— 2. خرج X: 1479.
— 3. pro حدث : 612, 827. Cfr. fieri,3.
— 4. - : *1188*, 2661. {*2522*}.
ita: 1. كذلك, هكذا : 54, 55, 2407.
— 2. ita etiam = ايضًا : 1363. Cfr. etiam,1.
— 3. 1961 (postquam ita est = اذ قد بان هذا), 1160-1 (-), 3280 (اذا كان هكذا).
itaque (ف): 512, 737, 782, 824, 925,; *1702* & 1898 (corr. ex: igitur). Cfr. ergo,1.
item: 1. و : 12, 37, 59, 59, 381, 386, 479, 487, 499, 515, 523, 523, 525, 526, 629,.
— 2. ف : 436, 505, 1289. Cfr. ergo,1.
— 3. ثمّ : 121, 415, 638, 655, 672, 673, 700, 728,. Cfr. deinde,1.
iterum: 1. ثمّ : 80, 102; 1630 (et iterum). Cfr. deinde,1.
— 2. ايضًا : 326. Cfr. etiam,1.
— 3. - : 51.
iudicium (حكم (خلّة)): 1023.
iungi: 1. عدل III: 40 (corr. ex: agregari; mut. in: equari). Cfr. equari,2.
— 2. جمع : 192 (que simul iuncta = فتجمعهما), 1351 (simul iuncta = مجموعين). Cfr. agregare,3.
Latus: 1. ضلع : 21, 22, 25, [134] (corr. in: radix), 3224, 3224.
— 2. جذر (de superficie quadrata): 23 (mut. in: radix).

— 3. ناحية : 3301.
— 4. - : 382, *383*.
levis esse (سهل ; على : D.): 1256, [1302], 3557.
liber (كتاب): (ipsius auctoris:) 1, 520-1, 536, 2263, 2266, 2549, 3558; (Algorismi:) 3, 72;
(Elementorum:) 69, 146, 190, 270, 302, 355, 462.
libet (راد IV): 982 (اذا اردنا = cum libuerit (mut. in: cum volemus)).
linea (خطّ): 25, 30, 32, 32, 133, 135, 139, 140,.
locus (منزلة): *578*, 620 (mut. in: pro), 803 (بمنزلة = loco). Cfr. gradus.
longitudo (طول): 23; 144, 188, 460, 512 (in longitudine (sua) = فى طوله).
longus (طول): 262, 3224, (longior, اطول). Cfr. magnus,4.
Magis esse (كَثَرَ): 3217. Cfr. plus,2.
magnus: 1. اعظم (maior): 295, 1067, 1080, 1111, 1113, 1122, 1128, 1129, 1138, 1148, 1168,.
— 2. كثير : 2167-2168, 2212-2217, 2230-2236, 2383, 2386, 2543-2546, 2558, 2567, 2608, 2610 (كثير = maior, maius); 244, 1237, 2554, 2555, (اكثر = maior); 2396 (اكثر = magnus); (ب : in / - :) 3253 / 3208, 3288. Cfr. plus.
— 3. كبير : 2216-2241 (شىء كبير et شىء صغير; v. res,1); 2927 (اكبر = magnus); 261, 1069, 1076, 1239, 1256, 1264, 1575, 1587, , 2599, (اكبر = maior).
— 4. اطول (maior): 297, 3232. Cfr. longus.
maior: Cfr. magnus.
manifestus: 1. (manifestum est:) بيّن / تبيّن : 133, 453, 469, 476, 503, 1005, 1650, 1960-1, 1974, 2037, 2112-3, / 307, 326, 333, 1349, 2081, 2144, 2939, 3251, 3304, 3341, 3426, 3499. Cfr. oculus; planus,1.
— 2. واضح : 1728 (- et planus); 3260.
— 3. واجب : 1870.
— 4. معلوم : 3194-5 (mut. in: scitum).
manus (ad manum (per)venire = جاء): Cfr. pervenire; venire.
medietas (نصف): 35, 53, 76, 83, 91, 92 (mut.), 99, 105, 108,. Cfr. medium,1.
medium, n.: 1. نصف : *92*, 367. Cfr. medietas.
— 2. - : Cfr. dividere,4.
medius, adj.: 1. نصف : 487 (radix et media = radix et medietas radicis, v. 489), 1072 (census et medius = 1 et $\frac{1}{2}$ census). Cfr. dimidius.
— 2. اوسط : 2921-2944, 2955-2966, 3004-3011, 3052, 3058, 3063-3101.
melior, melius (احسن): 1045, 1047, 1052.
memorare: 1. خبر IV: 520 (memorare et ordinare).
— 2. ذكر : 2. Cfr. dicere,2.
mensura (قدر): 3213. Cfr. proportio,2.
minor, minus: Cfr. parvus; plus.
minuere: 1. نقص : 599; (من : de/ex/ab:) 76, 212, 1178, 1297, / 99, 118, 211, 1296, 1533, / 3157, 3180; *2257* (منقوص, pro: (accepta) radice eorum *diminuta*). Cfr. diminuere,2; diminutus; prohicere,4; subtrahere,1.

— 2. سقط IV: 234, 1031, 1185, 1186, 1187, 1189, 1242, 1248, 1587,. Cfr. prohicere,1.
— 3. لقى IV: 1237, 1245, 1575, 2105. Cfr. prohicere,2.
modus: 1. امر : 566 (summa modi = جملة الامر).
— 2. مسمّى : 570 (mut. in: significatum).
— 3. عمل : 1480.
— 4. - : 3556.
moneta (فلس): [2012] (corr. in: bizantius), 2203-*2208*.
monstrare: 162 (porta autem monstratur ut = واما العلّة للباب).
(el) mukabale, mucabele: 527 (el mukabale, id est oppositio), 532.
multi (كثير): 2264, *3556*.
multiplicare: 1. ضرب (فى : in): 4, 79, 116, 121, 122, 210, 222, 229-30,; (مضروب = multiplicatus, cum multiplicatur:) [22] (corr. in: ductus), 135, 136, 3514; 3347 (quando multiplicas = ضربك). Cfr. ducere,3; extendere; percutere.
— 2. اخذ (= accipere): 800 (multiplicare in 2 radicem suam = اخذ جذريه).
multiplicatio: 1. ضرب : 140, 141, 144, 145, 146, 147, 148, 188, (189, bis: -), 191,; 2194 (multiplicatio inter se = ضرب احدهما فى الآخر). Cfr. ductus.
— 2. مضروب (sc.: multiplicatio A in B = ا مضروبًا فى ب): 179, 1089-90, 1090, 2327.
— 3. - (sc.: multiplicatio A in B = ا فى ب): 189, 189, 354, 396,.
Narrare: 1. ذكر : 71. Cfr. dicere,2.
— 2. وصف : 1024, 1026, 1039, 1043. Cfr. dicere,3; exponere,8.
necessarius est (يحتاج): 1, 1466, 1472.
necesse est: 1. ينبغى : 90. Cfr. conveniens.
— 2. يجب : 3436, 3441, 3445, 3455, 3461, 3550.
— 3. راد IV: *111* (e mut.).
nomen (اسم): 8, 569; *1902*, [3006] (-).
nominare: 1. لقى IV: 1045.
— 2. ورد (على : D.): 1257.
— 3. ذكر : 2265.
notus (معلوم): 799, 801, 804, 805, 808, 809, 810.
numerus: 1. عدد : 3, 7, 12, 13, 17, 42, 49, 50,; 22, 133-4, 424, 446, 454, 455, 485, 491, 498, 499, (من (ال)عدد = ex/de numero, numeris; v. dragma,2); 581-2 (omnis numerus ex numeris = كلّ عدد من الاعداد), 2804 (numerus numerorum = عدد من الاعداد ايّ عدد كان); 876, 876, 877 (mensura superficiei; v. 880-3).
— 2. عدّة : 626.
— 3. مقسوم عليه = numerus per quem dividitur: Cfr. dividens; partiens.

— 4. 843 (علم جذر اى مال هو = scire ipsa in numero), 966 (= scire cuius numeri sint radix).
— 5. - : (64), 75, 82, 98,.
nunc: 2956 (فى هذا الوقت); *604* (-).
Oculus (exponere ita ut oculis appareat, scire scientia subiecta oculis, exponere expositione oculis/oculo manifesta = بيّن عيانًا): 151, 276, 311, 591.
omnis: 1. كلّ : 239, 545, 581, *703*, 898, 967, 1279, 1321, 1349,; 4, 1416 (كلّ شىء = omne). Cfr. totus,1.
— 2. جميع : 126, *798* (mut. ex: totus), 943, 3002; [1751] (corr. in: totus). Cfr. totus,2.
operare: 1. عمل : 856. Cfr. intelligere. {*2522*}.
— 2. بدأ VIII: *562* (operari per ordinem).
operatio: 1. عمل : 3554 (عمل), 3557 (عمل فى حلّ). Cfr. opus.
— 2. - : 3481.
opponere (قبل III; ب , ه): *1864* (aliquid, cum); 1919 (ad, cum); 3156, 3201 (per, D.); 2118, (3285), 3522 (قابل ب = opponere in aliquo); (abs.) 3542. {*1310*}.
oppositio (مقابلة): *528*. Cfr. mukabale.
opus (عمل): 1256, [1302], 3530. Cfr. operatio,1.
ordinare: 1. وصف : 413 (= dicere et ordinare); 453 (=ordinare et exponere). Cfr. dicere,3.
— 2. خبر IV: 560 (ordinare et exponere); 520 (= memorare et ordinare).
ordo: 1. مثال : 813.
— 2. منزلة : *578*. Cfr. gradus.
— 3. Cfr. operare,2.
— 4. - : [2193].
ostendere: 1. رأى IV: 890.
— 2. - : 322.
Pannus (ثوب): 1741-1786.
pars: 1. قِسم : 29 (mut.), 608, 1063, 1064, 1066, 1067, 1069,.
— 2. قِسم = قَسم : saepe, ut in 915 (ما خرج من القَسم); v. 1355, 1756, 1914, 3472. Cfr. divisio,3.
— 3. جزء : [2843].
— 4. عدد : 1440.
partiens (مقسوم عليه): 915. Cfr. dividens; numerus,3.
partitus, partitum (مقسوم): 915 (partitus) & 918 (numerus partitus) mut. in: divisus. Cfr. divisus,1; dividendus.
parvus: 1. صغير : 2218-2241 (شىء صغير = incognita; cfr. res,1); (minor = اصغر :) 1080, 1112, 1113, 1122, 1127, 1129, 1148, 1168, , 2599,.

— 2. قليل : 2167-2169, 2212-2217, 2231-2236, 2384, 2542-2546, 2558, 2567, 2609, 2610 (قليل = minor, minus); 85, 223, 240, 293, 307, 330, 338, 443, 1237, [1489], 1538, 1631, 2554-2555, 2603, (اقلّ); (minus esse:) قَلَّ : 39, 3217, 3277.
— 3. minus = ناقص : 47, 54.
— 4. (non minus neque plus = لا زيادة ولا نقصان): 225-6.
— 5. minus = الّا : 2653; 3199 (2ª manu). Cfr. diminutus,4.
pecunia (عدد = census, summa): *1575* (mut. ex: census).
percutere (ضرب): 232, *642* (mut. ex: ducere), 707, 1211, 1661. Cfr. ducere,3.
permutare: [1362].
pervenire: 1. جاء : 944; (- ad manum:) 797. Cfr. manus.
— 2. اتى : (- ad manum:) 239.
planus: 1. planum est: بيّن / تبيّن : 413, 489, 1940, 1993, 3308, 3356 / 1347. Cfr. manifestus,1 & 2; verus.
— 2. قائم : 3350. Cfr. rectus; stans.
plurimus, -um (كثير): 527; 1257.
plus: 1. اكثر : 85, 240, 259, 291, 309, 331, 405, 419, 443,.
— 2. كَثُرَ = plus esse: 39 (vel plus vel minus = ما كثر او قلّ), 3277.
— 3. زائد : 47, 54 (vel plus vel minus... = جميع ... زائدها وناقصها).
ponderare (pro المؤونة): 1258.
ponere: 1. جعل : 20, *130*, 163, 258, 292, 307, 309, 330, 331, 344,; 262, 296, 1209 (posuimus = لانّا جعلنا); 1752, *1939* (-). Cfr. facere,1; comunis,1.
— 2. واضع n. ag. (- انا : ponam): 529, 590, 602, 897, 919, 1058, 1546, 1591, 1638.
— 3. ذهب : [*729*]. Cfr. abiicere,1.
porta: 1. باب : 95, 100, 114, 162, 203, 209, 422, 425, 429, 435, 447, 496, 572, 577, 596, 637, 654, 670, 699,; 222 (mut. in: capitulum), *241* (mut. ex: capitulum), 593 (mut. in: [expositio], regula), 1680 (corr. ex: capitulum). Cfr. capitulum; expositio,3; regula,1; via,1.
— 2. قول : 3516. Cfr. dicere,1; dictio.
positio (-): 3231, 3245 (secundum positionem = ex hypothesi).
possibilis: 1. (possibile est:) يجوز : 565.
— 2. (possibile est:) يمكن : 352, 362, 3240.
— 3. (non est possibile:) ليس / لا يمكن / ليس يمكن : 360 / 361 / 967, 971. Cfr. impossibilis.
— 4. - : 1018, 1022, 1035.
post (بعد): 1285, 2914.
postea: 1. ثمّ : 1758, 1783, 1810, 1890, 1946, 2049, 2091,; (et postea:) 1537 (corr. ex: deinde), 1686, 1715, 1884, 1922, 1950, 1958, 1963, 2032, 2052,. Cfr. deinde,1.

— 2. 2901 (fac in eis secundum algebram, et erunt postea = تقابل بها فيكون بعد المقابلة).
posterior (آخر): 1516, 1517, 1523, 1525, 1537-8, 1540, 1543, 1543-4, [1550] (corr. in: postremus), 1572. Cfr. postremus; secundus; ultimus.
postmodum (-): 1550.
postquam: 1. اذ : 1520, 1567, 1611, 1660, 1961.
— 2. بعد / بعد ان : 1867, 1879, 1901, 1902, / 1451, 3186, 3278.
— 3. اذا : 3197, 3280, 3473.
— 4. - : 1161.
postremus: 1. ثانى : 1510. Cfr. posterior.
— 2. آخر : 1550 (e corr.), 1552-3, 1554, 1564, *1583*, 1584, 1585,.
precedere (قدم V): 2548. Cfr. premittere.
predictus (-): 340, 1026, 1120, 1145, 1218 (e.g.: tertia specierum 6 predictarum = الضرب الثالث من الضروب الستّة).
premittere: 1789 (fac sicut premisimus et exposuimus superius = تعمل ايضًا كنحو ما تقدّم من عملنا); 3321-2 (sicut premisimus et exposuimus = كما قد قدّمنا بيانه); 3436, 3557 (id quod premisi(mus) = ما (قد) تقدّم); 3515 (id quod premisimus = ما قد تقدّم بيانه). Cfr. precedere.
preponere: 1. قدم II: 3431.
— 2. (- et dicere:) بان II: 1423.
preter: 1. غير : 565; 3073 (دون غير = et non preter); 1254, 3056 (على غير); 2549 (preter hanc viam = بغير هذا العمل, mut. in: aliâ viâ). Cfr. cadere,1 (*preterquam*).
— 2. الّا : 1043, [1053].
pretium (قيمة): [1741]-1782.
primus (اوّل): 1 (s.n.); 241, 445, 879, 928 (-), 1063, 1077, 1466, 1481, 1489, 1491,.
principium (اوّل): 1431, 1698, 2470, 2567.
prior (اوّل): 1479, 1515, 1521, 1524, 1539, 1540, 1543, 1555,
prius (اوّلاً): *247*, 351, 532, 3226. Cfr. redire,1.
pro (بمنزلة): *580* (e corr.), 594, 596, 600, 601, *620*,. Cfr. gradus.
probatio (برهان): 3320, 3363, 3404; [662] & [681] corr. in: demonstratio, q.v.
producere: 1. - / كان : 140, 185, 264, 271, 299, 303, 392, 568, *571*, , 1499 (e corr.) / 135, 136-7, 574, 1071-2, 1511, 1518, 1556, 1565,; 922 (producit = مثل); 649 (productum = وهو).
— 2. pro مبلغ : 338 (quod producetur = مبلغ ذلك). Cfr. provenire,3.
— 3. جمع VIII: (producitur:) 581, 970, 1116 (mut.),; (producit:) 904, 904, 906 (e corr.), 906,; *2384* (corr. ex: agregari). Cfr. fieri,4.
— 4. equalia produci = سوى VIII: Cfr. equalis,2.

— 5. (- lineam) خرج IV: 153, 312, 468, *486* (corr. ex: facere), 506, 610, [627] (corr. in: protrahere), 683, 773 (mut. in: protrahere), 819-20, 847,. Cfr. protrahere.
productum: 1. مضروب : 1065.
— 2. ما اجتمع : 1636, *1864*, *2326* & 2384 (corr. ex: agregatum), 2468, 2772 (-), 2805, 3398, *3492* (mut. ex: agregatum). Cfr. agregatum.
— 3. ضرب : *2504* (mut. ex: multiplicatio), 2962 (productum ex medio in se = ضرب الاوسط فى مثله).
pro(h)icere: 1. سقط IV (من : de/ab/ex): 82, 105, 124, 196, 656, 731, 790, 792, 1142, 1419, 1420, 1442, / 764 / 233, 235, 238, 1137, 1177, 1263, 1378, 1624, 1633, 1662, 1824,; 996 (mut. in: subtrahere); 1867, 1879 (postquam proiecisti = بعد اسقاطك). Cfr. abiicere,2; diminuere,1; minuere,2; subtrahere,2.
— 2. لقى IV: 1653, 1695; 1837, 1862, 2243, 2251, 2272, (de); 1708, 1846, 2101, 2224, 2256, (ex); *2481*, 2883, (3040), 3116, 3147, 3173 (comuniter aliquid cum aliquo: ب; v. seq.).
— 3. ذهب IV (aliquid cum : ب ه): 761. Cfr. abiicere,1.
— 4. نقص : 1854. Cfr. minuere,1.
— 5. رفع : 1901, 1902 (postquam proiecisti, proieceris = بعد رفع).
proponere: [*3006*] (-).
proportio: 1. نسبة : 871, 871, 872, 872, 873, 873, 874, 875, 1365, 1365, 1367, 1367, 1368, 1368.
— 2. قدر (من , عند : ad): 905, 906-7, 907, 907, 927, 927, 953, 958, 961, 962, 1361, 1362, 1363, 1729-30, 1730, 1734, 3381 seqq. Cfr. esse,4; mensura.
— 3. (- est:) عدّ : Cfr. esse,4.
proportionalis (متناسب): 879.
proportionari: 1. عدل : 11; 35 (corr. in: equari), 199-200 (mut. in: equari). Cfr. equari.
— 2. عدل III: 57 (mut. in: equari).
— 3. proportionatur ad = هو منسوب الى : 8.
propterea (+ quod, quia: لانّ): 182, 283, 313-4, 358, 365, 385, 463, 470, 477,. Cfr. cum,B,1; quia.
prosperare (-): 1713.
protrahere (lineam: خرج IV): 627 (corr. ex: producere), 629, 682, 711, 712-3, *773* (mut. ex: producere), 825, 826 (-), 846, 848, 850, 864, 866, 985, 1003. Cfr. producere,5.
provenire: 1. - / كان : 158, 632, 687, 742, 745, 775, / 26, *79* (corr. ex: esse), 81, 97, 101, 102, 103, 116, , *1171* (mut. ex: esse), [1711] & [2657] & [2680] (corr. in: esse).
— 2. صار : 1690.
— 3. بلغ (ه : ad): 89-90, 110, [1542], 1589; 223 (quod proveniet = مبلغ ذلك ; v. producere,2). Cfr. agregare,2; devenire.

— 4. جمع VIII: 290, 327, 415, 417, 419, 420, 588, 602, 805, 808, 899, 902, 903,; 3428 (quod provenit = المجتمع). Cfr. fieri,4.
— 5. عاد : 1888. Cfr. redire,1.
— 6. خرج : *3466* & *3512* (mut. ex: exire).
punctum (نقطة): 143, *143*, 153, 267, 268, 299, 300,. Cfr. versus.
Quadratum, n.: 1. في مثله , في نفسه (ضرب); من ضرب : 395, 397, 407, 409, 1164, 1164, 1165, 1166, 1195 seqq, 1230 seqq,; 1370 (e corr.).
— 2. مربّع : 283, 375, 382, 502, 665, 709, 828, 830, 830,.
— 3. سطح : 23, 26, 261, 291, 328, [709], 844 (post.), 1106 (post.), 1224,.
— 4. سطح مربّع : 312, 823, 842, 844 (prius), 863, 985, 1003,.
— 5. مال : 806-810 (errore).
quadratus, adj. (مربّع): 20, 131, 152, 165, 259, 277, 293, 346, 375, 684, 692, 718. Cfr. superficies,2.
quadruplus, quadruplum (n.) (اربعة امثال): 824-5, 1124,.
quando: 1. اذا : 195, 538, 542, 549, 553, 555, 557, 879, 1329, 1333, 1346,.
— 2. متى ما , متى : 1917, 2004, 2306, 2974. Cfr. cum,B,2.
quantitas: 1. قدر : 3329 (errore).
— 2. secundum/per quantitatem = بمثل : 291, 328.
quantus,-um (كم): (adv.) 571, 576, 592, 595, 619, 636, 653, 669, 697,; (adj.) 164, *1001*, 2643, 2657. Cfr. tantum,2; vicis,2.
quare: 1. ف : 25, 409, 464, 480, 493, 516,. Cfr. ergo,1.
— 2. لِمَ : 19.
— 3. وهو : 3288.
quasi (كانّه , فكانّه , وكانّه): 893, 940 (-), *1962*, 2089, 2146, 2163, 2228, 2242, 2246, 2539,. Cfr. sicut,2.
quemadmodum: 1. كنحو ما : 444.
— 2. على ما : 406.
quesitus: 1. مطلوب : 1098, 1101, 1103, 1109, 1217, 1831, 1883, 1907.
— 2. مسؤول عنه : 1875.
— 3. - : 2425.
questio: 1. مسئلة : 18, 65, 74, 203, 224, 227, 290, 326, 330, 338, 344, 421, 424, 444, 446, 530, 1058, 1077, 1079, 1088, 1098, 1111, 1119, 1127, 1145, 1170, 1190, 1199, 1217, 1257, 1285, 1426, 1431, 1465, 1465, 1466, 1471, 1479, 1480, 1481, 1481, 1545, 1679, 1698, 1728, *1729*, 1771, 1786, 1964, 2264, 2470-1, 2540, 2548, 2567, 2622, 2771, 3555, 3556; (- :) 112, 337, 430, 497, 2299, 2359, 2382.
— 2. سؤال : 1044, 2157.
— 3. ما القي : 1047.
— 4. باب : [222].

qui, quod: 1. هو , الذى , ذلك , ما : 1, 2, 3,.
— 2. (id) quod est...ex/de = ما ... من : 4, 5, 6, 9, 40, 48-9, 55, 86,.
— 3. 15, 96, 101, 130, (est quod = ان = est ut); 201 (quod est = ومعناه , mut. in: id est; cfr. expositio,1); 560 (hoc est quod = انّما ; cfr. cogere).
— 4. اىّ , qui(s): 822, 890, 892, 937, 938, 966, 982, 1001,.
quia: (لانّ :) *134*, 277, 569, 752, 989, 1009, 1010, (cfr. propterea); 988 (بانّ); 1018 (prius; فانّ); 1558 (وانّما ; cfr. cogere).
quicquid: 2405 (كلّ شىء معك = quicquid habes). Cfr. totus,1.
quidam (بعض): 17*42*, 2267; (- in quosdam, - quibusdam = بعضها فى بعض, بعضها الى بعض :) 533, 538, 540-1, 544, 551, 553, ; 796. Cfr. aliquantus; aliquot.
quilibet: 1. كلّ, كلّ واحد : 10, 22, 204, 530, , 865, 1106,; (corr. in: unusquisque:) [579], [1681], [2868].
— 2. واحد : 1487 seqq; 1491-2 (quilibet primorum / secundorum = الواحد الاوّل / الثانى).
quinarius (- خمسة): 27 (mut.). Cfr. quincuplus.
quincuplus (- خمسة, خمسة امثال; alicuius / ad aliquid): 16, *27*, 28, 93, 110,.
quisque (كلّ , كلّ واحد): 1751, 1779, 2535, 2553.
quivis (كلّ): 920, 946.
quod: Cfr. qui.
quomodo (كيف): 537, 540.
quoniam (لانّ): 158; 3233 (-).
quotiens (- est A in B, totiens est unitas in G: ا يعدّ ب بقدر ما فى ج من الآحاد): 952-3.
Radix: 1. جذر = res incognita: 3, 3 (def.), 6, 8, 12, 13, 14, 17, 19 (post.), 35,; 794 (mut. ex: res); 2333 (corr. in: res), 2638 (mut. in: res). Cfr. res.
— 2. جذر = radix numeri / superficiei (quadratae): 15, 16, 16, 18, 19 (prius), 32, 36 (post.), / *23*, 26, 27, 160, 263, 274,; (radix = مأخوذ جذر :) 1041, 1044, (cfr. accipere). Radix radicis: 2890 seqq.
— 3. ضلع : *134* (radix superficiei; corr. ex: latus).
— 4. شىء : 603, 744, 795, 1076, 1134, 1174, 1463, 1528, 1530, 1577, 1674, 1707,; (corr. in: res:) 2282, 2293, 2334, 2713, 2900, 3046, 3275; (mut. in: res:) 681, 2767. Comp. 1709 (medietas radicum = نصف الاشياء) & 1710 (= نصف الاجذار).
ratio: 1. انّما : 2406 (et ratio quod hoc sit ita = وانّما صار هذا كذلك). Cfr. cogere.
— 2. صورة (?): 603 (v. 617 (-), 634 (causa = معنًى)).
— 3. شكل : [335].
rectitudo: 1. على استقامه = secundum rectitudinem: 820, 826, 827, 846, 848, 985, 986, 1003-4,; [1004] (corr. in: in directum); 153 (= in rectitudine).

— 2. في طوله = in rectitudine: 1163. Cfr. longitudo.
rectus: rectorum angulorum = قائم الزوايا : 664. Cfr. planus; stans.
redire: 1. عاد : 915, 1810, 1885, 1891, 1918, 2005, 2026, 2116, 2249, 2307; 1390-1 (redibit sicut fuit prius). Cfr. provenire,5.
— 2. رجع : 1431, 1698, 1945, 1978, 2500, 2567, 2614, 3016. Cfr. revertere.
— 3. كان اوّلاً : *917*.
— 4. جمع VIII: *1116*.
reducere: 1. ردّ (الى : ad): 39, 47, 85, 86, 87, 90, 91, 240, 443, 1095, 1291, 1527, 162*0*, 1671, 1702, 1804, 240*4*, 2440, *2584*, 2654, 2695, 271*6*, 2718, 2831, 2903, 2994, 3042, 3092, 3523, 3542; 55 (تُرَدّ...وكذلك تفعل = reduces...ita reducendo...).
— 2. كمل IV: 1415 (reduc...ad censum unum = اكمل...حتّى يكون مالاً كاملاً).
regula: 1. باب : *593*. Cfr. porta.
— 2. قياس : 1584. Cfr. exemplum,2.
— 3. - : *1255*, *1257*.
reiicere, reiectus: 1. لقى IV: 1561 (mut. in: sublatus).
— 2. - : [*1838*].
reintegrare: 1. كمل IV: 111, 1*457*, *1848*, 2068, 2120, 2512; 1*458* (اكمالك = reintegrabis). {2139}. Cfr. restaurare,3.
— 2. كان : (1942).
— 3. جبر : *1073*, 2310 (ب : [ad] *[cum] per*), 2312 & 2510 (cum). Cfr. restaurare,1.
reintegratio: 1. جبر : 2264, 2268.
— 2. اكمال : 1849, 2513.
reliquus (آخر : (unus...) reliquus): 66, 1179, 1181, 1267, 1270, 1396, 1421, 1432, 1436, 1440,. Cfr. alius.
remanere (بقى): 77, 84, 107, 118, 125, 154-5, 197, 211-2, 213, 233, 235,; *2482*, 2582 (كان).
res (شىء): 1. = x (sc. res quaesita): 533, 533, 534, 542, 543,; (def.:) 538, 569. Res magna & res parva (sc. res quaesitae x_1, x_2): 2216-2226, 2234-2241.
— 2. جذر : 794 (mut. in: radix), *2333* (corr. ex: radix), 2479, *2638* (mut. ex: radix).
— 3. = aliquid: 3514.
— 4. - : 3197 (postquam sic est res = اذا كان هذا كذلك).
reservare (حفظ): 1274.
residuum, n.: 1. الذى يبقى, ما بقى / باقٍ : 1077, 1914, *2484*, 2599, 3282 / 994, 1044, 1179, 1181, 1247, 1436, 1450, 1474, , 1887, 1900, 1902, 2603, 2629; 236 (بقى = habebis residuum, scilicet).
— 2. فضل : [1970].
residuus: 1. باقٍ : 149, 161; 321, 400 (يبقى), 1746 (ما بقى من).

— 2. آخر : 1253, 2130. Cfr. alius.
responsio (جواب): 1045, 1046, 3056.
restaurare: 1. جبر (ب : cum/per): 1072 (mut. in: reintegrare), 1117, 1132, 1174, 1243, 1251, 1285, 1289, *1453*, 1455, 1669, 1705, 1815, 1843, 2334, 2364, 2388, 2435, 2437, 2459, 2475, 2479, 2494, 2508, 2578, 2580, 2693, 2825, 2826, 2886, 2987, 3036, 3038, 3089, 3110, 3114, 3144, 3146, 3523, 3542; 1801 (restaura = اجبر وقابل). {*1310*}, {*1315*}, {*2792*}.
— 2. قبل III (ب : in): 1807 (mut. in: facere); (restaurare et apponere:) 1394, 1407, 1413-4, 1429-30, 1435, 1829, 1930; 1439 (restaurare et opponere).
— 3. كمل IV: 162*2*, 1764, 1818; [1457], [1458] (corr. in: reintegrare); 1848 (mut. in: reintegrare). Cfr. reintegrare,1.
— 4. عمل : *1767* (mut. ex: facere).
— 5. لقى IV: [2481] (corr. in: prohice comuniter).
— 6. ردّ : [2404].
restauratio (جبر): *527*, 1060, 3259. Cfr. geber.
revertere, reverti (رجع): 2002, 2023, 2470. Cfr. redire,2.
Scientia: 1. - : Cfr. ducere,1; exire,3; oculus.
— 2. معرفة : 534.
scilicet (in codice semper .*s*. vel *silicet* scriptum): 1. - / الذى هو, وهى , وهو , التى هى : 69, *179*, *180*, *183*, *206*, (*344*), *561*, *576*, 640, 6*79*, *707*, *1041*, *1283-4*, 1425, *1652*, *1652*, / 29, 56, 76, 132, 446, 1181, 12*76*, 1449,.
— 2. اعنى : 335.
scire: 1. علم : 91, 164, 221, 568, 822, 843, 915, 966, 982, 1000, 1025,; 1869, 2600, 3188, *3195*, 3224 (scitum est = معلوم); e.g.: 292, 1896 (iam scivisti = قد علمت); 91 (iam scis = قد علمنا), 2086 (iam scivimus = قد علمنا); 2682 (scias = اعلم).
— 2. نظر : 890 & 937 (scire prius); 938 (scire).
— 3. عرف : (1256).
— 4. 275 (si volueris scire scientia subiecta oculis = ان احببت ان نبيّن عياناً); 966, 976, 993 (ut sciamus, scias = حتّى); ducere ad sciendum: Cfr. ducere,1.
se, se ipsum/-a: 1. نفس : 4, 7, 145, 145, 147, 148, 159, 189, 190, (in - : فى نفسه , نفسها); 7 (per - : ب).
— 2. مثل (in - : فى مثله , مثلها): 74, 78, 81, 97, 101, 103, 116, 120,.
secties (ستّ مرّات): 1090, 1093 (sexcuplum: 1104).
secundum: 1. كنحو ، على نحو (ما) : 2028, 3263, 3557. Cfr. sicut,5.
— 2. بقدر : 1364 (e corr.).
— 3. على (ما) : 813, 856, 944, 1039, 1333, 2045, *2472*,. Cfr. rectitudo,1.
— 4. 291 (secundum quantitatem = بمثل); 339 (secundum quod = لانّ); 2157 (secundum hanc questionem = لانّ السؤال كذلك); 2454 (secundum viam aliam

= بالوجه الآخر).
secundus: 1. ثانی : 880, 1088, 1099.
— 2. آخر : 1494, 1494, 1503 (mut. in: ultimus). Cfr. posterior.
sed: 1. (- tantum:) انّما : 967.
— 2. (ف...) وامّا = sed (...est): 5, 229, 1188; 446 (فامّا).
— 3. و : 8, 119, 308, 398, 399, 400, 408, 410 (prius), 411,.
semel (مرّة): 1066, 1071, 1084, 1084.
semis (نصفًا): 1066, 1071, 1075, 1084, 1085, 1085.
semper (ابدًا): 17, 1356, 1359, 2194, 2531, 3189, 3291, [3313], 3344, 3518; (- :) 240, 14*25*.
separatus (مفرد): 539. Cfr. divisus,3.
servare (حفظ): 977, 979, 995, 996, 1027, 1029, 1032, 1448. Cfr. custodire; reservare.
si: 1. ان : 151, 214, 237, 571, 809, 965, 993, 1034, 1037,.
— 2. اذا : 109, 202, 551, 1000, 1372, [2145],.
— 3. لو : Cfr. dicere,1.
sic (كذلك): 39, 40, 47, 48, 126, (296), 579, 598, 599,. Cfr. etiam; facere,7; similiter,2.
sicut: 1. مثل : 14, 17, 18, 27, 37, 63, 110, 145,; 2746 (sicut ipse = مثله); 87 (مثل ما). Cfr. equalis,3.
— 2. ك : 35, 43, 45, 50, *51*, (cfr. dicere,1); 92 (et erit sicut = فكانّه ; cfr. quasi); 871, 872, 873, 875, 958, 1365 seqq, (- (proportio): كنسبة, كقدر).
— 3. كما (- dixit, exposuit, narravit): 146, 190, 270, 302, 354, 461, 513 (post.), 597,.
— 4. على ما : 631, 890, 892, 937, 2210, 2244, 2621, 3473, 3485, 3533.
— 5. نحو, (ما) على نحو ، كنحو : 241, 1063, 1771, 1789, 1807, 3286. Cfr. dicere,1; secundum,1.
— 6. بمنزلة : 1786. Cfr. gradus.
— 7. - : 895.
— 8. 262, 296 (لانّا جعلناه كذلك = sicut posuimus (ipsum)); 878 (وقد بيّن ذلك اقليدس = sicut exposuit Euclides).
significatum (مسمًّى): *570* (mut. ex: modus).
signum (اشارة): 3295 (ex via signi). Cfr. innuitio.
similis (شبيه / (- esse:) شبه IV): 1480, 1680 / 1679 (ما اشبهها = similes istis); 601 (-), 3253 (- ; v. forma,2). Cfr. consimilis.
similiter: 1. ك : 53 (similiter item).
— 2. كذلك : 85, 587, 597, 812, 1012,; 972 (similiter etiam accidit). Cfr. sic.
— 3. ايضًا : 701, 905; 1050, 1755 (similiter etiam). Cfr. etiam,1.
— 4. - : 610, 650, 775, 1765.
simul: 84, 106, 125, 1144 (que simul (sunt) = وذلك); *2968* (-); cfr. iungi,2.

singillatim (كلّ واحد): *10* (e corr.).
singularis (-): 3555-6.
sive (او): 240, 240, 569. Cfr. aut.
solvere (-): 1744.
solutio: 1. - : 308.
— 2. قياس : *1490*. Cfr. expositio,2.
species: 1. ضرب (species aequationis): 2, 10, 56, 337, 337, 414, 415, 520, 530, 530, 1059, 1078, 1099, 1120, 1145, 1190, 1217, 1217, 2265, 2466.
— 2. جنس (potestas incognitae / species): 582, 582, 583, 589 / 1046.
— 3. نوع (species computationis): 798, 856, 914, 944, 945.
stans (قائم): 7. Cfr. planus; rectus.
sub (دون): 5.
subiectus: Cfr. oculus.
sublatus: 1. منقوص (من : de, ex): *2520* (e corr.), 2565, 2569, 2573, 2575, 2598, 2619, 2698, 2726, 2797, 3049, 3051, 3055, 3061, 3096, 3097,; [*2572*] & [*2576*], mut. ex: diminutus; [2596] (انقص). Cfr. diminutus,1; subtractus,2.
— 2. ناقص : *3167* (e mut.). Cfr. diminutus,2.
— 3. لقى IV: *388* (corr. ex: abiectus), *1561* (mut. ex: reiectus).
subtractio (نقصان): 972, 1051.
subtractus: 1. الّا : 997, 1032, 1044, 1052, 1129,. Cfr. diminutus.
— 2. منقوص : 2452, [*2614*], 2620, 2744, 3001, 3100, (3102), 3546. Cfr. diminutus,1; sublatus,1.
subtrahere: 1. نقص (من : de): 993, 999, 1031, 1042, 25*96*, 3205. Cfr. minuere,1.
— 2. سقط IV: *996*, 1474. Cfr. prohicere,1.
sufficere (مكن IV): 796.
sufficientia (كفاية): 914, 1056.
summa (جملة): 566, v. modus.
superficies: 1. سطح : 20, 23, 26, 26, 28, 28, 30, 33, 131, 133,.
— 2. مربّع = quadrata superficies: 178 (mut.).
superfluum: 1. فضل (بين : alicuius super aliquo, inter aliquid et aliquid): 1475, 1493, 1507, 1509, 1515, 1524, 1541, 1542, 1553, 1556,; *1970* (corr. ex: residuum). Cfr. differentia,1.
— 2. زيادة : 1584, 1612.
superior, superius: 1. الذى قبل : 1465, 1964.
— 2. فيما تقدّم : 2151. Cfr. premittere.
supra: 1. فوق : 4.
— 2. قبل : 1334.
Tamen: 1. قد + impf.: 565.
— 2. الّا ان : 566.
— 3. (-): [1477], 1743.

tantum: 1. sed...tantum = انّما : 967. Cfr. cogere.
— 2. tantum...quantum: 3234. {1153-1158}.
terminatio (حدّ): 528-9 (errore). Cfr. continentia.
tollere: *2990* (2ª manu).
totiens: Cfr. quotiens.
totus: 1. كلّ : 36, 159, 443, 493, 634, 812, 855-6; (كلّ شىء معك = totum quod habes, habemus:) 92, 1459, 1527, 1621, 1671, 1673, 1702, 1704, 1766,; (كلّ ما معك = totum quod habes:) 2410, 2832; (كلّ ما كان = totum quod erit:) 443. Cfr. omnis,1; quicquid.
— 2. جميع : 227, 255, *390*, 474, 479, 650 (= totum agregatum), 743, 798 (mut. in: omnis), 991, 1751 (corr. ex: omnis),; 1293 (جميع ما معك = totum quod habes). Cfr. omnis,2.
— 3. - : 297, 306, 325, 410, 465, 481, 518,.
trahere (خرج IV; الى : ad): 1190, 1217. Cfr. ducere,1; exire,3.
tractatus (مقالة = βιβλίον Elementorum): 146, 190, 270, 302, 355, 462, 513.
tunc: 1. حينئذ : 246, 543.
— 2. ف : 552. Cfr. ergo,1.
Ultimus: 1. آخر, sc. secundus: *1503* (e corr.), 1505, 1506, 1518, 1526, 1573, 1721,. Cfr. posterior.
— 2. ثانى : 1508.
unde: 1. و : 326 (و...ايضًا = unde...iterum).
— 2. وانّما : 1045. Cfr. cogere.
— 3. فان : [571].
unitas: 1. أحد : 4, 175, 180, 37*6*, 881 (bis), 883,; 624-5 (id quod est in ea de unitatibus = ما فيه من الآحاد), 880-1.
— 2. واحد : *927* (corr. ex: unum), 928, 929, 953, *956* & *956* (e mut. & corr.), 958,.
universalis (جامع): *897-8*, *919-20*, *945*. Cfr. comunis,2.
unus, unum: 1. (n., adj.) واحد : 4 (prius), 22, 25, 3*2*, 39, 45, 46, 47, 54, 86, 90, ,927 (mut. in: unitas), 2564 (unus); (كلّ واحد:) 1505, 1593 (e corr.), 1598; 1507-8, 1510, (الواحد الاوّل...الواحد الثانى / الآخر) = unus pri(m)orum...unus posteriorum (postremorum, secundorum, ultimorum)).
— 2. احد : 4 (post.), 6, 6, 9, , 376 (corr. in: unitates).
— 3. درهم : 1356 (mut.), 1425.
— 4. تامّ : 1765 (ut sit census unus = حتّى يصير مالًا تامًّا).
— 5. 601-2 (unum in aliud = بعضها فى بعض).
unusquisque (واحد / كلّ واحد / كلّ): 579, 601, 601, 631, 1237, 1746-7, 17*48*, / 614, 614, 12*68*, 1321, 1323, 13*50*, / 15*36*, 1537.
usque ad (الى): 313, 827 (= ad, v. 826).

ut: 72, 115, 372, 404, (est ut = est quod = ان (هو)).
— 2. حتّى : 966, 971, 975, 993, 1017, 1073, 1819, 1887, 2406, 2513.
— 3. - : 3018.
uterque: 1. هما جميعًا : 67.
— 2. كلّ واحد : 1263, *1280*, 1281, 1283,.
uti (عمل): 2549.
Vel: 1. ولا : 8.
— 2. او : 85, 85, 85, 351, 533, 539, 539, 539, 799,. Cfr. aut.
velle: 1. راد IV: 111 (mut.), 164, 799, 801, 806, 809, 814, 819, 822,; (et hoc est quod voluimus exponere = وذلك ما اردنا ان نبيّن :) 33, 161, 198, 289, 329, 370, 401, 412, 495, 519, 618, 635, 651, 668, 696, 723, 756, 783, 832, 855, 886, 913, 930, 964, 992, 1016, 1087, 1110, 1126, 1169, 1234, 1353, 1371, 1519, 1565, 1610, 3330, 3338, 3392, 3429; 1770 (et illud est quod voluimus exponere; v. 329); 1198 (وذلك ما اردنا بيانه); {*1343*}.
— 2. حبّ IV: 151, 275, 310, 1856, 1867, 2549, 2774, 3063, 3186.
— 3. شاء : 214, 237, 1180, 1254, [1298], 1372, 1430, 1430, 1437, 1627, 1724,.
venire (ad manum: جاء): 127, 856.
verbi gratia: 1. 87 (وهو مثل); 814 (مثال ذلك); 1443, [1490] (قياسه).
— 2. - : [61], 62, *200*, 585.
— 3. ثمّ (?): [587].
vero (و/ -): 7, 66, 158, 266, 552, 556, 691, / 767; (وامّا :) 435, 970.
versus: 351 (in alio puncto ab ipso versus ... vel versus = خارجًا منها (sc. نقطة) ... او داخلًا عنها ...); 404 (ex parte ... versus = داخلًا).
verum: 1. وامّا : 77, 129, 429.
— 2. ف : 1039.
— 3. و : 3232.
verus (صحيح): 3259, 3295; 3350 (verus et planus = صحيح قائم).
via: 1. باب : 65, 71, 77, 119, 129. Cfr. porta.
— 2. مثال : 944. Cfr. exemplum.
— 3. جهة : 1372, 2600 (viâ aliâ = على هذه الجهة الاخرى), 2776, 3139, 3181, 3204.
— 4. وجه : 1397, 1422, 1679, 1680, 1729, 1771, 1823, 2454, 2486.
— 5. سبيل : 561.
— 6. عمل : 2549, 2550.
— 7. فنّ : 2267.
— 8. طريق : 3259, 3295, 3305, 3308, 3351, 3370, 3556.
— 9. 3057 (et ducunt ad viam unam = وهما يودّيان الى شىء واحد); 3218 (secundum hanc viam = على ذلك).
vicis: 1. مرّة : 181, 375, 547, 1091, 1093, 1107.
— 2. امثال : 501 (تسعة امثال = 9 vicibus quantum est).

videlicet (-): *677*, *706*, 733, 734, 766,; 2432 (وهو).
videre (رأى): 562 (vidimus = رأينا), 631 (sicut vides = على ما ترى).
vir (-): 1748, *1758*.
vocare (سمى II): 1902, [3006].
Zoz, zuz (= quarta sicli; فلس): [2155]-2160. Cfr. bizantius.

Appendice

1. Introduction

La traduction latine de l'Algèbre d'Abū Kāmil contient —si l'on fait abstraction des deux lacunes majeures signalées par le traducteur lui-même— l'exposition de l'algèbre proprement dite et de ses applications à des polygones réguliers dans son intégralité. Ce n'est pas le cas de la dernière partie du traité originel (fol. 79ʳ-111ʳ du texte arabe conservé), laquelle traite d'abord des équations indéterminées de type diophantien puis présente, dès le fol. 95ʳ, des exemples de problèmes récréatifs. En effet, la traduction latine —à l'opposé de la traduction en hébreu— ne rapporte de cette partie que son tout début, puisqu'elle s'interrompt au milieu du deuxième problème (fol. 80ʳ,2 du texte arabe): le traducteur a cessé son travail au bas du fol. 97 *recto* du manuscrit de Paris.

L'édition du texte latin d'Abū Kāmil devait évidemment inclure ce petit fragment. En outre, l'introduction qu'on y trouve va permettre d'illustrer de fort belle manière le fait déjà mentionné que notre traducteur, s'il se sent assez à l'aise lorsqu'il s'agit d'un texte proprement mathématique, peine quelque peu lorsqu'il doit rendre une partie plus littéraire, un trait d'autant plus marqué ici qu'il paraît avoir singulièrement relâché ses efforts en cette fin de traduction. Nous pouvons même mettre de côté les euphémismes et affirmer sans ambages que sa version de l'introduction abonde en contresens au point d'en devenir çà et là incompréhensible voire burlesque. On dirait même qu'il ne lui chaut plus de savoir ce qui a été étudié ou le sera: deux *manifestum est* sont ainsi censés annoncer le programme de la dernière partie. Nous voilà bien éloignés du travail consciencieux de la première partie. Il est vrai que son auteur eût peut-être remédié à ces déficiences s'il avait mené son travail de traduction et de révision plus avant.

Les fantaisies de la traduction latine nous ont amené à la faire précéder d'une version française de l'introduction. Comme dans ce qui précède, les parenthèses angulaires enfermeront des parties omises par le traducteur en latin; quant aux chiffres entre parenthèses rondes, ils renvoient aux propositions de la deuxième partie (voir l'étude de M. Lorch).

«Au nom de Dieu clément et miséricordieux. ⟨Il n'y a de force et de puissance qu'en Lui.⟩

Nous en avons terminé avec ce qui fut en grande partie inaccessible aux mathématiciens de notre connaissance appartenant aux gens de notre génération ainsi qu'aux anciens sur lesquels des informations sont parvenues jusqu'à nous, à savoir: la recherche des cordes du pentagone et du décagone qu'un cercle connu contient respectivement (1, 2) ou qui eux-mêmes contiennent un cercle connu (3, 4); la recherche du diamètre du cercle qui contient un pentagone ou un décagone connus (5 & 7, 9) ou que contiennent un pentagone ou un décagone connus (6 & 8, 10); la recherche de la grandeur de la corde de la quinzième partie de la circonférence d'un cercle connu (11); la recherche de la grandeur du côté du pentagone et du décagone équilatéraux et équiangulaires de surfaces connues (16, 17), et des côtés de triangles de surface connue qui sont dans un pentagone ou un décagone équilatéral et équiangulaire (18-20); d'autres choses (encore), que nous avons exposées ⟨en suffisance⟩ dans notre présent livre.

Nous allons à présent exposer nombre de problèmes qui ne sont pas déterminés et que certains mathématiciens appellent fluides —ce qui signifie qu'ils trouvent leur issue en de nombreuses voies— à l'aide d'un raisonnement probant et d'une théorie claire. D'aucuns sont discutés par les mathématiciens qui y font usage de méthodes dépourvues d'un fondement solide; d'autres ont été résolus par moi à l'aide d'un principe correct et d'une démarche facile d'un grand profit. Loué soit Dieu qui en a facilité l'accès, Lui qui connaît le tréfonds de la pensée. Nous exposerons encore bien d'autres choses qu'ont présentées les mathématiciens dans leurs ouvrages et qu'ils traitent par les méthodes de l'algèbre, en telle manière que celui qui les lit et y réfléchit les comprenne correctement ⟨sans qu'il ne fasse que répéter servilement et suivre aveuglément l'auteur. Dieu en sera témoin.⟩».

2. Texte latin (fol. 97^r,6 - 97^r,bas)

N.B.: Les règles d'édition du texte sont les mêmes que celles qui ont été utilisées précédemment, la seule différence étant que les corrections à la traduction sont placées dans des notes en bas de page.

In nomine Pii et Misericordis. Iam applicuimus ad[1] inventionem illius quod erat longinquum multum ad sciendum[2]. Et erat longinqu⟨u⟩m in oculis illorum quos vidimus ex compotistis de hominibus nostre generationis, et [etiam] quorum famam audivimus ex antiquis, de corda pentagoni et decagoni circumdantium circulum *notum* in unoquoque eorum vel circumdatorum a

[1] Le texte a قد انتهينا من = *ad finem pervenimus*, et non قد انتهينا الى.

[2] استخراج ما كان يتعذر كثير منه; une expression semblable, ما كان استخراج كثير منه يتعذر, est rendue avec encore plus de fantaisie dans l'introduction de la partie précédente (voir lignes 14-15 du texte de M. Lorch).

circulo noto, sive de inventione diametri circuli circumdantis in pentagono vel decagono notis vel circumdati ab ipsis notis, et de inventione quantitatis corde quindecagoni in circulo noto, et de inventione quantitatis lateris pentagoni et decagoni equilaterorum et equiangulorum per mensuram eorum notam, et laterum triangulorum notorum mensura quando erunt in pe⟨n⟩tagono et decagono equilateris et equiangulis, et aliorum que exposuimus in libro nostro isto[3].

Et manifestum est nunc[4] multum ex questionibus que non sunt terminate, quas nominant quidam compotiste «questiones»[5]. Cuius expositio est quia est inventio viarum multarum[6] in [d-] gravitate dubiosa. Et intellige⟨n⟩tia manifesta[7] ex eis est id quod regirant [quod] compotiste in portis[8] preter causam stantem **immobilem, vel: rectam** et operantur in ea. Et ex illis *est* quod inveniunt[9] per radicem veram et causam rectam[10] multe utilitatis. Et laus sit [g-] Creatori qui iuvit **vel: causavit** nos in hoc, quia Ipse novit secreta. Et manifestum est etiam[11] [va-] satis ex eo quod perscrutati sunt compotiste in libris eorum et *quod* fecerunt illud in portis [in] restaurationis et oppositionis, propterea quod intelliga[n]t illud legens illud et intelligens in eo animadver[sio-]tentia vera[12].

Si dicemus tibi: Census est cui[us] radix[*ci*] ⟨est⟩ *et* add[itur]*e* super eum

---

[3]على ما فيه كفاية, ajoute l'arabe.

[4]Pour: *exponemus nunc* (وانّا نبيّن الآن); une confusion entre نبيّن et تبيّن apparaît aussi dans le texte de la première partie (ligne 3370).

[5]سيّالة = *fluidae, fluxiles*, de la racine سال et non سأل. Le traducteur ne pouvait connaître ce terme, tout à fait particulier, qualifiant en mathématique les problèmes indéterminés. M. O. Naises nous a obligeamment signalé que le même terme désignait, depuis le IXe siècle (ap. J.-C.), une ouverture fort prisée du jeu d'échecs.

[6]اعنى بها تخرج بصوابات كثيرة; le terme précédemment mentionné signifie donc que les problèmes indéterminés ont une multitude de solutions numériques —de fait une infinité, comme cela apparaîtra au début du premier problème (v. n. 14).

[7]بقياس مقنع ومذهب واضح. La traduction est incompréhensible, tant à cause de la coupure inappropriée de la phrase que de l'interprétation des deux premiers termes. Le traducteur a-t-il lu ثقل مقنّع? Ou interprété le premier terme comme اِلتباس?

[8]On a peine à imaginer ce qu'un lecteur, mis devant la phrase *id quod regirant compotiste in portis*, y verrait comme signification ou comme métaphore. Si le sens de *porta* est devenu familier à qui a lu la première partie, le verbe *regyrare* est un nouveau venu. L'arabe est ما يدور بين الحسّاب بالابواب, et l'origine de l'interprétation apparaît lorsque l'on sait que le sens principal de deux formes de دور est celui de *convertere*, alors qu'une a aussi le sens dérivé abstrait (comp. *conversare*) de *disputare*. N'accablons pas notre traducteur: il ne manque pas toujours de vocabulaire, comme le montre son choix subséquent de possibilités pour قائم.

[9]استخرجته, *inveni (illud)*.

[10]حيلة سهلة, plutôt *artem facilem*.

[11]ونبيّن ايضًا, cf. n. 4.

[12]ليفهمه مَن قرأه ونظر فيه فهمًا صحيحًا. Comme traduction mot à mot, on ne pourrait guère faire mieux.

5 dragmas et erit illi radix, [sicut] *quantus*[13] *est* cens[*ui*]us? Hec questio non est terminata, et exit ad eam ex viis quarum non narratur plurimum illarum[14]. Et nos complebimus in[15] duabus [rebus] viis ex ea et quomodo est opus in ea, propterea ut sit tibi leve questio[16] illius quod voluisti *i[d]llud* ex viis, cum adiutorio Dei. Et causa in inventione ⟨unius⟩ duarum viarum in hac questione est quod facias censum tuum censum; et illi est radix, que est res. Cui adde 5, et erit census et 5 dragme. Et necesse est quod sit illi radix. Et iam scivisti quia **(A79^{v})** radix eius est maior re, quia census solius radix est res. Fac ⟨ergo⟩ radicem eius rem et numerum, et erit ex numero id quod si duces in se erit minus ex dragmis que sunt cum censu, que in hac questione sunt 5 dragme. Faciamus illud [d-] rem et dragmam. Et ducamus e[a]um in se, et provenie⟨n⟩t census et dragma et 2 res, equales censui et 5 dragmis. Prohice censum et dragmam ex censu et 5 dragmis, et remanebunt 2 res, equales 4 rebus (sic). Igitur res equatur 2. Et hec est radix census, et census est 4. Cum ergo addes ei 5 dragmas, erunt 9 dragme, quorum radix est 3. Et si facies radicem eius rem et 2 dragmas, et duces eam in se, provenie(n)t census et 4 res et 4 dragme, equales censui et 5 dragmis. Prohice ergo censum et 4 ex censu et 5 dragmis, et remaneb(un)t 4 res, equales dragme. Ergo res equatur $\frac{1}{4}$ dragme. Que est radix census, et census est $\frac{e\ 1}{8\ 2}$. Et si addes illi 5 dragmas, erunt 5 dragme et $\frac{e\ 1}{8\ 2}$, cuius radix est 2 dragme et $\frac{1}{4}$. Et si [addes illi 5 dragmas] si (sic) volueris, fac radicem census rem et $\frac{1}{2}$ dragme, vel rem et dragmam et $\frac{1}{2}$, vel rem et dragmam et $\frac{1}{3}$, vel id quod volueris ex numeris dum modo sit quod proveniet ex multiplicatione eius quod addis ⟨ad⟩ rem in se minus 5. Et similiter fac totum quod ad te perveniet ex hac specie secundum hoc exemplum.

Et si dicemus tibi: Est census, cui est radix, ex quo minui 10 dragmas, et fuit residuo radix. Hec questio etiam est indeterminata. Et iam scivisti quia radix census est res et quia radix census diminutis 10 dragmis est minus re. Fac ergo illam rem diminuta dragma, vel diminutis 2 dragmis, vel diminuto eo quod volueris ex numero vel fractionibus vel numero et fractionibus. Et fac illud rem diminuta dragma. Et duc eum in se, et provenient census et dragma diminutis 2 rebus, equales censui diminutis 10 dragmis. Restaura censum cum 10 dragmis, et adde illas ad censum ⟨et dragmam⟩ diminutis 2 rebus, et erunt census et 11 dragme diminutis 2 rebus, equales censui. Restaura item eum *cum* 2 rebus, et adde illas ad censum, et prohice censum ex censu; et remaneb(un)t 11 dragme, equales 2 rebus. Ergo res equatur 5 et $\frac{1}{2}$. Que est radix census, et census est **(A80^{r})** 30 et $\frac{1}{4}$. Si minues ex eo 10 dragmas, remanebunt 20 dragme et $\frac{1}{4}$. Quorum radix est 4 et $\frac{1}{2}$. Et si facies

---

[13] *sicut* = كما, *quantus* = كم.

[14] ما لا يحصى كثرة, (*vie*) *multitudine innumerabiles*; le texte a bien *narratur* et pas *numeratur*.

[15] نخبر, pas نجبر, donc *trademus* ou *notificabimus*. Parmi l'infinité de manières de donner à l'expression x^2+5 une valeur carrée, Abū Kāmil va en choisir deux.

[16] طلب, donc *questio* ici dans le sens de *investigatio*.

radicem census tui diminutis 10 dragmis (**Hic desinit**).

3. Glossaire

Nous omettons de cette liste les termes géométriques déjà mentionnés par M. Lorch, la plupart des mots couramment trouvés dans la première partie, ainsi que certains des mots rencontrés dans les notes.

Antiqui: متقدّمون
audivit: انتهى اليه
causare (iuvare): سهل II
compotista: حاسب
dum modo: بعد ان
facere in: عمل ب
fama: خبر
fractio: كسر
generatio: عصر
homines: اهل
immobilis (rectus, stans): قائم
indeterminatus = non terminatus
intelligentia: فهم, مذهب
intelligere: فهم
inventio: استخراج
iuvare (causare): سهل II
leve esse alicui: سهل على
nominare: سمى II
notus: معلوم
nunc: الآن
operare in: عمل على
opus: عمل
perscrutare: رسم
pervenire: جاء
porta: باب
prohicere: لقى IV
quantitas: قدر
questio: مسئلة ; طلب
quidam (pl.): بعض
radix: اصل
rectus (immobilis, stans): قائم
restaurare cum: جبر ب
secreta: ذات الصدور

solus: وحده
species: نوع
stans (immobilis, rectus): قائم
(non) terminatus: (غير) محدود
utilitas: منفعة
verus: صحيح
via: صواب
videre: شهد III

Some 14th and 15th Century Texts on Practical Geometry

A. Simi and L. Toti Rigatelli

Until recently the main goal of the research carried out by the Centro Studi della Matematica Medioevale (Mediaeval Mathematics Study Centre) at the University of Siena was the reconstruction of the history of algebra in the period between the 13th – 16th centuries. Only very recently has the focus shifted to the study of practical geometry ([2] and [25]). Thus the authors are far from having an overall picture of this discipline and in this paper will limit themselves to giving some information regarding the contents of some 14th and 15th century texts they have examined. The reader should consult W. Van Egmond's catalogue ([27]) for morphological descriptions of the codices cited.

As far as we know today, there were few treatises dedicated entirely to practical geometry, while it is not unusual to find a chapter on geometrical problems in treatises on the abacus. Of the manuscripts dedicated exclusively to geometry we have examined the following:

1) Ms. 205. I in the University of Bologna Library (d. 1430);
2) L. IV. 18 in the Town Library of Siena (c. 1460);
3) Palat. 577 in the National Library of Florence (c. 1460);
4) Moreni 130 in the Riccardiana Library of Florence (c. 1464).

* * *

The first codex is entitled *Geometria, vulgarmente arte de mexura*. The author, anonymous, was almost certainly a master of the abacus from Northern Italy, to judge by the dialect he uses in the text. The only exact geographic reference made is to the city of Ferrara, on f. 29r. The treatise deals with calculating the area of plane figures with the idea of applying the results to the measurement of agricultural land (*mexurare terre*). The plane figures in the texts are always considered as composed of six basic figures. These figures are:

a) the circle (*tondo*);
b) the semicircle (*mezzotondo*);
c) the segment of a circle (*forma di tondo*);
d) the square (*quadro*);
e) the rectangle (*forma di quadro or quadrangolo simplize*);
f) the right-angled triangle (*triangolo simplize*).

Interestingly, the author gives three different formulas for the area of the circle:

$$A_1 = d^2 \cdot 11/14 \qquad A_2 = d^2 - 3/14 \cdot d^2 \qquad A_3 = d \cdot C \cdot 1/4$$

where d represents the diameter of the circle and C its circumference.

The author proposes completely original formulas for calculating the area of the segment and the arc of a circle. Indicating the chord with c and the sagitta with s, the formulas are, respectively:

$$A = \{[c - 3/14 \cdot (c - 2s)] \cdot s\} \cdot 11/14$$

$$a = \pi/3 \cdot (c + 2s^2/c) \ .$$

The first gives a result slightly less than the correct one, and the second slightly in excess of it, although both are very near the mark.

The part between ff. 36v. – 43 r., on the subdivision of triangles, trapezia and quadrilaterals, is directly inspired by the fourth section of Leonardo Pisano's *Practica Geometriae* (1220). Here the author demonstrates a fine algebraic ability ([13]).

In the second part of the manuscript, on units of measure, the author uses those then in use in the cities of Ferrara and Bologna. It should be pointed out that the units of measure for surfaces are made to correspond (for practical purposes) to units of capacity for grain. For example, 100 square perches are considered equivalent to one and half bushels of wheat.

The codex has not been transcribed.

* * *

The second manuscript, also anonymous, is entitled *Trattato di geometria pratica*. This is rather a long text which, unlike the others, begins with an extensive theoretical introduction before going on to present specific problems. The author begins by defining point, line, segment, surface, plane and angle, and then goes on to define the principal plane and solid figures. Some lemmata on proportion precede the formulas for calculating the area of plane surfaces. Among the properties of proportion, let us look at the following:

> If four numbers are in continuous proportion, the product of the square of the first by the fourth is equal to the cube of the second and the product of the square of the fourth by the first is equal to the cube of the third.

The author then goes on to some propositions on the triangle: he enunciates Heron's theorem and makes use of the theorems of Pythagoras and Euclid without explicitly citing these authors.

Furthermore, in quite a number of problems he uses a formula which, as the reader will recognize, expresses Carnot's theorem, which is also present in Euclid's *Elements*. Given that the triangle is ABC and D the projection of C on AB the author claims that:

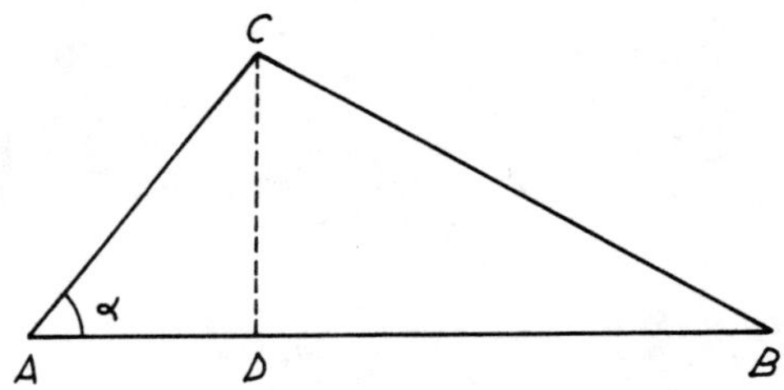

Fig. 1, $AD = (AB^2 + AC^2 - BC^2)/2AB$

This is followed by twelve different methods for calculating the area of a circle. The particular case of a circle with a diameter of seven is considered; all twelve formulas are always valid only in this latter case. Usually at least some of the formulas are not valid. The following list includes the formulas which are not always valid:

1) $A = d^2 \cdot 11/14$
2) $A = (C \cdot d)/4$
3) $A = C/2 \cdot d/2$
4) $A = C/4 \cdot d$
5) $A = C^2 \cdot d/88$ not always valid
6) $A = C^2/(12 + 4/7)$
7) $A = 3/4 \cdot C + C$ not always valid
8) $A = d \cdot 11/14 \cdot d$
9) $A = d^2/2 \cdot d \cdot C/2$
10) $A = [(C/4)^2 : C/2] \cdot 3 + (C/4)^2$ not always valid
11) $A = d^2 - d^2/7 - d^2/14$
12) $A = (C/4)^2 + 3/11 \cdot (C/4)^2$

(where A, C and d are the area, the circumference and the diameter of the circle respectively.)

We remind the reader that the inscription of three circles in a triangle, each one being tangent to the other two and to two sides of the triangle, is known as Malfatti's problem (named after the Italian mathematician who, in 1803, solved it) ([19]). Although the earliest formulation of the problem known to us can be found in a late-14th century codex, which we will discuss further on, the first correct solution to the problem, for the particular case of an equilateral triangle, appears in the Ms. L. IV. 18.

The author begins by examining the problem which has been formulated in the following manner:

> Given a triangle with sides 6, $\sqrt{12}$ and $\sqrt{48}$, work out the length of the radius of the semicircle which can be inscribed in the triangle.

Before solving the problem, the author states that it derives from another problem which, given an equilateral triangle with sides of 12 yards, demands the inscription of three circles (with the aforementioned conditions).

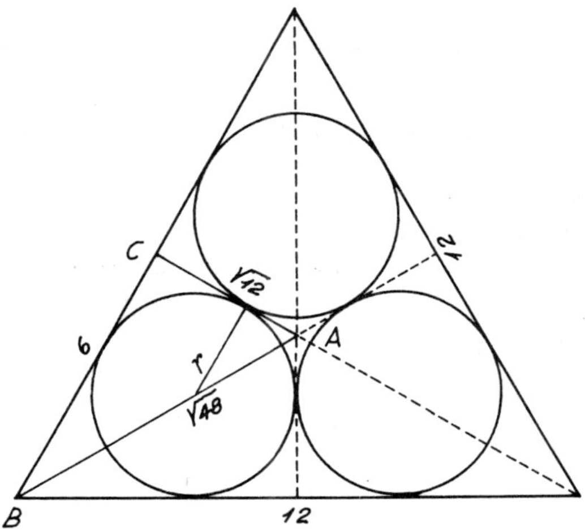

Fig. 2

The rather original steps which the author uses in solving the problem are as follows:

a) Firstly he observes that ... *sempre, nella magior faccia, capir magior tondo* ..., (the longest side will always be able to hold the greatest circle), therefore he takes the line AB as the base of the triangle.

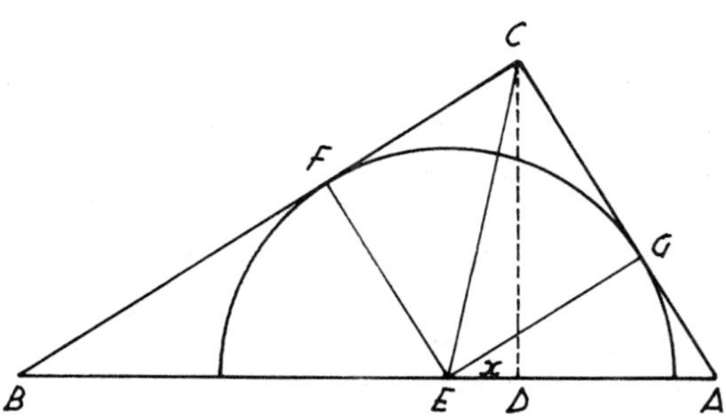

Fig. 3

b) By applying a formula equivalent to Pythagoras' generalized theorem to the triangle ABC, the author determines AD, where D is the foot of the height relative to the base AB. Then, by subtraction, he calculates $BD = AB - AD$.

c) Applying Pythagoras' theorem to the right-angled triangle ACD, he obtains the height CD.

d) Then he fixes a point E on the base AB, between D and B (because $AD < BD$), and puts $DE = x$. The aim is to determine DE in such a way that E becomes the centre of the required semicircle.

e) Applying Pythagoras' theorem to the right-angled triangle CDE, the author determines the length of the line CE as a function of x.

f) He then determines the height GE of the triangle ACE relative to the base AC by applying, first, Pythagoras' generalized theorem to the triangle ACE, with the aim of calculating GC and AG, and then Pythagoras' theorem to the right-angled triangle CEG. Using the same method, he calculates the height EF of the triangle EBC, relative to the base BC.

g) Finally, seeing that the segments EF and EG are both radii of the required semicircle, the author puts $EF(x) = EG(x)$. This equality gives rise to an equation of the following type:

$$ax^2 + bx = c\ .$$

Among the many points worthy of note in the Codex L. IV. 18, of particular interest is the reference to the golden section of a segment used to solve the problem of calculating the area of a regular pentagon of which the side is known. By tracing two diagonals of the polygon, both from the same vertex, the author dissects the figure into three isosceles triangles, two of which are of equal size. The author uses the concept of the golden section to determine the length of the diagonals.

To calculate the area of a nine-sided polygon inscribed in a circle, the author affirms that if the side of the polygon is 6, then the radius of the circle will be 9, and from this he deduces a general rule:

> The radius of a circle with a polygon of side 6 inscribed in it is equal to the number of sides of the polygon itself.

Clearly the rule is false, even if it gives, as the table below shows, approximate values, too large from the triangle to the pentagon and too small from the heptagon to the decagon.

	Exact	Approximate ($l = 6$)
Triangle	$r = l/\sqrt{3}$	$r = 3.46$
Square	$r = l/\sqrt{2}$	$r = 4.25$
Pentagon	$r = 2l/\sqrt{10 - 2\sqrt{5}}$	$r = 5.10$
Hexagon	$r = l$	$r = 6$
Heptagon	$r = l/2\sin(25°43')$	$r = 6.97$
Octagon	$r = l/\sqrt{2 - \sqrt{2}}$	$r = 7.83$
Nonagon	$r = l/2\sin 20°$	$r = 8.82$
Decagon	$r = 2l/(\sqrt{5} - 1)$	$r = 9.75$.

In solid geometry, the following problem seems worthy of note:

> If we know that a well with a depth of n yards costs L_n lire, determine the cost L_m of another well with a depth of m yards, in the same ratio.

With this in mind, the author first determines the sum of all the natural numbers between 1 and n and 1 and m, using the formulas: $n(n+1)/2$ and $m(m+1)/2$. From the proportion $n(n+1)/2 : m(m+1)/2 = L_n : L_m$, he gets the value $L_m = 2L_n/[n(n+1)] \cdot m(m+1)/2$, where $2L_n/[n(n+1)]$ is the "ratio of the well", that is to say the cost of digging the first yard.

The practical nature of the treatise is stressed by the questions discussed between ff. 76v. and 81v. on the so-called *a ochio* measurements (based on observation). Among the authors who wrote on this subject in the 15th century, which was also treated in the seventh section of Fibonacci's *Practica Geometriae*, are the Florentine scholar Gratia de' Castellani and Leon Battista Alberti ([4], [13] and [14]). The first wrote a work in Latin entitled *De Visu*, an extract of which was translated into the Tuscan dialect and can be found in the Codex Ottoboniano 3307 in the Apostolic Library in the Vatican (c. 1465). The second wrote the short treatise *Ludi Rerum Mathematicarum* between 1450 and 1452; many of the problems discussed by Alberti deal with measurements based on observation.

Some of the instruments used by the author of Ms. L. IV. 18 are different from those used by Gratia de' Castellani and Leon Battista Alberti. These were the plumb-line, the quadrant and mirrors. Our author also made use of rods and the sun's shadow. The surveying methods were usually based on the similarity of triangles.

The plumb-line is used in a problem that involves measuring the height of a mountain. The instrument is made of a right-angled isosceles triangle with a thread hanging from the vertex of the right angle. If the hypotenuse is placed on the plane to be measured, if this is horizontal, the plumb-bob

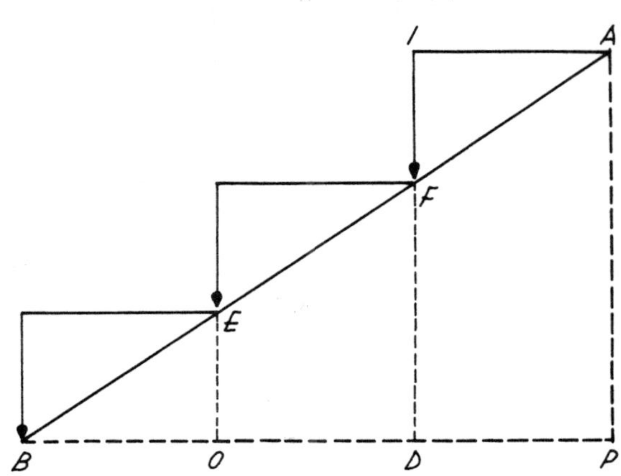

Fig. 4

will fall over its centre, while if the plane is not level it will not fall over the centre. Fig. 4 illustrates the procedure for measuring AP. The method consist of taking a cane AI, of length l, with a thread hanging from its end I.

The plumb-line is used to place the cane in a horizontal position. The length of the thread must be adjusted each time, so that the plumb-bob touches the segment AB at a point F; after moving the cane to F, the procedure is repeated. It will be repeated for the number of times n necessary to reach the foot B of the mountain; since $BP = n \cdot l$, Pythagoras' theorem can be applied.

There are 11 problems concerning measurements made using the quadrant. This instrument is made up of two rigid rods of the same length, AO and AB, which limit a circular sector of 90°. On the rod OB, at O and B, are two holes through which the object whose height or distance is to be measured can be sighted. The arc AB is subdivided in degrees as in Figure 5.

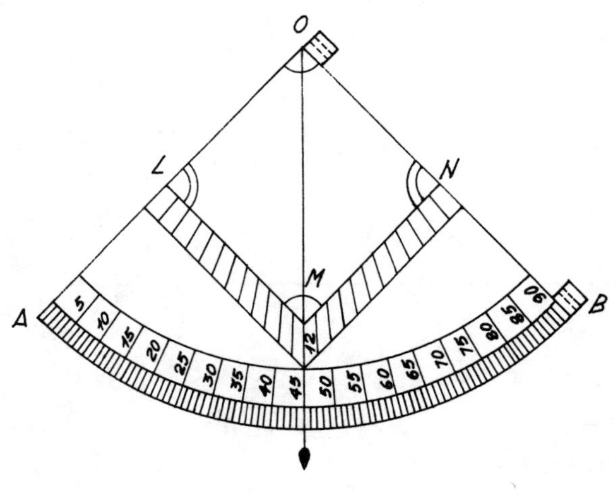

Fig. 5

The thread is attached to O. Inside the sector there is a set-square LMN, mounted in such a way that $L = N = 90°$; and each of the sides is divided into 12 sections. The so-called *recta* shadow is measured from L to M (as will be discussed in more detail later) and the *versa* shadow from M to N. The shadow produced by an opaque body placed perpendicularly to the horizon was called a *recta* shadow, while the shadow projected onto an ideal surface perpendicular to the horizon was called a *versa* shadow.

The problem regarding the measurement of the height of an object $A'B'$ (obtained by using the *recta* shadow scale) seems to be sufficiently clear in the use of the quadrant.

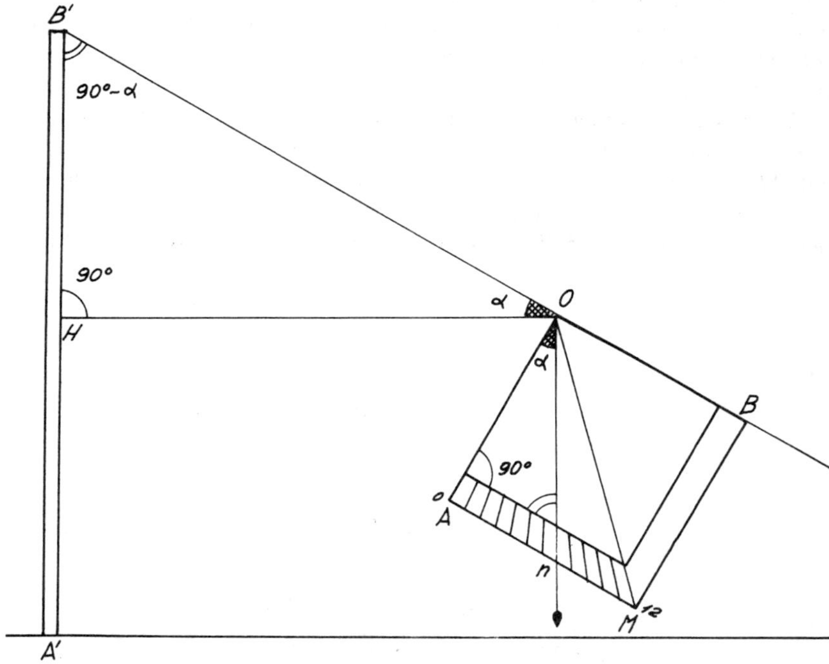

Fig. 6

In figure 6 the observer's eye is at B and the length of the segment OB is negligible with respect to the length of BB'. The dimensions have been changed to clarify the use of the instrument; thus the observer's height is $A'H$. Hence the following proportion is valid:
$B'H : AN = OH : AO$, or, that is, $B'H : AN = OH : 12$, thus $B'H = OH \cdot n/12$, where OH is the observer–object distance.

Having solved some even more complex problems, the author thinks it opportune to offer further clarifications regarding the concepts of *recta* and *versa* shadows and with the aim of determining for both the two subdivisions in twelve parts, he recommends taking the points of the shadow every hour. He goes on to say that it is very important to know how to go from the *recta* shadow to the corresponding *versa* shadow and vice versa. Supposing that the *recta* shadow is oR and the corresponding *versa* shadow is oV, both can be determined by the proportion:

$AB : OC = AO : EC$, that is, $oR : 12 = 12 : oV$, thus $oR = 144/oV$ and $oV = 144/oR$.

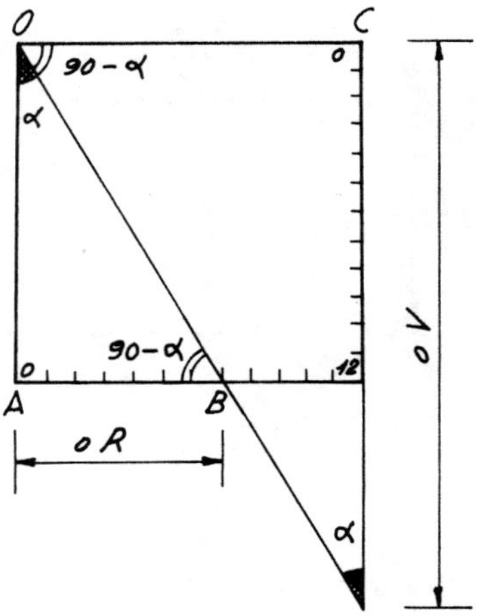

Fig. 7

Regarding the use of mirrors we have only provided illustrative diagrams (lacking in the author's text), but not the solutions to the problems based on the similarity of triangles.

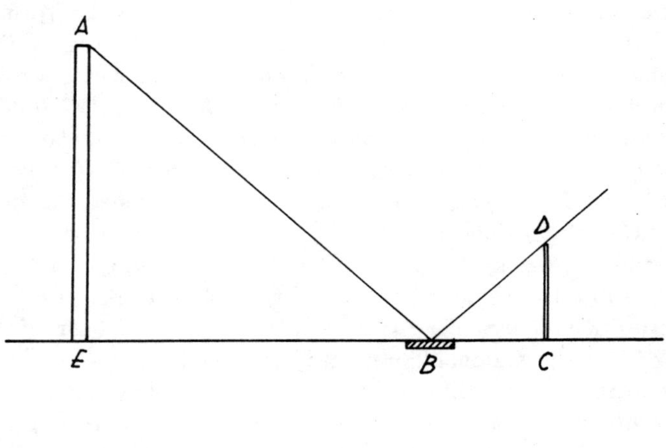

Fig. 8

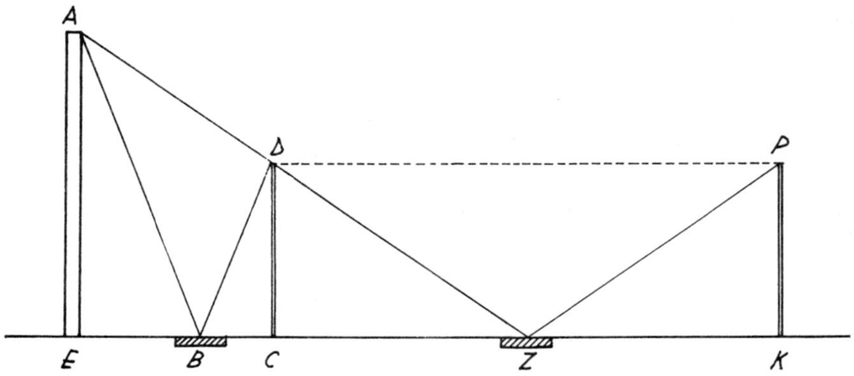

Fig. 9

Figure 8 refers to the problem of determining the height of AE, with the observer's eye at D and the distances BE and BC known. Figure 9 refers to the case where it is not possible to measure the distance between the mirror and the object.

The codex was transcribed by Monica Sani and Annalisa Simi ([20] and [22]).

* * *

The Palatino Codex 577 is of particular interest. The first part, from sheet 1r. to f. 241r., contains a *Tractato di praticha di geometria*, by an anonymous author. According to Ettore Picutti, it was written by Maestro Benedetto da Firenze, author of the well-known Codex L. IV. 21 in the Town Library of Siena. However a careful reading of the text would seem to disprove this attribution. At the end of f. 11v., with reference to an earlier work of his, the author states: *Nella praticha d'arismetricha diciemo delle 10 dimostrazioni [di proposizioni del II libro di Euclide] e chonclusioni, inchominciando alla prima e per insino alla decima e questo fu nel primo chapitolo dell'ottava parte [...]* ("In the *praticha d'arismetricha* we discussed the 10 prooofs [of propositions from Book II of Euclid] and conclusions, from the first to the tenth and this was in the first chapter of the eight part [...]").

The subject of chapter I of the eighth book of L. IV. 21 is different and relates to *ragioni di meriti e di schonti* ([3]), while in f. 317r. of the Palatino Codex 573 in the National Library of Florence, reference is made to the proofs of the Euclidean propositions. There we find the following:

La ottava parte di questo tractato, che chontiene el travagliamento delle radicie, ... divideremo questa parte in 11 chapitoli: nel primo chapitolo diremo alchune dimostrazioni le quali nel sechondo di Euclide sono dimostrate [...] ("The eighth part of this treatise, which contains the determination of the roots, ... we will divide this part in 11 chapters: in the second chapter we will discuss some proofs which are given in Euclid's second [...]").

Therefore it seems reasonable to suppose that the two authors, of the codices Palatino 577 and Palatino 573, are one and the same person. Gino Arrighi has identified the author of the second codex as a pupil of the Florentine Domenicho d'Agostino Vaiaio ([6]). Obviously, Maestro Benedetto and Vaiaio's pupil could well be the same person.

As far as the contents of the treatise are concerned, after careful analysis, Ettore Picutti states that: "All the 'geometria' of the *Summa*, from the beginning on page 59v. (119 folios), is the transcription of the first 241 folios of the Codex Palatino 577", confirming once again Luca Pacioli's inclination for plagiarism ([16] and [18]).

* * *

Manuscript no. 4 contains a treatise composed of 67 numbered folios entitled *Regole di geometria pratica*. The author's name, Orbetano da Montepulciano, has come down to us thanks to an annotation on f. 40r.:

Ora te voglio admagiestrare del modo de fare lu ssimo nobele demenutivo nostro, trovato per Mastro Paulo gieometro da Fiorenza e datoglie la regola per me da Orbetano da Montepulciano.

This annotation also strengthens the hypothesis that the Moreni manuscript 130 was inspired by the mathematical culture of the Florentine abacus schools, which in the 14th and 15th centuries were famous enough to make their influence felt throughout Tuscany. We believe that the '*Maestro Paulo gieometro da Fiorenza*' mentioned by the author could be the well-known Paolo dell'Abaco who founded the *Bottega d'abaco* of *Santa Trinita* in Florence ([15] and [26]).

The text, probably unfinished, is neither structured nor homogenous. It is a miscellany of 141 problems presented without any criteria of order. Among the parts most worthy of note are those dedicated to calculating the quantity of wine contained in full or partially full casks, an interesting subject which has been little studied thus far ([7], [8], [9], [10] and [24]).

Among the public officials of the *maestri d'abaco* in the Senese area in the 15th century were the supervisors of water carriers and road maintenance and also the officials of an office called *martelino*, who controlled the amount of wine sold to hotels. The owners of the latter had to pay a tax on the wine they sold. This explains the great care shown by Senese authors of the period in explaining the calculation of casks and barrels.

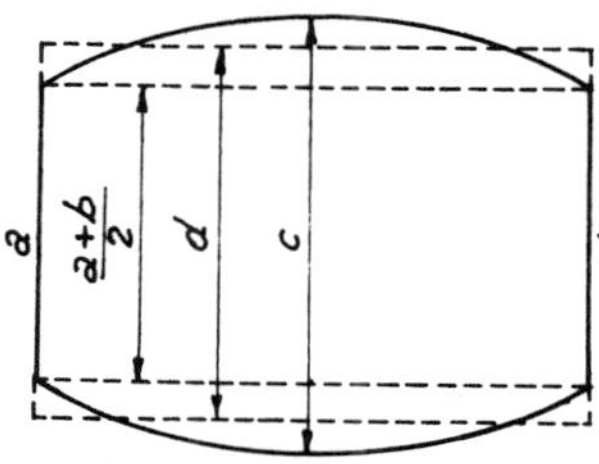

Fig. 10

Orbetano gives two formulas for calculating the capacity of casks. These can be rendered as follows:

$$V = d^2 \cdot l \qquad V = d^2 \cdot l \cdot 11/14 = \pi \cdot (d/2)^2 \cdot l$$

obtained, respectively, by supposing that the cask was approximately equivalent to a square–based prism or a cylinder. In these formulas, l represents the length of the cask, and d the so-called *diametro perfetto* (perfect diameter) or, that is, with reference to figure 10, the size obtained using the formula:

$$d = [(a + b)/2 + c]/2 ,$$

where a and b are the mean values between the x-axis and the y-axis with respect to the front and back base of the cask (the latter being approximated to ellipses) and c the height of the inspection hole (known as the *cocchiume*).

Calculating the amount of wine missing in a cask means measuring how much the level has declined, with respect to the maximum level, by inserting a dipstick into the inspection hole. It is then a question of going from linear measure to capacity by way of some appropriate formula and the use of tables. Orbetano solves the problem by proposing four different rules, with as many tables for the *scemi* (amounts of missing wine):

i) the rule *del 60* (f. 40r.), which, according to the author, can be attributed to Paolo dell'Abaco;

ii) the rule *per blaccio e per oncia* (f. 63r.);

iii) the rule *del garofano d'oro* (f. 42r., f. 50r., f. 64r.);

iv) the rule on f. 62v.

Of these four rules only ii) and iii) are completely original and not mentioned by other Senese authors. We will only examine the rule of the *garofano d'oro* (golden carnation) in any detail since, according to the author, this is the most precise and exact in the art of geometry.

n	n′
1	8
2	21
3	38
4	58
5	79
6	101
7	126
8	150
9	175
10	200

The table associated with this rule is composed of two columns, in the first of which the number n can be found, using the formula:

$$n = s_n/d \cdot k$$

where $s_n = s - 1/4 \cdot [c - (a+b)/2]$ represents the so-called *scemo netto* (net amount of missing wine) and s the height of the wine missing measured with the dipstick through the inspection hole, a and b are the mean values between the x-axis and the y-axis with respect to the front and back base of the cask, c the height of the inspection hole, d the perfect diameter and k an appropriate coefficient, which we will come back to shortly.

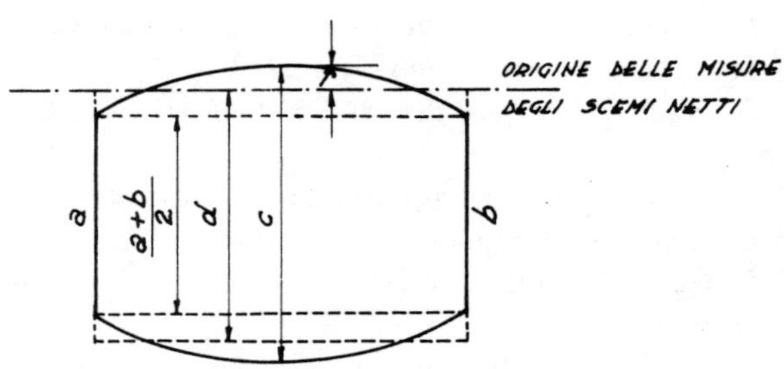

Fig. 11

By the way, note that it is necessary to pass from the *scemo* (height of the wine missing) s to the *scemo netto* (net amount of wine missing) $s_n = s - r$ so that the origin of the measure of the *scemi* coincides with the upper extremity of the perfect diameter d of the cylinder to which the author approximates the cask.

The number n', in the second column, divided by k, tells us how empty the cask is for each barrel of its total capacity. Hence, to work out the capacity of the *scemo* it is necessary to multiply the quotient n'/k by the number of barrels that the cask would contain if it were full.

Having given the *garofano d'oro* table, Orbetano goes on to make an observation about the general validity of the rule. Its application is the same no matter what units of measurement are in use in a given county or area; the only difference is that the coefficient k by which to multiply and divide, must represent the number of the sub-units in use in the county or area in which it is applied.

The codex was transcribed by Annalisa Simi ([15] and [23]).

* * *

The manuscripts containing a chapter concerning geometric problems that we examined were:

a) Pluteo 30, 26 in the Mediceo Laurenziana Library of Florence (d. 1370);
b) L. IX. 28 in the Town Library of Siena (d. 1384);
c) Ms. 1. XI in the Library of the University of Bologna (14th century);
d) Palat. 575 in the National Library of Florence (c. 1460);
e) Ms. 1612 in the Library of the University of Bologna (d. 1464).

Codex a) contains, from f. 37v. to f. 61v., a treatise on the art of geometry (*Tractato de l'arte de la geumetria*), compiled by Giovanni de' Danti, which is drawn from the much larger treatise by a certain Maestro Magrobono, according to a note on f. 38r. Of the 40 chapters of Magrobono's text the author here offers only 16. On f. 37v. he provides the following list of the chapter titles, which gives us a fairly clear idea of the manuscript's contents:

1) *La regola dei vocabuli dei talgli ed ogni altra cosa che potesse concorrere*;
2) *La regola de mesurare e tondi per tucte maniere*;
3) *La regola de fare del quadro cubico palla ritonda*;
4) *Le regole degli sciemi tondi e de le loro diferentie*;
5) *Le regole di mectere dentro e di fuori ai tondi, cio quadri, trianguli, altre misure*;
6) *Le regole de' quadri, bislonghi, per ogni modo che po' concorrere*;
7) *Le regole de' trianguli, scudi, paratelle e altre forme*;
8) *Le regole dei difitij, cio torri, case, arche, bocti e ogni altra forma che concorresse*;
9) *Le regole dei modelli sestati, co lli mesurare ogni capacit*;
10) *Le regole de quantit de mesura del vino e de le terre depte di Toscana*;
11) *Le regole degli sciemi quadrati*;
12) *Le regole degli sciemi quadrati per modo de propositione*;
13) *Le regole degli sciemi, tracto di tucto quadrato*;
14) *Le regole de fare la corda de mesurare bocti*;
15) *El libro del tractato del quadrante il quale contiene in s la regola e il modo de mesurare ogni cosa nella quale si contiene misure con uno bello ordine*;
16) *Le regole delle taule del bisexto nella quale si pu sempre in che luogo il sole*.

It is interesting to note that in chapter 2), on the area of a circle, the author gives the formula $A = C/2 \cdot d/2$, the validity of which he attempts to prove by reducing the circle to an equivalent rectangle, with one dimension equal to the radius and the other equal to the semicircumference, by way of suitable cuts.

In chapter 4) the author proposes three completely original formulas for calculating the area of a circular segment. Indicating the chord with c, the sagitta with s, the arc with a and the diameter of the circle with d, the formulas are:

$A_1 = (c^2 - 3/4 \cdot c^2)/s$

$A_2 = \sqrt{s^2 \cdot \{d^2 - [(d-2s)^2 + s \cdot (d-2s)]\}}$

$A_3 = d/2 \cdot a/2 \pm c/2 \cdot (d/2 - s)$ $\quad$ $>$ or $<$ a half circle.

The first of these is wrong, the second gives a result in excess, while the third is exact.

Chapters 9) to 14) are about calculating the capacity of casks and *scemi* (amount of wine missing). The author uses methods different from those given by Orbetano, but without explaining the use of the tables given. In chapter 10) the author lists a number of different units of measure of capacity (for wine) and length then in use in various areas of Tuscany and Umbria.

The codex has not been transcribed.

* * *

Gilio di Cecco da Montepulciano, Maestro d'abaco in Siena in 1374, is the author of codex b). His treatise does not seem to have been meant for didactic purposes, but rather to have been a notebook for his personal use. The manuscript measures just 80 × 62 mm and easily fits into the palm of a hand.

The part relating to geometry is composed of 35 folios and deals mainly with plane geometry. Of particular interest are the problems which require the inscription of a number of tangent circles in different figures, such as the triangle, the square and the circle.

It is in Gilio's work that we find the earliest formulation of Malfatti's problem (see above). In the particular case of the equilateral triangle:

> Uno scudo ch' per ogni faccia bracia 12, vogliovi mettare dentro 3 tondi magiori che si pu e sieno uguali; domando quanto sarano ampi, cio quant' per lo diametro loro.

To solve this problem Gilio proposes the following formula (without giving any explanation):

$$d = \sqrt{3/4 \cdot l^2 + l/2} - l/2$$

where d stands for the diameter of the circle to be determined and l for the side of the equilateral triangle. The result obtained using this formula is an approximation, greater than the correct value.

The codex has been transcribed in part by Raffaella Franci and Annalisa Simi ([12] and [21]).

* * *

Composed of 26 folios, the part dealing with geometry in codex c) is entitled *Le regolle de la geometria versibelle*. The text is by an anonymous author who uses a Northern Italian vernacular.

Among the many interesting questions regarding plane geometry in the codex, we would like to examine a curious rule for calculating the area of an irregular pentagon. Let n_1, n_2, n_3, n_4 and n_5 be the lengths of the sides, with $n_1 > n_2 > n_3 > n_4 > n_5$. The rule can be summarized in the formula:

$$A = [(n_1 + n_2)^2 + (n_3 + n_4)^2 - n_5^2]/4 ,$$

which obviously gives a result which is an approximation of the correct one.

This text also has a part dealing with calculating the capacity of casks and how full or empty they are. It is significant that the author gives a series of equivalences which allow us to pass from units of measure for volume to units of measure for capacity.

The codex has not been transcribed.

* * *

Manuscript d) contains a treatise on the abacus, comprising 180 folios, by an anonymous Florentine author. The work has not been put into any specific order, and the questions on geometry are scattered throughout the text.

Among the interesting problems in plane geometry is one in which the author determines the area of a regular octagon with side l, using the following formula:

$$A = 2l^2 + \sqrt{2(2l^2)^2} .$$

The three methods the author suggests for calculating the area of a quadrilateral are also of interest. The first consists in dissecting the figure into triangles and rectangles whose areas are easy to calculate. The second involves calculating the product between the half-sum of the opposite sides (a method used already by Babylonians). The third is applied in the particular case of the quadrilateral in figure 12.

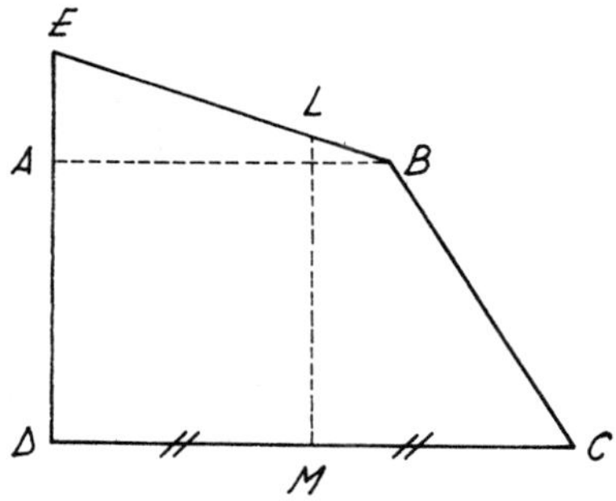

Fig. 12

The formula for calculating the area is $A = (AB+CD)/2 \cdot LM$, where M indicates the mean point of CD. Obviously the last two methods give results which are only approximations of the correct one.

Annalisa Simi has transcribed the parts of this codex dealing with arithmetic and algebra ([1]).

* * *

Piero Jachomo di Maestro Antonio de Chapelam, of Bologna, is the author of codex e). This work, which perfectly fits into the category of abacus treatises, is entitled *Libro del Arixmetricha e Zemetria* and is composed of a miscellany of 438 numbered problems. The part dealing with geometry is to be found between ff. 94v.–120v. and 140v.–141v.

Among the problems on plane geometry, of particular interest is the formula which, given the area A of an equilateral triangle allows us to determine the side l:

$$l = \sqrt{\sqrt{A^2 \cdot (5 + 1/3)}} \, .$$

In solid geometry the author proposes an incorrect, although not entirely senseless, formula for the measure S of the total surface of a sphere: $S = (C/2)^2$, where C is the circumference of a great circle of the sphere. The same formula is also given by the authors of codexes Pluteo 30, 26 and Palat. 575.

The codex has not been transcribed.

* * *

In this paper the authors have concentrated on the numerous approximative (and rather strange) formulas given by the various authors of the codexes, rather than considering often more complex problems which were solved in a more usual way. The aim has been to show the practical nature of geometrical studies in the 14th and 15th centuries.

Bibliography

[1] Anonimo Fiorentino: *Regole di geometria e della cosa* (from Palatino 575 codex in the National Library of Florence). Edited and with an introduction by Annalisa Simi. *Quaderni del Centro Studi della Matematica Medioevale*, no. 20, University of Siena, 1992.

[2] Anonimo Senese: *Differenze di geometria e misure a ochio* (from the Plimpton Ms. 194 in the Columbia University Library). Edited and with an introduction by Maria Teresa Rivolo. *Quaderni del Centro Studi della Matematica Medioevale*, no. 13, University of Siena, 1986.

[3] Arrighi G.: Il Codice L. IV. 21 della Biblioteca degli Intronati di Siena e la Bottega dell'abaco a Santa Trinita in Firenze. *Physis* **7** (1965), 369-400.

[4] Arrighi G.: Un estratto del "De visu" di M°Grazia de' Castellani. *Acts of the Giorgio Ronchi Foundation* **22** (1967), 44-58.

[5] Arrighi G.: Il trattato di geometria e la volgarizzazione del "Liber Quadratorum" di Leonardo Pisano del Codice Palatino 577 (sec. XV) della Biblioteca Nazionale di Firenze. *Acts of the Giorgio Ronchi Foundation* **22** (1967), 760-775.

[6] Arrighi G.: Nuovi contributi per la storia della matematica in Firenze nell'et di mezzo. (Il Codice Palatino 573 della Biblioteca Nazionale di Firenze). *Proceedings of the Istituto Lombardo di Scienze, Lettere ed Arti, Classe di Scienze (A)*, **101** (1967), 395-437.

[7] Arrighi G.: La tenuta delle botti ed il calcolo degli scemi in un'opera del senese Tommaso della Gazzaia. *Storia dell'Agricoltura magazine* **3**, 1967, 2-24.

[8] Arrighi G.: Il Codice L. IV. 16 (sec. XV) della Biblioteca degli Intronati di Siena. Gli "Elementi" di Euclide, un vocabolario di abbreviature e gli "scemi di botti". *Accademie e Biblioteche d'Italia* **38**, no. 1, Rome, 1970, 34-41.

[9] Arrighi G.: Le tavole di Antonio di Marchionne (sec. XVI) per la tenuta delle botti e gli scemi. *Storia dell'Agricoltura magazine*, no. 2, August 1973, 129-139.

[10] Arrighi G.: Misura delle botti e calcolo degli scemi secondo il Petriboni (sec. XV). *Physis* **25**, fasc. 1, 1983, 171-175.

[11] Franci R. and Toti Rigatelli L.: *Introduzione all'aritmetica mercantile del Medioevo e del Rinascimento*. Edizioni Quattroventi, Urbino, 1982.

[12] Gilio: *Questioni d'algebra* (from the L. IX. 28 codex in the Town Library of Siena). Edited and with an introduction by Raffaella Franci. Quaderni del Centro Studi della Matematica Medioevale, no. 6, University of Siena, 1983.

[13] Leonardo Pisano: *Practica geometriae ed opuscoli*. Works by Leonardo Pisano published by Baldassarre Boncompagni, vol. II, Rome, 1862.

[14] Leon Battista Alberti: *Ludi Matematici*. Edited by Raffaele Rinaldi, with a preface by Ludovico Geymonat. Guanda, Milan, 1980.

[15] Orbetano da Montepulciano: *Regole di Geometria Pratica* (from the Moreni Ms. 130 in the Riccardiana Library in Florence). Edited and with an introduction by Annalisa Simi. Quaderni del Centro Studi della Matematica Medioevale, no. 19, University of Siena, 1991.

[16] Pacioli Luca: *Summa de arithmetica, geometria, proportioni et proportionalit*. Venice, 1495.

[17] Picutti E.: Il libro dei quadrati di Leonardo Pisano e i problemi di analisi indeterminata nel Codice Palatino 577 della Biblioteca Nazionale di Firenze. *Physis* **1 – 4** (1979), 195-339.

[18] Picutti E.: Sui plagi matematici di frate Luca Pacioli. *Le Scienze* **246** (1989), 72-79.

[19] Procissi A.: Di alcuni problemi anteriori ed analoghi a quello di Malfatti. *Acts of the conference on G. Malfatti*, Ferrara, (1981).

[20] Sani M.: *Il Codice L. IV. 18 della Biblioteca Comunale di Siena (Parte prima)*. Thesis for a degree in mathematics. University of Siena, 1988 – 1989.

[21] Simi A.: *I problemi di geometria dal Codice L. IX. 28 della Biblioteca Comunale di Siena*. Thesis for a degree in mathematics. University of Siena, 1987 – 1988.

[22] Simi A.: *Trascrizione ed analisi di parte del Codice L. IV. 18 della Biblioteca Comunale di Siena (sec. XV)*. Rapporto Matematico no. 214, Mathematics Department of the University of Siena, 1989.

[23] Simi A.: *Trascrizione ed analisi del Codice Moreni 130 della Biblioteca Riccardiana di Firenze*. Rapporto Matematico no. 229, Mathematics Department of the University of Siena, 1991

[24] Simi A.: *La tenuta delle botti ed il calcolo degli scemi nelle opere di alcuni abacisti senesi dei secoli XV e XVI*. Acts of the Fisiocritici Academy in Siena, 15, tomo 11 (1992), 1-8.

[25] Tommaso della Gazzaia: *Praticha geometria e tutte misure di terre* (from the C. III. 23 codex in the Town Library of Siena). Transcribed by Cinzia Nanni and with an introduction by Gino Arrighi. Quaderni del Centro Studi della Matematica Medioevale, no. 1, University of Siena, 1982.

[26] Toti Rigatelli L.: Matematici fiorentini del Tre – Quattrocento. *Symposia Mathematica* **27** (1986), 3-21.

[27] Van Egmond W.: *Practical Mathematics in the Italian Renaissance: A catalog of Italian Abbacus manuscripts and printed books to 1600*. Istituto e Museo di Storia della Scienza, Florence, 1980.

Addresses of the authors

Prof. Dr. P. Bockstaele,
Graetboslaan 9,
B 3050 Oud Heverlee
Belgium

Prof. Dr. H.J.M. Bos
Department of Mathematics, University of Utrecht
P.O.Box 30.010
NL 3508 TA Utrecht
Netherlands

Dr. S. Brentjes
Karl-Sudhoff-Institut für Geschichte der Naturwissenschaften und Medizin
Augustusplatz
D 04109 Leipzig
Germany

Dr. C.S.M. Burnett
Warburg Institute
Woburn Square
London WC1H 0AB
United Kingdom

Dr. A. Djebbar
Ministre de l'Éducation Nationale
Ministère de l'Éducation Nationale
Ben Aknoun
Alger
Algeria

Dr. Y. Dold-Samplonius
Türkenlouisweg 14
D 69151 Neckargemünd
Germany

Prof. Dr. M. Folkerts
Institut für Geschichte der Naturwissenschaften der Universität München
Deutsches Museum
D 80306 München
Germany

Dr. J.P. Hogendijk
Department of Mathematics, University of Utrecht
P.O.Box 30.010
NL 3508 TA Utrecht

Prof. Dr. B. Hughes O.F.M.
Department of secondary and adult education
California State University
18111 Nordhoff Street
Northridge Ca. 91330
USA

Prof. Dr. W. Kaunzner
Zollerstraße 9
D 93053 Regensburg
Germany

Prof. Dr. P. Kunitzsch
Davidstraße 17
D 81927 München
Germany

Dr. R. Lorch
Brahmsstraße 8
D 82110 Germering
Germany

Dr. J. van Maanen
Department of Mathematics, University of Groningen
P.O. Box 800
NL 9700 XX Groningen
Netherlands

Dr. G. Molland
Department of History and Philosophy of Science
University of Aberdeen
Kings College
Aberdeen AB9 2UB
United Kingdom

Prof. Dr. B.A. Rosenfeld
409 Martin Terrace
State College, Pa. 16803
USA

Prof. Dr. C.J. Scriba
Institut für Geschichte der Naturwissenschaften, Mathematik und Technik
Universität Hamburg
Bundesstraße 55
D 20146 Hamburg
Germany

Dr. J. Sesiano
Département des Mathématiques
École Polytechnique Fédérale
MA (Ecublens)
CH 1015 Lausanne
Switzerland

Prof. Dr. A. Simi
Prof. Dr. L. Toti Rigatelli
Università di Siena
Via del Capitano, 15
I 53100 Siena
Italy